Advanced High School Statistics

Fourth Edition

Leah Dorazio
Statistics and Computer Science Teacher
San Francisco University High School

David Diez
Data Scientist
OpenIntro

Mine Çetinkaya-Rundel
Associate Professor of the Practice, Duke University
Professional Educator, RStudio

Christopher D Barr
Investment Analyst
Varadero Capital

Table of Contents

Preface

Advanced High School Statistics covers a first course in statistics, providing an introduction to applied statistics that is clear, concise, and accessible. This book was written to align with the new AP® Statistics Course Description[1], but it is also popular in non-AP courses and community colleges.

This book may be downloaded as a free PDF at **openintro.org/ahss**.

We hope readers will take away three ideas from this book in addition to forming a foundation of statistical thinking and methods.

(1) Statistics is an applied field with a wide range of practical applications.

(2) You don't have to be a math guru to learn from real, interesting data.

(3) Data are messy, and statistical tools are imperfect. But, when you understand the strengths and weaknesses of these tools, you can use them to learn about the real world.

Textbook overview

The chapters of this book are as follows:

1. **Exploring one-variable data and collecting data.** Data structures, variables; data summaries and graphs; data collection principles; sampling methods and experimental designs.
2. **Probability, random variables, and probability distributions.** The basic principles of probability and random variables; binomial and normal distributions; sampling distributions and the central limit theorem.
3. **Inference for categorical data: proportions.** Sampling distributions for a sample proportion and a difference in sample proportions; introduction to statistical inference; inference for population proportions and contingency tables using the normal and chi-square distributions.
4. **Inference for numerical data: means.** Sampling distributions for a sample mean and difference in sample means; inference for a population mean and a difference in population means using the t-distribution.
5. **Regression analysis.** An introduction to linear correlation and regression with two variables.

Online resources

OpenIntro is focused on increasing access to education by developing free, high-quality education materials. In addition to textbooks, we provide the following free resources on our website to help teachers and students be successful.

- Video overviews for each section of the textbook
- Lecture slides for each section of the textbook
- Casio and TI calculator tutorials
- Video solutions for selected section and chapter exercises
- Statistical software labs
- A small but growing number of Desmos activities[2]

[1] AP® is a trademark registered and owned by the College Board, which was not involved in the production of, and does not endorse, this product. apcentral.collegeboard.org/media/pdf/ap-statistics-revised-course-framework.pdf

[2] openintro.org/ahss/desmos

All of these resources can be found at:

openintro.org/ahss

We also have improved the ability to access data in this book through the addition of Appendix B, which provides additional information for each of the data sets used in the main text. Online guides to each of these data sets are also provided at **openintro.org/data** and through a companion R package.

Examples and exercises

Many examples are provided to establish an understanding of how to apply methods.

E

EXAMPLE 0.1

This is an example.

Full solutions to examples are provided here, within the example.

When we think the reader should be ready to try an example problem on their own, we frame it as Guided Practice.

G

GUIDED PRACTICE 0.2

The reader may check or learn the answer to any Guided Practice problem by reviewing the full solution in a footnote.[3]

Exercises are also provided at the end of each section and each chapter for practice or homework assignments. Solutions for odd-numbered exercises are given in Appendix A.

Getting involved

We encourage anyone learning or teaching statistics to visit **openintro.org** and get involved. We also value your feedback. Please provide feedback, report typos, and review known typos at

openintro.org/ahss/feedback

Acknowledgements

This project would not be possible without the passion and dedication of all those involved. The authors would like to thank the OpenIntro Staff for their involvement and ongoing contributions. We are also very grateful to the hundreds of students and instructors who have provided us with valuable feedback since we first started working on this project in 2009. A special thank you to Stephen Miller and Juan Gomez for reviewing and providing feedback on the fourth edition of AHSS.

[3]Guided Practice solutions are always located down here!

Chapter 1

Exploring one-variable data and collecting data

Scientists seek to answer questions using rigorous methods and careful observations. These observations – collected from the likes of field notes, surveys, and experiments – form the backbone of a statistical investigation and are called **data**. Statistics is the study of how best to collect, analyze, and draw conclusions from data.

In this chapter, we begin by introducing the four-step statistical process and some important data basics that will lay the groundwork for our statistical investigations. We look at ways to verbally, graphically, and numerically summarize data once it is collected. Then we investigate techniques for collecting data and common sources of bias that arise during data collection. After finishing this chapter, you will have the tools for identifying strengths and weaknesses in data-based conclusions, tools that are essential to be an informed citizen and a savvy consumer of information.

For videos, slides, and other resources, please visit
www.openintro.org/ahss

1.1 Introduction to data

You collect data on dozens of questions from all of the students at your school. How would you organize all of this data? Effective presentation and description of data is a first step in most analyses. This section introduces one structure for organizing data as well as some terminology that will be used throughout this book. We also introduce loan data from Lending Club which will be used extensively in this chapter.

Learning objectives

1. Identify components within a statistical study and distinguish between a population and a sample.
2. Determine an investigative question within a statistical study.
3. Identify observational units, variables, parameters, and statistics from a statistical study or data set.
4. Identify variables as categorical or numerical and numerical variables as discrete or continuous.

1.1.1 Why do we collect data?

Researchers from a wide array of fields have questions or problems that require the collection and analysis of data. Let's consider three examples.

- Climate scientists: how will the global temperature change over the next 100 years?
- Psychology: can a simple reminder about saving money cause students to spend less?
- Political science: what fraction of adults in the United States approve of the job Congress is doing?

What questions from current events or from your own life can you think of that could be answered by collecting and analyzing data?

Before diving into statistical terminology and methods, it is helpful to put statistics in the context of a general process of investigation:

1. Identify a statistical question or problem.
2. Collect relevant data on the topic.
3. Analyze the data.
4. Form a conclusion.

Questions whose answers are already known or questions that cannot be address with the collection of data are not *statistical questions*. Statistics as a subject has three primary components: How best can we collect data? How should it be analyzed? And what can we infer from the analysis?

We should think of this process as a feedback loop; one conclusion may prompt a follow-up question – or possibly many follow-up questions. When posing a statistical question, it is important that it be well-formed so that a clear plan for data collection can be made. Moreover, the research question should not change based on the data analysis or results. Avoiding bias and maintaining integrity in data collection and analysis is fundamental to the statistical process.

1.1.2 Populations and samples

Consider the following research questions:

1. What is the average mercury content in swordfish in the Atlantic Ocean?
2. What proportion of eligible voters plan to actually vote in the next United States presidential election?
3. Does a new drug reduce the number of deaths in patients with severe heart disease?

Each research question refers to a target **population**, or population of interest. In the first question, the target population is all swordfish in the Atlantic ocean, and each fish represents an individual. Often times, it is too expensive to collect data for every individual in a population. Instead, a sample is taken. A **sample** represents a subset of the individuals and is often a small fraction of the population. For instance, 60 swordfish (or some other number) in the population might be selected, and this sample data may be used to provide an estimate of the population average in order to answer the research question. We use a sample size n from the entire population size N. Here, n is 60 and the total number of swordfish in the Atlantic ocean (N) is unknown. When data are collected from a sample to answer an investigative question about a larger population, we call that investigation a **statistical study**.

G

GUIDED PRACTICE 1.1

For the second and third research questions above, identify the population of interest. For these Guided Practice questions, you can check your answer in the footnote.[1]

1.1.3 Observational units, variables, parameters, and statistics

We collect a sample of data to better understand the characteristics of a population. A **variable** is a characteristic we measure for each individual of the population. We call each individual of the population an **observational unit** or a **case**. We may be interested in estimating a mean, median, proportion, or some other summary of a population. These population values are called **parameters**. More specifically, a parameter is a numerical attribute or summary of a variable of interest for a population. Such summary values we calculate from a particular sample are called **statistics**. In general, we use a calculated statistic to estimate an unknown parameter.

E

EXAMPLE 1.2

Earlier we asked the question: what is the average mercury content in swordfish in the Atlantic Ocean? Identify the variable to be measured, the observational unit, the parameter of interest, and the corresponding statistic.

The variable is the level of mercury content in swordfish in the Atlantic Ocean. It will be measured for each individual swordfish, which is the observational unit. The parameter of interest is the average mercury content in *all* swordfish in the Atlantic Ocean, while the statistic will be the average mercury content in the swordfish in our sample.

[1](2) The population of interest is eligible voters in the United States. (3) The population of interest includes all people with severe heart disease.

G

GUIDED PRACTICE 1.3

For the second question regarding the proportion of eligible voters that plan to vote in the next United States presidential election, identify the observational unit, the variable to be measured, the parameter of interest, and the statistic.[2]

E

EXAMPLE 1.4

Based on the definitions of statistic and parameter, which is a variable and which is a fixed number?

A statistic can vary from sample to sample, so it is a variable. A parameter is a numerical attribute or summary of a population, so at any one point in time, it is a fixed quantity (which is often unknown).

1.1.4 Data matrices

The information that we collect on each individual is called a **datum** (singular of **data**). A collection of data is called a **data set**. Figure 1.1 displays rows 1, 2, 3, and 50 of a data set for 50 randomly sampled loans offered through Lending Club, which is a peer-to-peer lending company. These observations will be referred to as the `loan50` data set.

Each row in the table represents a single loan. Each row corresponds to an observational unit. An **observational unit** or **case** is an item or individual from which data is collected. The columns represent characteristics, called variables, for each of the loans. A **variable** is a characteristic that may change from one observational unit to another. For example, the first row represents a loan of $7,500 with an interest rate of 7.34%, where the borrower is based in Maryland (MD) and has an income of $70,000.

G

GUIDED PRACTICE 1.5

What is the grade of the first loan in Figure 1.1? And what is the home ownership status of the borrower for that first loan? [3]

In practice, it is especially important to ask clarifying questions to ensure important aspects of the data are understood. For instance, it is always important to be sure we know what each variable means and the units of measurement. Descriptions of the `loan50` variables are given in Figure 1.2.

	loan_amount	interest_rate	term	grade	state	total_income	homeownership
1	7500	7.34	36	A	MD	70000	rent
2	25000	9.43	60	B	OH	254000	mortgage
3	14500	6.08	36	A	MO	80000	mortgage
⋮	⋮	⋮	⋮	⋮	⋮	⋮	⋮
50	3000	7.96	36	A	CA	34000	rent

Figure 1.1: Four rows from the `loan50` data matrix.

[2]The observational unit is an individual eligible voter. The variable, or characteristic that we record on each eligible voter, is whether or not they plan to vote in the next US presidential election. The parameter of interest is the proportion of all eligible voters that plan to vote in the next US presidential election, and the statistic will be the proportion in our sample that plan to vote in the next US presidential election.

[3]The loan's grade is A, and the borrower rents their residence.

variable	description
`loan_amount`	Amount of the loan received, in US dollars.
`interest_rate`	Interest rate on the loan, in an annual percentage.
`term`	The length of the loan, which is always set as a whole number of months.
`grade`	Loan grade, which takes values A through G and represents the quality of the loan and its likelihood of being repaid.
`state`	US state where the borrower resides.
`total_income`	Borrower's total income, including any second income, in US dollars.
`homeownership`	Indicates whether the person owns, owns but has a mortgage, or rents.

Figure 1.2: Variables and their descriptions for the `loan50` data set.

The data in Figure 1.1 represent a **data matrix**, which is a convenient and common way to organize data, especially if collecting data in a spreadsheet. Each row of a data matrix corresponds to a unique case (observational unit), and each column corresponds to a variable.

When recording data, use a data matrix unless you have a very good reason to use a different structure. This structure allows new cases to be added as rows or new variables as new columns.

GUIDED PRACTICE 1.6

The grades for assignments, quizzes, and exams in a course are often recorded in a gradebook that takes the form of a data matrix. How might you organize grade data using a data matrix?[4]

1.1.5 Types of variables

Examine the `interest_rate`, `term`, `state`, and `grade` variables in the `loan50` data set. Each of these variables is inherently different from the other three, yet some share certain characteristics.

First consider `interest_rate`, which is said to be a **numerical** variable because it can take a wide range of numerical values, and it is sensible to add, subtract, or take averages with those values. On the other hand, we would not classify a variable reporting telephone area codes as numerical because the average, sum, and difference of area codes doesn't have any clear meaning.

The `term` variable is also numerical, although it seems to be a little different than `interest_rate`. This variable represents number of months and can only take whole non-negative numbers such as 36 and 60. For this reason, the `term` variable is said to be **discrete** because it can only take numerical values with jumps. On the other hand, the `interest_rate` variable is said to be **continuous**.

The variable `state` can take up to 51 values after accounting for Washington, DC: `AL`, `AK`, ..., and `WY`. Because the responses themselves are categories, `state` is called a **categorical** variable, and the possible values are called the variable's **levels**.

Finally, consider the `grade` variable, which represents the quality of the loan and takes on values A through G. This variable seems to be a hybrid: it is a categorical variable but the levels have a natural ordering. A variable with these properties is called an **ordinal** variable, while a regular categorical variable without this type of special ordering is called a **nominal** variable. To simplify analyses, any ordinal variable in this book will be treated as a nominal (unordered) categorical variable.

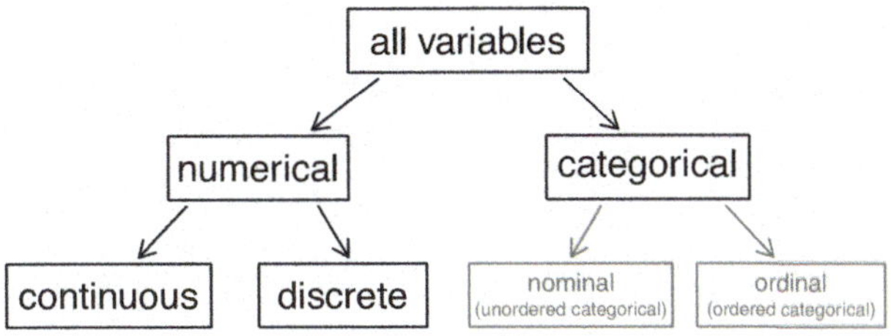

Figure 1.3: Breakdown of variables into their respective types.

E

EXAMPLE 1.7

Data were collected about students in a statistics course. Three variables were recorded for each student: number of pets, student height, and whether the student had previously taken a statistics course. Classify each of the variables as continuous numerical, discrete numerical, or categorical.

The number of pets and student height represent numerical variables. Because the number of pets is a count, it is discrete. Height varies continuously, so it is a continuous numerical variable. The last variable classifies students into two categories – those who have and those who have not taken a statistics course – which makes this variable categorical.

G

GUIDED PRACTICE 1.8

An experiment is evaluating the effectiveness of a new drug in treating migraines. A `group` variable is used to indicate the experiment group for each patient: treatment or control. The `num_migraines` variable represents the number of migraines the patient experienced during a 3-month period. Classify each variable as either numerical or categorical.[5]

---

[5]The `group` variable can take just one of two group names, making it categorical. The `num_migraines` variable describes a count of the number of migraines, which is an outcome where basic arithmetic is sensible, which means this is a numerical outcome; more specifically, because it represents a count, `num_migraines` is a discrete numerical variable.

Section summary

- Statistics can be understood as an investigative process that involves asking a valid statistical research question, collecting data, analyzing the data, and forming a conclusion.
- A **population** is the entire group of interest. The population size is represented by N.
- A **sample** is the subset of the population that a researcher collects data on. The number of item in the sample, called the sample size, is represented by n.
- When data are collected from a sample to answer an investigative question about a population, we call this a **statistical study**. Statistical studies are necessary when it is too difficult or expensive to collect data from every individual in the population.
- A **datum** (singular form of data) is a piece of information about an individual. A collection of data is called a **data set**.
- An investigative question for a statistical study should have a defined purpose, should not change based on the data analysis or results, and should be posed so that the required data can be collected and analyzed.
- Researchers often summarize data in a table, where the rows correspond to individuals, also called **observational units** or **cases**, and the columns correspond to the **variables**, the values of which are recorded for each individual.
- A **parameter** is a numerical attribute or summary of a variable of interest for the entire population. Researchers take a sample from the population to estimate this unknown quantity. The estimate calculated from the sample is called the **statistic**. A statistic can vary from sample to sample, while a parameter is a fixed number.
- Variables can be **numerical** (measured on a numerical scale) or **categorical** (taking on levels, such as low/medium/high). Numerical variables can be **continuous**, where all values within a range are possible, or **discrete**, where only specific values, usually integer values, are possible.

Exercises

1.1 Air pollution and birth outcomes, study components. Researchers collected data to examine the relationship between air pollutants and preterm births in Southern California. During the study air pollution levels were measured by air quality monitoring stations. Specifically, levels of carbon monoxide were recorded in parts per million, nitrogen dioxide and ozone in parts per hundred million, and coarse particulate matter (PM_{10}) in $\mu g/m^3$. Length of gestation data were collected on 143,196 births between the years 1989 and 1993, and air pollution exposure during gestation was calculated for each birth. The analysis suggested that increased ambient PM_{10} and, to a lesser degree, CO concentrations may be associated with the occurrence of preterm births.[6]

(a) Identify the main research question of the study.

(b) Who are the subjects in this study, and how many are included?

(c) What are the variables in the study? Identify each variable as numerical or categorical. If numerical, state whether the variable is discrete or continuous. If categorical, state whether the variable is ordinal.

1.2 Buteyko method, study components. The Buteyko method is a shallow breathing technique developed by Konstantin Buteyko, a Russian doctor, in 1952. Anecdotal evidence suggests that the Buteyko method can reduce asthma symptoms and improve quality of life. In a scientific study to determine the effectiveness of this method, researchers recruited 600 asthma patients aged 18-69 who relied on medication for asthma treatment. These patients were randomnly split into two research groups: one practiced the Buteyko method and the other did not. Patients were scored on quality of life, activity, asthma symptoms, and medication reduction on a scale from 0 to 10. On average, the participants in the Buteyko group experienced a significant reduction in asthma symptoms and an improvement in quality of life.[7]

(a) Identify the main research question of the study.

(b) Who are the subjects in this study, and how many are included?

(c) What are the variables in the study? Identify each variable as numerical or categorical. If numerical, state whether the variable is discrete or continuous. If categorical, state whether the variable is ordinal.

1.3 Cheaters, study components. Researchers studying the relationship between honesty, age and self-control conducted an experiment on 160 children between the ages of 5 and 15. Participants reported their age, sex, and whether they were an only child or not. The researchers asked each child to toss a fair coin in private and to record the outcome (white or black) on a paper sheet, and said they would only reward children who report white.[8]

(a) Identify the main research question of the study.

(b) Who are the subjects in this study, and how many are included?

(c) The study's findings can be summarized as follows: "Half the students were explicitly told not to cheat and the others were not given any explicit instructions. In the no instruction group probability of cheating was found to be uniform across groups based on child's characteristics. In the group that was explicitly told to not cheat, girls were less likely to cheat, and while rate of cheating didn't vary by age for boys, it decreased with age for girls." How many variables were recorded for each subject in the study in order to conclude these findings? State the variables and their types.

1.4 Stealers, study components. In a study of the relationship between socio-economic class and unethical behavior, 129 University of California undergraduates at Berkeley were asked to identify themselves as having low or high social-class by comparing themselves to others with the most (least) money, most (least) education, and most (least) respected jobs. They were also presented with a jar of individually wrapped candies and informed that the candies were for children in a nearby laboratory, but that they could take some if they wanted. After completing some unrelated tasks, participants reported the number of candies they had taken.[9]

(a) Identify the main research question of the study.

See parts (b) and (c) on the following page.

[6] B. Ritz et al. "Effect of air pollution on preterm birth among children born in Southern California between 1989 and 1993". In: *Epidemiology* 11.5 (2000), pp. 502–511.

[7] J. McGowan. "Health Education: Does the Buteyko Institute Method make a difference?" In: *Thorax* 58 (2003).

[8] Alessandro Bucciol and Marco Piovesan. "Luck or cheating? A field experiment on honesty with children". In: *Journal of Economic Psychology* 32.1 (2011), pp. 73–78.

[9] P.K. Piff et al. "Higher social class predicts increased unethical behavior". In: *Proceedings of the National Academy of Sciences* (2012).

(b) Who are the subjects in this study, and how many are included?

(c) The study found that students who were identified as upper-class took more candy than others. How many variables were recorded for each subject in the study in order to conclude these findings? State the variables and their types.

1.5 Relaxing after work. The General Social Survey asked the question, "After an average work day, about how many hours do you have to relax or pursue activities that you enjoy?" to a random sample of 1,155 Americans. The average relaxing time was found to be 1.65 hours. Determine which of the following is an observation, a variable, a sample statistic (value calculated based on the observed sample), or a population parameter.

(a) An American in the sample.

(b) Number of hours spent relaxing after an average work day.

(c) 1.65.

(d) Average number of hours all Americans spend relaxing after an average work day.

1.6 Cats on YouTube. Suppose you want to estimate the percentage of videos on YouTube that are cat videos. It is impossible for you to watch all videos on YouTube so you use a random video picker to select 1000 videos for you. You find that 2% of these videos are cat videos. Determine which of the following is an observation, a variable, a sample statistic (value calculated based on the observed sample), or a population parameter.

(a) Percentage of all videos on YouTube that are cat videos.

(b) 2%.

(c) A video in your sample.

(d) Whether or not a video is a cat video.

1.7 Fisher's irises. Sir Ronald Aylmer Fisher was an English statistician, evolutionary biologist, and geneticist who worked on a data set that contained sepal length and width, and petal length and width from three species of iris flowers (*setosa*, *versicolor* and *virginica*). There were 50 flowers from each species in the data set.[10]

(a) How many cases were included in the data?

(b) How many numerical variables are included in the data? Indicate what they are, and if they are continuous or discrete.

(c) How many categorical variables are included in the data, and what are they? List the corresponding levels (categories).

Photo by Ryan Claussen (http://flic.kr/p/6QTcuX) CC BY-SA 2.0 license

1.8 Smoking habits of UK residents. A survey was conducted to study the smoking habits of UK residents. Below is a data matrix displaying a portion of the data collected in this survey. Note that "£" stands for British Pounds Sterling, "cig" stands for cigarettes, and "N/A" refers to a missing component of the data.[11]

	sex	age	marital	grossIncome	smoke	amtWeekends	amtWeekdays
1	Female	42	Single	Under £2,600	Yes	12 cig/day	12 cig/day
2	Male	44	Single	£10,400 to £15,600	No	N/A	N/A
3	Male	53	Married	Above £36,400	Yes	6 cig/day	6 cig/day
⋮	⋮	⋮	⋮	⋮	⋮	⋮	⋮
1691	Male	40	Single	£2,600 to £5,200	Yes	8 cig/day	8 cig/day

(a) What does each row of the data matrix represent?

(b) How many participants were included in the survey?

(c) Indicate whether each variable in the study is numerical or categorical. If numerical, identify as continuous or discrete. If categorical, indicate if the variable is ordinal.

[10] R.A Fisher. "The Use of Multiple Measurements in Taxonomic Problems". In: *Annals of Eugenics* 7 (1936), pp. 179–188.

[11] National STEM Centre, Large Datasets from stats4schools.

1.2 Representing categorical data

How do we visualize and summarize categorical data? In this section, we will introduce tables and other basic tools for categorical data that are used throughout this book.

Learning objectives

1. Construct and interpret tabular representations (one-way tables) for a categorical variable.
2. Describe and summarize a categorical variable with counts, proportions, ratios, and percents.
3. Interpret graphical representations of a categorical variable.
4. Justify a claim using tabular and graphical representations of a categorical variable.
5. Use tabular and graphical representations to compare two or more data sets in terms of the same categorical variable.

1.2.1 Tabular representations of a categorical variable

Recall the `loan50` data set we encountered in the previous section. This data set represents a sample from a much larger loan data set called `loans_full_schema`. This larger data set contains information on 10,000 loans made through Lending Club. One of the variables in the `loans_full_schema` data set is `homeownership`. This is a categorical variable with three levels: rent, mortgage, own. Consider Figure 1.4 for the `homeownership` variable, where each count in the table represents the number of times a particular variable outcome occurred in the data set. For example, the value 3858 corresponds to the number of loans in the data set where the borrower rents their home. A table like this that summarizes counts for each value type of a categorical variable is called a **frequency table**.

`homeownership`	Count
mortgage	4789
own	1353
rent	3858
Total	10000

Figure 1.4: A table summarizing the frequencies of each value for the `homeownership` variable.

Sometimes it is more helpful to record the proportion of times a particular variable outcome occurred, as in Figure 1.5. In this case, we can use a **relative frequency table**. The total for a relative frequency table always adds to 1, because all cases are represented on the table.

homeownership	Relative frequency
mortgage	0.4789
own	0.1353
rent	0.3858
Total	1

Figure 1.5: A table summarizing the relative frequencies of each value for the `homeownership` variable.

1.2.2 Bar charts and pie charts

A **bar chart** (also called **bar plot** or **bar graph**) is a common way to display a single categorical variable. In a bar chart, each bar represents a category of a categorical variable and the height of each bar corresponds to the frequency (count) or relative frequency (proportion) for that category. The left panel of Figure 1.6 shows a frequency bar chart for the `homeownership` variable. In the right panel, the counts are converted into proportions, showing the proportion of observations that are in each level (e.g. 3858/10000 = 0.3858 for `rent`).

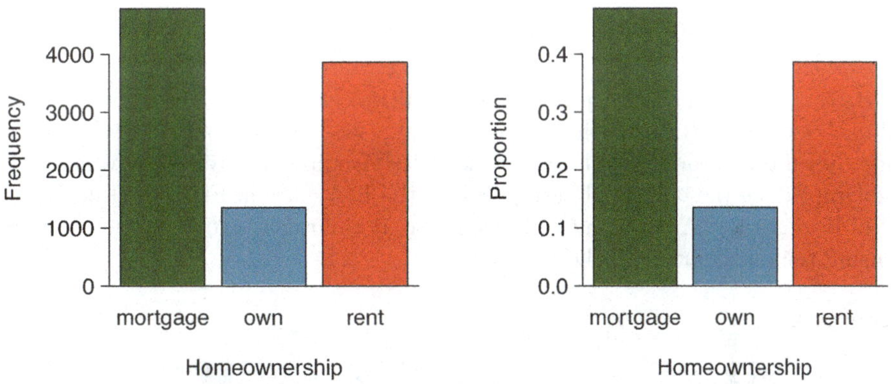

Figure 1.6: Two bar charts of `homeownership`. The left panel shows the counts, and the right panel shows the proportions in each group.

A pie chart is shown in Figure 1.7 representing the same information in the bar charts in Figure 1.6. Each slice of a pie chart represents a category of the categorical variable of interest. The area of each slice, as a fraction of the total area, corresponds to the relative frequency of observational units falling within each category. The sum of the slices' areas together will equal 1, or 100% of the total area. Pie charts can be useful for giving a high-level overview to show how a set of cases break down. However, it is also difficult to decipher certain details in a pie chart. For example, it's not immediately obvious that there are more loans where the borrower has a mortgage than rent when looking at the pie chart, while this detail is very obvious in the bar chart.

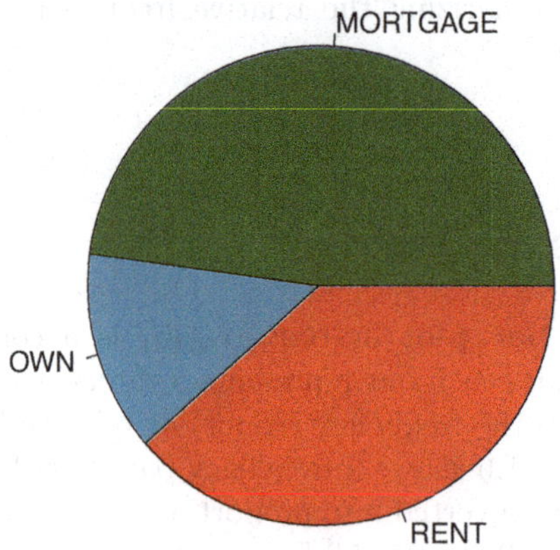

Figure 1.7: A pie chart of `homeownership`.

Pie charts can work well when the goal is to visualize a categorical variable with very few levels, especially if each level represents a simple fraction (e.g., one-half, one-quarter, etc.). However, they can be quite difficult to read when they are used to visualize a categorical variable with many levels. For example, the pie chart in Figure 1.8 and the bar chart in Figure 1.9 both represent the distribution of loan grades (A through G). In this case, it is far easier to compare the counts of each loan grade using the bar chart than the pie chart.

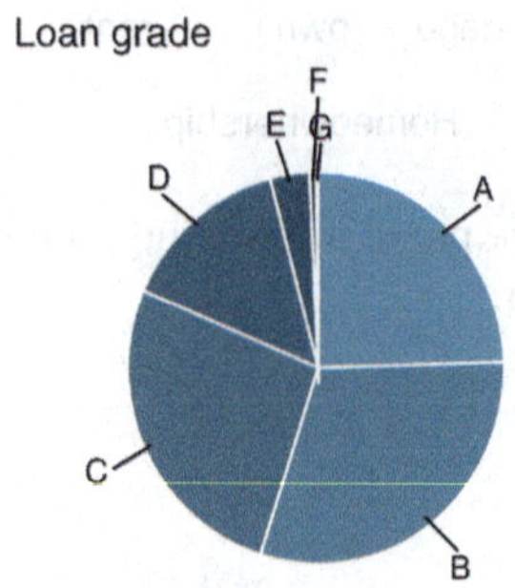

Figure 1.8: (a) Pie chart

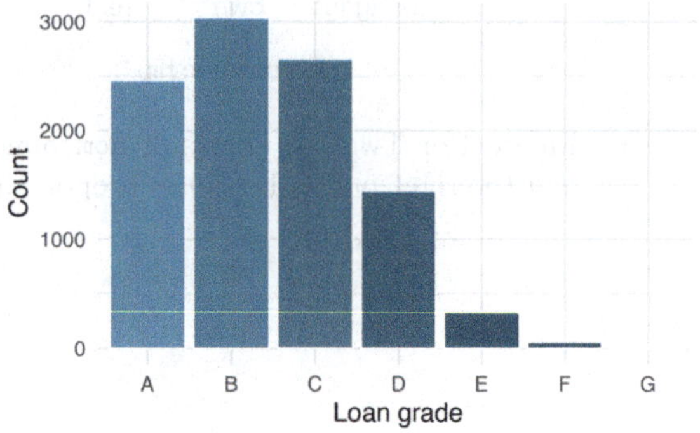

Figure 1.9: (b) Bar chart

Figure 1.10: A pie chart and bar chart of loan grades.

1.2.3 Comparing data sets in terms of a categorical variable

We can use tabular and graphical summaries of a categorical variable to compare two or more groups or data sets. Here, we compare `homeownership` for individual loans and for joint loans using relative frequency tables and bar charts. Because the number of individual loans and the number of joint loans are quite different, it makes more sense to compare proportions rather than numbers.

Homeownership for Individual loans	Relative frequency	Homeownership for Joint loans	Relative frequency
mortgage	0.4514	mortgage	0.6355
own	0.1376	own	0.1224
rent	0.4110	rent	0.2421

Figure 1.11: Two tables summarizing the relative frequencies of each value for the `homeownership` variable. The table on the left includes only individual loans and the table on the right includes only joint loans.

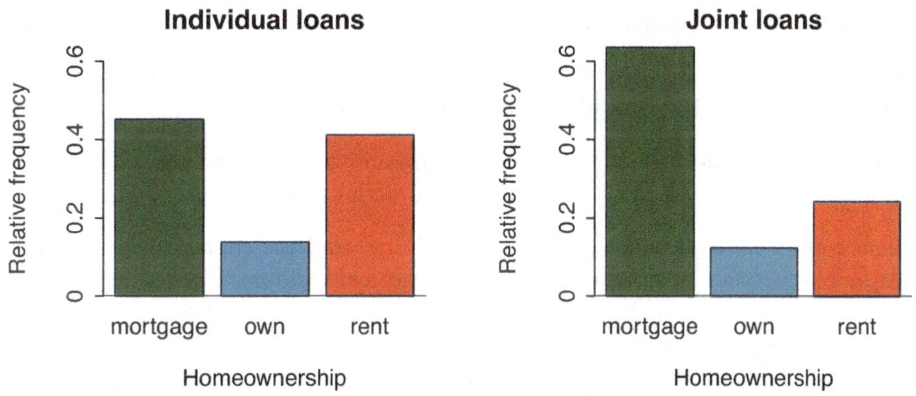

Figure 1.12: Two bar charts summarizing the relative frequencies of each value for the `homeownership` variable. The bar chart on the left includes only individual loans and the bar chart on the right includes only joint loans.

EXAMPLE 1.9

Using the bar charts in Figure 1.12, compare homeownership type for those with individual versus joint loans. What differences can be observed?

A higher percent of joint loans than individual loans are for mortgage (a little over 60% versus a little over 40%), whereas a higher percent of individual loans than joint loans are for rent (about 40% versus about 25%).

Section summary

- Tabular representations of a single categorical variable include a (one-way) frequency table and a relative frequency table. A **frequency table** shows the **number** or **count** of observational units in each category of a categorical variable, while a **relative frequency table shows** the **proportion** or **percent**.
- Percentages, relative frequencies, and ratios all provide the same information as proportions.
- **Bar charts**, also called bar graphs, display frequencies (counts) or relative frequencies (proportions) for the categories of a single categorical variable. Each bar on a bar chart represents a category of the categorical variable of interest. The height or length of each bar corresponds to the frequency.
- **Pie charts** are used to display frequencies (counts) or relative frequencies (proportions) for categorical data. Each slice on a pie chart represents a category of the categorical variable of interest. The area of each slice, as a fraction of the total area, corresponds to the relative frequency of observational units falling within each category. The sum of the slices' areas together will equal 1, or 100% of the total area.
- Pie charts can be more difficult to read and bar charts are generally a better option.
- Frequency and relative frequency tables, bar charts, and pie charts can be used to compare two or more data sets in terms of the same categorical variable.
- Counts, relative frequencies, and graphical representations of categorical variables reveal information that can be used to justify claims about the variables in context.
- Frequency and relative frequency tables, bar charts, and pie charts can be used to compare two or more data sets in terms of the same categorical variable.

Exercises

1.9 Antibiotic use in children. The bar plot and the pie chart below show the distribution of pre-existing medical conditions of children involved in a study on the optimal duration of antibiotic use in treatment of tracheitis, which is an upper respiratory infection.

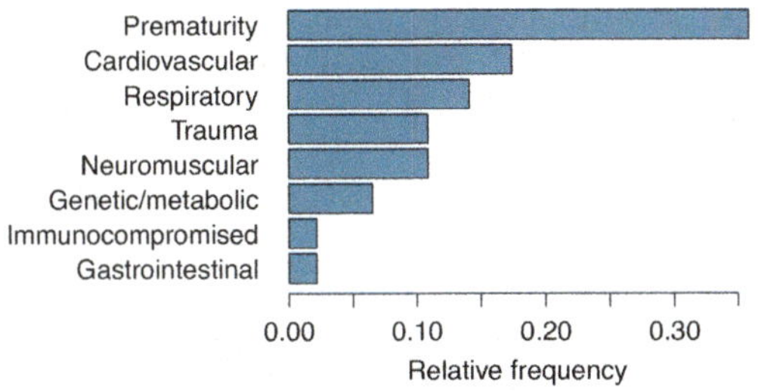

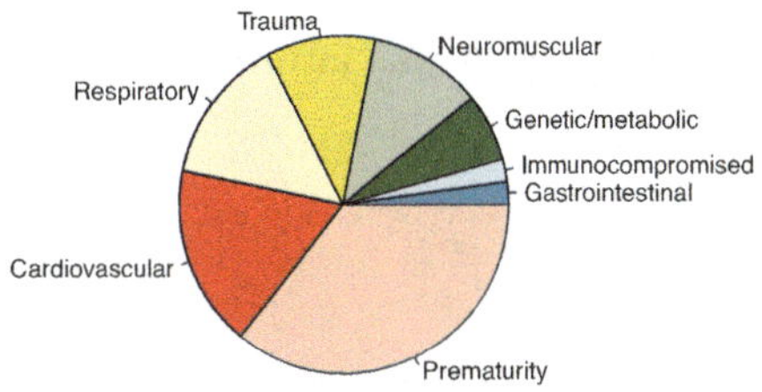

(a) What features are apparent in the bar plot but not in the pie chart?

(b) What features are apparent in the pie chart but not in the bar plot?

(c) Which graph would you prefer to use for displaying these categorical data?

1.10 Health coverage, frequencies. The Behavioral Risk Factor Surveillance System (BRFSS) is an annual telephone survey designed to identify risk factors in the adult population and report emerging health trends. The following table summarizes the respondents reported health status:[12]

Health Status	
Excellent	4,657
Very good	6,972
Good	5,675
Fair	2,019
Poor	677
Total	20,000

(a) Report the relative frequency of respondents with Excellent health status as a ratio, a proportion, and a percent.

(b) Report the relative frequency of respondents with Fair or Poor health status as a ratio, a proportion, and a percent.

(c) True or False: Most people reported Very good health status.

[12]Office of Surveillance, Epidemiology, and Laboratory Services Behavioral Risk Factor Surveillance System, BRFSS 2010 Survey Data.

1.3 Representing numerical data with graphs

How do we visualize and describe the distribution of household income for counties within the United States? What shape would the distribution have? What other features might be important to notice? In this section, we will explore techniques for summarizing numerical variables using the `loan50` data set, which was introduced in Section 1.1.

Learning objectives

1. Understand what the term *distribution* means and how to summarize it in a table or graph.
2. Create and interpret stem-and-leaf plots, dot plots, and histograms for visualizing the distribution of a numerical variable.
3. Describe the shape of a distribution as approximately symmetric, right skewed, or left skewed, and identify a distribution as unimodal, bimodal, multimodal, or uniform.
4. Summarize a distribution with respect to center, spread, shape, gaps, clusters, and outliers.
5. Justify a claim using graphical representations of a quantitative variable.

1.3.1 Stem-and-leaf plots and dot plots

Here we revisit the `loan50` data set, which includes data on 50 randomly sampled loans offered through Lending Club, a peer-to-peer lending company, and we look at the variable `interest_rate`. We would like to visualize and summarize the distribution of this numerical variable. The term **distribution** refers to the values that a variable takes and the frequency of these values. To simplify the `interest_rate` data, we round the values to the nearest percent. For example, 10.9% is recorded as 11.

11	10	26	10	9	10	17	6	8	13
17	5	7	5	8	25	18	10	8	19
14	20	9	10	11	5	7	15	12	13
11	9	10	7	18	17	8	6	7	7
13	16	11	10	9	10	21	11	9	6

Figure 1.13: The interest rate, in %, for 50 loans from Lending Club.

Rather than look at the data as a list of numbers, which makes the distribution difficult to discern, we will organize it into a table called a **stem-and-leaf plot** shown in Figure 1.14. In a stem-and-leaf plot, each number is broken into two parts. The first part is called the **stem** and consists of the beginning digit(s). The second part is called the **leaf** and consists of the final digit(s). The stems are written in a column in ascending order, and the leaves that match up with those stems are written on the corresponding row, with the leaf values getting larger as they are farther from the stem. Figure 1.14 shows a stem-and-leaf plot of the interest rate for 50 loans from Lending Club. The stem represents the tens place and the leaf represents the ones place. For example, `0 | 5` corresponds to 5% and `2 | 6` corresponds to 26%. When making a stem-and-leaf plot, remember to include a legend that describes what the stem and what the leaf represent. Without this, there is no way of knowing if 2 | 6 represents 2.6, 26, 260, 2600, etc.

```
0 | 5556667777788889999
1 | 000000001111123333456777889
2 | 0156


Legend: 2 | 0 = 20%
```

Figure 1.14: A stem-and-leaf plot of the interest rate in 50 loans.

G

GUIDED PRACTICE 1.10

There are only four numbers on the bottom row. Why is this the case?[13]

When there are too many numbers on one row or there are only a few stems, we *split* each row into two halves, with the leaves from 0-4 on the first half and the leaves from 5-9 on the second half. The resulting graph is called a **split stem-and-leaf plot**. Figure 1.15 shows the previous stem-and-leaf redone as a split stem-and-leaf.

```
0 |
0 | 5556667777788889999
1 | 000000001111123334
1 | 56777889
2 | 01
2 | 56


Legend: 2 | 5 = 25%
```

Figure 1.15: A split stem-and-leaf.

G

GUIDED PRACTICE 1.11

Rounding to the nearest percent, what is the lowest interest rate in the `loan50` data set? What is the largest?[14]

Another simple graph for univariate numerical data is a dot plot. A **dot plot** represents each value of a numerical variable with a dot and can be oriented vertically or, more commonly, horizontally. When oriented horizontally, each dot is placed above the horizontal axis at the location corresponding to the value of the observation, with nearly identical values stacked on top of each other. The higher the stack of dots, the greater the number of occurrences there are at those values. An example using the same data set, interest rate from 50 loans, is shown in Figure 1.16.

Graphs such as this make it easy to observe important features of the data, such as frequency, the location of clusters, and the presence of gaps.

E

EXAMPLE 1.12

Based on the dot plot, which value has the highest frequency, and are there any gaps in interest rate for the `loan50` data set?

The highest frequency is at 10%. There is a small gap between 21% and 25%.

[13]There are only 4 loans with interest rates in the 20% range.
[14]The lowest interest rate is 5%, and the largest is 26%. That is a big range!

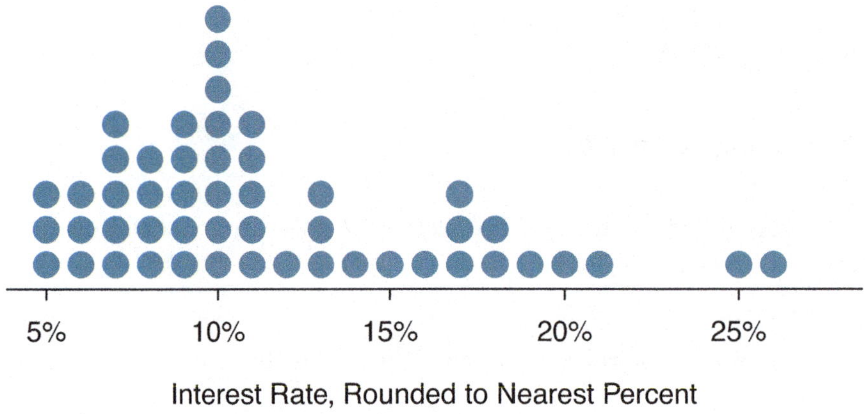

Figure 1.16: A dot plot of `interest_rate` for the `loan50` data set.

Additionally, we can easily identify any observations that appear to be unusually distant from the rest of the data. Unusually distant observations are called **outliers**. Later in this chapter we will provide numerical rules of thumb for identifying outliers. For now, it is sufficient to identify them by observing gaps in the graph. In this case, the values 25% and 26% could be classified as outliers because they are numerically distant from most of the data.

OUTLIERS ARE EXTREME

An **outlier** is an observation that is unusually small or large relative to the rest of the data.

WHY IT IS IMPORTANT TO LOOK FOR OUTLIERS

Examination of data for possible outliers serves many useful purposes, including

1. Identifying asymmetry in the distribution.
2. Identifying data collection or entry errors. For instance, we re-examined the loans purported to have an interest rate of 25% and 26% to ensure these value were accurate.
3. Providing insight into interesting properties of the data.

GUIDED PRACTICE 1.13

G

Consider a data set that consists of the following numbers: 12, 12, 12, 12, 12, 13, 13, 14, 14, 15, 19. Which graph would better illustrate the data: a stem-and-leaf plot or a dot plot? Explain.[15]

[15]Because all the values begin with 1, there would be only one stem (or two in a split stem-and-leaf). This would not provide a good sense of the distribution. For example, the gap between 15 and 19 would not be visually apparent. A dot plot would be better here.

1.3.2 Histograms

Stem-and-leaf plots and dot plots are ideal for displaying data from small samples because they show the exact values of the observations and how frequently they occur. However, they are impractical for larger samples. For larger samples, rather than showing the value of each observation, we prefer to think of the value as belonging to a *bin*. For example, in the `loan50` data set, we created a table of counts for the number of loans with interest rates between 5.0% and 7.5%, then the number of loans with rates between 7.5% and 10.0%, and so on. Observations that fall on the boundary of a bin (e.g. 10.00%) are allocated to the lower bin. This tabulation is shown in Figure 1.17. These binned counts are plotted as bars in Figure 1.19(a) into what is called a **histogram**, which resembles a more heavily binned version of the stacked dot plot shown in Figure 1.16.

Interest Rate	5.0% - 7.5%	7.5% - 10.0%	10.0% - 12.5%	12.5% - 15.0%	···	25.0% - 27.5%
Count	11	15	8	4	···	1

Figure 1.17: Counts for the binned `interest_rate` data.

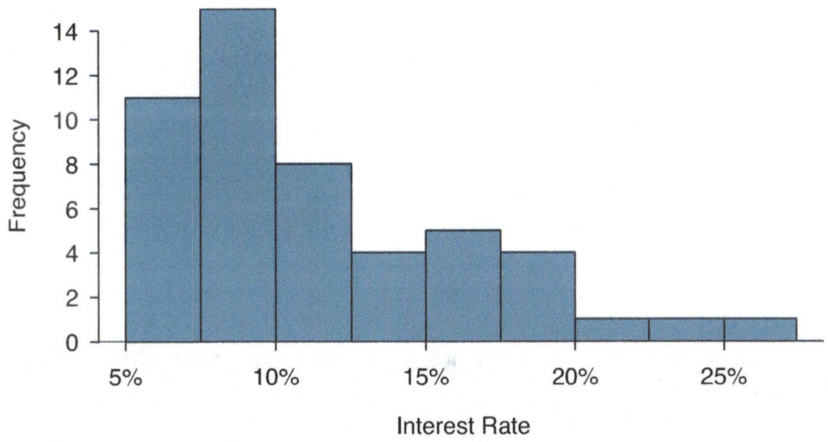

Figure 1.18: A histogram of `interest_rate`. This distribution is strongly skewed to the right.

G

GUIDED PRACTICE 1.14

What can you see in the dot plot and stem-and-leaf plot that you cannot see in the frequency histogram?[16]

DRAWING HISTOGRAMS

1. The variable should be placed on the horizontal axis. Before drawing the histogram, label both axes and draw a scale for each.

2. Draw bars such that the width of the bar corresponds to the bin width and the height of the bar is the frequency of that bin.

Histograms provide a view of the **data density**. Higher bars represent where the data are relatively more common. For instance, there are more loans with interest rates between 7.5%-10% than between 10%-12.5%. The bars make it easy to see how the density of the data changes relative to the number of loans.

[16] Interest rates for individual loans.

EXAMPLE 1.15

How many loans had interest rates less than 10%?

The height of the bars corresponds to frequency. There were 11 cases from 5 to less than 7.5 and 15 cases from 7.5 to less than 10, so there were 11 + 15 = 26 loans that had interest rates less than 10%.

EXAMPLE 1.16

Approximately how many loans had an interest rate less than 6%?

Based just on this histogram, we cannot know the exact answer to this question. We only know that 11 loans had interest rates between 5 and 7.5 percent. If the number of loans is evenly distribution on this interval, then we can estimate that approximately $11/2.5 \approx 4$ loans fell in the range between 5 and 6 percent.

EXAMPLE 1.17

What *percent* of the loans had interest rates greater than 10%?

From the first example, we know that 26 loans had interest rates less than 10%. Because there are 50 loans in total, there must be 24 loans that had interest rates greater than 10%. To find the percent, compute 24/50 = 0.48 = 48%.

Just as we constructed a frequency table and frequency histogram, we can construct a relative frequency table and relative frequency histogram where we represent the proportion rather than the number within each bin.

GUIDED PRACTICE 1.18

How will the appearance of a relative frequency histogram differ from that of the corresponding frequency histogram?[17]

We see in Figure 1.19, the frequency and relative frequency histograms differ only in the scale on the vertical axes. Choose one or the other depending whether it is more important to be able to quickly read off the number or proportion in each bin. When comparing distributions, it is generally better to use relative frequency histograms, especially when the number of observations represented in each graph are very different.

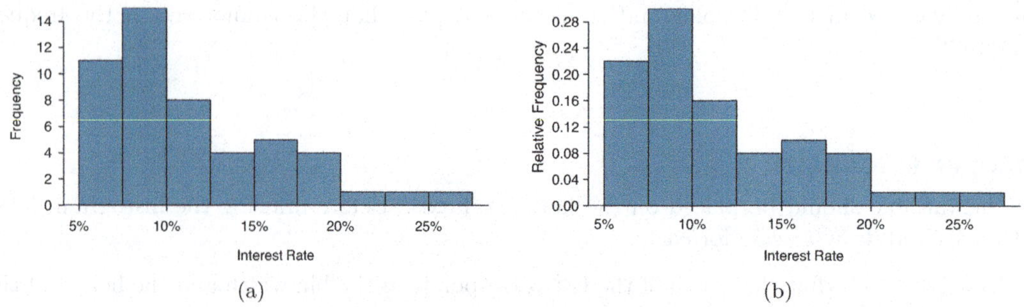

Figure 1.19: A comparison of a frequency histogram with its corresponding relative frequency histogram.

---

[17]Changing from frequency to relative frequency involves dividing all the frequencies by the same number, so only the vertical scale (the numbers on the y-axis) changes. The appearance of the histogram remains the same.

1.3.3 Describing shape

Histograms are especially convenient for describing the **shape** of the data distribution. Figure 1.19(a) shows that more loans have a lower interest rate, while fewer loans have a large interest rate. When data trail off to the right in this way and have a longer right tail, the shape is said to be **right skewed**.[18] Data sets with the reverse characteristic – a longer, thin tail to the left – are said to be **left skewed**. Data sets that show roughly equal trailing off in both directions may be **approximately symmetric**.

LONG TAILS TO IDENTIFY SKEW

When data trail off in one direction, the distribution has a **long tail**. If a distribution has a longer left tail, it is left skewed. If a distribution has a longer right tail, it is right skewed.

G **GUIDED PRACTICE 1.19**

Take a look at the dot plot in Figure 1.16. Can you see the skew in the data? Is it easier to see the skew in the frequency histogram, the dot plot, or the stem-and-leaf plot?[19]

G **GUIDED PRACTICE 1.20**

Would you expect the distribution of number of pets per household to be right skewed, left skewed, or approximately symmetric? Explain.[20]

In addition to looking at whether a distribution is skewed or symmetric, histograms, stem-and-leaf plots, and dot plots can be used to identify modes. A **mode** is represented by a prominent peak in the distribution.[21] There is only one prominent peak in the histogram of `interest_rate`.

Figure 1.20 shows histograms that have one, two, or three prominent peaks. Such distributions are called **unimodal**, **bimodal**, and **multimodal**, respectively. Any distribution with more than 2 prominent peaks is called multimodal. Notice that in Figure 1.19(a) there was one prominent peak in the unimodal distribution with a second less prominent peak that was not counted, because it only differs from its neighboring bins by a few observations.

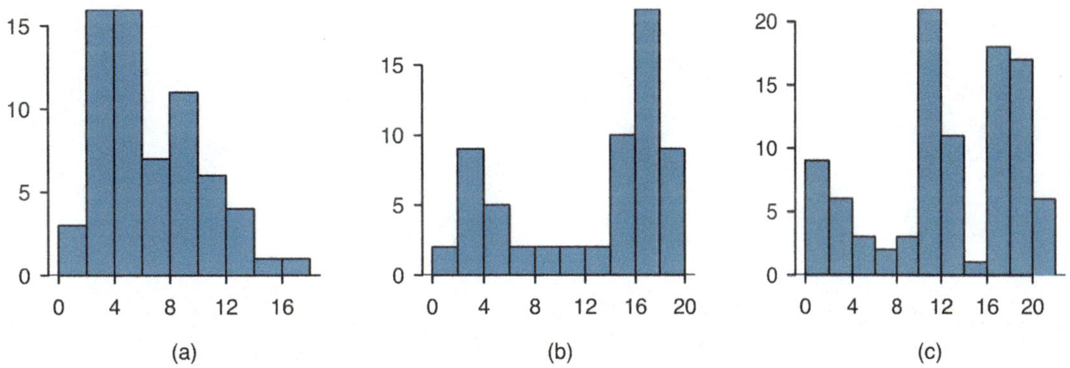

Figure 1.20: Counting only prominent peaks, the distributions are (left to right) unimodal, bimodal, and multimodal.

[18] Other ways to describe data that are right skewed: **skewed to the right**, **skewed to the high end**, or **skewed to the positive end**.

[19] The skew is visible in all three plots.

[20] We suspect most households would have 0, 1, or 2 pets but that a smaller number of households will have 3, 4, 5, or more pets. Based on this, we would expect there to be greater density over the small numbers, suggesting the distribution will have a long right tail and be right skewed.

[21] Another definition of mode, which is not typically used in statistics, is the value with the most occurrences. It is common to have *no* observations with the same value in a data set, which makes this other definition useless for many real data sets.

When a distribution has no prominent peaks and when the frequency of each value is approximately the same, we call the distribution **uniform**, or uniformly distributed.

GUIDED PRACTICE 1.21

Height measurements of young students and adult teachers at a K-3 elementary school were taken. Would you anticipate the distribution of height for this data set to be uniform, unimodal, bimodal or multimodal?[22]

1.3.4 Summarizing distributions

We can now summarize the shape of distributions of a numerical variable in terms of skew and modes. We would also like to have some sense of the center and the variability, or spread, of a distribution. We will quantify these terms in the next section. For now let us develop an intuitive understanding of the terms center and spread by comparing histograms (a) and (b) in Figure 1.20.

GUIDED PRACTICE 1.22

Compare histograms (a) and (b) in Figure 1.20. Which distribution has a larger center? Which has a larger spread?

When summarizing the distribution of a numerical variable, it is helpful to comment on center, spread, and shape, as well as note the presence of cluster, gaps, or outliers. For example, in graph (b) of Figure 1.20, we observe that there are two peaks and two clusters near the values of 4 and 18 and the distribution of the variable is somewhat left skewed. There are no apparent gaps or outliers.

The graphical summaries of this section and the numerical summaries of the next section fall into the realm of **descriptive statistics**. Descriptive statistics is about describing or summarizing data; it does not attribute properties of the data to a larger population. **Inferential statistics**, on the other hand, uses samples to generalize or to infer something about a larger population. We will delve into inferential statistics in Chapter 3 and Chapter 4.

[22]There might be two height groups visible in the data set: one of the students and one of the adults. That is, the data are probably bimodal.

[22]Note that the horiztonal axis is not the same for graphs (a) and (b). Visually, we would say that distribution (b) has a larger center. If we very roughly estimate the center of distribution (a) at around 8, we can see that the center of distribution (b) is certainly higher than 8. Also, distribution (b) has larger variability/spread as it has a larger range and less concentration in the middle.

Section summary

- When looking at a **univariate** numerical display, researchers want to understand the distribution of the variable. The term **distribution** refers to the values that a variable takes and the frequency of those values.
- Two graphs that are useful for showing the distribution of a small number of observations are the stem-and-leaf plot and dot plot. These graphs are ideal for displaying data from small samples because they show the exact values of the observations and how frequently they occur. For larger data sets it is common to use a histogram to display the distribution of a variable. A histogram shows the frequency or relative frequency of values or intervals of values.
- A **histogram** places the observed values of the numerical variable into ordered intervals, or bins, along the horizontal axis. Each bar represents an interval or bin, and the height of each bar shows the frequency or relative frequency of the observations within that interval. Altering the interval widths, or bin widths, can change the appearance of the histogram. Alternatively, a histogram can be constructed with bins on the vertical axis with bars appearing horizontally.
- A **stem-and-leaf plot** splits each value of the numerical variable into two parts: a "stem" (the first digit or digits) and a "leaf" (usually the single digit after the stem digit or digits). Both stems and leaves are ordered from smallest to largest.
- A **dot plot** represents each value of the numerical variable by a dot. Each dot is placed above the horizontal or beside the vertical axis corresponding to the value of that observation, with nearly identical values stacked on top of each other.
- Descriptions of the distribution of one numerical variable should include shape, center, and variability (spread) as well as any unusual features such as outliers, gaps, or clusters in context.
- Distributions may be **symmetric** or they may have a long tail. If a distribution has a long left tail (with greater concentration over the higher numbers), it is **left skewed**. If a distribution has a long right tail (with greater concentration over the smaller numbers), it is **right skewed**.
- Distributions may be **unimodal**, **bimodal**, or **multimodal**, depending on the number of main peaks. A distribution with no prominent peaks, where the frequency of each value is approximately the same is called **uniform**.
- An **outlier** is an observation that is unusually small or large relative to the rest of the data.
- A **gap** is a region in a distribution between two values in which there are no observed data.
- **Clusters** are concentrations of values usually separated by gaps.
- Graphical representations of a quantitative variable may reveal information that can be used to justify claims about the variable in context.

Exercises

1.11 Fiber in your cereal . The Cereal FACTS report provides information on nutrition content of cereals as well as who they are targeted for (adults, children, families). We have selected a random sample of 20 cereals from the data provided in this report. Shown below are the fiber contents (percentage of fiber per gram of cereal) for these cereals.[23]

	Brand	Fiber %		Brand	Fiber %
1	Pebbles Fruity	0.0%	11	Cinnamon Toast Crunch	3.3%
2	Rice Krispies Treats	0.0%	12	Reese's Puffs	3.4%
3	Pebbles Cocoa	0.0%	13	Cheerios Honey Nut	7.1%
4	Pebbles Marshmallow	0.0%	14	Lucky Charms	7.4%
5	Frosted Rice Krispies	0.0%	15	Pebbles Boulders Chocolate PB	7.4%
6	Rice Krispies	3.0%	16	Corn Pops	9.4%
7	Trix	3.1%	17	Frosted Flakes Reduced Sugar	10.0%
8	Honey Comb	3.1%	18	Clifford Crunch	10.0%
9	Rice Krispies Gluten Free	3.3%	19	Apple Jacks	10.7%
10	Frosted Flakes	3.3%	20	Dora the Explorer	11.1%

(a) Create a stem and leaf plot of the distribution of the fiber content of these cereals.

(b) Create a dot plot of the fiber content of these cereals.

(c) Create a histogram and a relative frequency histogram of the fiber content of these cereals.

(d) What percent of cereals contain more than 7% fiber?

1.12 Sugar in your cereal. The Cereal FACTS report from Exercise 1.11 also provides information on sugar content of cereals. We have selected a random sample of 20 cereals from the data provided in this report. Shown below are the sugar contents (percentage of sugar per gram of cereal) for these cereals.

	Brand	Sugar %		Brand	Sugar %
1	Rice Krispies Gluten Free	3%	11	Corn Pops	31%
2	Rice Krispies	12%	12	Cheerios Honey Nut	32%
3	Dora the Explorer	22%	13	Reese's Puffs	34%
4	Frosted Flakes Red. Sugar	27%	14	Pebbles Fruity	37%
5	Clifford Crunch	27%	15	Pebbles Cocoa	37%
6	Rice Krispies Treats	30%	16	Lucky Charms	37%
7	Pebbles Boulders Choc. PB	30%	17	Frosted Flakes	37%
8	Cinnamon Toast Crunch	30%	18	Pebbles Marshmallow	37%
9	Trix	31%	19	Frosted Rice Krispies	40%
10	Honey Comb	31%	20	Apple Jacks	43%

(a) Create a stem and leaf plot of the distribution of the sugar content of these cereals.

(b) Create a dot plot of the sugar content of these cereals.

(c) Create a histogram and a relative frequency histogram of the sugar content of these cereals.

(d) What percent of cereals contain more than 30% sugar?

[23] JL Harris et al. "Cereal FACTS 2012: Limited progress in the nutrition quality and marketing of children's cereals". In: *Rudd Center for Food Policy & Obesity.* 12 (2012).

1.4 Numerical summaries and box plots

What are the different ways to measure the center of a distribution, and why is there more than one way to measure the center? How do you know if a value is "far" from the center? What does it mean to be an outlier? We will continue with the `loan50` data set and investigate multiple quantitative summarizes for numerical data.

Learning objectives

1. Calculate, interpret, and compare the two measures of center (mean and median).
2. Calculate the five-number summary and interpret a percentile.
3. Calculate and interpret the three measures of spread (standard deviation, interquartile range, and range).
4. Describe the effect that changing units has on each of the summary quantities.
5. Identify and apply the two rules of thumb for calculating outliers.
6. Justify the selection of a summary statistic for describing quantitative data.
7. Construct and interpret a box plot, which provides a graphical representation of the five-number summary.
8. Understand how a distribution's shape affects the relationship between the mean and the median.
9. Compare the distribution of a numerical variable across groups, using multiple dot plots or histograms with the same scale, back-to-back stem-and-leaf plots, or parallel box plots.
10. Summarize and compare distributions of a numerical variable by commenting on center, spread, shape, and noticing the presence of cluster, gaps, and outliers.
11. Justify a claim using multiple graphical representations or summary statistics of a quantitative variable.
12. Calculate Z-scores with population parameters and interpret them in context.
13. Compare Z-scores as measures of relative position for distributions.

1.4.1 Measures of center

In the previous section, we saw that modes can occur anywhere in a data set. Therefore, mode is not a measure of center. We understand the term *center* intuitively, but there are multiple ways to formally define what is the center. Here we will focus on the two most common: the mean and median.

The **mean**, often called the **average**, is a common way to measure the center of a distribution of data. To compute the mean interest rate, we add up all the interest rates and divide by the number of observations:

$$\bar{x} = \frac{10.90\% + 9.92\% + 26.30\% + \cdots + 6.08\%}{50} = 11.57\%$$

The sample mean is often labeled $\bar{x}$. The letter x is being used as a generic placeholder for the variable of interest, `interest_rate`, and the bar over the x communicates we're looking at the average interest rate, which for these 50 loans was 11.57%. It is useful to think of the mean as the balancing point of the distribution, and it's shown as a triangle in Figure 1.21.

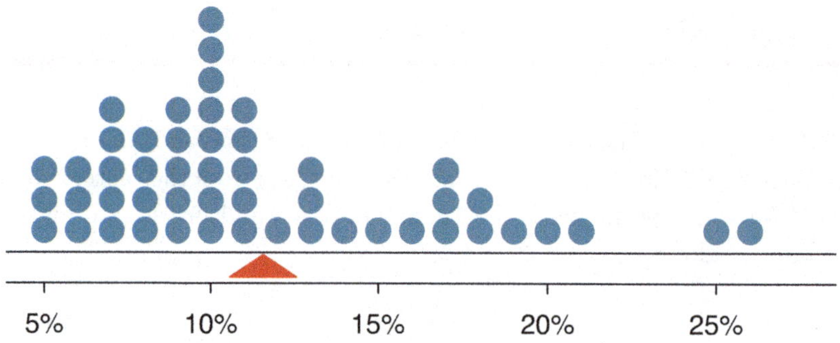

Figure 1.21: A dot plot of `interest_rate` for the `loan50` data set. The distribution's mean is shown as a red triangle.

MEAN

The sample mean can be computed as the sum of the observed values divided by the number of observations:

$$\bar{x} = \frac{x_1 + x_2 + \cdots + x_n}{n}$$

where $x_1, x_2, \ldots, x_n$ represent the n observed values.

GUIDED PRACTICE 1.23

Examine the equation for the mean. What does x_1 correspond to? And x_2? Can you infer a general meaning to what x_i might represent?[24]

GUIDED PRACTICE 1.24

What was n in this sample of loans?[25]

The `loan50` data set represents a sample from a larger population of loans made through Lending Club. We could compute a mean for this population in the same way as the sample mean. However, the population mean has a special label: μ. The symbol μ is the Greek letter *mu* and represents the average of all observations in the population. Sometimes a subscript, such as $_x$, is used to represent which variable the population mean refers to, e.g. μ_x. Often times it is too expensive to measure the population mean precisely, so we often estimate μ using the sample mean, $\bar{x}$.

EXAMPLE 1.25

The average interest rate across all loans in the population can be estimated using the sample data. Based on the sample of 50 loans, what would be a reasonable estimate of μ_x, the mean interest rate for all loans in the full data set?

The sample mean, 11.57%, provides a rough estimate of μ_x. While it's not perfect, this is our single best guess of the average interest rate of all the loans in the population under study.

In Chapter 3 and beyond, we will develop tools to characterize the accuracy of *point estimates* like the sample mean. As you might have guessed, point estimates based on larger samples tend to be more accurate than those based on smaller samples.

[24] x_1 corresponds to the interest rate for the first loan in the sample (10.90%), x_2 to the second loan's interest rate (9.92%), and x_i corresponds to the interest rate for the ith loan in the data set. For example, if $i = 4$, then we're examining x_4, which refers to the fourth observation in the data set.

[25] The sample size was $n = 50$.

The median provides another measure of center. The **median** splits an ordered data set in half. There are 50 interest rates in the `loan50` data set (an even number) so the data are perfectly split into two groups of 25. We take the median in this case to be the average of the two middle observations: $(9.93 + 9.93)/2 = 9.93$. When there are an odd number of observations, there will be exactly one observation that splits the data into two halves, and in this case that observation is the median (no average needed).

MEDIAN: THE NUMBER IN THE MIDDLE

In an ordered data set, the **median** is the observation right in the middle. If there are an even number of observations, the median is the average of the two middle values.

Graphically, we can think of the mean as the balancing point. To estimate the median from a histogram, we try to split the *area* in half so that 50% of the area is to the left and 50% of the area is to the right.

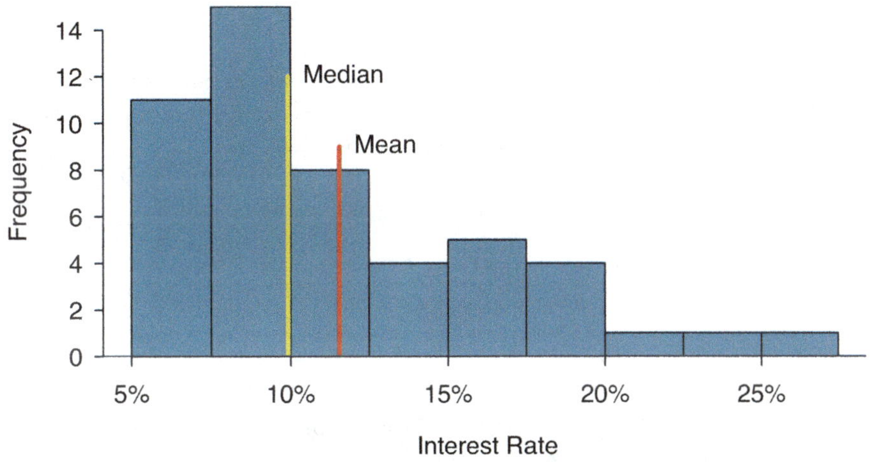

Figure 1.22: A histogram of `interest_rate` with its mean and median shown.

E

EXAMPLE 1.26

Based on the data, why is the mean greater than the median in this data set?

Consider the two largest values which are 25% and 26% . These values drag up the mean because they increase the sum (the total). However, they do not drag up the median because their magnitude does not change the location of the middle value.

THE MEAN FOLLOWS THE TAIL

In a right skewed distribution, the mean is greater than the median.
In a left skewed distribution, the mean is less than the median.
In a symmetric distribution, the mean and median are approximately equal.

G

GUIDED PRACTICE 1.27

Consider the distribution of individual income in the United States. Which is greater: the mean or median? Why?[26]

[26]Because a small percent of individuals earn extremely large amounts of money while the majority earn a modest amount, the distribution is skewed to the right. Therefore, the mean is greater than the median.

1.4.2 Standard deviation and IQR as measures of spread

The mean interest rate in the `loan50` data set is 11.57%. Say that you are offered an interest rate of 15%. Should we consider this value to be far from the mean? In order to answer this question, it is not enough to know the center of the data set and its **range** (computed as: maximum value - minimum value). We must know about the variability of the data set within that range. Low variability or small spread means that the values tend to be more clustered together. High variability or large spread means that the values tend to be far apart.

EXAMPLE 1.28

Is it possible for two data sets to have the same range but different spread? If so, give an example. If not, explain why not.

Yes. An example is: 1, 1, 1, 1, 1, 9, 9, 9, 9, 9 and 1, 5, 5, 5, 5, 5, 5, 5, 5, 5, 9.

The first data set has a larger spread because values tend to be farther away from each other, while in the second data set, most values are clustered together at the mean.

Here, we introduce the standard deviation as a measure of spread. Though its formula is a bit tedious to calculate by hand, the standard deviation is very useful in data analysis and roughly describes how far away, on average, the observations are from the mean.

We call the distance of an observation from its mean its **deviation**. Below are the deviations for the 1st, 2nd, 3rd, and 50th observations in the `interest_rate` variable:

$$\begin{aligned} x_1 - \bar{x} &= 10.90 - 11.57 = -0.67 \\ x_2 - \bar{x} &= 9.92 - 11.57 = -1.65 \\ x_3 - \bar{x} &= 26.30 - 11.57 = 14.73 \\ &\vdots \\ x_{50} - \bar{x} &= 6.08 - 11.57 = -5.49 \end{aligned}$$

If we square these deviations and then take an average, the result is equal to the sample **variance**, denoted by s^2:

$$\begin{aligned} s^2 &= \frac{(-0.67)^2 + (-1.65)^2 + (14.73)^2 + \cdots + (-5.49)^2}{50 - 1} \\ &= \frac{0.45 + 2.72 + 216.97 + \cdots + 30.14}{49} \\ &= 25.52 \end{aligned}$$

We divide by $n - 1$, rather than dividing by n, when computing a sample's variance; there's some mathematical nuance here, but the end result is that doing this makes this statistic slightly more reliable and useful.

Notice that squaring the deviations does two things. First, it makes large values relatively much larger, seen by comparing $(-0.67)^2$, $(-1.65)^2$, $(14.73)^2$, and $(-5.49)^2$. Second, it gets rid of any negative signs.

The **standard deviation** is defined as the square root of the variance:

$$s = \sqrt{25.52} = 5.05$$

While often omitted, a subscript of $_x$ may be added to the variance and standard deviation, i.e. s_x^2 and s_x, if it is useful as a reminder that these are the variance and standard deviation of the observations represented by x_1, x_2, ..., x_n.

CALCULATING THE STANDARD DEVIATION

The standard deviation is the square root of the variance. It is roughly the "typical" distance of the observations from the mean.

$$s_X = \sqrt{\frac{1}{n-1}\sum (x_i - \bar{x})^2}$$

The variance is useful for mathematical reasons, but the standard deviation is easier to interpret because it has the same units as the data set. The units for variance will be the units squared (e.g. meters^2). Formulas and methods used to compute the variance and standard deviation for a population are similar to those used for a sample.[27] However, like the mean, the population values have special symbols: σ^2 for the variance and σ for the standard deviation. The symbol σ is the Greek letter *sigma*.

THINKING ABOUT THE STANDARD DEVIATION

It is useful to think of the standard deviation as the "typical" or "average" distance that observations fall from the mean.

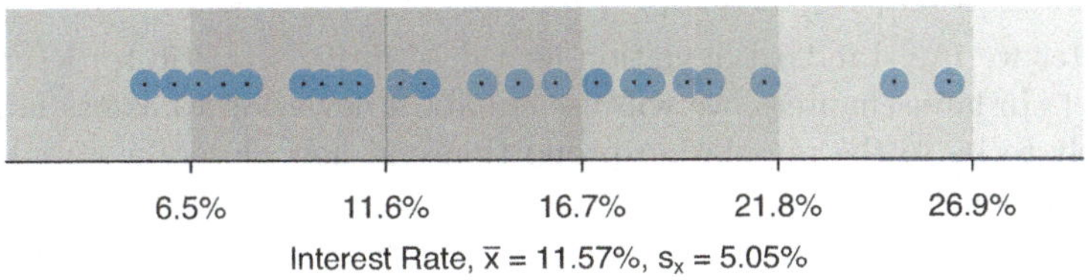

Figure 1.23: For the `interest_rate` variable, 34 of the 50 loans (68%) had interest rates within 1 standard deviation of the mean, and 48 of the 50 loans (96%) had rates within 2 standard deviations. Usually about 70% of the data are within 1 standard deviation of the mean and 95% within 2 standard deviations, though this is far from a hard rule.

E

EXAMPLE 1.29

Given that the average interest rate for the `loan50` data set is 11.57%, with a standard deviation of 5.05%, is an interest rate of 15% especially far from the mean?

Because 15% is within one standard deviation of the mean, it is not especially far from the mean. If the value were more than 2 standard deviations away from the mean, we would consider it far from the mean.

In Chapter 2, we encounter a bell-shaped distribution known as the *normal distribution*. The **empirical rule** tells us that for nearly normal distributions, about 68% of the data will be within one standard deviation of the mean, about 95% will be within two standard deviations of the mean, and about 99.7% will be within three standard deviations of the mean. However, as seen in Figure 1.24, these percentages generally do not hold if the distribution is not bell-shaped.

G

GUIDED PRACTICE 1.30

On page 31, the concept of shape of a distribution was introduced. A good description of the shape of a distribution should include modality and whether the distribution is symmetric or skewed to one side. Using Figure 1.24 as an example, explain why such a description is important.[28]

---

[27] The only difference is that the population variance has a division by n instead of $n-1$.

[28] Figure 1.24 shows three distributions that look quite different, but all have the same mean, variance, and standard deviation. Using modality, we can distinguish between the first plot (bimodal) and the last two (unimodal). Using skewness, we can distinguish between the last plot (right skewed) and the first two. While a graph tells a more complete story, we can use modality and shape (symmetry/skew) to characterize basic information about a distribution.

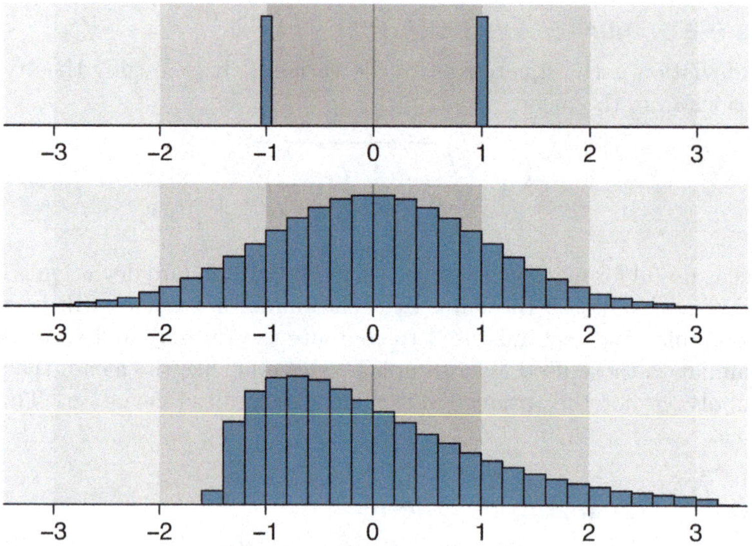

Figure 1.24: Three very different population distributions with the same mean $\mu = 0$ and standard deviation $\sigma = 1$.

In this chapter we use standard deviation as a descriptive statistic to describe the variability in a given data set. In later chapters, we will use standard deviation to assess how closely a sample proportion is likely to be to the population proportion and how closely a sample mean is likely to be to the population mean.

Another measure of spread involves looking at the range of the middle 50% of the data in a data set. Q_1 represents the **first quartile**, which is the 25th percentile, and is the median of the smaller half of the data set. Q_3 represents the **third quartile**, or 75th percentile, and is the median of the larger half of the data set. We calculate the variability in the data using the range of the middle 50% of the data: $Q_3 - Q_1$. This quantity is called the **interquartile range** (IQR, for short). It, like the standard deviation, is a measure of variability or **spread** in data. The more variable the data, the larger the standard deviation and IQR tend to be.

In the `loan50` data set, there are 50 values, so there are 25 values in the lower and in the upper half of the data set. Q_1 is the median of the lower 25 values, and so is the middle or 13th value. In this case, Q_1 is 7.96%. Q_3 is the median of the upper 25 values, and is the 38th value, which is 13.72%.

INTERQUARTILE RANGE (IQR)

The IQR is computed as

$$IQR = Q_3 - Q_1$$

where Q_1 and Q_3 are the 25th and 75th percentiles.

EXAMPLE 1.31

Calculate the IQR of interest rate for the `loan50` data set. How does this compare to the previously calculated standard deviation of interest rate?

The $IQR = Q_3 - Q_1 = 13.72 - 7.96 = 5.76$. The IQR of interest rate is 5.76%, while the standard deviation of interest rate was 5.05%. In general, these two measures of spread will yield different values.

1.4.3 Linear transformations of data and changing units

EXAMPLE 1.32

Begin with the following list: 1, 1, 5, 5. Multiply all of the numbers by 10. What happens to the mean? What happens to the standard deviation? How do these compare to the mean and the standard deviation of the original list?

E

The original list has a mean of 3 and a standard deviation of 2. The new list: 10, 10, 50, 50 has a mean of 30 with a standard deviation of 20. Because all of the values were multiplied by 10, both the mean and the standard deviation were multiplied by 10. [29]

EXAMPLE 1.33

Start with the following list: 1, 1, 5, 5. Multiply all of the numbers by -0.5. What happens to the mean? What happens to the standard deviation? How do these compare to the mean and the standard deviation of the original list?

E

The new list: -0.5, -0.5, -2.5, -2.5 has a mean of -1.5 with a standard deviation of 1. Because all of the values were multiplied by -0.5, the mean was multiplied by -0.5. Multiplying all of the values by a negative flipped the sign of numbers, which affects the location of the center, but not the spread. Multiplying all of the values by -0.5 multiplied the standard deviation by +0.5 because the standard deviation cannot be negative.

EXAMPLE 1.34

Again, start with the following list: 1, 1, 5, 5. Add 100 to every entry. How do the new mean and standard deviation compare to the original mean and standard deviation?

E

The new list is: 101, 101, 105, 105. The new mean of 103 is 100 greater than the original mean of 3. The new standard deviation of 2 is the *same* as the original standard deviation of 2. Adding a constant to every entry shifted the values, but did not stretch or contract them.

Suppose that a researcher is looking at a list of 500 temperatures recorded in Celsius (C). The mean of the temperatures listed is given as 27°C with a standard deviation of 3°C. Because she is not familiar with the Celsius scale, she would like to convert these summary statistics into Fahrenheit (F). To convert from Celsius to Fahrenheit, we use the following conversion:

$$x_F = \frac{9}{5}x_C + 32$$

Fortunately, she does not need to convert each of the 500 temperatures to Fahrenheit and then recalculate the mean and the standard deviation. The unit conversion above is a linear transformation of the following form, where $a = 9/5$ and $b = 32$:

$$aX + b$$

[29]Here, the population standard deviation was used in the calculation. These properties can be proven mathematically using properties of sigma (summation).

Using the examples as a guide, we can solve this temperature-conversion problem. The mean was 27°C and the standard deviation was 3°C. To convert to Fahrenheit, we multiply all of the values by 9/5, which multiplies both the mean and the standard deviation by 9/5. Then we add 32 to all of the values which adds 32 to the mean but does not change the standard deviation further.

$$\begin{aligned} \bar{x}_F &= \frac{9}{5}\bar{x}_C + 32 \\ &= \frac{5}{9}(27) + 32 \\ &= 80.6 \end{aligned} \qquad \begin{aligned} \sigma_F &= \frac{9}{5}\sigma_C \\ &= \frac{9}{5}(3) \\ &= 5.4 \end{aligned}$$

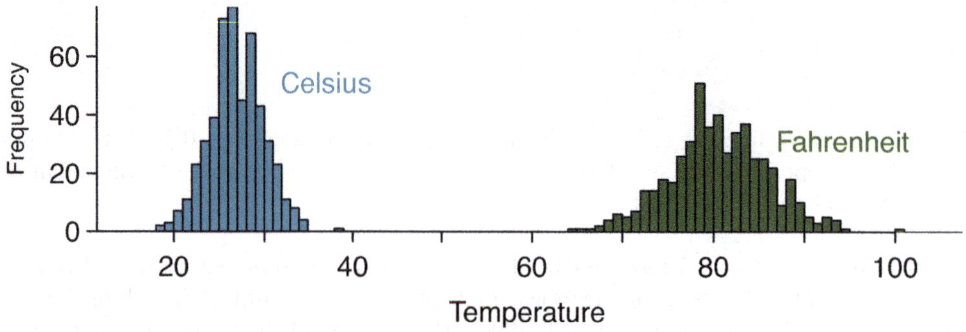

Figure 1.25: 500 temperatures shown in both Celsius and Fahrenheit.

ADDING SHIFTS THE VALUES, MULTIPLYING STRETCHES OR CONTRACTS THEM

Adding a constant to every value in a data set shifts the mean but does not affect the standard deviation. Multiplying the values in a data set by a constant will change the mean and the standard deviation by the same multiple, except that the standard deviation will always remain positive.

EXAMPLE 1.35

Consider the temperature example. How would converting from Celsuis to Fahrenheit affect the median? The IQR?

The median is affected in the same way as the mean and the IQR is affected in the same way as the standard deviation. To get the new median, multiply the old median by 9/5 and add 32. The IQR is computed by subtracting Q_1 from Q_3. While Q_1 and Q_3 are each affected in the same way as the median, the additional 32 added to each will cancel when we take $Q_3 - Q_1$. That is, the IQR will be increase by a factor of 9/5 but will be unaffected by the addition of 32.

For a more mathematical explanation of the IQR calculation, see the footnote.[30]

---

[30] New IQR $= \left(\frac{9}{5}Q_3 + 32\right) - \left(\frac{9}{5}Q_1 + 32\right) = \frac{9}{5}\left(Q_3 - Q_1\right) = \frac{9}{5} \times \text{(old IQR)}$.

1.4.4 Outliers and robust statistics

Why do we have two measures of spread? As with median, the IQR is more robust, that is, more resistant to outliers. While the IQR only looks at the 25th and 75th percentile values, the standard deviation calculation involves all the data points and so can be inflated by extremely small or extremely large values.

Here we present two common rules of thumb for identifying outliers.

RULES OF THUMB FOR IDENTIFYING OUTLIERS

There are two rules of thumb for identifying outliers in a numerical data set:

- More than 1.5× IQR below Q_1 or above Q_3
- More than 2 standard deviations above or below the mean.

Both are important for the AP exam. In practice, consider these to be only rough guidelines.

GUIDED PRACTICE 1.36

For the `loan50` data set, $Q_1 = 7.96$ and $Q_3 = 13.72$. $\bar{x} = 11.57$ and $s = 5.05$. What values would be considered an outlier on the low end using each rule?[31]

GUIDED PRACTICE 1.37

Examine the distribution of the interest rates in the `loan50` data set. What does the fact that there are outliers on the high end but not on the low end suggestion?[32]

How are the sample statistics of the `interest_rate` data set affected by the observation, 26.3%? What would have happened if this loan had instead been only 15%? What would happen to these summary statistics if the observation at 26.3% had been even larger, say 35%? These scenarios are plotted alongside the original data in Figure 1.26, and sample statistics are computed under each scenario in Figure 1.27.

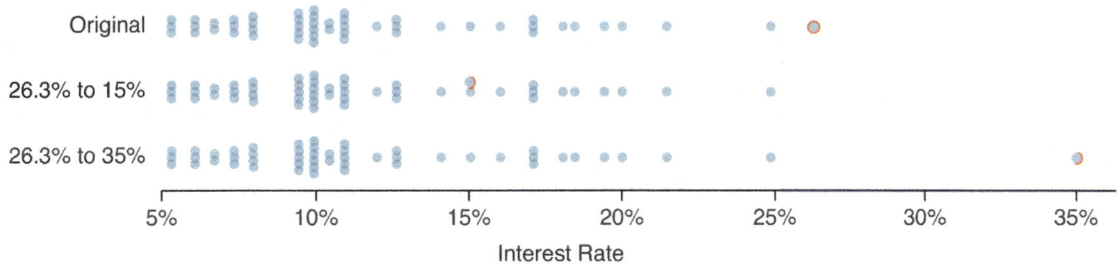

Figure 1.26: Dot plots of the original interest rate data and two modified data sets.

[31] $Q_1 - 1.5 \times IQR = 7.96 - 1.5 \times (13.72 - 7.96) = 5.76$, so values less than -1.22% would be considered an outlier using the first rule of thumb. Using the second rule of thumb, a value less than $\bar{x} - 2 \times s = 11.57 - 2 \times 5.05 = 1.47$ would be considered an outlier. Note that these are just rules of thumb and usually yield different values.

[32] It suggests that the distribution has a right hand tail, that is, that it is right skewed.

	robust		not robust	
scenario	median	IQR	$\bar{x}$	s
original interest_rate data	9.93%	5.76%	11.57%	5.05%
move 26.3% → 15%	9.93%	5.76%	11.34%	4.61%
move 26.3% → 35%	9.93%	5.76%	11.74%	5.68%

Figure 1.27: A comparison of how the median, IQR, mean ($\bar{x}$), and standard deviation (s) change had an extreme observations from the interest_rate variable been different.

GUIDED PRACTICE 1.38

(a) Which is more affected by extreme observations, the mean or median? Figure 1.27 may be helpful. (b) Is the standard deviation or IQR more affected by extreme observations?[33]

The median and IQR are called **robust statistics** because extreme observations have little effect on their values: moving the most extreme value generally has little influence on these statistics. On the other hand, the mean and standard deviation are more heavily influenced by changes in extreme observations, which can be important in some situations.

EXAMPLE 1.39

The median and IQR did not change under the three scenarios in Figure 1.27. Why might this be the case?

The median and IQR are only sensitive to numbers near Q_1, the median, and Q_3. Because values in these regions are stable in the three data sets, the median and IQR estimates are also stable.

GUIDED PRACTICE 1.40

The distribution of loan amounts in the loan50 data set is right skewed, with a few large loans lingering out into the right tail. If you were wanting to understand the typical loan size, should you be more interested in the mean or median?[34]

---

[33](a) Mean is affected more. (b) Standard deviation is affected more. Complete explanations are provided in the material following Guided Practice 1.38.

[34]Answers will vary! If we're looking to simply understand what a typical individual loan looks like, the median is probably more useful. However, if the goal is to understand something that scales well, such as the total amount of money we might need to have on hand if we were to offer 1,000 loans, then the mean would be more useful.

1.4.5 Box plots

A **box plot** summarizes a data set using five statistics while also plotting unusual observations. Figure 1.28 provides a vertical dot plot alongside a box plot of the `interest_rate` variable from the `loan50` data set.

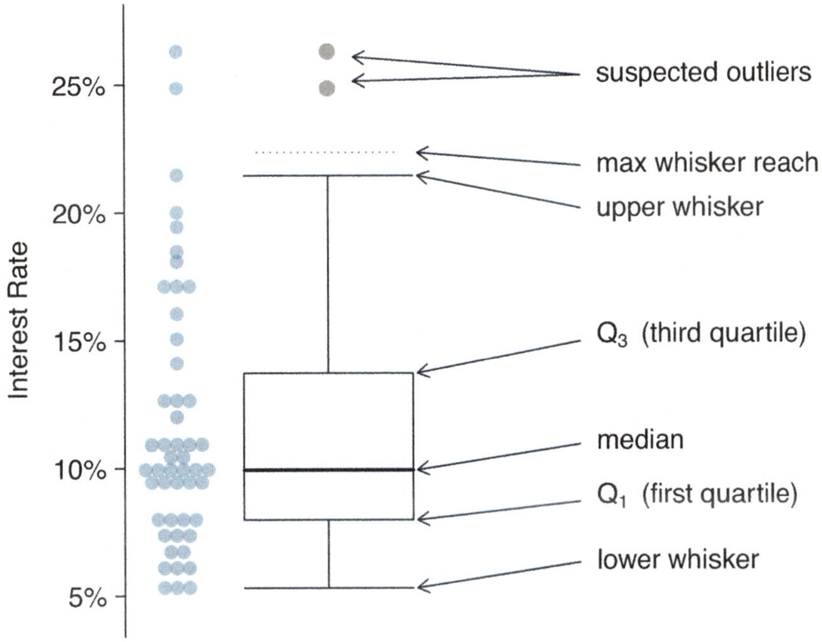

Figure 1.28: A vertical dot plot, where points have been horizontally stacked, next to a labeled box plot for the interest rates of the 50 loans.

The five summary statistics used in a box plot are known as the **five-number summary**, which consists of the minimum, the maximum, and the three quartiles (Q_1, Q_2, Q_3) of the data set being studied. Recall that Q_1 represents the 25th percentile and Q_3 represents the 75th percentile. Q_2 represents the 50th percentile, also known as the median.

To build a box plot, draw an axis (vertical or horizontal) and draw a scale. Draw a dark line denoting Q_2, the median. Next, draw a line at Q_1 and at Q_3. Connect the Q_1 and Q_3 lines to form a rectangle. The width of the rectangle corresponds to the IQR and the middle 50% of the data is in this interval.

OUTLIERS IN THE CONTEXT OF A BOX PLOT

When in the context of a box plot, define an **outlier** as an observation that is more than $1.5 \times IQR$ above Q_3 or $1.5 \times IQR$ below Q_1. Such points are marked using a dot or asterisk in a box plot.

Extending out from the rectangle, the **whiskers** attempt to capture all of the data remaining outside of the box, except outliers. In Figure 1.28, the upper whisker does not extend to the last two points, which are beyond $Q_3 + 1.5 \times IQR$ and are outliers, so it extends only to the last point below this limit.[35] The lower whisker stops at the lowest value, 5, because there are no additional data to reach. Outliers are each marked with a dot or asterisk. In a sense, the box is like the body of the box plot and the whiskers are like its arms trying to reach the rest of the data.

Box plots may be oriented vertically as seen in Figure 1.28 or horizontally as seen in Figure 1.29. In both cases, the box plot and the information it tells us is identical, with the only difference being the axis that the variable is graphed along.

[35] You might wonder, isn't the choice of $1.5 \times IQR$ for defining an outlier arbitrary? It is! In practical data analyses, we tend to avoid a strict definition since what is an unusual observation is highly dependent on the context of the data.

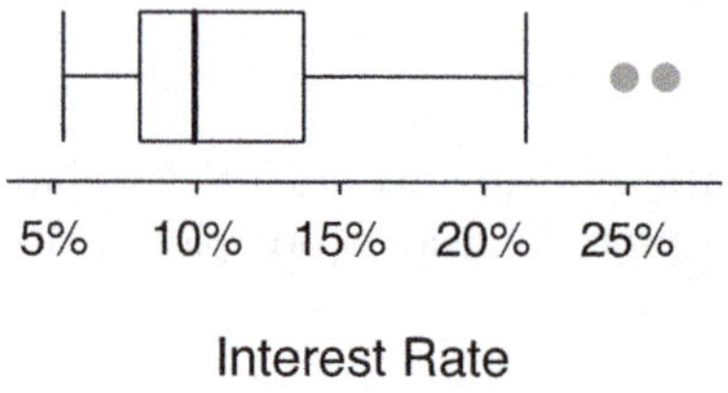

Figure 1.29: A box plot for interest rate, with the variable along the horizontal axis.

GUIDED PRACTICE 1.41

Using Figure 1.29, estimate the following values for `interest_rate` in the `loan50` data set: (a) Q_1, (b) Q_3, and (c) IQR.[36]

EXAMPLE 1.42

Compare the box plot to the graphs previously discussed: stem-and-leaf plot, dot plot, frequency and relative frequency histogram. What can we learn more easily from a box plot? What can we learn more easily from the other graphs?

It is easier to immediately identify the quartiles from a box plot. The box plot also more prominently highlights outliers. However, a box plot, unlike the other graphs, does not show the *distribution* of the data. For example, we cannot generally identify modes using a box plot.

EXAMPLE 1.43

Is it possible to identify skew from the box plot?

Yes. Looking at the lower and upper whiskers of this box plot, we see that the lower 25% of the data is squished into a shorter distance than the upper 25% of the data, implying that there is greater density in the low values and a tail trailing to the upper values. This box plot is right skewed.

GUIDED PRACTICE 1.44

True or false: there is more data between the median and Q_3 than between Q_1 and the median.[37]

---

[36] These visual estimates will vary a little from one person to the next: $Q_1 = 8\%$, $Q_3 = 14\%$, $\text{IQR} = Q_3 - Q_1 = 6\%$. (The true values: $Q_1 = 7.96\%$, $Q_3 = 13.72\%$, $\text{IQR} = 5.76\%$.)

[37] False. Because Q_1 is the 25th percentile and the median is the 50th percentile, 25% of the data fall between Q_1 and the median. Similarly, 25% of the data fall between Q_2 and the median. The distance between the median and Q_3 is larger because that 25% of the data is more spread out.

EXAMPLE 1.45

Consider the following ordered data set.

5 5 9 10 15 16 30 40 80

Find the five-number summary and identify how small or large a value would need to be to be considered an outlier. Are there any outliers in this data set?

There are nine numbers in this data set. Because n is odd, the median is the middle number: 15. When finding Q_1, we find the median of the lower half of the data, which in this case includes 4 numbers (we do not include the 15 as belonging to either half of the data set). Q_1 then is the average of 5 and 9, which is $Q_1 = 7$, and Q_3 is the average of 30 and 40, so $Q_3 = 35$. The min is 5 and the max is 80. To see how small a number needs to be to be an outlier on the low end we do:

$$\begin{aligned} Q_1 - 1.5 \times IQR &= Q_1 - 1.5 \times (Q_3 - Q_1) \\ &= 7 - 1.5 \times (35 - 7) \\ &= -35 \end{aligned}$$

On the high end we need:

$$\begin{aligned} Q_3 + 1.5 \times IQR &= Q_3 + 1.5 \times (Q_3 - Q_1) \\ &= 35 + 1.5 \times (35 - 7) \\ &= 77 \end{aligned}$$

There are no numbers less than -41, so there are no outliers on the low end. The observation at 80 is greater than 77, so 80 is an outlier on the high end.

1.4.6 Comparing numerical data across groups

Some of the more interesting investigations can be considered by examining numerical data across groups. The methods required here aren't really new. All that is required is to make a numerical plot for each group. To make a direct comparison between two groups, create a pair of dot plots or a pair of histograms drawn using the same scales. It is also common to use back-to-back stem-and-leaf plots and parallel box plots, all of which are explored here.

The `loan50` data set is a sample from the much larger data set `loans_full_schema`. To compare interest rates across the individual and joint application types, we take a look at a random sample of 50 individual loans and a random sample of 40 joint loans from the `loans_full_schema` data set. The interest rates for each loan are shown in Figure 1.30, separated by application type.

Interest rate for randomly sampled loans

Individual					Joint			
14	14	20	14	14	7	6	12	15
9	13	19	11	13	20	20	8	22
7	11	14	14	18	9	10	10	5
15	13	19	16	12	13	16	8	17
17	6	15	14	17	10	23	9	15
15	14	12	6	7	27	12	15	15
8	14	17	15	16	15	7	26	6
12	7	31	12	7	14	18	21	14
13	17	12	10	16	21	10	12	31
12	7	12	7	16	30	14	15	12

Figure 1.30: Left: interest rate for a random sample of 50 individual loan. Right: interest rate for a random sample of 40 joint loans.

In order to compare the distribution of interest rate for each loan type, we start with a **back-to-back stem-and-leaf plot**, shown in Figure 1.31. As with a stem-and-leaf plot, a back-to-back stem-and-leaf plot is useful when the data sets have a small number of values. The leaf values should get larger as you move farther away from the central stem. Here, the stem represents the tens place, the leaf represents the ones place, and we split the stems to stretch them out.

```
               Individual loans          Joint loans

                         9877777766| 0 |566778899
           444333333222211111111100| 1 |000022223444
                   998777766665555| 1 |555555678
                                 0| 2 |001123
                                  | 2 |67
                                 0| 3 |01

          Legend: 3 |1 = 31% interest rate
```

Figure 1.31: Back-to-back stem-and-leaf plot for interest rate, split by whether the loan type was individual or joint.

In addition to a back-to-back stem-and-leaf plot, we can use two dot plots or two histograms vertically stacked on the same scale for easy comparison. These are shown in Figure 1.32. Another option is to use parallel box plots, as shown in Figure 1.33.

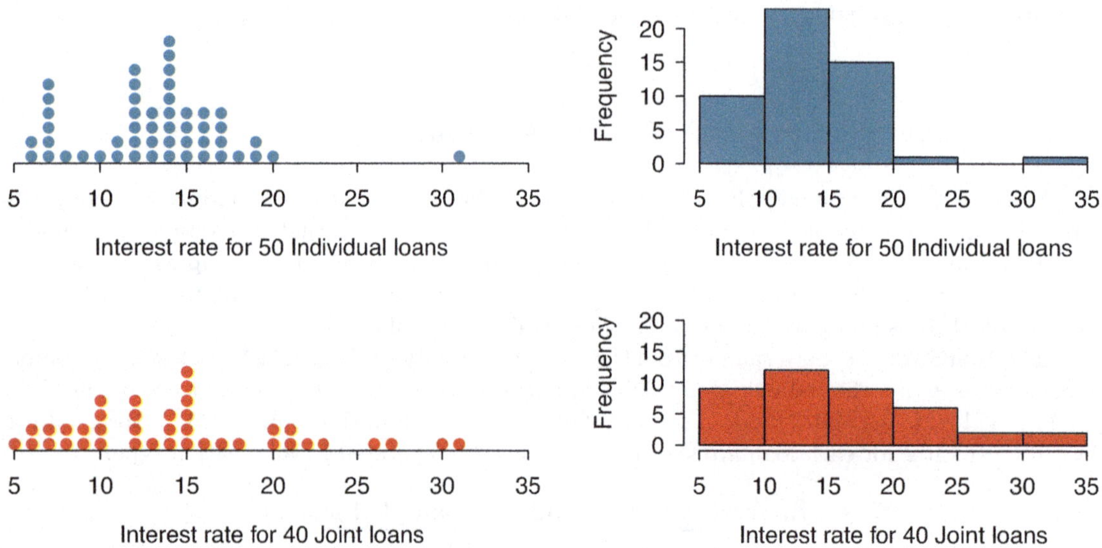

Figure 1.32: Vertically stacked graphs comparing distribution of interest rate for individual versus joint loans. Left: stacked dot plots; Right: stacked histograms.

EXAMPLE 1.46

What are benefits or drawbacks to using a back-to-back stem-and-leaf, stacked dot plots, stacked histograms, or parallel box plots to compare a variable across groups?

The back-to-back stem-and-leaf and the stacked dot plots give us the most detail, because we can see every individual value. However, these are only practical for small data sets. The stacked histograms are good for comparing the distributions of larger data sets. Parallel box plots do not show us the details of the distributions, but they are good for quickly comparing the five-number summaries and are particularly convenient for comparing across more than two groups.

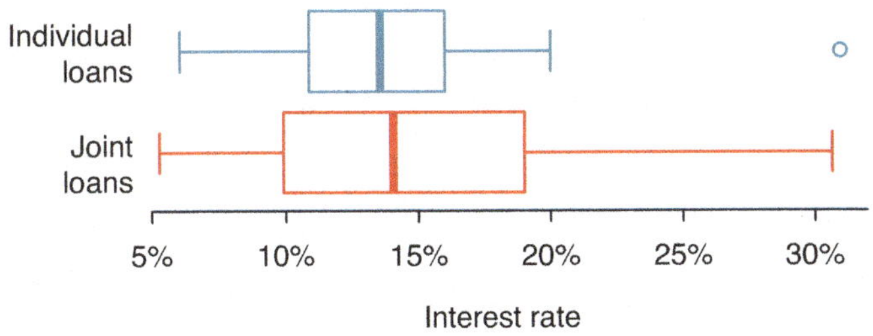

Figure 1.33: Parallel box plots comparing 5-number summary of interest rate for individual versus joint loans.

EXAMPLE 1.47

Use the parallel box plots to compare the distribution of interest rate between individual and joint loans.

Ⓔ The distributions of interest rate for individual and joint loans have a similar median (center), with the median interest rate for joint loans being just slightly higher. The ranges are similar, but the interest rate for joint loans has a larger IQR, meaning more spread in the middle 50% of the data. Both distributions are right skewed. The interest rates for individual loans has a gap between about 20% and 30%, with the largest value being an outlier. The interest rates for joint loans have no outliers.

COMPARING DISTRIBUTIONS

When comparing distributions, compare them with respect to center, spread, and shape as well as any unusual observations. Also, include contextual information about the variable.

1.4.7 Z-scores

In the previous section, we compared the *distribution* of a numerical variable across groups using graphs and summary statistics. Here we investigate a quantity for comparing the relative position of *individual* values between distributions or within a distribution.

We are often interested in knowing and comparing how far values are from the mean. However, because we may want to compare values that are in different units, such as degrees Celsius and degrees Fahrenheit, it will be helpful to have a standardized way to compare distance from the mean. Measuring the *number of standard deviations* values are from the mean will allow us to compare relative distance from the mean even when values are in different units.

The mean interest rate for the `loan50` data set was calculated as 11.57%, with a standard deviation of 5.05%. The highest interest rate is about 26%. Is that value unusually high relative to the rest of the data set? The value 26% is about 14.5% above the mean, so it is between 2 and 3 standard deviations above the mean, meaning it is quite high relative to the rest of the data set. This can be found by doing

$$\frac{26 - 11.57}{5.05} = 2.86$$

The number of standard deviations a value is above or below the mean is known as the **Z-score**. A Z-score has no units, and therefore is sometimes also called *standard units*.

THE Z-SCORE

The Z-score of an observation is the number of standard deviations it falls above or below the mean. We compute the Z-score for an observation x that follows a distribution with mean μ and standard deviation σ using

$$Z = \frac{x - \mu}{\sigma}$$

Observations above the mean always have positive Z-scores, while those below the mean always have negative Z-scores. If an observation is equal to the mean, then the Z-score is 0.

EXAMPLE 1.48

Head lengths of brushtail possums have a mean of 92.6 mm and standard deviation 3.6 mm. Compute the Z-scores for possums with head lengths of 95.4 mm and 85.8 mm.

For $x_1 = 95.4$ mm:

$$Z_1 = \frac{x_1 - \mu}{\sigma} = \frac{95.4 - 92.6}{3.6} = 0.78$$

For $x_2 = 85.8$ mm:

$$Z_2 = \frac{85.8 - 92.6}{3.6} = -1.89$$

We can use Z-scores to roughly identify which observations are more unusual than others. An observation x_1 is said to be more unusual than another observation x_2 if the absolute value of its Z-score is larger than the absolute value of the other observation's Z-score: $|Z_1| > |Z_2|$. This technique is especially insightful when a distribution is symmetric.

GUIDED PRACTICE 1.49

Which of the observations in Example 1.48 is more unusual?[38]

GUIDED PRACTICE 1.50

Let X represent a random variable from a distribution with $\mu = 3$ and $\sigma = 2$, and suppose we observe $x = 5.19$.
(a) Find the Z-score of x.
(b) Interpret the Z-score.[39]

Because Z-scores have no units, they are useful for comparing distance to the mean for distributions that have different standard deviations or different units.

---

[38] Because the *absolute value* of Z-score for the second observation ($x_2 = 85.8$ mm $\rightarrow Z_2 = -1.89$) is larger than that of the first ($x_1 = 95.4$ mm $\rightarrow Z_1 = 0.78$), the second observation has a more unusual head length.

[39] (a) Its Z-score is given by $Z = \frac{x-\mu}{\sigma} = \frac{5.19-3}{2} = \frac{2.19}{2} = 1.095$. (b) The observation x is 1.095 standard deviations *above* the mean. We know it must be above the mean, because Z is positive.

EXAMPLE 1.51

The average daily high temperature in June in LA is 77°F with a standard deviation of 5°F. The average daily high temperature in June in Iceland is 13°C with a standard deviation of 3°C. Which would be considered more unusual: an 83°F day in June in LA or a 19°C day in June in Iceland?

Both values are 6° above the mean. However, they are not the same number of standard deviations above the mean. 83 is $\frac{83-77}{5} = 1.2$ standard deviations above the mean, while 19 is $\frac{19-13}{3} = 2$ standard deviations above the mean. Therefore, a 19°C day in June in Iceland would be more unusual than an 83°F day in June in LA, as the $|Z|$ for 19°C is greater than for 83°F ($|2| > |1.2|$).

1.4.8 Technology: summarizing a single variable

Download the `loan50` CSV file from openintro.org/data. Open it and graphically and numerically summarize the `loan_amount` variable.

Desmos: desmos.com/calculator

1. Open the CSV file. Copy the column that corresponds to the `loan_amount` variable.
2. In left panel of Desmos, click in a cell and type: `a =`
   Paste the column that you copied after the = sign and then hit return on your keyboard. You can also manually enter data between the [ ] brackets separated by a comma.
3. Use `histogram(data, bin width)`, replacing `data` and `bin width` with the desired values, as illustrated below. Here, we named our data `a`, but you can name it whatever you like. You can also use `boxplot(data)` and `dotplot(data)`. Click the magnifying glass to Zoom Fit the graphing window and click the wrench icon to add labels to the axes.
4. Use `stats(data)`, `mean(data)`, and `stdev(data)` to numerically summarize the variable as illustrated below.

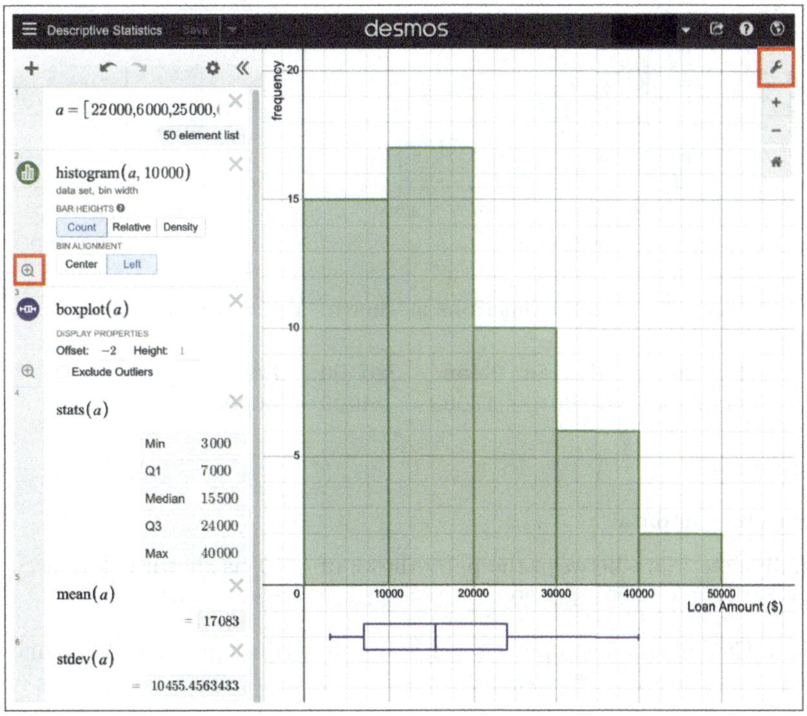

In this book we will also provide some basic R code to address the technology questions posed. In many cases, there will be similarities between the Desmos syntax and the R syntax. The goal is to provide the interested reader a small glimpse into analyzing data using a real world statistical software language. These sections can be skipped. For a more comprehensive introduction to statistical data analysis, we encourage you to explore OpenIntro's Statistical Software labs at openintro.org/book/statlabs. Here you will find labs in R (Base), R (Tidyverse), Rguroo, Jamovi, JASP, Python, SAS, and Stata.

R: You will need to first download R and RStudio from: `posit.co/download/rstudio-desktop`.

1. Open the CSV file. Copy the data that corresponds to the `loan_amount` variable. Do not include the header/variable name.
2. In RStudio, at the > prompt, type: `a = scan()` and hit return. Paste the copied data, then hit return and return again.
```
> a = scan()
1:  22000
2:  6000
...
49:  20000
50:  15000
51:
Read 50 items
```
3. Use `hist()` and `boxplot()` as shown below. Replace labels with appropriate values.
```
> hist(a, main = "Loan Amounts for 50 loans made through Lending Club",
xlab = "Loan Amount ($)", ylab = "frequency", col = "lightblue")

> boxplot(a, main = "Loan Amounts for 50 loans made through Lending Club", xlab
= "Loan Amount ($)", horizontal = TRUE, col = "lightblue")
```

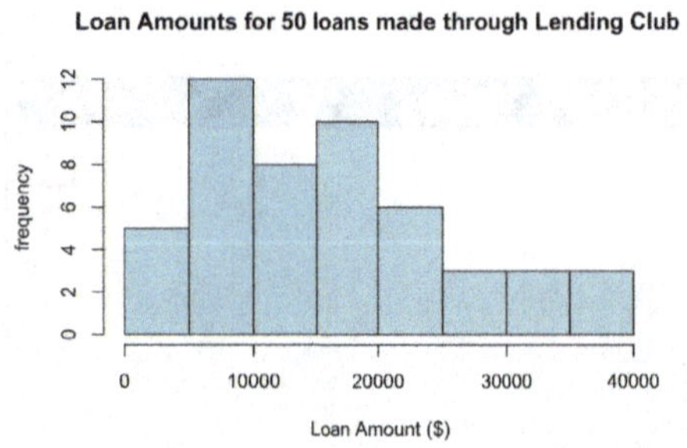

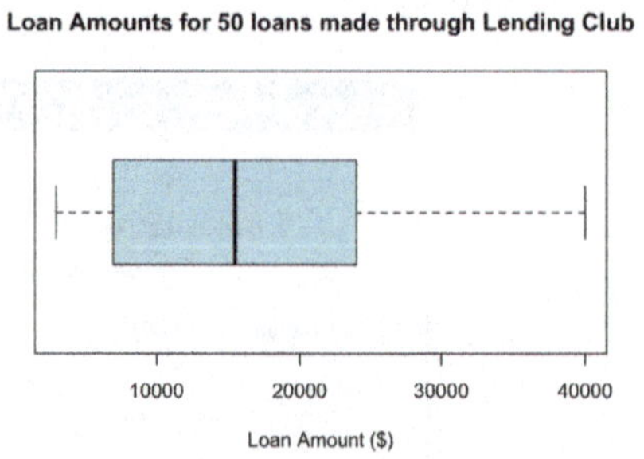

4. Use `summary()`,[40] and `sd()` functions as shown. You can ignore the `[1]`.
```
> summary(a)
Min.     1st Qu.     Median   Mean    3rd Qu.   Max.
3000     7125        15500    17083   24000     40000
> sd(a)
[1] 10455.46
```

Other ways to read in data:

Manually enter data between the parentheses of `c( )` as shown below.
```
> a = c(22000, 6000, 25000, ..., 5825, 20000, 15000)
```

Access a CSV file on a computer or on the web. Adjust path and file name as necessary.
```
> loan50 = read.csv("Users/FirstnameLastname/Desktop/loan50.csv")
```

Install the `openintro` package to access *all* data sets found at openintro.org/data.
```
> install.packages("openintro")
> library(openintro)
```

[40]There are multiple definitions for computing quartiles. To ensure values match the definition used here, for small data sets of length n, you may need to add `quantile.type=6` when n is odd or `quantile.type=2` when n is even.

If using the `openintro` package or reading from a CSV file, use the structure `dataset$variable` to access a particular variable. For example, replace `a` from above with `loan50$loan_amount`.

```
> sd(loan50$loan_amount)
[1] 10455.46
```

Now explore data sets, for example `loan50`, with these useful functions:
`View(loan50)`, `str(loan50)`, `summary(loan50)`, `head(loan50)`.

Calculator: Your teacher may give you data files. Alternately, manually enter data into a list as described here. Instructions for the open-source NumWorks calculator (numworks.com/simulator) are included along with example output. For the TI-83/84 and Casio calculators, general instructions are provided and worked example videos can be accessed via the icon or at **openintro.org/ti** and **openintro.org/casio**.

NUMWORKS: SUMMARIZING DATA

Use `OK` or `EXE` to make a selection.

1. Click the yellow Home button (above the black Power button) to get to the home screen and select `Statistics`.
2. Make sure you have `Data` at the top highlighted, then enter your values. Use the `x` button near the upper right to delete an entry. To clear a whole list, press the up arrow until the list name is highlighted, then press the `x` button, right arrow, and choose `Confirm`.
3. Use the arrows to choose `Graph` and choose the desired graph. Note that you can use the arrows to get to `Settings` to change the Bin width and X start for a Histogram. Using arrows on Boxplot will show each value of the 5-number summary.

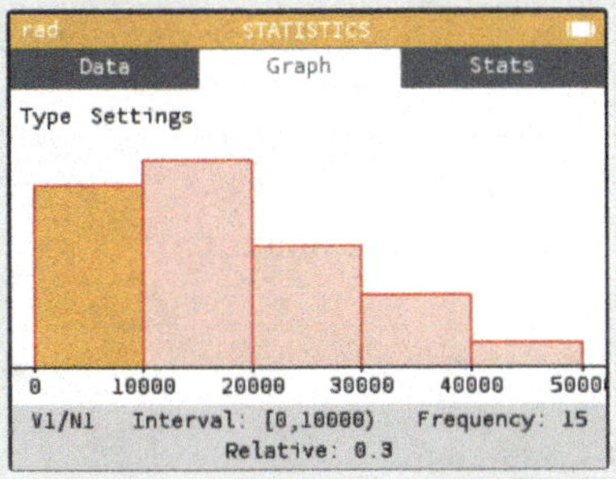

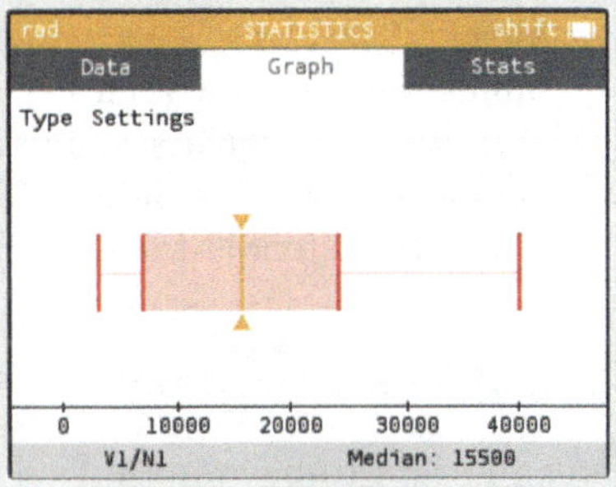

4. Use the arrows to choose `Stats` to see the summary statistics. Use the down arrow to see more summary statistics.

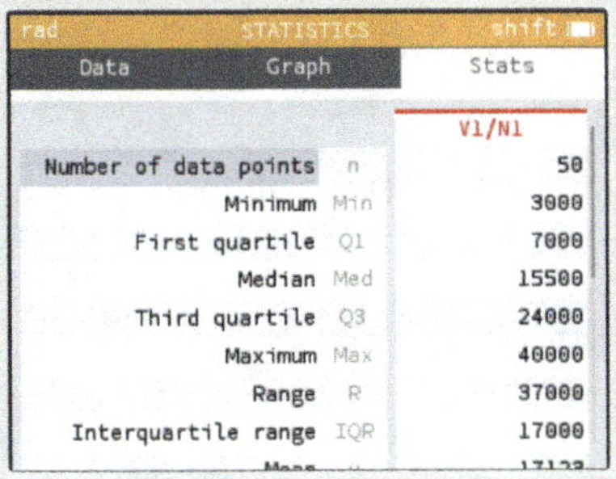

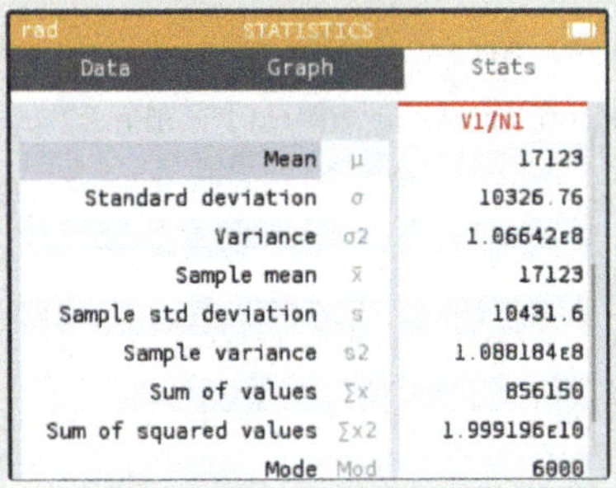

TI-83/84: ENTERING DATA

The first step in summarizing data or making a graph is to enter the data set into a list. Use `STAT`, `Edit`.

1. Press `STAT`.
2. Choose `1:Edit`.
3. Enter data into `L1` or another list.

TI-84: CALCULATING SUMMARY STATISTICS

Use the `STAT`, `CALC`, `1-Var Stats` command to find summary statistics such as mean, standard deviation, and quartiles.

1. Enter the data as described previously.
2. Press `STAT`.
3. Right arrow to `CALC`.
4. Choose `1:1-Var Stats`.
5. Enter `L1` (i.e. `2ND 1`) for List. If the data is in a list other than `L1`, type the name of that list.
6. Leave `FreqList` blank.
7. Choose `Calculate` and hit `ENTER`.

TI-83: Do steps 1-4, then type `L1` (i.e. `2nd 1`) or the list's name and hit `ENTER`.

Calculating the summary statistics will return the following information. It will be necessary to hit the down arrow to see all of the summary statistics.

$\bar{x}$	Mean	n	Sample size or # of data points
Σx	Sum of all the data values	minX	Minimum
Σx^2	Sum of all the squared data values	Q_1	First quartile
Sx	Sample standard deviation	Med	Median
σx	Population standard deviation	maxX	Maximum

TI-83/84: DRAWING A BOX PLOT

1. Enter the data to be graphed as described previously.
2. Hit `2ND Y=` (i.e. `STAT PLOT`).
3. Hit `ENTER` (to choose the first plot).
4. Hit `ENTER` to choose `ON`.
5. Down arrow and then right arrow three times to select box plot with outliers.
6. Down arrow again and make `Xlist: L1` and `Freq: 1`.
7. Choose `ZOOM` and then `9:ZoomStat` to get a good viewing window.

TI-83/84: WHAT TO DO IF YOU CANNOT FIND L1 OR ANOTHER LIST

Restore lists `L1-L6` using the following steps:

1. Press `STAT`.
2. Choose `5:SetUpEditor`.
3. Hit `ENTER`.

CASIO FX-9750GII: ENTERING DATA

1. Navigate to `STAT` (`MENU` button, then hit the `2` button or select `STAT`).
2. Optional: use the left or right arrows to select a particular list.
3. Enter each numerical value and hit `EXE`.

CASIO FX-9750GII: DRAWING A BOX PLOT AND 1-VARIABLE STATISTICS

1. Navigate to `STAT` (`MENU`, then hit `2`) and enter the data into a list.
2. Go to `GRPH` (`F1`).
3. Next go to `SET` (`F6`) to set the graphing parameters.
4. To use the 2nd or 3rd graph instead of `GPH1`, select `F2` or `F3`.
5. Move down to `Graph Type` and select the ▷ (`F6`) option to see more graphing options, then select `Box` (`F2`).
6. If `XList` does not show the list where you entered the data, hit `LIST` (`F1`) and enter the correct list number.
7. Leave `Frequency` at `1`.
8. For `Outliers`, choose `On` (`F1`).
9. Hit `EXE` and then choose the graph where you set the parameters `F1` (most common), `F2`, or `F3`.
10. If desired, explore 1-variable statistics by selecting `1-Var` (`F1`).

CASIO FX-9750GII: DELETING A DATA LIST

1. Navigate to `STAT` (`MENU`, then hit `2`).
2. Use the arrow buttons to navigate to the list you would like to delete.
3. Select ▷ (`F6`) to see more options.
4. Select `DEL-A` (`F4`) and then `F1` to confirm.

Section summary

- The **mean** and the **median** are measures of **center**.
  - The **mean** is the sum of all the observations divided by the number of observations, n. $\bar{x} = \frac{1}{n}\sum x_i = \frac{\sum x_i}{n} = \frac{x_1+x_2+\ldots+x_n}{n}$
  - In an ordered data set, the **median** is the middle number when n is odd. When n is even, the median is the average of the two middle numbers.
- The mean follows the tail. In a **right skewed** distribution, the mean is greater than the median. Analogously, in a **left skewed** distribution, the mean is less than the median.
- The pth **percentile** of a data set is the value that has p% of the data less than or equal to it when the data set is ordered from smallest to largest. The first quartile (Q_1) is the 25th percentile, the median (Q_2) is the 50th percentile, and the third quartile (Q_3) is the 75th percentile.
- **Standard deviation (SD)**and **Interquartile range (IQR)** are measures of spread. SD measures the typical spread from the mean, whereas IQR measures the spread of the middle 50% of the data. **Range** is also sometimes used as a measure of spread.
  - Standard deviation is the square root of the variance. $s_X = \sqrt{\frac{1}{n-1}\sum(x_i - \bar{x})^2}$
  - IQR $= Q_3 - Q_1$, i.e. the difference between the third quartile and the first quartile.
  - Range $= max - min$, i.e. the difference between the maximum value and minimum value.
- **Changing units**. Adding a constant to every value in a data set shifts the mean but does not affect the standard deviation. Multiplying the values in a data set by a constant multiplies the mean and the standard deviation by that constant, except that the standard deviation must always remain positive.
- **Outliers** are observations that are extreme relative to the rest of the data. Two rules of thumb for identifying observations as outliers are:
  - more than 2 standard deviations above or below the mean
  - more than $1.5 \times IQR$ below Q_1 or above Q_3
- Mean, standard deviation, and range are sensitive to outliers and are considered nonresistant (non-robust) measures. Median and IQR are more robust and less sensitive to outliers.
- Summary statistics of a quantitative variable may reveal information that can be used to justify claims about the variable in context
- **Box plots** provide a graphical representation of the **five-number summary**, which consists of: min, Q_1, Q_2, Q_3, max. While a box plot does not indicate frequency or modes, it can show skew and outliers.
- Histograms and dot plots on the same scale, back-to-back stem-and-leaf plots, and parallel box plots can be used to compare important features between two or more distributions of the same numerical variable.
- A comparison of graphical representations for two or more distributions should include a comparison of center, spread, shape, outliers, and any unusual features. Put descriptions in *context*, identifying the variable(s) being summarized by name and including relevant units.
- Multiple quantitative one-variable graphical representations may reveal information that can be used to justify claims about the variable in context.
- A **Z-score** represents the number of standard deviations a value in a data set is above or below the mean. To calculate a Z-score use: $Z = \frac{x - \text{mean}}{SD}$. *Z-scores do not depend on units* and are useful for comparing relative positions of individual values within a distribution or between distributions

Exercises

1.13 Smoking habits of UK residents, Part I. A survey was conducted to study the smoking habits of UK residents. The histograms below display the distributions of the number of cigarettes smoked on weekdays and weekends, and they exclude data from people who identified themselves as non-smokers. Describe the two distributions and compare them.[41]

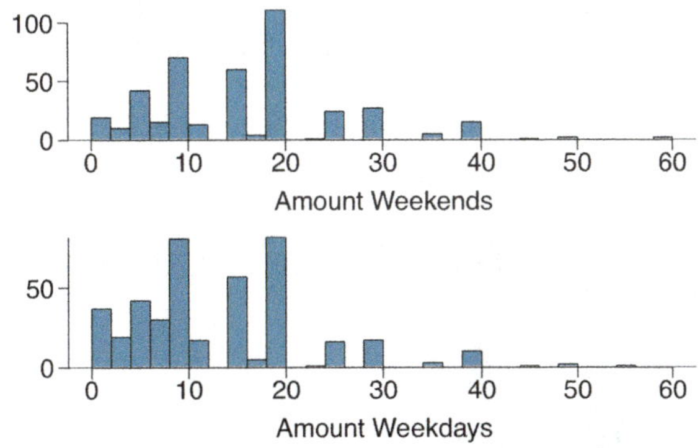

1.14 Stats scores, Part I. Below are the final exam scores of twenty introductory statistics students.

79, 83, 57, 82, 94, 83, 72, 74, 73, 71, 66, 89, 78, 81, 78, 81, 88, 69, 77, 79

Draw a histogram of these data and describe the distribution.

1.15 Smoking habits of UK residents, Part II. A random sample of 5 smokers from the data set discussed in Exercise 1.13 is provided below.

gender	age	maritalStatus	grossIncome	smoke	amtWeekends	amtWeekdays
Female	51	Married	£2,600 to £5,200	Yes	20 cig/day	20 cig/day
Male	24	Single	£10,400 to £15,600	Yes	20 cig/day	15 cig/day
Female	33	Married	£10,400 to £15,600	Yes	20 cig/day	10 cig/day
Female	17	Single	£5,200 to £10,400	Yes	20 cig/day	15 cig/day
Female	76	Widowed	£5,200 to £10,400	Yes	20 cig/day	20 cig/day

(a) Find the mean amount of cigarettes smoked on weekdays and weekends by these 5 respondents.

(b) Find the standard deviation of the amount of cigarettes smoked on weekdays and on weekends by these 5 respondents. Is the variability higher on weekends or on weekdays?

1.16 Factory defective rate. A factory quality control manager decides to investigate the percentage of defective items produced each day. Within a given work week (Monday through Friday) the percentage of defective items produced was 2%, 1.4%, 4%, 3%, 2.2%.

(a) Calculate the mean for these data.

(b) Calculate the standard deviation for these data, showing each step in detail.

1.17 Days off at a mining plant. Workers at a particular mining site receive an average of 35 days paid vacation, which is lower than the national average. The manager of this plant is under pressure from a local union to increase the amount of paid time off. However, he does not want to give more days off to the workers because that would be costly. Instead he decides he should fire 10 employees in such a way as to raise the average number of days off that are reported by his employees. In order to achieve this goal, should he fire employees who have the most number of days off, least number of days off, or those who have about the average number of days off?

1.18 Medians and IQRs. For each part, compare distributions (1) and (2) based on their medians and IQRs. You do not need to calculate these statistics; simply state how the medians and IQRs compare. Make sure to explain your reasoning.

(a) (1) 3, 5, 6, 7, 9
(2) 3, 5, 6, 7, 20

(b) (1) 3, 5, 6, 7, 9
(2) 3, 5, 7, 8, 9

(c) (1) 1, 2, 3, 4, 5
(2) 6, 7, 8, 9, 10

(d) (1) 0, 10, 50, 60, 100
(2) 0, 100, 500, 600, 1000

[41] National STEM Centre, Large Datasets from stats4schools.

1.19 Means and SDs. For each part, compare distributions (1) and (2) based on their means and standard deviations. You do not need to calculate these statistics; simply state how the means and the standard deviations compare. Make sure to explain your reasoning. *Hint:* It may be useful to sketch dot plots of the distributions.

(a) (1) 3, 5, 5, 5, 8, 11, 11, 11, 13
(2) 3, 5, 5, 5, 8, 11, 11, 11, 20

(b) (1) -20, 0, 0, 0, 15, 25, 30, 30
(2) -40, 0, 0, 0, 15, 25, 30, 30

(c) (1) 0, 2, 4, 6, 8, 10
(2) 20, 22, 24, 26, 28, 30

(d) (1) 100, 200, 300, 400, 500
(2) 0, 50, 300, 550, 600

1.20 Mix-and-match. Describe the distribution in the histograms below and match them to the box plots.

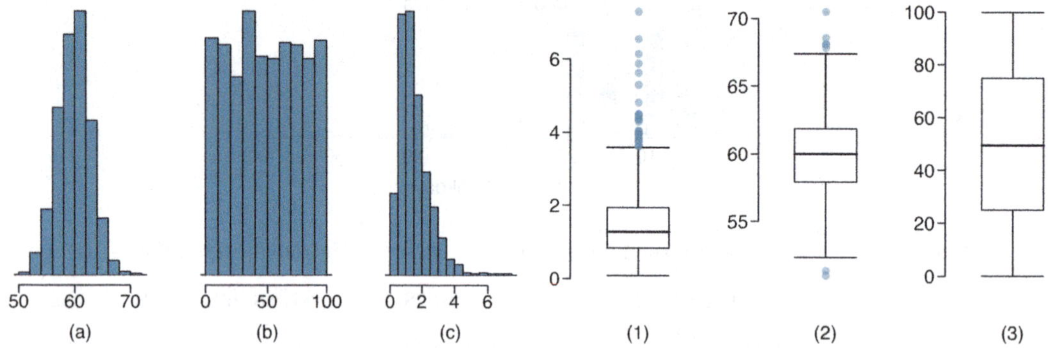

1.21 Air quality. Daily air quality is measured by the air quality index (AQI) reported by the Environmental Protection Agency. This index reports the pollution level and what associated health effects might be a concern. The index is calculated for five major air pollutants regulated by the Clean Air Act and takes values from 0 to 300, where a higher value indicates lower air quality. AQI was reported for a sample of 91 days in 2011 in Durham, NC. The relative frequency histogram below shows the distribution of the AQI values on these days.[42]

(a) Estimate the median AQI value of this sample.
(b) Would you expect the mean AQI value of this sample to be higher or lower than the median? Explain your reasoning.
(c) Estimate Q1, Q3, and IQR for the distribution.
(d) Would any of the days in this sample be considered to have an unusually low or high AQI? Explain your reasoning.

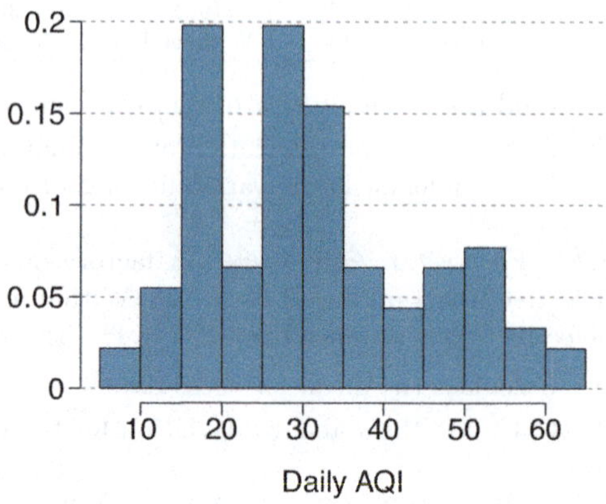

1.22 Median vs. mean. Estimate the median for the 400 observations shown in the histogram, and note whether you expect the mean to be higher or lower than the median.

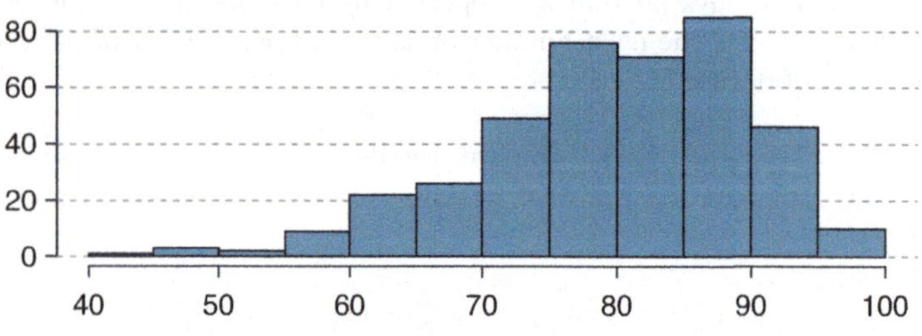

[42] US Environmental Protection Agency, AirData, 2011.

1.23 Histograms vs. box plots. Compare the two plots below. What characteristics of the distribution are apparent in the histogram and not in the box plot? What characteristics are apparent in the box plot but not in the histogram?

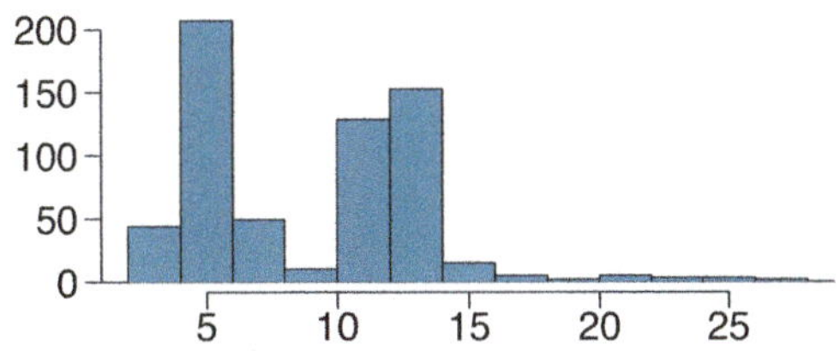

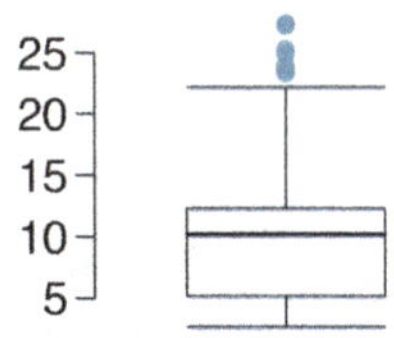

1.24 Facebook friends. Facebook data indicate that 50% of Facebook users have 100 or more friends, and that the average friend count of users is 190. What do these findings suggest about the shape of the distribution of number of friends of Facebook users?[43]

1.25 Distributions and appropriate statistics, Part I. For each of the following, state whether you expect the distribution to be symmetric, right skewed, or left skewed. Also specify whether the mean or median would best represent a typical observation in the data, and whether the variability of observations would be best represented using the standard deviation or IQR. Explain your reasoning.

(a) Number of pets per household.

(b) Distance to work, i.e. number of miles between work and home.

(c) Heights of adult males.

1.26 Distributions and appropriate statistics, Part II. For each of the following, state whether you expect the distribution to be symmetric, right skewed, or left skewed. Also specify whether the mean or median would best represent a typical observation in the data, and whether the variability of observations would be best represented using the standard deviation or IQR. Explain your reasoning.

(a) Housing prices in a country where 25% of the houses cost below \$350,000, 50% of the houses cost below \$450,000, 75% of the houses cost below \$1,000,000 and there are a meaningful number of houses that cost more than \$6,000,000.

(b) Housing prices in a country where 25% of the houses cost below \$300,000, 50% of the houses cost below \$600,000, 75% of the houses cost below \$900,000 and very few houses that cost more than \$1,200,000.

(c) Number of alcoholic drinks consumed by college students in a given week. Assume that most of these students don't drink since they are under 21 years old, and only a few drink excessively.

(d) Annual salaries of the employees at a Fortune 500 company where only a few high level executives earn much higher salaries than all the other employees.

[43] Lars Backstrom. "Anatomy of Facebook". In: *Facebook Data Team's Notes* (2011).

1.27 Income at the coffee shop. The first histogram below shows the distribution of the yearly incomes of 40 patrons at a college coffee shop. Suppose two new people walk into the coffee shop: one making \$225,000 and the other \$250,000. The second histogram shows the new income distribution. Summary statistics are also provided.

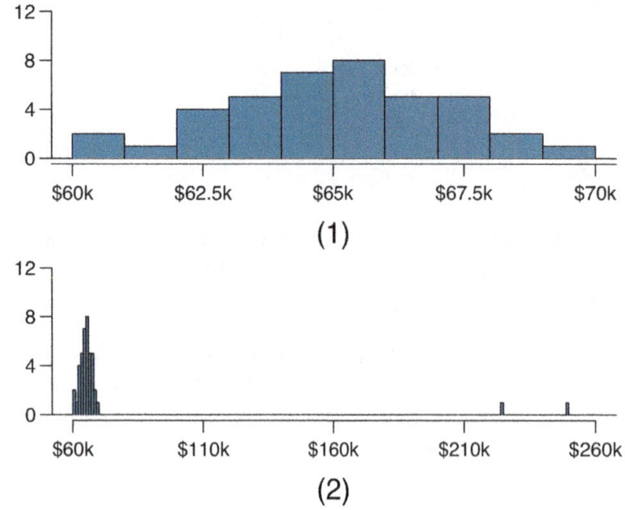

	(1)	(2)
n	40	42
Min.	60,680	60,680
1st Qu.	63,620	63,710
Median	65,240	65,350
Mean	65,090	73,300
3rd Qu.	66,160	66,540
Max.	69,890	250,000
SD	2,122	37,321

(a) Would the mean or the median best represent what we might think of as a typical income for the 42 patrons at this coffee shop? What does this say about the robustness of the two measures?

(b) Would the standard deviation or the IQR best represent the amount of variability in the incomes of the 42 patrons at this coffee shop? What does this say about the robustness of the two measures?

1.28 Midrange. The *midrange* of a distribution is defined as the average of the maximum and the minimum of that distribution. Is this statistic robust to outliers and extreme skew? Explain your reasoning

1.29 Commute times. The US census collects data on time it takes Americans to commute to work, among many other variables. The histogram below shows the distribution of average commute times in 3,144 US counties in 2023. Describe the numerical distribution for commute times.

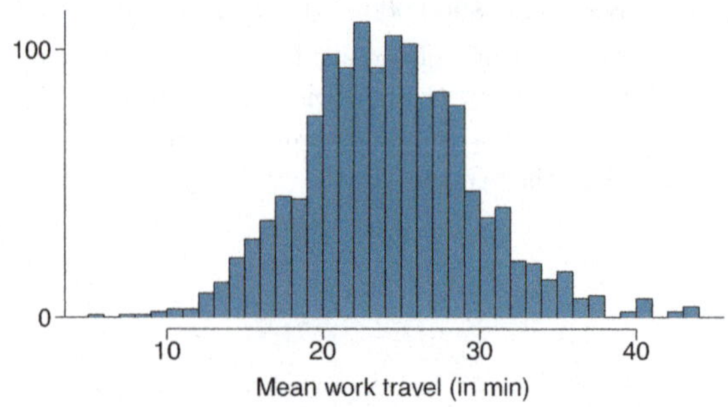

1.30 Hispanic/Latine population. The US census collects data on race and ethnicity of Americans, among many other variables. The histogram below shows the distribution of the percentage of the population that is Hispanic/Latine in 3,144 counties in the US in 2023.

(a) Describe the shape of the distribution.

(b) Is the mean of this distribution greater than or less than the median?

(c) Provide a range for the median percentage of the population that is Hispanic/Latine for counties in the US.

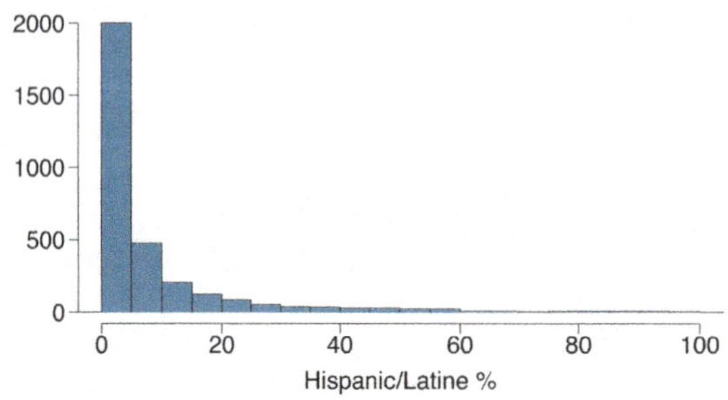

1.5 Overview of data collection principles

How do researchers collect data? Why are the results of some studies more reliable than others? The way a researcher collects data depends upon the research goals. In this section, we look at different methods of collecting data and consider the types of conclusions that can be drawn from those methods.

Learning objectives

1. Identify the main types of data collection: census, experiment, and observation study, and determine which one is appropriate for a given investigative question.
2. Classify a study as observational or experimental, and determine when a study's results can be generalized to the population and when a causal relationship can be drawn.
3. Identify and explain possible confounding factors within a study.

1.5.1 The investigative question and types of conclusions

In Section 1.1, we saw that the first step of the investigative process is to identify a research question. An investigative question for a specific study should have a defined purpose. There are two types of conclusions that can be drawn from statistical studies. The first is a *causal conclusion* about the effect of a treatment and the second is a *generalization* or inference from a sample to a larger population.

If the purpose of the study is to draw a causal conclusion, an experiment with random assignment of treatments should be implemented. For example, if researchers want to see if a certain drug *causes* a reduction of blood pressure, they would carry out an experiment and randomly assign some people to take the drug and others to take a placebo (fake treatment).

If, on the other hand, the purpose is to generalize the results of a sample to a larger population, an observational study with random sampling should be used. For example, researchers often want to estimate a parameter or quantity about a population, such as the average household size or proportion of registered voters that will vote for a certain candidate for mayor. Here, it is not an experiment that is needed, but rather a random sample from the target population, in this case registered voters in the city. Taking a random sample from the population will allow the researcher to generalize and say that the estimate from the sample (the statistic) is a reasonable estimate for the population parameter.

Sometimes we want to draw a causal conclusion *and* generalize that causal conclusion to a larger population. To be able to draw both types of conclusions, we need a random sample of individuals and then a randomized experiment implemented on those individuals. Here we contrasted the primary goal of a random sample with the primary goal of an experiment. We explore sampling techniques and experimental design more thoroughly later in this chapter.

The second step of the investigative process is data collection. The nature of the research question will determine whether an experiment or observational study is appropriate. The third step of the statistical process is analysis. A researcher should have a well-define variable of interest and a clearly stated parameter of interest. We will consider two main types of analysis in this textbook: confidence intervals and hypothesis tests, both of which will be introduced in Chapter 3. With confidence intervals, we seek to estimate the parameter within a range of reasonable values. With hypothesis tests, we seek to determine how much evidence there is that the parameter is greater than, less than, or different from a certain hypothesized value.

Figure 1.34: In February 2010, some media pundits cited one large snow storm as valid evidence against global warming. As comedian Jon Stewart pointed out, "It's one storm, in one region, of one country."

February 10th, 2010.

Lastly, it is important that the investigative question indicates the type(s) of conclusion(s) applicable from the study. The investigative question should provide the population to which the conclusions will be applicable and, in the case of an experiment that uses random assignment, a cause-and-effect conclusion.

1.5.2 Anecdotal evidence

Consider the following possible responses to the three research questions:

1. A man on the news got mercury poisoning from eating swordfish, so the average mercury concentration in swordfish must be dangerously high.
2. I met two students who took more than 7 years to graduate from Duke, so it must take longer to graduate at Duke than at many other colleges.
3. My friend's dad had a heart attack and died after they gave him a new heart disease drug, so the drug must not work.

Each conclusion is based on data. However, there are two problems. First, the data only represent one or two cases. Second, and more importantly, it is unclear whether these cases are actually representative of the population. Data collected in this haphazard fashion are called **anecdotal evidence**.

ANECDOTAL EVIDENCE

Be careful of making inferences based on anecdotal evidence. Such evidence may be true and verifiable, but it may only represent extraordinary cases. The majority of cases and the average case may in fact be very different.

Anecdotal evidence typically is composed of unusual cases that we recall based on their striking characteristics. For instance, we may vividly remember the time when our friend bought a lottery ticket and won $250 but forget most the times she bought one and lost. Instead of focusing on the most unusual cases, we should examine a representative sample of many cases.

1.5.3 Explanatory and response variables

When we ask questions about the relationship between two variables, we sometimes also want to determine if the change in one variable causes a change in the other. Consider the following rephrasing of an earlier question about the `county` data set:

> *If there is an increase in the median household income in a county, does this drive an increase in its population?*

In this question, we are asking whether one variable affects another. If this is our underlying belief, then *median household income* is the **explanatory** variable and the *population change* is the **response** variable in the hypothesized relationship.[44]

EXPLANATORY AND RESPONSE VARIABLES

When we suspect one variable might causally affect another, we label the first variable the explanatory variable and the second the response variable.

explanatory variable —might affect→ response variable

For many pairs of variables, there is no hypothesized relationship, and these labels would not be applied to either variable in such cases.

ASSOCIATION DOES NOT IMPLY CAUSATION

Labeling variables as *explanatory* and *response* does not guarantee the relationship between the two is actually causal, even if there is an association identified between the two variables. We use these labels only to keep track of which variable we suspect affects the other.

In many cases, the relationship is complex or unknown. It may be unclear whether variable A explains variable B or whether variable B explains variable A. For example, it is now known that a particular protein called REST is much depleted in people suffering from Alzheimer's disease. While this raises hopes of a possible approach for treating Alzheimer's, it is still unknown whether the lack of the protein causes brain deterioration, whether brain deterioration causes depletion in the REST protein, or whether some third variable causes both brain deterioration and REST depletion. That is, we do not know if the lack of the protein is an explanatory variable or a response variable. Perhaps it is both.[45]

[44] Sometimes the explanatory variable is called the **independent** variable and the response variable is called the **dependent** variable. However, this becomes confusing since a *pair* of variables might be independent or dependent, so we avoid this language.

[45] nytimes.com/2014/03/20/health/fetal-gene-may-protect-brain-from-alzheimers-study-finds.html

1.5.4 Introduction to experiments

There are two primary types of data collection: experiments and observational studies.

When researchers want to investigate the possibility of a causal connection, they conduct an **experiment**. For all experiments, the researchers must impose a treatment. For most studies there will be both an explanatory and a response variable. For instance, we may suspect administering a drug will reduce mortality in heart attack patients over the following year. To check if there really is a causal connection between the explanatory variable and the response, researchers will collect a sample of individuals and split them into groups. The individuals in each group are *assigned* a treatment. When individuals are randomly assigned to a group, the experiment is called a **randomized experiment**. For example, each heart attack patient in a drug trial could be randomly assigned into one of two groups: the first group receives a **placebo** (fake treatment) and the second group receives the drug. See the case study in Section 1.7.1 for another example of an experiment.

In an experiment, a treatment is applied to an observational unit called an **experimental unit**. When the experimental units are people, we commonly refer to them as subjects or participants rather than experimental units. In our example, each heart attack patient received a placebo or a drug, so the experimental unit or subject in this study is a heart attack patient. At the end of the study, the **response variable** is measured on each experimental unit. Here, there response variable could be whether or not a heart attack patient survived at least 5 years after receiving the drug or placebo. Reachers will compare the proportion in each group that survived at least 5 years. If the difference is *significant*, and if the experiment is well-designed, then the researchers will be able to draw a causal conclusion about the effect of the drug.

Some experiments study more than one factor (explanatory variable) at a time, and each of these factors may have two or more levels (possible values). For example, suppose a researcher plans to investigate how the type and volume of music affect a person's performance on a particular video game. Because these two factors, `type` and `volume`, could interact in interesting ways, we do not want to test one factor at a time. Instead, we want to do an experiment in which we test all *combinations* of the factors. Let's say that `volume` has two levels (soft and loud) and that `type` has three levels (dance, classical, and punk). Then, we would want to have experiment groups for each of the six (2 x 3 = 6) combinations: soft dance, soft classical, soft punk, loud dance, loud classical, loud punk. Each combination is a **treatment**. Therefore, this experiment will have 2 factors and 6 treatments. To replicate each treatment 10 times, one would need to play the game 60 times.

GUIDED PRACTICE 1.52

G A researcher wants to compare the effectiveness of four different drugs. She also wants to test each of the drugs at two doses: low and high. Describe the factors, levels, and treatments of this experiment.[46]

As the number of factors and levels increases, the number of treatments become large and the analysis of the resulting data becomes more complex, requiring the use of advanced statistical methods. We will investigate only one factor at a time in this book.

[46]There are two factors: type of drug, which has four levels, and dose, which has 2 levels. There will be 4 x 2 = 8 treatments: drug 1 at low dose, drug 1 at high dose, drug 2 at low dose, and so on.

1.5.5 Observational studies

Researchers perform an **observational study** when they collect data without interfering with how the data arise. For instance, researchers may collect information via surveys, review medical or company records, or follow a **cohort** of many similar individuals to study why certain diseases might develop. In each of these situations, researchers merely observe or take measurements of things that arise naturally.

Observational studies come in two forms: prospective and retrospective studies. A **prospective study** identifies individuals and collects information as events unfold. For instance, medical researchers may identify and follow a group of similar individuals over many years to assess the possible influences of behavior on cancer risk. One example of such a study is The Nurses' Health Study, started in 1976 and expanded in 1989. This prospective study recruits registered nurses and then collects data from them using questionnaires. **Retrospective studies** collect data after events have taken place, e.g. researchers may review past events in medical records. Some data sets, such as `county`, may contain both prospectively- and retrospectively-collected variables. Local governments prospectively collect some variables as events unfolded (e.g. retails sales) while the federal government retrospectively collected others during the 2010 census (e.g. county population counts).

E

EXAMPLE 1.53

Suppose that a researcher is interested in the average tip customers at a particular restaurant give. Should she carry out an observational study or an experiment?

In addressing this question, we ask, "Will the researcher be imposing any treatment?" Because there is no treatment or interference that would be applicable here, it will be an observational study. Additionally, one consideration the researcher should be aware of is that, if customers know their tips are being recorded, it could change their behavior, making the results of the study inaccurate.

Making causal conclusions based on experiments is often reasonable. However, making the same causal conclusions based on observational data is treacherous and is not recommended. Observational studies are generally only sufficient to show associations.

G

GUIDED PRACTICE 1.54

Suppose an observational study tracked sunscreen use and skin cancer, and it was found people who use sunscreen are more likely to get skin cancer than people who do not use sunscreen. Does this mean sunscreen *causes* skin cancer?[47]

Some previous research tells us that using sunscreen actually reduces skin cancer risk, so maybe there is another variable that can explain this hypothetical association between sunscreen usage and skin cancer. One important piece of information that is absent is sun exposure. Sun exposure is what is called a **confounding variable**, **confounding factor**, or a **confounder**).

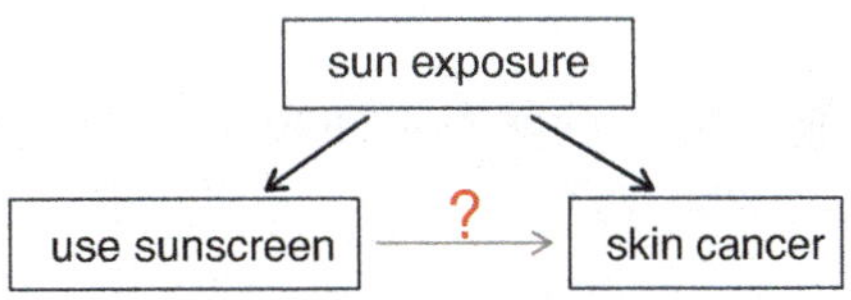

CONFOUNDING VARIABLE

A confounding variable is a variable that is associated with both the explanatory *and* response variables. Because of the confounding variable's association with both variables, we do not know if the response is due to the explanatory variable or due to the confounding variable.

[47] No. See the paragraph following the exercise for an explanation.

Sun exposure is a confounding factor because it is associated with both the use of sunscreen and the development of skin cancer. People who are out in the sun all day are more likely to use sunscreen, and people who are out in the sun all day are more likely to get skin cancer. Research shows us the development of skin cancer is due to the sun exposure. The variables of sunscreen usage and sun exposure are **confounded**, and without this research, we would have no way of knowing which one was the true cause of skin cancer.

EXAMPLE 1.55

In a study that followed 1,169 non-diabetic adults who had been hospitalized for a first heart attack, the people that reported eating chocolate had increased survival rate over the next 8 years than those that reported not eating chocolate. Also, those who ate more chocolate tended to live longer on average. The researchers controlled for several confounding factors, such as age, physical activity, smoking, and many other factors. Can we conclude that the consumption of chocolate caused the people to live longer?

This is an observational study, not a controlled randomized experiment. Even though the researchers controlled for many possible variables, there may still be other confounding factors. (Can you think of any that weren't mentioned?) While it is possible that the chocolate had an effect, this study cannot prove that chocolate increased the survival rate of patients.

EXAMPLE 1.56

The authors who conducted the study did warn in the article that additional studies would be necessary to determine whether the correlation between chocolate consumption and survival translates to any causal relationship. That is, they acknowledged that there may be confounding factors. One possible confounding factor not considered was mental health. In context, explain what it would mean for mental health to be a confounding factor in this study.

Mental health would be a confounding factor if, for example, people with better mental health tended to eat more chocolate, and those with better mental health *also* were less likely to die within the 8 year study period. Notice that if better mental health were not associated with eating more chocolate, it would not be considered a confounding factor because it wouldn't explain the observed associated between eating chocolate and having a better survival rate. If better mental health were associated only with eating chocolate and not with a better survival rate, then it would also not be confounding for the same reason. Only if a variable that is associated with both the explanatory variable of interest (chocolate) and the outcome variable in the study (survival during the 8 year study period) can it be considered a confounding factor.

While one method to justify making causal conclusions from observational studies is to exhaust the search for confounding variables, there is no guarantee that all confounding variables can be examined or measured.

ASSOCIATION ≠ CAUSATION

In general, association does not imply causation, and causation can only be inferred from a randomized experiment.

Section summary

- The nature and purpose of the investigative question should guide the data collection process and should be phrased in terms of the variable(s) of interest in the study. If the purpose is to draw a causal conclusion an experiment should be used. If the purpose is to make a generalization from a sample to a larger population, a random sample from the population of interest should be taken.
- Depending on the investigative question, one of two main data analysis methods may be used: confidence intervals or hypothesis tests. With confidence intervals, we seek to estimate a stated parameter within a range of reasonable values. With hypothesis tests, we seek to determine how much evidence there is that the stated parameter is greater than, less than, or different from a certain hypothesized value.
- An investigative question should indicate the type(s) of conclusion(s) applicable from the study. The investigative question should provide the population to which the conclusions will be applicable and, in the case of an experiment that uses random assignment, a cause-and-effect conclusion.
- A **census** consists of recording information from all items or individuals in a population.
- An **experiment** is a study in which a researcher assigns conditions, or treatments, to experimental units to explore an investigative question of interest. The goal is to draw a *causal* conclusion about the effect of the treatment(s).
- The experimental unit is the observational unit to which the treatment is assigned. When experimental units consist of people, they are sometimes referred to as subjects or participants.
- In an experiment, an **explanatory variable**, or factor, is a variable whose different categories, or levels, are imposed on the experimental units. The different categories, or levels, of the explanatory variable are called treatments. When there is more than one explanatory variable, the combinations of the categories, or levels, of the explanatory variables are called treatments.
- A **response variable** is an outcome measured on each experimental unit after the treatment has been administered.
- An **observational study** is a study where treatments are not imposed. The researcher records the values of the variables of interest in order to explore an investigative question of interest.
- To be able to genera
- A **prospective study** is one in which the observational units of study are selected at a point in time, and data are gathered both at that time and into the future.
- A **retrospective study** is one in which the observational units of study are selected at a point in time and data from the past are gathered.
- A **survey** is an observational study in which the data are collected from humans using a standard set of questions.
- A **confounding variable** in an observational study provides an alternative explanation for the observed relationship between the explanatory and response variables determined in the study, thereby lowering the credibility of the assertion of a causal relationship between the explanatory and response variables of interest. To be a confounding variable, a variable must be associated with both the explanatory variable and the response variable.
- A sample is considered random when all observational units in the sample have an equal chance of being selected from the population. A random mechanism is any resource used to select the observational units to be included in the sample.
- It is appropriate to generalize from a sample to the population of individuals from which the sample was selected only when the individuals in the sample are randomly selected from the population. Do not generalize from anecdotal evidence or from volunteer or other types of non-random samples.
- When observational units, or experimental units, in a sample are not randomly selected from a population, it is appropriate to make generalizations only about a population of individuals that are similar to those used in the study.

Exercises

1.31 Air pollution and birth outcomes, scope of inference. Exercise 1.1 introduces a study where researchers collected data to examine the relationship between air pollutants and preterm births in Southern California. During the study air pollution levels were measured by air quality monitoring stations. Length of gestation data were collected on 143,196 births between the years 1989 and 1993, and air pollution exposure during gestation was calculated for each birth.

(a) Identify the population of interest and the sample in this study.

(b) Comment on whether or not the results of the study can be generalized to the population, and if the findings of the study can be used to establish causal relationships.

1.32 Cheaters, scope of inference. Exercise 1.3 introduces a study where researchers studying the relationship between honesty, age, and self-control conducted an experiment on 160 children between the ages of 5 and 15. The researchers asked each child to toss a fair coin in private and to record the outcome (white or black) on a paper sheet, and said they would only reward children who report white. Half the students were explicitly told not to cheat and the others were not given any explicit instructions. Differences were observed in the cheating rates in the instruction and no instruction groups, as well as some differences across children's characteristics within each group.

(a) Identify the population of interest and the sample in this study.

(b) Comment on whether or not the results of the study can be generalized to the population, and if the findings of the study can be used to establish causal relationships.

1.33 Buteyko method, scope of inference. Exercise 1.2 introduces a study on using the Buteyko shallow breathing technique to reduce asthma symptoms and improve quality of life. As part of this study 600 asthma patients aged 18-69 who relied on medication for asthma treatment were recruited and randomly assigned to two groups: one practiced the Buteyko method and the other did not. Those in the Buteyko group experienced, on average, a significant reduction in asthma symptoms and an improvement in quality of life.

(a) Identify the population of interest and the sample in this study.

(b) Comment on whether or not the results of the study can be generalized to the population, and if the findings of the study can be used to establish causal relationships.

1.34 Stressed out, Part I. A study that surveyed a random sample of otherwise healthy high school students found that they are more likely to get muscle cramps when they are stressed. The study also noted that students drink more coffee and sleep less when they are stressed.

(a) What type of study is this?

(b) Can this study be used to conclude a causal relationship between increased stress and muscle cramps?

(c) State possible confounding variables that might explain the observed relationship between increased stress and muscle cramps.

1.6 Sampling methods and sources of bias

How do opinion polls work? How do research organizations collect the data, and what types of bias should we look out for? You have probably read or heard claims from many studies and polls. A background in statistical reasoning will help you assess the validity of such claims.

Learning objectives

1. Identify and describe how to implement different random sampling methods, including simple, stratified, cluster, and systematic.
2. Justify the appropriateness of a sampling method.
3. Identify potential sources of bias in sampling methods.
4. Understand when it is valid to generalize and to what population that generalization can be made.

1.6.1 Sources of bias when sampling from a population

We might try to estimate the time to graduation for Duke undergraduates in the last 5 years by collecting a sample of students. All graduates in the last 5 years represent the *population*, and graduates who are selected for review are collectively called the *sample*. The goal is to use information from the sample to generalize or make an inference to the population. In order to be able to generalize, we must *randomly* select a sample from the population of interest. The most basic type of random selection is equivalent to how raffles are conducted. For example, in selecting graduates, we could write each graduate's name on a raffle ticket and draw 100 tickets. The selected names would represent a random sample of 100 graduates.

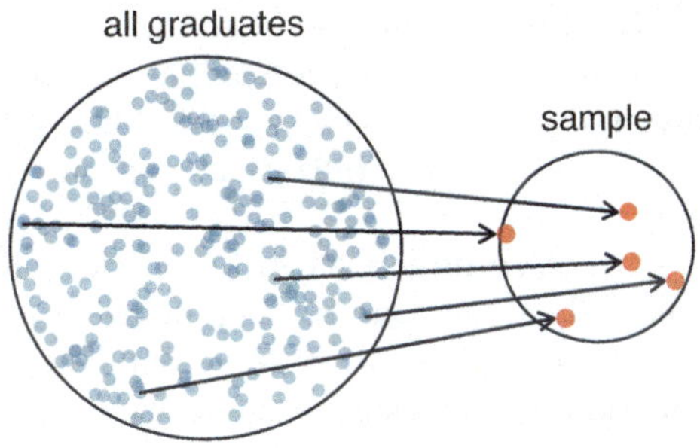

Figure 1.35: In this graphic, five graduates are randomly selected from the population to be included in the sample.

Why pick a sample randomly? Why not just pick a sample by hand? Consider the following scenario.

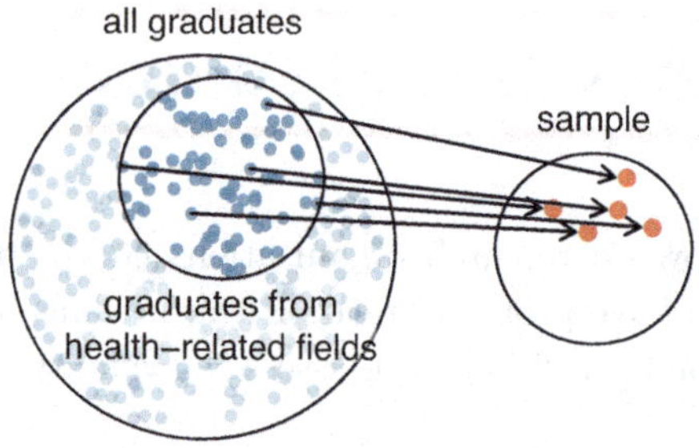

Figure 1.36: Instead of sampling from all graduates equally, a nutrition major might inadvertently pick graduates with health-related majors disproportionately often.

EXAMPLE 1.57

Suppose we ask a student who happens to be majoring in nutrition to select several graduates for the study. What kind of students do you think she might select? Do you think her sample would be representative of all graduates?

Perhaps she would pick a disproportionate number of graduates from health-related fields. Or perhaps her selection would be well-representative of the population. When selecting samples by hand, we run the risk of picking a *biased* sample, even if that bias is unintentional or difficult to discern.

Bias in a sampling method is a systematic error in the sampling procedure that results in a statistic being consistently larger or consistently smaller than the parameter the statistic is used to estimate. There are multiple types of bias when sampling. If the student majoring in nutrition picked a disproportionate number of graduates from health-related fields, this would introduce undercoverage bias into the sample. **Undercoverage bias** occurs when some individuals of the population are inherently less likely to be included in the sample than others, making the sample not representative of the population. In the example, this bias creates a problem because a degree in health-related fields might take more or less time to complete than a degree in other fields. Suppose that it takes longer. Since graduates from other fields would be less likely to be in the sample, the undercoverage bias would cause her to *overestimate* the parameter.

Sampling randomly resolves the problem of undercoverage bias, *if the sample is randomly selected from the entire population of interest.* If the sample is randomly selected from only a subset of the population, say, only graduates from health-related fields, then the sample will not be representative of the population of interest. Generalizations can only be made to the population from which the sample is randomly selected.

Another common downfall is a **convenience sample**, where individuals who are easily accessible are more likely to be included in the sample. For instance, if a political survey is done by stopping people walking in the Bronx, this will not represent all of New York City. It is often difficult to discern what sub-population a convenience sample represents.

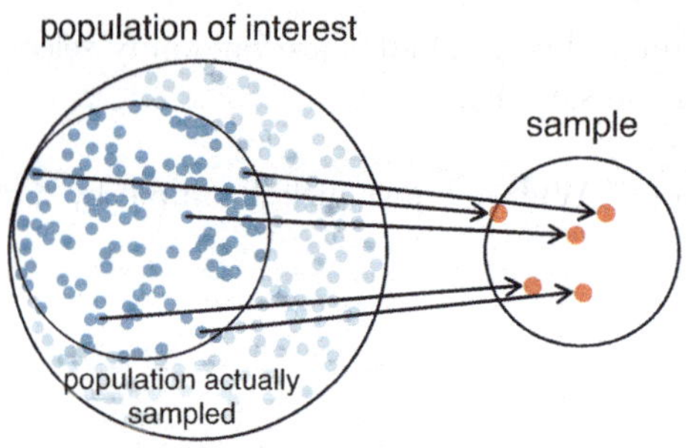

Figure 1.37: Due to the possibility of non-response, surveys studies may only reach a certain group within the population. It is difficult, and often times impossible, to completely fix this problem.

Similarly, a volunteer sample is one in which people's responses are solicited and those who choose to participate, respond. This introduces **voluntary response bias**, which is a problem because those who choose to participate may tend to have different opinions than the rest of the population, resulting in a biased sample.

G

GUIDED PRACTICE 1.58

We can easily access ratings for products, sellers, and companies through websites. These ratings are based only on those people who go out of their way to provide a rating. If 50% of online reviews for a product are negative, do you think this means that 50% of buyers are dissatisfied with the product?[48]

The act of taking a random sample helps minimize bias; however, bias can crop up in other ways. Even when people are picked at random, e.g. for surveys, caution must be exercised if the **non-response** is high. For instance, if only 30% of the people randomly sampled for a survey actually respond, then it is unclear whether the results are **representative** of the entire population. This **non-response bias** can skew results.

Even if a sample has no undercoverage bias and no non-response bias, there is an additional type of bias that often crops up and undermines the validity of results, known as response bias. **Response bias** refers to a broad range of factors that influence how a person responds, such as question wording, question order, and influence of the interviewer. This type of bias can be present even when we collect data from an entire population in what is called a **census**. Because response bias is often subtle, one must pay careful attention to how questions were asked when attempting to draw conclusions from the data.

E

EXAMPLE 1.59

Suppose a high school student wants to investigate the student body's opinions on the food in the cafeteria. Let's assume that she manages to survey every student in the school. How might response bias arise in this context?

There are many possible correct answers to this question. For example, students might respond differently depending upon who asks the question, such as a school friend or someone who works in the cafeteria. The wording of the question could introduce response bias. Students would likely respond differently if asked "Do you like the food in the cafeteria?" versus "The food in the cafeteria is pretty bad, don't you think?"

WATCH OUT FOR BIAS

Undercoverage bias, non-response bias, and response bias can still exist within a random sample. Always determine how a sample was chosen, ask what proportion of people failed to respond, and critically examine the wording of the questions.

When there is no bias in a sample, increasing the sample size tends to increase the precision and reliability of the estimate. When a sample is biased, it may be impossible to decipher helpful information from the data, even if the sample is very large.

G

GUIDED PRACTICE 1.60

A researcher sends out questionnaires to 50 randomly selected households in a particular town asking whether or not they support the addition of a traffic light in their neighborhood. Because only 20% of the questionnaires are returned, she decides to mail questionnaires to 50 more randomly selected households in the same neighborhood. Comment on the usefulness of this approach.[49]

---

[48] Answers will vary. From our own anecdotal experiences, we believe people tend to rant more about products that fell below expectations than rave about those that perform as expected. For this reason, we suspect there is a negative bias in product ratings on sites like Amazon. However, because our experiences may not be representative, we also keep an open mind.

[49] The researcher should be concerned about non-response bias, and sampling more people will not eliminate this issue. The same type of people that did not respond to the first survey are likely not going to respond to the second survey. Instead, she should make an effort to reach out to the households from the original sample that did not respond and solicit their feedback, possibly by going door-to-door.

1.6.2 Simple, systematic, stratified, and cluster sampling

Almost all statistical methods for observational data rely on a sample being random and unbiased. When a sample is collected using a nonrandom sampling method, such as a convenience or voluntary response sample, potential bias is introduced and a generalization to the population of interest is not warranted.

The most basic random sample is called a **simple random sample**, which is equivalent to using a raffle to select cases. This means that each case in the population has an equal chance of being included and there is no implied connection between the cases in the sample. Three other common random sampling methods include: systematic, stratified, and cluster. Figure 1.38 provides a graphical representation of simple versus systematic sampling while Figure 1.39 provides a graphical representation of stratified and cluster sampling.

Simple random sampling is probably the most intuitive form of random sampling. Consider the salaries of Major League Baseball (MLB) players, where each player is a member of one of the league's 30 teams. We will use N to represent the population size. Here N is the total number of players during regular season, which is 750. To take a simple random sample of $n = 120$ of these baseball players and their salaries, we could number each player from 1 to 750. Then we could randomly select 120 numbers between 1 and 750 (without replacement) using a random number generator or random digit table. The players with the selected numbers would comprise our sample.

Two properties are always true in a simple random sample:

1. Each case in the population has an equal chance of being included in the sample.
2. Each *group* of n cases has an equal chance of making up the sample.

The statistical methods in this book focus on data collected using simple random sampling. Note that Property 2 – that each group of n cases has an equal chance making up the sample – is not true for the remaining four sampling techniques. As you read each one, consider why.

Though less common than simple random sampling, **systematic sampling** is sometimes used when there exists a convenient list of all of the individuals of the population. Suppose we have a roster with the names of all the MLB players for a particular season. To take a systematic random sample, number them from 1 to 750. Select one random number between 1 and 750 and let that player be the first individual in the sample. Then, depending on the desired sample size, select every 10th number or 20th number, for example, to arrive at the sample.[50] If there are no patterns in the salaries based on the numbering then this could be a reasonable method.

EXAMPLE 1.61

A systematic sample is not the same as a simple random sample. Provide an example of a sample that can come from a simple random sample but not from a systematic random sample.

Answers can vary. If we take a sample of size 3, then it is possible that we could sample players numbered 1, 2, and 3 in a simple random sample. Such a sample would be impossible from a systematic sample. Property 2 of simple random samples does not hold for other types of random samples.

Sometimes there is a variable that is known to be associated with the quantity we want to estimate. In this case, a stratified random sample might be selected. **Stratified sampling** is a divide-and-conquer sampling strategy. The population is divided into groups called **strata**. The strata are chosen so that similar cases are grouped together and a sampling method, usually simple random sampling, is employed to select a certain number or a certain proportion of the whole within each stratum. In the baseball salary example, the 30 teams could represent the strata; some teams have a lot more money (we're looking at you, Yankees).

[50] If we want a sample of size $n = 150$, it would make sense to select every 5th player because $750/150 = 5$. Suppose we randomly select the number 741. Then player 741, 746, 1, 6, 11, $\cdots$, 731, and 736 would make up the sample.

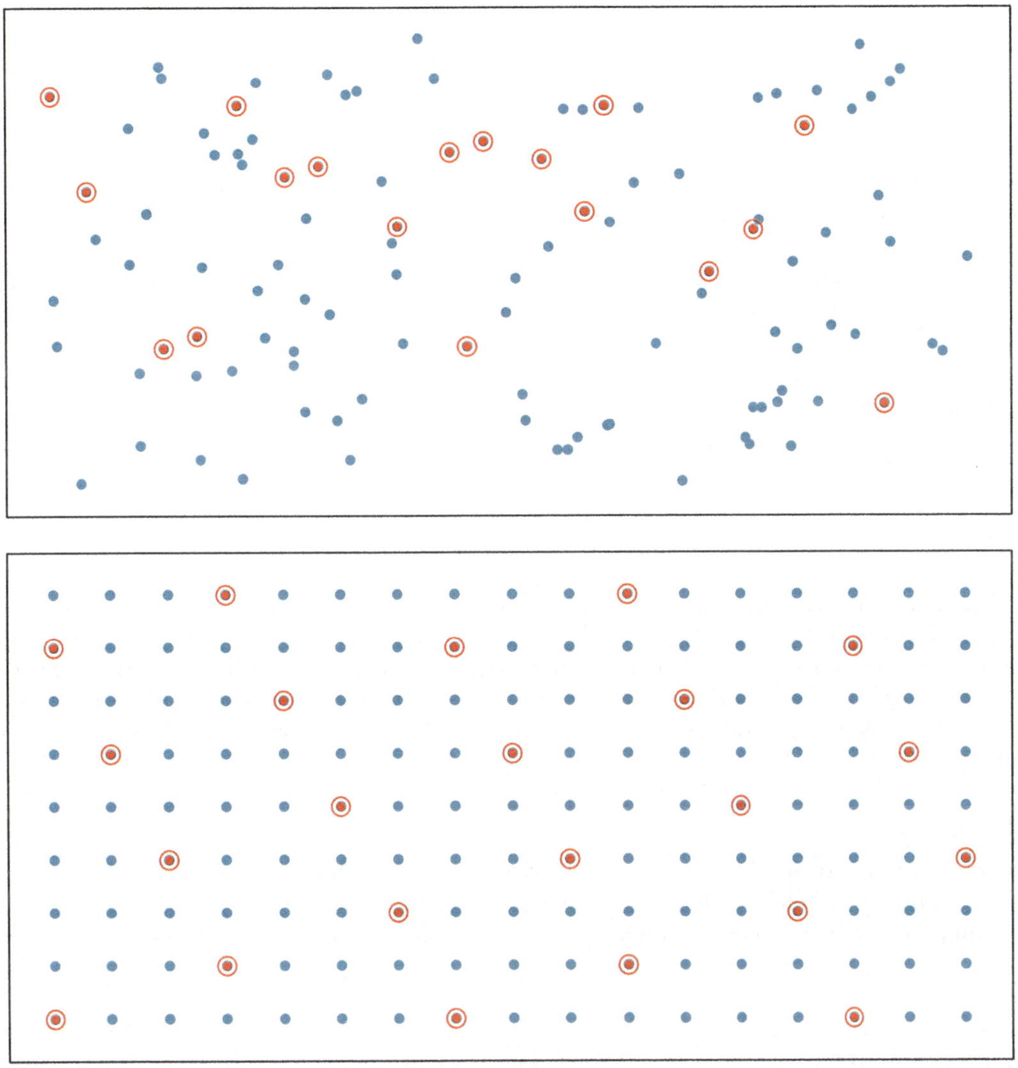

Figure 1.38: Examples of simple random sampling and systematic sampling. In the top panel, simple random sampling was used to randomly select 18 cases. In the lower panel, systematic random sampling was used to select every 7th individual.

EXAMPLE 1.62

For this baseball example, briefly explain how to select a stratified random sample of size $n = 120$.

Each team can serve as a stratum, and we could take a simple random sample of 4 players from each of the 30 teams, yielding a sample of 120 players.

Stratified sampling is inherently different than simple random sampling. For example, the stratified sampling approach described would make it impossible for the entire Yankees team to be included in the sample.

EXAMPLE 1.63

Stratified sampling is especially useful when the cases in each stratum are very similar *with respect to the outcome of interest*. Why is it good for cases within each stratum to be very similar?

We should get a more stable estimate for the subpopulation in a stratum if the cases are very similar. These improved estimates for each subpopulation will help us build a reliable estimate for the full population. For example, in a simple random sample, it is possible that just by random chance we could end up with proportionally too many Yankees players in our sample, thus overestimating the true average salary of all MLB players. A stratified random sample can assure proportional representation from each team.

Next, let's consider a sampling technique that randomly selects groups of people. **Cluster sampling** is much like simple random sampling, but instead of randomly selecting *individuals*, we randomly select groups or **clusters**. Unlike stratified sampling, cluster sampling is most helpful when there is a lot of case-to-case variability within a cluster but the clusters themselves don't look very different from one another. That is, we expect individual strata to be **homogeneous** (self-similar), while we expect individual clusters to be **heterogeneous** (diverse) with respect to the variable of interest.

Sometimes cluster sampling can be a more economical random sampling technique than the alternatives. For example, if neighborhoods represented clusters, this sampling method works best when each neighborhood is very diverse. Because each neighborhood itself encompasses diversity, a cluster sample can reduce the time and cost associated with data collection, because the interviewer would need only go to some of the neighborhoods rather than to all parts of a city, in order to collect a useful sample.

EXAMPLE 1.64

Suppose we are interested in estimating the proportion of students at a certain school that have part-time jobs. It is believed that older students are more likely to work than younger students. What sampling method should be employed? Describe how to collect such a sample to get a sample size of 60.

Because grade level affects the likelihood of having a part-time job, we should take a stratified random sample. To do this, we can take a simple random sample of 15 students from each grade. This will give us equal representation from each grade. Note: in a simple random sample, just by random chance we might get too many students who are older or younger, which could make the estimate too high or too low. Also, there are no well-defined clusters in this example. We wouldn't want to use the grades as clusters and sample everyone from a couple of the grades. This would create too large a sample and would not give us the nice representation from each grade afforded by the stratified random sample.

ADVANCED SAMPLING TECHNIQUES REQUIRE ADVANCED METHODS

The methods of inference covered in this book generally only apply to simple random samples. More advanced analysis techniques are required for systematic, stratified, and cluster random sampling.

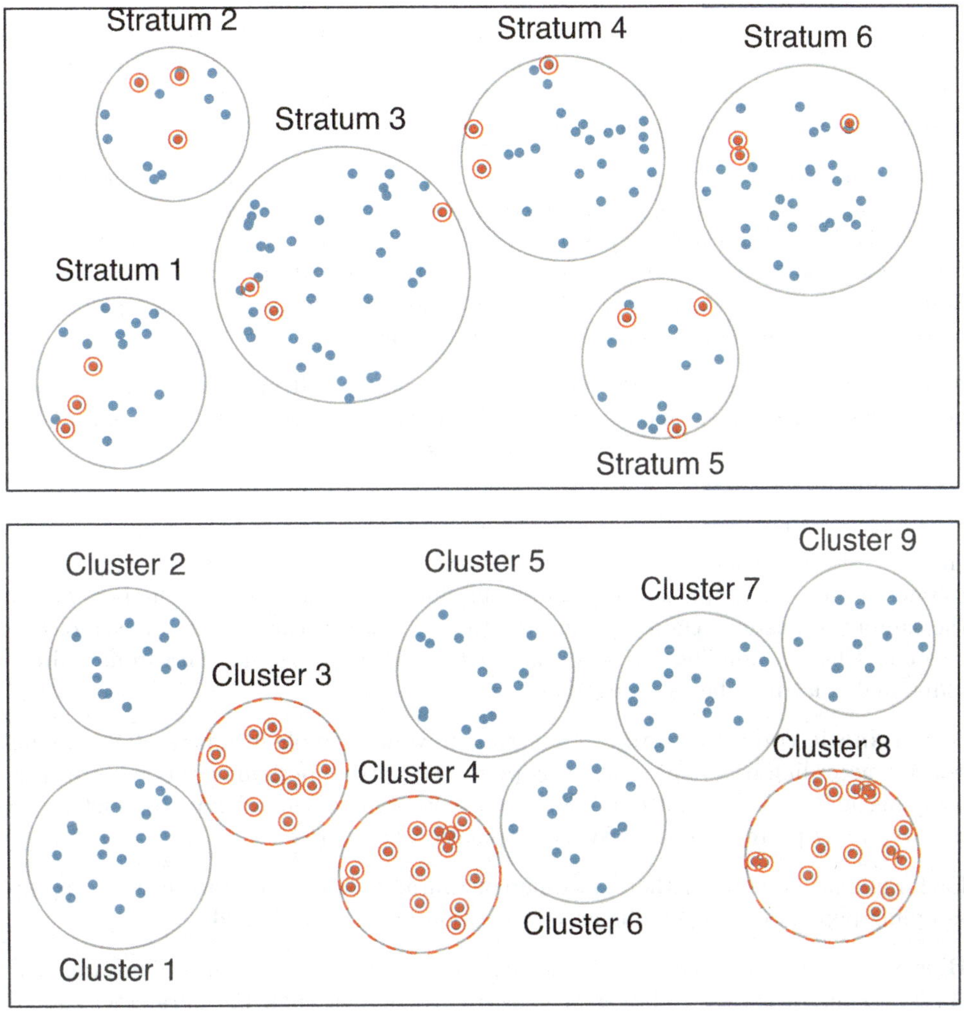

Figure 1.39: Examples of stratified and cluster sampling. In the top panel, stratified sampling was used: cases were grouped into strata, and then simple random sampling was employed within each stratum. In the bottom panel, cluster sampling was used, where data were binned into nine cluster and three clusters were randomly selected.

Section summary

- **Generalizations** from a sample can be made to a population only if the sample is random. Furthermore, the generalization can be made only to the population from which the sample was randomly selected, not to a larger or different population.
- Sampling *without* replacement is a sampling strategy in which an observational unit from a population can be selected only once. The observational unit is not returned to the population before subsequent selections of observational units are made, so there is no chance that the observational unit can be selected again.
- Sampling *with* replacement is a sampling strategy in which an observational unit from the population can be selected more than once. The observational unit is returned to the population before subsequent selections of observational units are made, so it is possible that the observational unit could be selected again.
- In a **simple random sample** (SRS) of size n, every individual has the same chance of being selected and every sample of the size n has the same chance of being the sample selected. This method is the basis for many types of sampling mechanisms. A common way to select a simple random sample is to number each individual of the population from 1 to N. Using a random number generator, integers from 1 to N are randomly selected *without* replacement and the corresponding individuals become part of the sample.
- A **stratified random sample** involves the division of all individuals in a population into non-overlapping groups, called strata, that are similar in some what that might affect their responses. Within each stratum a simple random sample is selected, and the selected individuals are combined to form one sample.
- A **cluster random sample** involves the division of a population into smaller groups, called clusters. Ideally, each cluster mirrors the heterogeneity of the population, with clusters similar to one another. A simple random sample of clusters is selected from the population to form the sample of clusters. Data are collected from all observational units in each of the selected clusters. Cluster sampling can save time and money by collecting data from entire groups of individuals that may be close together.
- A **systematic random sample** is a method in which sample members from a population are selected according to a random starting point and a fixed, periodic interval between successive sampling units. A systematic random sample is sometimes easier to conduct but one must beware of any patterns in the way the population is ordered.
- Each random sampling method has different characteristics that make it more appropriate for sampling populations depending on the question being investigated.
- **Bias** in a sampling method is a systematic error in the sampling procedure that results in a statistic being consistently larger or consistently smaller than the parameter the statistic is used to estimate.
- **Voluntary response bias** is a bias that may occur when a sample consists entirely of volunteers.
- **Undercoverage bias** may occur when the sampling method fails to include part of the population or a part of the population is less likely to be selected based on the sampling method.
- **Non-response bias** may occur because of a failure to obtain responses from some individuals chosen to be sampled. The respondents and nonrespondents could differ significantly in ways that are important for the study.
- **Response bias** may occur when responses to a survey or measurements of observational units tend to differ from the "true" value in one direction. Examples include questions that are confusing or leading (question wording bias) or self-reported responses.
- Nonrandom sampling methods (e.g., samples chosen by convenience or voluntary response) introduce potential bias because they do not use random chance to select the individuals.

Exercises

1.35 Course satisfaction across sections. A large college class has 160 students. All 160 students attend the lectures together, but the students are divided into 4 groups, each of 40 students, for lab sections administered by different teaching assistants. The professor wants to conduct a survey about how satisfied the students are with the course, and he believes that the lab section a student is in might affect the student's overall satisfaction with the course.

(a) What type of study is this?

(b) Suggest a sampling strategy for carrying out this study.

1.36 Housing proposal across dorms. On a large college campus first-year students and sophomores live in dorms located on the eastern part of the campus and juniors and seniors live in dorms located on the western part of the campus. Suppose you want to collect student opinions on a new housing structure the college administration is proposing and you want to make sure your survey equally represents opinions from students from all years.

(a) What type of study is this?

(b) Suggest a sampling strategy for carrying out this study.

1.37 Evaluate sampling methods. A university wants to determine what fraction of its undergraduate student body support a new $25 annual fee to improve the student union. For each proposed method below, indicate whether the method is reasonable or not.

(a) Survey a simple random sample of 500 students.

(b) Stratify students by their field of study, then sample 10% of students from each stratum.

(c) Cluster students by their ages (e.g. 18 years old in one cluster, 19 years old in one cluster, etc.), then randomly sample three clusters and survey all students in those clusters.

1.38 Random digit dialing. The Gallup Poll uses a procedure called random digit dialing, which creates phone numbers based on a list of all area codes in America in conjunction with the associated number of residential households in each area code. Give a possible reason the Gallup Poll chooses to use random digit dialing instead of picking phone numbers from the phone book.

1.39 Haters are gonna hate, study confirms. A study published in the *Journal of Personality and Social Psychology* asked a group of 200 randomly sampled men and women to evaluate how they felt about various subjects, such as camping, health care, architecture, taxidermy, crossword puzzles, and Japan in order to measure their attitude towards mostly independent stimuli. Then, they presented the participants with information about a new product: a microwave oven. This microwave oven does not exist, but the participants didn't know this, and were given three positive and three negative fake reviews. People who reacted positively to the subjects on the dispositional attitude measurement also tended to react positively to the microwave oven, and those who reacted negatively tended to react negatively to it. Researchers concluded that "some people tend to like things, whereas others tend to dislike things, and a more thorough understanding of this tendency will lead to a more thorough understanding of the psychology of attitudes."[51]

(a) What are the cases?

(b) What is (are) the response variable(s) in this study?

(c) What is (are) the explanatory variable(s) in this study?

(d) Does the study employ random sampling?

(e) Is this an observational study or an experiment? Explain your reasoning.

(f) Can we establish a causal link between the explanatory and response variables?

(g) Can the results of the study be generalized to the population at large?

1.40 Family size. Suppose we want to estimate household size, where a "household" is defined as people living together in the same dwelling, and sharing living accommodations. If we select students at random at an elementary school and ask them what their family size is, will this be a good measure of household size? Or will our average be biased? If so, will it overestimate or underestimate the true value?

[51] Justin Hepler and Dolores Albarracín. "Attitudes without objects - Evidence for a dispositional attitude, its measurement, and its consequences". In: *Journal of personality and social psychology* 104.6 (2013), p. 1060.

1.41 Sampling strategies. A statistics student who is curious about the relationship between the amount of time students spend on social networking sites and their performance at school decides to conduct a survey. Various research strategies for collecting data are described below. In each, name the sampling method proposed and any bias you might expect.

(a) He randomly samples 40 students from the study's population, gives them the survey, asks them to fill it out and bring it back the next day.

(b) He gives out the survey only to his friends, making sure each one of them fills out the survey.

(c) He posts a link to an online survey on Facebook and asks his friends to fill out the survey.

(d) He randomly samples 5 classes and asks a random sample of students from those classes to fill out the survey.

1.42 Reading the paper. Below are excerpts from two articles published in the *NY Times*:

(a) An article titled *Risks: Smokers Found More Prone to Dementia* states the following:[52]

> "Researchers analyzed data from 23,123 health plan members who participated in a voluntary exam and health behavior survey from 1978 to 1985, when they were 50-60 years old. 23 years later, about 25% of the group had dementia, including 1,136 with Alzheimer's disease and 416 with vascular dementia. After adjusting for other factors, the researchers concluded that pack-a-day smokers were 37% more likely than nonsmokers to develop dementia, and the risks went up with increased smoking; 44% for one to two packs a day; and twice the risk for more than two packs."

Based on this study, can we conclude that smoking causes dementia later in life? Explain your reasoning.

(b) Another article titled *The School Bully Is Sleepy* states the following:[53]

> "The University of Michigan study, collected survey data from parents on each child's sleep habits and asked both parents and teachers to assess behavioral concerns. About a third of the students studied were identified by parents or teachers as having problems with disruptive behavior or bullying. The researchers found that children who had behavioral issues and those who were identified as bullies were twice as likely to have shown symptoms of sleep disorders."

A friend of yours who read the article says, "The study shows that sleep disorders lead to bullying in school children." Is this statement justified? If not, how best can you describe the conclusion that can be drawn from this study?

[52] R.C. Rabin. "Risks: Smokers Found More Prone to Dementia". In: *New York Times* (2010).

[53] T. Parker-Pope. "The School Bully Is Sleepy". In: *New York Times* (2011).

1.7 Experimental design

Does the use of stents reduce the risk of stroke? What are different ways to design an experiment to answer this question? What are possible sources of bias, and how can we try to minimize them? If we do not incorporate principles of good experimental design, we will not be able to draw a causal conclusion about the effectiveness of stents. This is why it is crucial to start with a well-designed experiment.

Learning objectives

1. Identify elements of a well-designed experiment.
2. Identify experimental designs.
3. Justify the appropriateness of a particular experimental design.
4. Justify the appropriateness of conclusions based on a well-designed experiment.

1.7.1 Case study: using stents to prevent strokes

Here, we introduce a classic challenge in statistics: evaluating the effectiveness of a medical treatment. Terms introduced here will be more explicitly defined later in the section. The goal for now is simply to get a sense of the role statistics can play in practice.

In this section we will consider an experiment that studies effectiveness of stents in treating patients at risk of stroke. Stents are devices put inside blood vessels that assist in patient recovery after cardiac events and reduce the risk of an additional heart attack or death. Many doctors have hoped that there would be similar benefits for patients at risk of stroke. We start by writing the principal question the researchers hope to answer:

> Does the use of stents reduce the risk of stroke?

The researchers who asked this question collected data on 451 at-risk patients. Each volunteer patient was randomly assigned to one of two groups:

> **Treatment group**. Patients in the treatment group received a stent and medical management. The medical management included medications, management of risk factors, and help in lifestyle modification.
>
> **Control group**. Patients in the control group received the same medical management as the treatment group, but they did not receive stents.

Researchers randomly assigned 224 patients to the treatment group and 227 to the control group. In this study, the control group provides a reference point against which we can measure the medical impact of stents in the treatment group.

Researchers studied the effect of stents at two time points: 30 days after enrollment and 365 days after enrollment. The results of 5 patients are summarized in Figure 1.40. Patient outcomes are recorded as "stroke" or "no event", representing whether or not the patient had a stroke at the end of a time period.

Considering data from each patient individually would be a long, cumbersome path towards answering the original research question. Instead, performing a statistical data analysis allows us to consider all of the data at once. Figure 1.41 summarizes the raw data in a more helpful way. In this table, we can quickly see what happened over the entire study. For instance, to identify the number of patients in the treatment group who had a stroke within 30 days, we look on the left-side of the table at the intersection of the treatment and stroke: 33.

Patient	group	0-30 days	0-365 days
1	treatment	no event	no event
2	treatment	stroke	stroke
3	treatment	no event	no event
⋮	⋮	⋮	
450	control	no event	no event
451	control	no event	no event

Figure 1.40: Results for five patients from the stent study.

	0-30 days		0-365 days	
	stroke	no event	stroke	no event
treatment	33	191	45	179
control	13	214	28	199
Total	46	405	73	378

Figure 1.41: Descriptive statistics for the stent study.

G

GUIDED PRACTICE 1.65

What proportion of the patients in the treatment group had no stroke within the first 30 days of the study? (Please note: answers to all Guided Practice exercises are provided using footnotes.)[54]

We can compute summary statistics from the table. A **summary statistic** is a single number summarizing a large amount of data.[55] For instance, the primary results of the study after 1 year could be described by two summary statistics: the proportion of people who had a stroke in the treatment and control groups.

Proportion who had a stroke in the treatment (stent) group: $45/224 = 0.20 = 20\%$.

Proportion who had a stroke in the control group: $28/227 = 0.12 = 12\%$.

These two summary statistics are useful in looking for differences in the groups, and we are in for a surprise: an additional 8% of patients in the treatment group had a stroke! This is important for two reasons. First, it is contrary to what doctors expected, which was that stents would *reduce* the rate of strokes. Second, it leads to a statistical question: do the data show a "real" difference between the groups?

This second question is subtle. Suppose you flip a coin 100 times. While the chance a coin lands heads in any given coin flip is 50%, we probably won't observe exactly 50 heads. This type of fluctuation is part of almost any type of data generating process. It is possible that the 8% difference in the stent study is due to this natural variation. However, the larger the difference we observe (for a particular sample size), the less believable it is that the difference is due to chance. So what we are really asking is whether the difference is **statistically significant**, that is, whether the difference so large that we should reject the notion that it was due to chance. To answer this question, we will need to carry out an inference procedure called a hypothesis test, which is introduced in Chapter 3.

While we don't yet have the statistical tools to fully address this question on our own, we can comprehend the conclusions of the published analysis: there was compelling evidence of harm by stents in this study of stroke patients.

Be careful: do not generalize the results of this study to all patients and all stents. This study looked at patients with very specific characteristics who volunteered to be a part of this study and who may not be representative of all stroke patients. In addition, there are many types of stents and this study only considered the self-expanding Wingspan stent (Boston Scientific). However, this study does leave us with an important lesson: we should keep our eyes open for surprises.

[54]There were 191 patients in the treatment group that had no stroke in the first 30 days. There were $33 + 191 = 224$ total patients in the treatment group, so the proportion is $191/224 = 0.85$.

[55]Formally, a summary statistic is a value computed from the data. Some summary statistics are more useful than others.

1.7.2 Reducing bias in human experiments

Just as random selection is essential in sampling in order to avoid selection bias, **random assignment** is essential in the context of experiments to determine which subjects will receive which treatments. If the researcher chooses which patients are in the treatment and control groups, she may unintentionally place sicker patients in the treatment group, biasing the experiment against the treatment.

Randomized experiments are essential for investigating cause and effect relationships, but they do not ensure an unbiased perspective in all cases. Human studies are perfect examples where bias can unintentionally arise. Here we reconsider a study where a new drug was used to treat heart attack patients. In particular, researchers wanted to know if the drug reduced deaths in patients.

These researchers designed a randomized experiment because they wanted to draw causal conclusions about the drug's effect. Study volunteers[56] were randomly placed into two study groups. One group, the **treatment group**, received the drug. The other group, called the **control group**, did not receive any drug treatment. In an experiment, the explanatory variable is also called a **factor**. Here the factor is receiving the drug treatment. It has two **levels**: yes and no, thus it is categorical. The response variable is whether or not patients died within the time frame of the study. It is also categorical.

Put yourself in the place of a person in the study. If you are in the treatment group, you are given a fancy new drug that you anticipate will help you. On the other hand, a person in the other group doesn't receive the drug and sits idly, hoping her participation doesn't increase her risk of death. These perspectives suggest there are actually two effects: the one of interest is the effectiveness of the drug, and the second is an emotional effect that is difficult to quantify.

Researchers aren't usually interested in the emotional effect, which might bias the study. To circumvent this problem, researchers do not want patients to know which group they are in. When researchers keep the patients uninformed about their treatment, the study is said to be **blind** or **single-blind**. But there is one problem: if a patient doesn't receive a treatment, she will know she is in the control group. The solution to this problem is to give fake treatments to patients in the control group. A fake treatment is called a **placebo**, and an effective placebo is the key to making a study truly blind. A classic example of a placebo is a sugar pill that is made to look like the actual treatment pill. Often times, a placebo results in a slight but real improvement in patients. This effect has been dubbed the **placebo effect**.

The patients are not the only ones who should be blinded: doctors and researchers can accidentally bias a study. When a doctor knows a patient has been given the real treatment, she might inadvertently give that patient more attention or care than a patient that she knows is on the placebo. To guard against this bias, which again has been found to have a measurable effect in some instances, most modern studies employ a **double-blind** setup where researchers who interact with subjects and are responsible for measuring the response variable are, just like the subjects, unaware of who is or is not receiving the treatment.[57]

G **GUIDED PRACTICE 1.66**

Look back to the study in subsection 1.7.1 where researchers were testing whether stents were effective at reducing strokes in at-risk patients. Is this an experiment? Was the study blinded? Was it double-blinded?[58]

---

[56] Human subjects are often called **patients**, **volunteers**, or **study participants**.

[57] There are always some researchers involved in the study who do know which patients are receiving which treatment. However, they do not interact with the study's patients and do not tell the blinded health care professionals who is receiving which treatment.

[58] The researchers assigned the patients into their treatment groups, so this study was an experiment. However, the patients could distinguish what treatment they received, so this study was not blind. The study could not be double-blind because it was not blind.

1.7.3 Elements of a well-designed experiment

The goal of an experiment is to be able to draw a *causal* conclusion about the effect of a treatment. Well-designed experiments are built on four main principles.

Comparison of Treatment Groups. In order to determine if a treatment is effective, there should be at least two treatment groups, one of which could be a control group. At the end of an experiment, the groups will be *compared* to determine if one treatment is, overall, more effective than another treatment or a placebo. Where possible the experiment should be **double-blind**, with neither the subjects nor the researchers knowing who is receiving which treatment.

Random Assignment. Subjects/experimental units should be randomly assigned to treatment groups (or treatments randomly assigned to subjects/experimental units). The purpose of the random assignment is to make the treatment groups as similar as possible with respect to **extraneous variables** – variables that could be associated with or impact the response variable. If random assignment is successful, the distributions of extraneous variables will be approximately the same for the treatments groups. For example, some subjects may be more susceptible to a disease than others due to dietary or genetic factors. Randomizing subjects into treatment groups helps *even out* the effects of such variables, and it also prevents accidental bias that result from choosing which treatment each subject/experimental unit receives.

Replication. Replication within an experiment means multiple subjects/experimental units are assigned to each treatment. The more cases researchers observe, the more accurately they can estimate the effect of the explanatory variable on the response. Also, in an experiment with six subjects, even if there is randomization, it is quite possible for the three healthiest people to be in the same treatment group. In a randomized experiment with 100 people, it is virtually impossible for the healthiest 50 people to end up in the same treatment group. Thus replication makes it more likely that the randomization will be successful in evening out the effects of extraneous variables that are associated with or impact the response variable.

Direct Control. Direct control in an experiment means keeping the settings of certain potential extraneous sources of variation in the response variable the same from experimental unit to experimental unit. For example, when patients take a drug in pill form, some patients take the pill with only a sip of water while others may have it with an entire glass of water. To control for the effect of water consumption, a doctor may ask all patients to drink a 12 ounce glass of water with the pill. A researcher can directly control the appearance of the treatment, the time of day it is taken, etc. She cannot directly control variables such as gender or age. To control for these types of variables, she might consider blocking, which is described in Section 1.7.4.

Sometimes, at the end of an experiment, a researcher may find that even though these four design elements were used, an extraneous variable, such as whether or not a subject smokes, was still unevenly distributed between the treatment groups. In this case, this extraneous variable of smoking is a **confounding variable**, as it is associated with the response variable *and* is more present in one treatment group than another (making it associated with the explanatory variable as well). This is why researchers often record information on extraneous variables – it allows them to determine if the randomization was effective at making the treatment groups similar with respect to those extraneous variables. In the next section we will consider three types of experimental design.

1.7.4 Completely randomized, blocked, and matched pairs design

A **completely randomized experiment** is one in which the subjects or experimental units are randomly assigned to each treatment group in the experiment. Suppose we have two treatments, one of which may be a placebo, and 300 subjects. To carry out a completely randomized design, we could assign each subject a unique integer from 1 to 300. Then we could use a random number generator to randomly generate 150 integers from 1 to 300 *without replacement*. The subjects assigned those integers would go in the first treatment group. The remaining subjects would go in the second treatment group. Note that this method of randomly assigning subjects to treatments is not equivalent to taking a simple random sample. Here we are not sampling a subset of a population; we are randomly *splitting* subjects into groups.

Researchers sometimes know or suspect that another variable, other than the treatment, influences the response. Under these circumstances, they may carry out a **blocked experiment**. In this design, they first group individuals into **blocks** based on the identified extraneous variable and then randomize subjects within each block to the treatment groups. This strategy is referred to as **blocking**. For instance, if we are looking at the effect of a drug on heart attacks, we might first split patients in the study into low-risk and high-risk blocks. Then we can randomly assign half the patients from each block to the control group and the other half to the treatment group, as shown in Figure 1.42. At the end of the experiment, we would incorporate this blocking into the analysis. By blocking by risk of patient, we control for this possible confounding factor. Additionally, by randomizing subjects to treatments within each block, we attempt to even out the effect of variables that we cannot block or directly control.

E

EXAMPLE 1.67

An experiment will be conducted to compare the effectiveness of two methods for quitting smoking. Identify a variable that the researcher might wish to use for blocking and describe how she would carry out a blocked experiment.

The researcher should choose the variable that is most likely to influence the response variable - whether or not a smoker will quit. A reasonable variable, therefore, would be the number of years that the smoker has been smoking. The subjects could be separated into three blocks based on number of years of smoking and each block randomly divided into the two treatment groups.

Even in a blocked experiment with randomization, other variables that affect the response can be distributed unevenly among the treatment groups, thus biasing the experiment in one direction. A third type of design, known as **matched pairs** addresses this problem. In a matched pairs experiment, pairs of people are matched on as many variables as possible, so that the comparison happens between very similar cases. This is actually a special type of blocked experiment, where the blocks are of size two.

An alternate form of matched pairs involves each subject receiving *both* treatments. Randomization can be incorporated by randomly selecting half the subjects to receive treatment 1 first, followed by treatment 2, while the other half receives treatment 2 first, followed by treatment.

G

GUIDED PRACTICE 1.68

How and why should randomization be incorporated into a matched pairs design?[59]

G

GUIDED PRACTICE 1.69

Matched pairs sometimes involves each subject receiving both treatments at the same time. For example, if a hand lotion was being tested, half of the subjects could be randomly assigned to put Lotion A on the left hand and Lotion B on the right hand, while the other half of the subjects would put Lotion B on the left hand and Lotion A on the right hand. Why would this be a better design than a completely randomized experiment in which half of the subjects put Lotion A on both hands and the other half put Lotion B on both hands?[60]

Because it is essential to identify the type of data collection method used when choosing an appropriate inference procedure, we will revisit sampling techniques and experiment design in the subsequent chapters on inference.

---

[59] Assume that all subjects received treatment 1 first, followed by treatment 2. If the variable being measured happens to increase naturally over the course of time, it would appear as though treatment 2 had a greater effect than it really did.

[60] The dryness of people's skins varies from person to person, but probably less so from one person's right hand to left hand. With the matched pairs design, we are able control for this variability by comparing each person's right hand to her left hand, rather than comparing some people's hands to other people's hands (as you would in a completely randomized experiment).

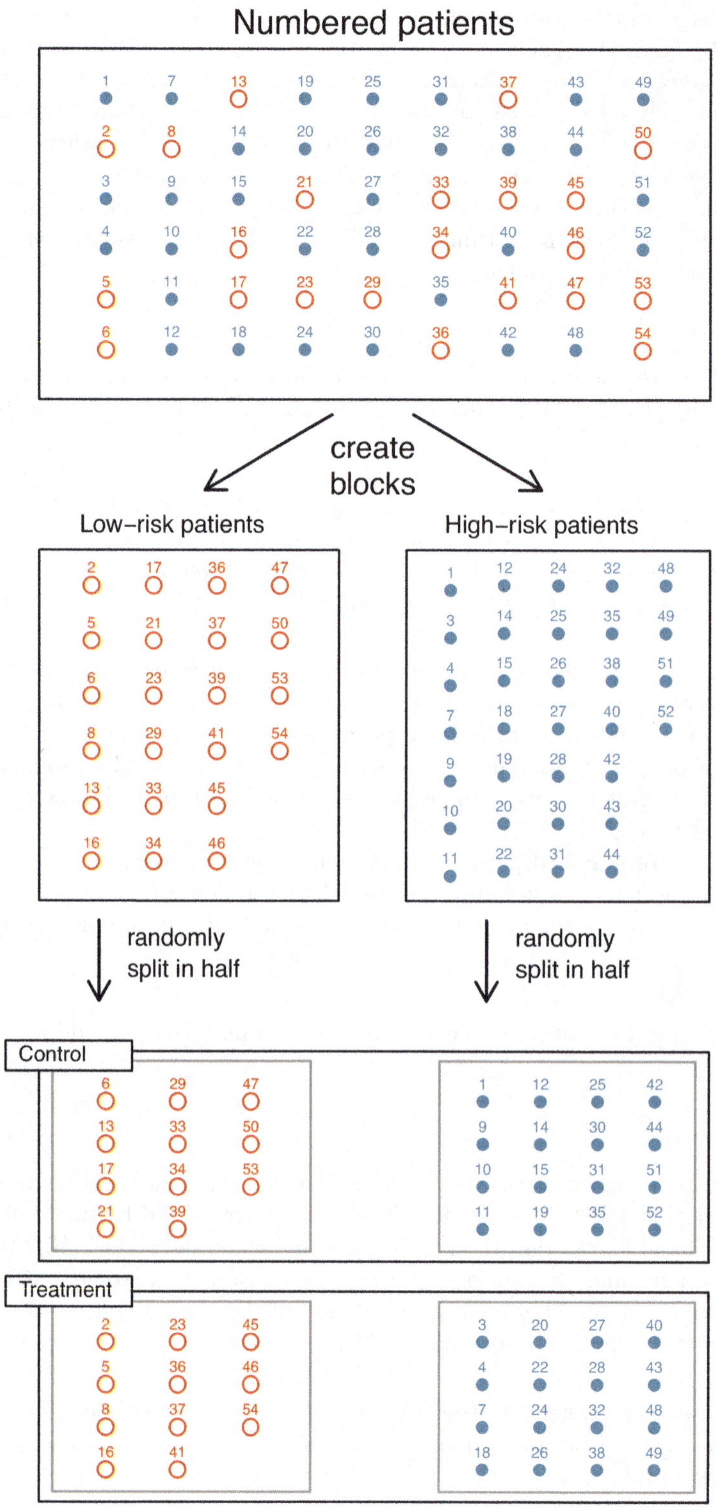

Figure 1.42: Blocking using a variable depicting patient risk. Patients are first divided into low-risk and high-risk blocks, then each block is evenly separated into the treatment groups using randomization. This strategy ensures an equal representation of patients in each treatment group from both the low-risk and high-risk categories.

1.7.5 Testing more than one variable at a time

Some experiments study more than one factor (explanatory variable) at a time, and each of these factors may have two or more levels (possible values). For example, suppose a researcher plans to investigate how the type and volume of music affect a person's performance on a particular video game. Because these two factors, `type` and `volume`, could interact in interesting ways, we do not want to test one factor at a time. Instead, we want to do an experiment in which we test all *combinations* of the factors. Let's say that `volume` has two levels (soft and loud) and that `type` has three levels (dance, classical, and punk). Then, we would want to have experiment groups for each of the six (2 x 3 = 6) combinations: soft dance, soft classical, soft punk, loud dance, loud classical, loud punk. Each combination is a **treatment**. Therefore, this experiment will have 2 factors and 6 treatments. To replicate each treatment 10 times, one would need to play the game 60 times.

GUIDED PRACTICE 1.70

Ⓖ A researcher wants to compare the effectiveness of four different drugs. She also wants to test each of the drugs at two doses: low and high. Describe the factors, levels, and treatments of this experiment.[61]

As the number of factors and levels increases, the number of treatments become large and the analysis of the resulting data becomes more complex, requiring the use of advanced statistical methods. We will investigate only one factor at a time in this book.

1.7.6 Drawing conclusions based on experiments

The goal in an experiment is for the treatment groups to be as similar as possible *except for the treatment*, so that at the end of the experiment any difference in response between the groups can be attributed to the treatment and not to some confounding variable. Using random assignment of subjects/experimental units to treatment groups allows for a cause-and-effect conclusion between the explanatory and response variables because the potential for confounding variables is reduced.

To be able to generalize this cause-and-effect relationship to a larger population, a random sample is needed. However, for ethical and practical reasons, subjects are rarely randomly sampled from a larger population; instead, most subjects are volunteers. For this reason, the precise population for which the conclusion applies may be unclear. For example, if an experiment to determine the most effective means to encourage individuals to vote is carried out only on college students, we may not be able to generalize the conclusions of the experiment to all adults in the population. While technically a generalization to a larger population is not valid without a random sample, in practice we say that the results of an experiment can be generalized to the population of subjects/experimental units similar to those that participated in the experiment. You will see this language later in the book when we investigate hypothesis testing for completely randomized experiments and matched pairs experiments.

[61] There are two factors: type of drug, which has four levels, and dose, which has 2 levels. There will be 4 x 2 = 8 treatments: drug 1 at low dose, drug 1 at high dose, drug 2 at low dose, and so on.

Section summary

- In an **experiment**, researchers impose a **treatment** to test its effects, with the goal being to draw a causal conclusion between the type of treatment (explanatory variable) and the response variable. In order for observed differences in the response to be attributed to the treatment and not to some other factor, it is important to make the treatment groups and the conditions for the treatment groups as similar as possible.

- A well-designed experiment should include comparison of at least two treatment groups, random assignment, replication, and direct control.

- A **control group** is a collection of experimental units that are created for comparison. A control group may be given a treatment different from the treatment of interest to determine if the treatment of interest has an effect (e.g., a treatment with an inactive substance, a placebo, may be given).

- The **placebo effect** is the difference between the average response to a placebo and the average response to no treatment.

- In a **single-blind**, also called single-masked, experiment, participants do not know which treatment they are receiving, but members of the research team who interact with them know which treatment each participant is receiving, or vice versa

- In a **double-blind**, also called double-masked, experiment, neither the participants nor the members of the research team who interact with them know which treatment each participant is receiving.

- An extraneous source of variation, also referred to as an **extraneous variable**, is a variable that is known (or believed) to affect the response but is not an explanatory variable being studied.

- The purpose of **random assignment** is to create treatment groups that are as similar as possible with respect to extraneous sources of variation. If random assignment is successful, the respective distributions of each extraneous variable will be approximately the same for all the treatment groups, thus *evening out* the effects of extraneous variables.

- A **confounding variable** in an experiment is a variable that is distributed differently among treatment groups *and* affects the response variable.

- **Replication**, or imposing the treatments on multiple subjects or experimental units, provides more data for comparison and decreases the likelihood that the treatment groups differ on some characteristic due to chance alone (i.e. in spite of the randomization).

- **Direct control** in an experiment means keeping variables that are within a researchers power to control (e.g. testing conditions) the *same* between treatment groups and from experimental unit to experimental unit. Direct control in an experiment helps to control for potential extraneous variables and sources of variation in the response variable.

- In a **completely randomized design**, subjects or experimental units are randomly assigned to different treatment groups. To carry out the random assignment, first number the units with unique integers from 1 to N. Then, use a random number generator to randomly choose integers from 1 to N without replacement and assign the subjects/units corresponding to those integers to a treatment group. Repeat as needed. Do this in such a way that the treatment group sizes are balanced, unless there exists a good reason to make one treatment group larger than another.

- In a **blocked design**, experimental units are first separated by a blocking variable thought to be associated with the response variable. Units are separated into groups or **blocks**, with each block being homogeneous or alike with respect to the blocking variable (e.g. block subjects by smoking status and create a smoking block and a nonsmoking block, when smoking is thought to be associated with the response variable being measured). Within *each* block, subjects are randomly assigned to the treatment groups, allowing the researcher to compare like to like within each block. Blocking allows for more precise comparisons of the response across the treatments, because within each block the treatments can be compared without having to worry about variation in the response being due to differences in the blocking variable.

- A **matched pairs design** uses a randomized block design where each block consists of a pair of experimental units that are as similar as possible. Each pair receives both treatments by randomly assigning one treatment to one member of the pair and the other treatment to the second member of the pair. Alternatively, instead of each pair receiving one treatment, a matched pairs experiment may involve each experimental unit getting *both* treatments, with the order of the treatments being randomized. A matched pairs design allows for the best comparison of like to like.

- A completely randomized, blocked, or matched pairs design may be more appropriate depending on the goals of the experiment, the characteristics of the subjects or experimental units, and the variables involved.

- Using random assignment of treatments to experimental units allows for cause-and-effect conclusions between the explanatory and the response variables because the potential for confounding variables is reduced.

- Depending on the experimental unit, it may be unethical or difficult to randomly select subjects or experimental units to participate in an experiment. In that case, the study's experimental units are obtained from volunteers and will represent the population of experimental units similar to those who participated in the study.

Exercises

1.43 Light and exam performance. A study is designed to test the effect of light level on exam performance of students. The researcher believes that light levels might have different effects on males and females, so wants to make sure both are equally represented in each treatment. The treatments are fluorescent overhead lighting, yellow overhead lighting, no overhead lighting (only desk lamps).

(a) What is the response variable?

(b) What is the explanatory variable? What are its levels?

(c) What is the blocking variable? What are its levels?

1.44 Vitamin supplements. To assess the effectiveness of taking large doses of vitamin C in reducing the duration of the common cold, researchers recruited 400 healthy volunteers from staff and students at a university. A quarter of the patients were assigned a placebo, and the rest were evenly divided between 1g Vitamin C, 3g Vitamin C, or 3g Vitamin C plus additives to be taken at onset of a cold for the following two days. All tablets had identical appearance and packaging. The nurses who handed the prescribed pills to the patients knew which patient received which treatment, but the researchers assessing the patients when they were sick did not. No significant differences were observed in any measure of cold duration or severity between the four groups, and the placebo group had the shortest duration of symptoms.[62]

(a) Was this an experiment or an observational study? Why?

(b) What are the explanatory and response variables in this study?

(c) Were the patients blinded to their treatment?

(d) Was this study double-blind?

(e) Participants are ultimately able to choose whether or not to use the pills prescribed to them. We might expect that not all of them will adhere and take their pills. Does this introduce a confounding variable to the study? Explain your reasoning.

1.45 Light, noise, and exam performance. A study is designed to test the effect of light level and noise level on exam performance of students. The researcher believes that light and noise levels might have different effects on graduate and undergraduate students, so wants to make sure both are equally represented in each treatment. The light treatments considered are fluorescent overhead lighting, yellow overhead lighting, no overhead lighting (only desk lamps). The noise treatments considered are no noise, construction noise, and human chatter noise.

(a) What is the response variable?

(b) How many factors are considered in this study? Identify them, and describe their levels.

(c) What is the role of the program type (graduate versus undergraduate) variable in this study?

1.46 Music and learning. You would like to conduct an experiment in class to see if students learn better if they study without any music, with music that has no lyrics (instrumental), or with music that has lyrics. Briefly outline a design for this study.

1.47 Soda preference. You would like to conduct an experiment in class to see if your classmates prefer the taste of regular Coke or Diet Coke. Briefly outline a design for this study.

1.48 Exercise and mental health. A researcher is interested in the effects of exercise on mental health and he proposes the following study: Use stratified random sampling to ensure representative proportions of 18-30, 31-40 and 41- 55 year olds from the population. Next, randomly assign half the subjects from each age group to exercise twice a week, and instruct the rest not to exercise. Conduct a mental health exam at the beginning and at the end of the study, and compare the results.

(a) What type of study is this?

(b) What are the treatment and control groups in this study?

(c) Does this study make use of blocking? If so, what is the blocking variable?

(d) Does this study make use of blinding?

(e) Comment on whether or not the results of the study can be used to establish a causal relationship between exercise and mental health, and indicate whether or not the conclusions can be generalized to the population at large.

(f) Suppose you are given the task of determining if this proposed study should get funding. Would you have any reservations about the study proposal?

---

[62] C. Audera et al. "Mega-dose vitamin C in treatment of the common cold: a randomised controlled trial". In: *Medical Journal of Australia* 175.7 (2001), pp. 359–362.

Chapter highlights

A raw data matrix/table may have thousands of rows. The data need to be summarized in order to make sense of all the information. In this chapter, we looked at ways to summarize data graphically, numerically, and verbally. We also investigated various ways that researchers collect data. The key concepts are the difference between a sample and an experiment, the role that randomization plays in each, and the types of conclusions that can be drawn.

Categorical data

- A single **categorical variable** is summarized with **counts** or **proportions** in a **frequency table** or **relative frequency table**. A **bar chart** is used to show the frequency or relative frequency of the categories that the variable takes on.

Numerical data

- When looking at a single **numerical variable**, we try to understand the **distribution** of the variable. The distribution of a variable can be represented with a frequency or relative frequency table and with a graph, such as a **stem-and-leaf plot** or **dot plot** for small data sets, or a **histogram** for larger data sets. If only a summary is desired, a **box plot** may be used to graph the **five-number summary**.
- The **distribution** of a variable can be described and summarized with **center** (mean or median), **spread** (SD or IQR), and **shape** (right skewed, left skewed, approximately symmetric). It is also helpful to note any unusual features such as outliers, gaps, or clusters.
- **Z-scores** and **percentiles** are useful for identifying a data point's relative position within a data set.
- **Outliers** are values that appear extreme relative to the rest of the data. Investigating outliers can provide insight into properties of the data or may reveal data collection/entry errors.
- When **comparing the distribution** of two numerical variables, use two dot plots, two histograms, a back-to-back stem-and-leaf, or parallel box plots.

Collecting data

- Researchers take a **random sample** in order to draw an **inference** to the larger population from which they sampled. When examining observational data, even if the individuals were randomly sampled, a correlation does not imply a causal link.
- In an **experiment**, researchers impose a treatment and use **random assignment** in order to make a **comparison** and draw **causal conclusions** about the effects of the treatment. While often implied, inferences to a larger population may not be valid if the subjects were not also *randomly sampled* from that population.
- **Stratifying vs Blocking**. Stratifying is used when sampling, where the purpose is to *sample* a subgroup from each stratum in order to arrive at a better *estimate* for the parameter of interest. Blocking is used in an experiment to *separate* subjects into blocks and then *compare* responses within those blocks. All subjects in a block are used in the experiment, not just a sample of them.
- Always be mindful of possible **confounding factors** when interpreting the results of observation studies.

It is the role of the researcher or data scientist to ask questions, to identify patterns and departure from patterns, and to make sense of this in the context of the data. Strong writing and presentation skills are critical for being able to communicate the methods and results to a wider audience.

Chapter exercises

1.49 Make-up exam. In a class of 25 students, 24 of them took an exam in class and 1 student took a make-up exam the following day. The professor graded the first batch of 24 exams and found an average score of 74 points with a standard deviation of 8.9 points. The student who took the make-up the following day scored 64 points on the exam.

(a) Does the new student's score increase or decrease the average score?

(b) What is the new average?

(c) Does the new student's score increase or decrease the standard deviation of the scores?

1.50 Infant mortality. The infant mortality rate is defined as the number of infant deaths per 1,000 live births. This rate is often used as an indicator of the level of health in a country. The relative frequency histogram below shows the distribution of estimated infant death rates for 224 countries for which such data were available in 2014.[63]

(a) Estimate Q1, the median, and Q3 from the histogram.

(b) Would you expect the mean of this data set to be smaller or larger than the median? Explain your reasoning.

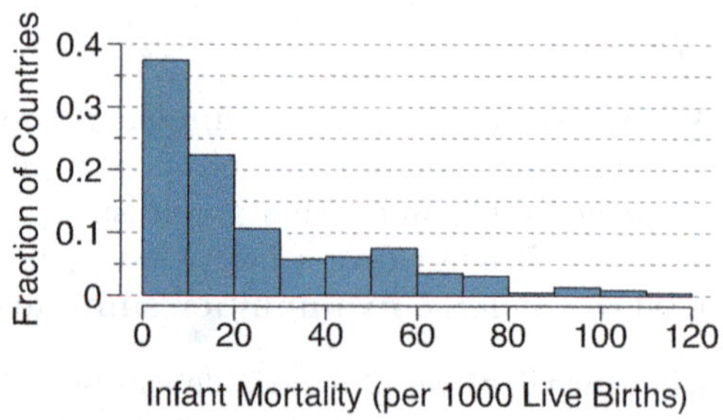

1.51 TV watchers. Students in an AP Statistics class were asked how many hours of television they watch per week (including online streaming). This sample yielded an average of 4.71 hours, with a standard deviation of 4.18 hours. Is the distribution of number of hours students watch television weekly symmetric? If not, what shape would you expect this distribution to have? Explain your reasoning.

1.52 A new statistic. The statistic $\frac{\bar{x}}{median}$ can be used as a measure of skewness. Suppose we have a distribution where all observations are greater than 0, $x_i > 0$. What is the expected shape of the distribution under the following conditions? Explain your reasoning.

(a) $\frac{\bar{x}}{median} = 1$

(b) $\frac{\bar{x}}{median} < 1$

(c) $\frac{\bar{x}}{median} > 1$

1.53 Oscar winners. The first Oscar awards for best actor and best actress were given out in 1929. The histograms below show the age distribution for all of the best actor and best actress winners from 1929 to 2018. Summary statistics for these distributions are also provided. Compare the distributions of ages of best actor and actress winners.[64]

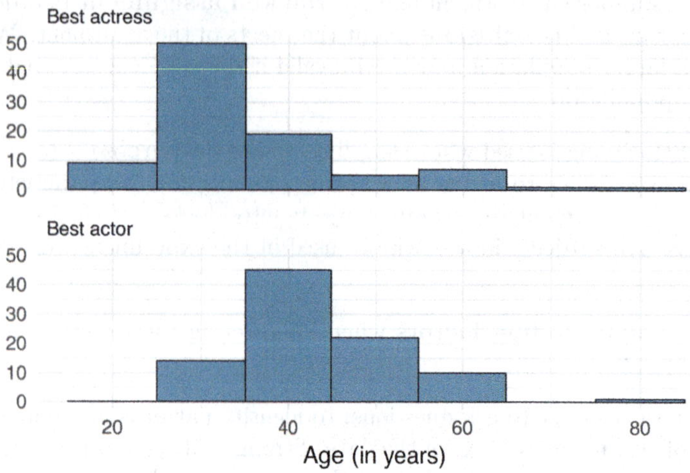

	Best Actress
Mean	36.2
SD	11.9
n	92

	Best Actor
Mean	43.8
SD	8.83
n	92

[63] CIA Factbook, Country Comparisons, 2014.

[64] Oscar winners from 1929 – 2012, data up to 2009 from the Journal of Statistics Education data archive and more current data from wikipedia.org.

1.54 Exam scores. The average on a history exam (scored out of 100 points) was 85, with a standard deviation of 15. Is the distribution of the scores on this exam symmetric? If not, what shape would you expect this distribution to have? Explain your reasoning.

1.55 Stats scores. Below are the final exam scores of twenty introductory statistics students.

57, 66, 69, 71, 72, 73, 74, 77, 78, 78, 79, 79, 81, 81, 82, 83, 83, 88, 89, 94

Create a box plot of the distribution of these scores. The five number summary provided below may be useful.

Min	Q1	Q2 (Median)	Q3	Max
57	72.5	78.5	82.5	94

1.56 Marathon winners. The histogram and box plots below show the distribution of finishing times for male and female winners of the New York Marathon between 1970 and 1999.

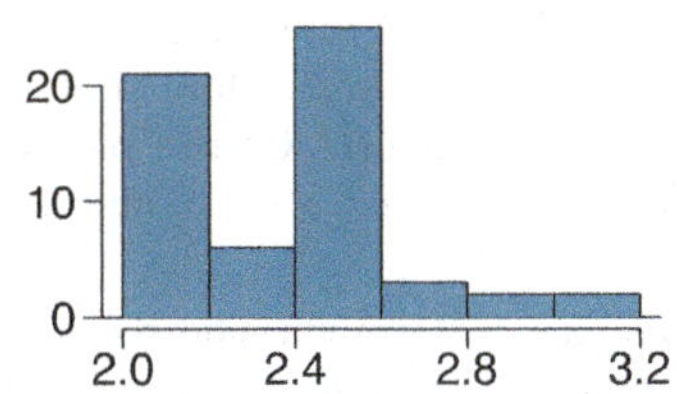

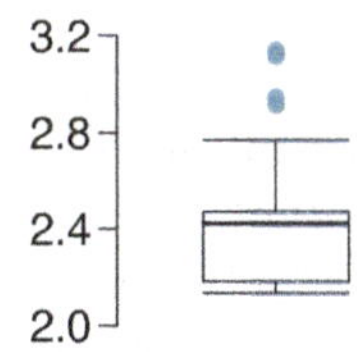

(a) What features of the distribution are apparent in the histogram and not the box plot? What features are apparent in the box plot but not in the histogram?

(b) What may be the reason for the bimodal distribution? Explain.

(c) Compare the distribution of marathon times for men and women based on the box plot shown below.

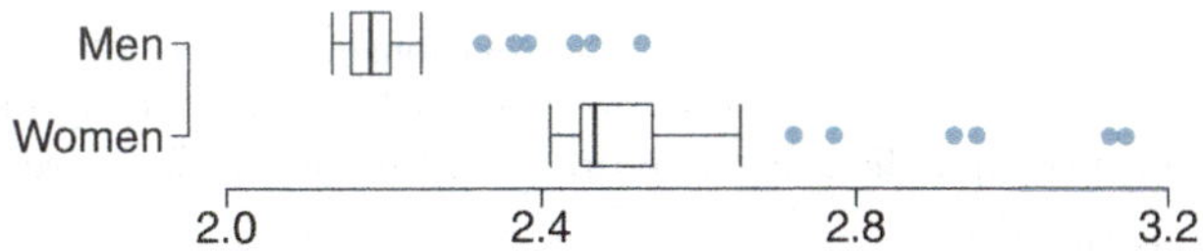

(d) The time series plot shown below is another way to look at these data. Describe what is visible in this plot but not in the others.

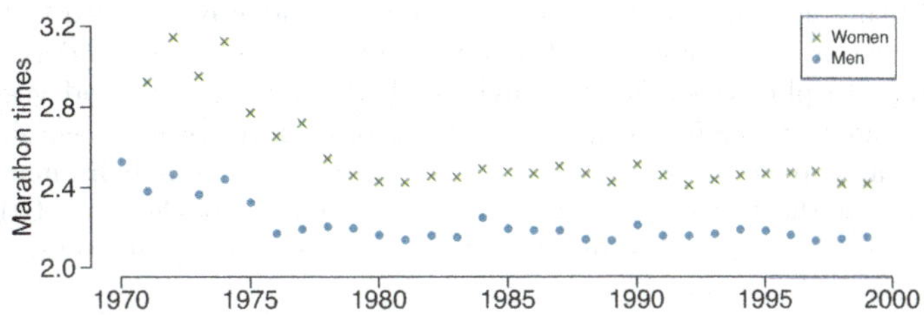

1.57 Chia seeds and weight loss. Chia Pets – those terra-cotta figurines that sprout fuzzy green hair – made the chia plant a household name. But chia has gained an entirely new reputation as a diet supplement. In one 2009 study, a team of researchers recruited 38 men and divided them randomly into two groups: treatment or control. They also recruited 38 women, and they randomly placed half of these participants into the treatment group and the other half into the control group. One group was given 25 grams of chia seeds twice a day, and the other was given a placebo. The subjects volunteered to be a part of the study. After 12 weeks, the scientists found no significant difference between the groups in appetite or weight loss.[65]

(a) What type of study is this?

(b) What are the experimental and control treatments in this study?

(c) Has blocking been used in this study? If so, what is the blocking variable?

(d) Has blinding been used in this study?

(e) Comment on whether or not we can make a causal statement, and indicate whether or not we can generalize the conclusion to the population at large.

[65]D.C. Nieman et al. "Chia seed does not promote weight loss or alter disease risk factors in overweight adults". In: *Nutrition Research* 29.6 (2009), pp. 414–418.

1.58 City council survey. A city council has requested a household survey be conducted in a suburban area of their city. The area is broken into many distinct and unique neighborhoods, some including large homes, some with only apartments, and others a diverse mixture of housing structures. For each part below, identify the sampling methods described, and describe the statistical pros and cons of the method in the city's context.

(a) Randomly sample 200 households from the city.

(b) Divide the city into 20 neighborhoods, and sample 10 households from each neighborhood.

(c) Divide the city into 20 neighborhoods, randomly sample 3 neighborhoods, and then sample all households from those 3 neighborhoods.

(d) Divide the city into 20 neighborhoods, randomly sample 8 neighborhoods, and then randomly sample 50 households from those neighborhoods.

(e) Sample the 200 households closest to the city council offices.

1.59 Flawed reasoning. Identify the flaw(s) in reasoning in the following scenarios. Explain what the individuals in the study should have done differently if they wanted to make such strong conclusions.

(a) Students at an elementary school are given a questionnaire that they are asked to return after their parents have completed it. One of the questions asked is, "Do you find that your work schedule makes it difficult for you to spend time with your kids after school?" Of the parents who replied, 85% said "no". Based on these results, the school officials conclude that a great majority of the parents have no difficulty spending time with their kids after school.

(b) A survey is conducted on a simple random sample of 1,000 women who recently gave birth, asking them about whether or not they smoked during pregnancy. A follow-up survey asking if the children have respiratory problems is conducted 3 years later. However, only 567 of these women are reached at the same address. The researcher reports that these 567 women are representative of all mothers.

(c) An orthopedist administers a questionnaire to 30 of his patients who do not have any joint problems and finds that 20 of them regularly go running. He concludes that running decreases the risk of joint problems.

1.60 Stressed out, Part II. In a study evaluating the relationship between stress and muscle cramps, half the subjects are randomly assigned to be exposed to increased stress by being placed into an elevator that falls rapidly and stops abruptly and the other half are left at no or baseline stress.

(a) What type of study is this?

(b) Can this study be used to conclude a causal relationship between increased stress and muscle cramps?

1.61 Eat better, feel better? In a public health study on the effects of consumption of fruits and vegetables on psychological well-being in young adults, participants were randomly assigned to three groups: (1) diet-as-usual, (2) an ecological momentary intervention involving text message reminders to increase their fruits and vegetable consumption plus a voucher to purchase them, or (3) a fruit and vegetable intervention in which participants were given two additional daily servings of fresh fruits and vegetables to consume on top of their normal diet. Participants were asked to take a nightly survey on their smartphones. Participants were student volunteers at the University of Otago, New Zealand. At the end of the 14-day study, only participants in the third group showed improvements to their psychological well-being across the 14-days relative to the other groups.[66]

(a) What type of study is this?

(b) Identify the explanatory and response variables.

(c) Comment on whether the results of the study can be generalized to the population.

(d) Comment on whether the results of the study can be used to establish causal relationships.

(e) A newspaper article reporting on the study states, "The results of this study provide proof that giving young adults fresh fruits and vegetables to eat can have psychological benefits, even over a brief period of time." How would you suggest revising this statement so that it can be supported by the study?

[66] Tamlin S Conner et al. "Let them eat fruit! The effect of fruit and vegetable consumption on psychological well-being in young adults: A randomized controlled trial". In: *PloS one* 12.2 (2017), e0171206.

1.62 Screens, teens, and psychological well-being. In a study of three nationally representative large-scale data sets from Ireland, the United States, and the United Kingdom (n = 17,247), teenagers between the ages of 12 to 15 were asked to keep a diary of their screen time and answer questions about how they felt or acted. The answers to these questions were then used to compute a psychological well-being score. Additional data were collected and included in the analysis, such as each child's sex and age, and on the mother's education, ethnicity, psychological distress, and employment. The study concluded that there is little clear-cut evidence that screen time decreases adolescent well-being.[67]

(a) What type of study is this?
(b) Identify the explanatory variables.
(c) Identify the response variable.
(d) Comment on whether the results of the study can be generalized to the population, and why.
(e) Comment on whether the results of the study can be used to establish causal relationships.

1.63 Stanford Open Policing. The Stanford Open Policing project gathers, analyzes, and releases records from traffic stops by law enforcement agencies across the United States. Their goal is to help researchers, journalists, and policymakers investigate and improve interactions between police and the public.[68] The following is an excerpt from a summary table created based off of the data collected as part of this project.

County	State	Driver's race	No. of stops per year	% of stopped cars searched	% of stopped drivers arrested
Apaice County	Arizona	Black	266	0.08	0.02
Apaice County	Arizona	Hispanic	1008	0.05	0.02
Apaice County	Arizona	White	6322	0.02	0.01
Cochise County	Arizona	Black	1169	0.05	0.01
Cochise County	Arizona	Hispanic	9453	0.04	0.01
Cochise County	Arizona	White	10826	0.02	0.01
...	...	...	...	...	...
Wood County	Wisconsin	Black	16	0.24	0.10
Wood County	Wisconsin	Hispanic	27	0.04	0.03
Wood County	Wisconsin	White	1157	0.03	0.03

(a) What variables were collected on each individual traffic stop in order to create to the summary table above?
(b) State whether each variable is numerical or categorical. If numerical, state whether it is continuous or discrete. If categorical, state whether it is ordinal or not.
(c) Suppose we wanted to evaluate whether vehicle search rates are different for drivers of different races. In this analysis, which variable would be the response variable and which variable would be the explanatory variable?

1.64 Space launches. The following summary table shows the number of space launches in the US by the type of launching agency and the outcome of the launch (success or failure).[69]

	1957 - 1999		2000 - 2018	
	Failure	Success	Failure	Success
Private	13	295	10	562
State	281	3751	33	711
Startup	-	-	5	65

(a) What variables were collected on each launch in order to create to the summary table above?
(b) State whether each variable is numerical or categorical. If numerical, state whether it is continuous or discrete. If categorical, state whether it is ordinal or not.
(c) Suppose we wanted to study how the success rate of launches vary between launching agencies and over time. In this analysis, which variable would be the response variable and which variable would be the explanatory variable?

---

[67] Amy Orben and AK Baukney-Przybylski. "Screens, Teens and Psychological Well-Being: Evidence from three time-use diary studies". In: *Psychological Science* (2018).
[68] Emma Pierson et al. "A large-scale analysis of racial disparities in police stops across the United States". In: *arXiv preprint arXiv:1706.05678* (2017).
[69] JSR Launch Vehicle Database, A comprehensive list of suborbital space launches, 2019 Feb 10 Edition.

Chapter 2

Probability, random variables, and probability distributions

2.1 Relationships between two categorical variables

2.2 Probability basics

2.3 Conditional probability, intersections, and unions

2.4 Discrete random variables

2.5 Binomial distributions

2.6 Normal distributions

2.7 Sampling distributions and the central limit theorem

Probability forms a foundation of statistics, and you're probably already aware of many of the ideas. However, formalization of the concepts is new for most. We begin by introducing probability concepts through examples that will be familiar to most people. Then we analyze and apply discrete probability distributions, including the binomial distribution, and the most important continuous distribution in statistics - the normal distribution. Finally, we introduce the concept of a sampling distribution which will lay the groundwork for the next two chapters.

For videos, slides, and other resources, please visit
www.openintro.org/ahss

2.1 Relationships between two categorical variables

How do we visualize and summarize categorical data? How can we see the relationship between two categorical variables? For example, is there an association between the categorical variables of homeownership type and application type? Does email type provide any useful value in classifying email as spam or not spam? In this section, we will introduce tables and other basic tools for categorical data that are used throughout this book.

Learning objectives

1. Compare tabular and graphical representations for the relationship between two categorical variables.
2. Justify a claim using tabular and graphical representations for the distributions of two categorical variables.
3. Calculate summary statistics from two-way tables.
4. Compare summary statistics for two categorical variables.
5. Justify a claim using summary statistics for two categorical variables.

2.1.1 Introduction to two-way tables

In Section 1.2 we summarized `homeownership`, a categorical variable from the `loan` data set that has three levels, using a one-way frequency table and a one-way relative frequency table. Those tables are reproduced here.

`homeownership`	Count
rent	3858
mortgage	4789
own	1353
Total	10000

`homeownership`	Relative Frequency
rent	0.3858
mortgage	0.4789
own	0.1353
Total	1.000

Sometimes we to represent the frequencies split among *two* categorical variables. Figure 2.1 summarizes two variables: `app_type` and `homeownership`. A table that summarizes data for two categorical variables in this way is called a **two-way table** or a **contingency table**. In a two-way frequency table, each value in the table represents the count or number of times a particular combination of outcomes occurred. For example, the value 3496 corresponds to the number of loans in the data set where the borrower rents and the application type was individual.

Row and column totals are also included. The **row totals** provide the total counts across each row (e.g. $3496 + 3839 + 1170 = 8505$), and **column totals** are total counts down each column.

		`homeownership`			
		rent	mortgage	own	Total
`app_type`	individual	3496	3839	1170	8505
	joint	362	950	183	1495
	Total	3858	4789	1353	10000

Figure 2.1: A two-way frequency table for `app_type` and `homeownership`.

We can also create a two-way relative frequency table that shows the proportion for each combination of categories. These proportions are called **joint relative frequencies**. To calculate these, we divide every count in the frequency table by the entire table total of 10,000. For example, we calculate the proportion of the loans that are classified as `own` *and* `individual` as 1170/10,000 = 0.1170.

The row and column totals are called the **marginal relative frequencies**. The marginal relative frequency for the first row is 0.8505, which found by taking the `individual` row total divided by the total for the entire table. The marginal relative frequency for the second column is 0.4789, which is found by taking the `mortgage` column total divided by the total for the entire table.

		homeownership			
		rent	mortgage	own	Total
app_type	individual	0.3496	0.3839	0.1170	0.8505
	joint	0.0362	0.0950	0.0183	0.1495
	Total	0.3858	0.4789	0.1353	1.000

Figure 2.2: A two-way relative frequency table for `app_type` and `homeownership`.

2.1.2 Graphical representations for two categorical variables

Bar charts and mosaic plots provide a way to visualize and compare the distributions of categorical variables. A **segmented bar chart**, or stacked bar chart, is a graphical display of two-way table information. For example, a segmented bar chart is shown in Figure 2.3(a), where we have first created a bar chart using the `homeownership` variable and then divided each group by the levels of `app_type`.

One related visualization to the segmented bar chart is the **side-by-side bar chart**. An example is shown in Figure 2.3(b).

Figures 2.3(c) and 2.3(d) show a standardize segmented bar chart and a standardized side-by-side bar chart. These visualizations are helpful for understanding the proportion of individual or joint loan applications for borrowers in each level of `homeownership`. Additionally, because the proportions of `joint` and `individual` vary across the groups, we can conclude that the two variables are associated.

EXAMPLE 2.1

Examine the four bar charts in Figure 2.3. When is the segmented, side-by-side, standardized segmented bar chart, or standardized side-by-side the most useful?

The segmented bar chart is most useful when it's reasonable to assign one variable as the explanatory variable and the other variable as the response, since we are effectively grouping by one variable first and then breaking it down by the others.

Side-by-side bar charts are more agnostic in their display about which variable, if any, represents the explanatory and which the response variable. It is also easy to discern the number of cases in of the six different group combinations. However, one downside is that it tends to require more horizontal space; the narrowness of Figure 2.3(b) makes the plot feel a bit cramped. Additionally, when two groups are of very different sizes, as we see in the `own` group relative to either of the other two groups, it is difficult to discern if there is an association between the variables.

The standardized segmented bar chart is helpful if the primary variable in the segmented bar chart is relatively imbalanced, e.g. the `own` category has only a third of the observations in the `mortgage` category, making the simple segmented bar chart less useful for checking for an association. The major downside of the standardized version is that we lose all sense of how many cases each of the bars represents.

The last plot is a standardized side-by-side bar chart. It shows the joint and individual groups as proportions within each level of homeownership, and it offers similar benefits and tradeoffs to the standardized version of the stacked bar plot.

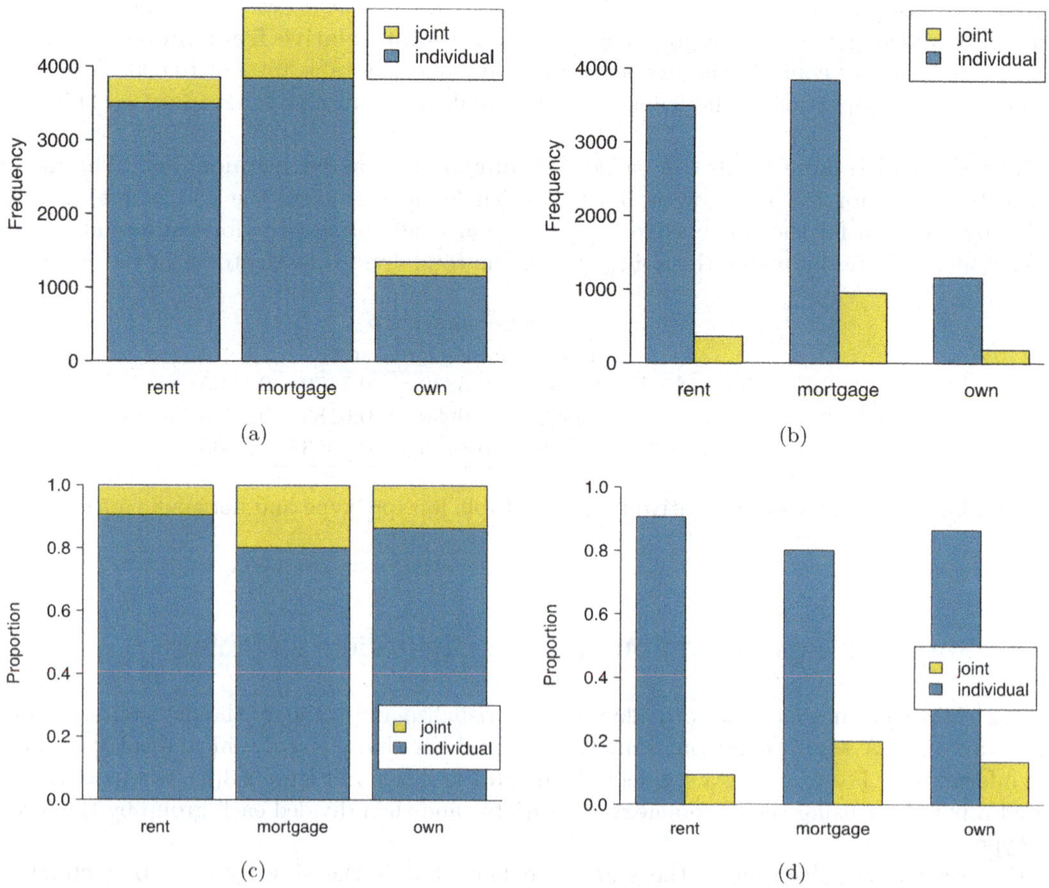

Figure 2.3: (a) segmented bar chart for `homeownership`, where the counts have been further broken down by `app_type`. (b) Side-by-side bar chart for `homeownership` and `app_type`. (c) Standardized version of the segmented bar chart. (d) Standardized side-by-side bar chart.

A **mosaic plot** is a visualization technique suitable for two-way tables that resembles a standardized segmented bar chart with the added benefit that we still see the relative group sizes of the primary variable as well.

To get started in creating our first mosaic plot, we'll break a square into columns for each category of the `homeownership` variable, with the result shown in Figure 2.4(a). Each column represents a level of `homeownership`, and the column widths correspond to the proportion of loans in each of those categories. For instance, there are fewer loans where the borrower is an owner than where the borrower has a mortgage. In general, mosaic plots use box *areas* to represent the number of cases in each category.

To finish the completed mosaic plot, the single-variable mosaic plot is further divided into pieces in Figure 2.4(b) using the `app_type` variable. As with the standardized segmented bar chart in Figure 2.3(c), each column is split proportional to the number of loans from individual and joint borrowers. For example, the second column represents loans where the borrower has a mortgage, and it was divided into individual loans (upper) and joint loans (lower). As another example, the bottom segment of the third column represents loans where the borrower owns their home and applied jointly, while the upper segment of this column represents borrowers who are homeowners and filed individually. We can again use this plot to see that the `homeownership` and `app_type` variables are associated, since some columns are divided in different vertical locations than others, which was the same technique used for checking an association in the standardized segmented bar chart.

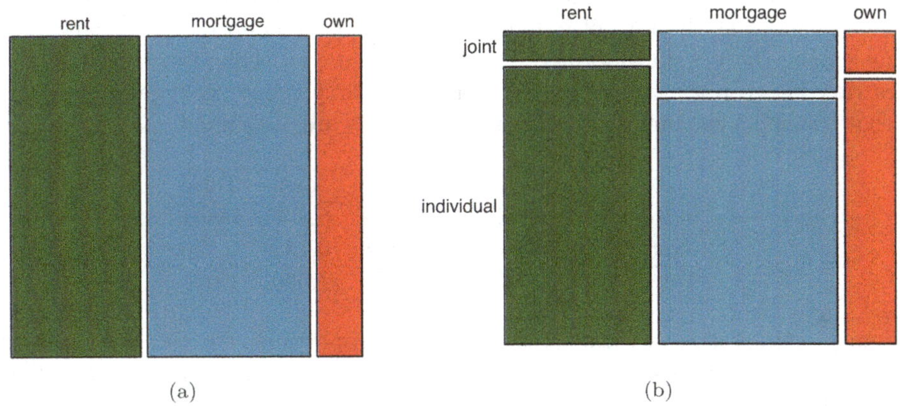

Figure 2.4: (a) The one-variable mosaic plot for `homeownership`. (b) Two-variable mosaic plot for both `homeownership` and `app_type`.

In Figure 2.4(b), we chose to first split by the homeowner status of the borrower. However, we could have instead first split by the application type, as in Figure 2.5. Like with the bar charts, it's common to use the explanatory variable to represent the first split in a mosaic plot, and then for the response to break up each level of the explanatory variable.

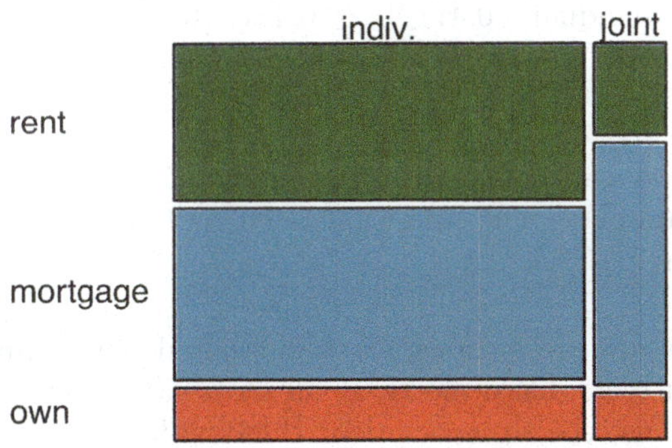

Figure 2.5: Mosaic plot where loans are grouped by the `homeownership` variable after they've been divided into the `individual` and `joint` application types.

2.1.3 Conditional relative frequencies

The bar charts and mosaic plots encountered in the previous section provide visual representations of counts or proportions in two-way tables. We will find these summaries useful when looking for evidence of an association between two categorical variables, such as `app_type` and `homeownership`. For instance, suppose we are interested in the following: for applicants who `rent`, what proportion apply as an individual and what proportion apply jointly? Also, how does this compare to applicants whose `homeownership` type is `mortgage`?

We can calculate the proportion of individual or joint application types *within each* `homeownership` type and compare them. These proportions are called **conditional relative freqencies**. Looking at Figure 2.6, we will condition on `homeownership` type and compute *column* proportions: we take each cell frequency and divide it by the column total to see what proportion of the column variable it represents. Notice that these proportions can be seen in the standardized stacked and segmented bar graphs shown in Figure 2.3 and in the mosaic plot in Figure 2.4(b).

In Figure 2.6, the value 0.906 indicates that 90.6% of renters applied as individuals. This is equivalent to saying that the rate of individual applications among renters is 90.6%. This rate is higher than among those with mortgages (80.2%) or who own their own home (85.1%). Because these rates vary between the three levels of `homeownership` (`rent`, `mortgage`, `own`), this provides evidence that the `app_type` and `homeownership` variables are associated.

	rent	mortgage	own	Total
individual	0.906	0.802	0.865	0.851
joint	0.094	0.198	0.135	0.150
Total	1.000	1.000	1.000	1.000

Figure 2.6: A two-way table with column proportions for the `app_type` and `homeownership` variables. The total for the last column is off by 0.001 due to a rounding error.

We can also ask a different question. What proportion of `individual` applicants rent, have a mortgage, or own? And how does this compare to `joint` applicants? Here, we condition on `app_type` and we calculate row proportions: we take each cell frequency and divide it by the row total to see what proportion of the row variable it represents, as shown in Figure 2.7. To calculate the row proportion at the intersection of `individual` and `rent`, we divide 3496 by the row total of 8505, which equals 0.411 and represents the proportion of individual applicants who rent.

	rent	mortgage	own	Total
individual	0.411	0.451	0.138	1.000
joint	0.242	0.635	0.122	1.000
Total	0.386	0.479	0.135	1.000

Figure 2.7: A two-way table with row proportions for the `app_type` and `homeownership` variables. The row total is off by 0.001 for the `joint` row due to a rounding error.

When comparing these row proportions, we would look down columns to see if the proportion of loans where the borrower rents, has a mortgage, or owns varied across the `individual` to `joint` application types. Here we also see an association: those with a `joint` application are more likely to have a mortgage and less likely to rent than those with an `individual` application. We can see these row proportions in the mosaic plot in Figure 2.5, while we can see the column proportions in the mosaic plot in Figure 2.4(b).

G

GUIDED PRACTICE 2.2

(a) What does 0.802 represent in Figure 2.6? (b) What does 0.451 represent in Figure 2.7? [1]

G

GUIDED PRACTICE 2.3

(a) What does 0.135 represent in the Figure 2.6? (b) What does 0.122 at the intersection of `joint` and `own` represent in Figure 2.7?[2]

[1](a) 0.802 represents the fraction of applicants with mortgages who applied as individuals. (b) 0.451 represents the proportion of individual applicants who have a mortgage.

[2](a) 0.135 represents the proportion of home-owning borrowers who had a joint application for the loan. (b) 0.122 represents the proportion of joint borrowers who own their home.

EXAMPLE 2.4

Data scientists use statistics to filter spam from incoming email messages. By noting specific characteristics of an email, a data scientist may be able to classify some emails as spam or not spam with high accuracy. One such characteristic is whether the email contains no numbers, small numbers, or big numbers. Another characteristic is the email format, which indicates whether or not an email has any HTML content, such as bolded text. We'll focus on email format and spam status using the `email` data set, and these variables are summarized in a two-way table in Figure 2.8. Which would be more helpful to someone hoping to classify email as spam or regular email for this table: row or column proportions?

Ⓔ

A data scientist would be interested in how the proportion of spam changes within each email format. This corresponds to column proportions: the proportion of spam in plain text emails and the proportion of spam in HTML emails.

If we generate the column proportions, we can see that a higher proportion of plain text emails are spam (209/1195 = 17.5%) than compared to HTML emails (158/2726 = 5.8%). This information on its own is insufficient to classify an email as spam or not spam, as over 80% of plain text emails are not spam. Yet, when we carefully combine this information with many other characteristics, we stand a reasonable chance of being able to classify some emails as spam or not spam with confidence.

	text	HTML	Total
spam	209	158	367
not spam	986	2568	3554
Total	1195	2726	3921

Figure 2.8: A two-way table for `spam` and `format`.

Example 2.4 points out that row and column proportions are not equivalent. Before settling on one form for a table, it is important to consider each to ensure that the most useful table is constructed.

We will revisit this idea of "conditioning" on a row or column variable to look for association between two variables in Section 2.3 when we investigate conditional probability. To determine of the association is "significant" or not, we will have to wait for Chapter 3, Inference for categorical data: proportions.

EXAMPLE 2.5

Look back to the mosaic plots in Figures 2.4(b) and 2.5 and the tables in Figures 2.7 and 2.6. Are there any obvious scenarios where one might be more useful than the other?

Ⓔ

None that we thought were obvious! What is distinct about `app_type` and `homeownership` vs the email example is that these two variables don't have a clear explanatory-response variable relationship that we might hypothesize (see Section 1.5.3 for these terms). Usually it is most useful to "condition" on the explanatory variable. For instance, in the email example, the email format was seen as a possible explanatory variable of whether the message was spam, so we would find it more interesting to compute the relative frequencies (proportions) for each email format.

Section summary

- A **two-way table**, also called a contingency table, can be used to summarize and compare data for two categorical variables. The entries in the cells of the table can be frequencies (i.e., counts) or relative frequencies (i.e., proportions).
- **Side-by-side bar charts**, **segmented bar charts**, and **mosaic plots** are examples of graphs used to display the relationship between two categorical variables. In these graphs, the frequency or relative frequency of each category, or level, of one of the categorical variables is displayed for each category of the other categorical variable.
- Graphical representations of two categorical variables can be used to compare the relationship of one categorical variable across the levels of the other categorical variable and determine whether the two variables are associated.
- Tabular and graphical representations for the distributions of two categorical variables may reveal information that can be used to justify claims about the variables in context.
- A **joint relative frequency** in a two-way table is a cell frequency divided by the total for the entire table.
- A **marginal relative frequency** in a two-way table is a row total divided by the total for the entire table or a column total divided by the total for the entire table.
- A **conditional relative frequency** is a relative frequency computed by restricting to a particular level, or category of interest. A conditional relative frequency can be a cell frequency in a column divided by the total for that column or it can be a cell frequency in a row divided by the total for that row.
- Summary statistics for two categorical variables can be used to compare distributions for evidence of an association between the two variables and may reveal information that can be used to justify claims about the variable in context.

Exercises

2.1 Views on the DREAM Act. A random sample of registered voters from Tampa, FL were asked if they support the DREAM Act, a proposed law which would provide a path to citizenship for people brought illegally to the US as children. The survey also collected information on the political ideology of the respondents. Based on the mosaic plot shown below, do views on the DREAM Act and political ideology appear to be independent? Explain your reasoning.[3]

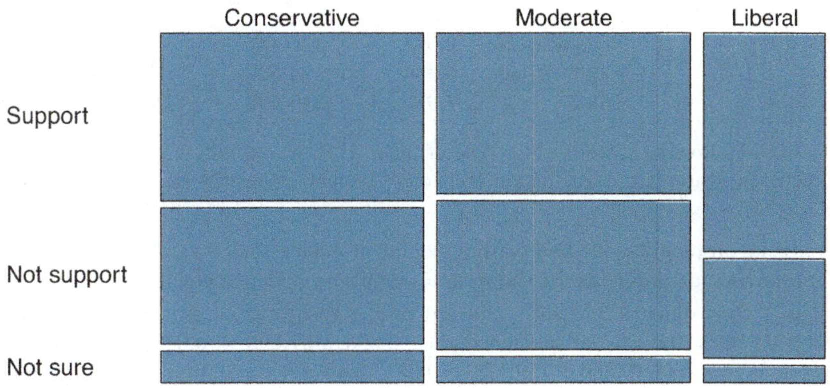

2.2 Raise taxes. A random sample of registered voters nationally were asked whether they think it's better to raise taxes on the rich or raise taxes on the poor. The survey also collected information on the political party affiliation of the respondents. Based on the mosaic plot shown below, do views on raising taxes and political affiliation appear to be independent? Explain your reasoning.[4]

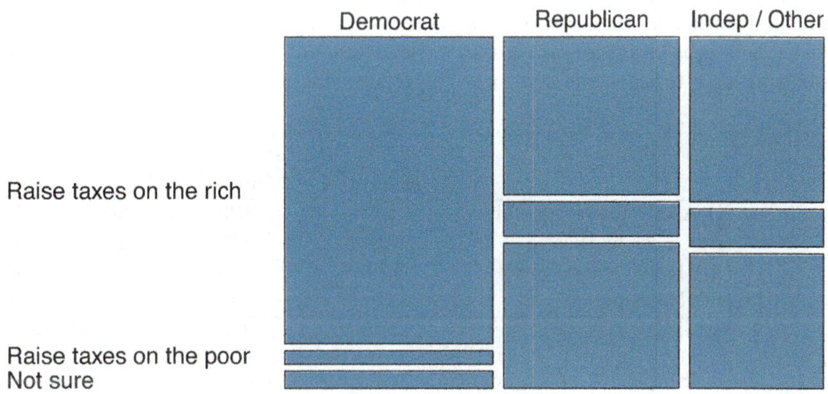

[3]SurveyUSA, News Poll #18927, data collected Jan 27-29, 2012.

[4]Public Policy Polling, Americans on College Degrees, Classic Literature, the Seasons, and More, data collected Feb 20-22, 2015.

2.3 Side effects of Avandia, Part I. Rosiglitazone is the active ingredient in the controversial type 2 diabetes medicine Avandia and has been linked to an increased risk of serious cardiovascular problems such as stroke, heart failure, and death. A common alternative treatment is pioglitazone, the active ingredient in a diabetes medicine called Actos. In a nationwide retrospective observational study of 227,571 Medicare beneficiaries aged 65 years or older, it was found that 2,593 of the 67,593 patients using rosiglitazone and 5,386 of the 159,978 using pioglitazone had serious cardiovascular problems. These data are summarized in the contingency table below.[5]

		Cardiovascular problems		
		Yes	No	Total
Treatment	Rosiglitazone	2,593	65,000	67,593
	Pioglitazone	5,386	154,592	159,978
	Total	7,979	219,592	227,571

Determine if each of the following statements is true or false. If false, explain why. *Be careful:* The reasoning may be wrong even if the statement's conclusion is correct. In such cases, the statement should be considered false.

(a) Since more patients on pioglitazone had cardiovascular problems (5,386 vs. 2,593), we can conclude that the rate of cardiovascular problems for those on a pioglitazone treatment is higher.

(b) The data suggest that diabetic patients who are taking rosiglitazone are more likely to have cardiovascular problems since the rate of incidence was (2,593 / 67,593 = 0.038) 3.8% for patients on this treatment, while it was only (5,386 / 159,978 = 0.034) 3.4% for patients on pioglitazone.

(c) The fact that the rate of incidence is higher for the rosiglitazone group proves that rosiglitazone causes serious cardiovascular problems.

(d) Based on the information provided so far, we cannot tell if the difference between the rates of incidences is due to a relationship between the two variables or due to chance.

2.4 Views on immigration. 910 randomly sampled registered voters from Tampa, FL were asked if they thought undocumented workers in the US should be (i) allowed to keep their jobs and apply for US citizenship, (ii) allowed to keep their jobs as temporary guest workers but not allowed to apply for US citizenship, or (iii) lose their jobs and have to leave the country. The results of the survey by political ideology are shown below.[6]

		Political ideology			
		Conservative	Moderate	Liberal	Total
Response	(i) Apply for citizenship	57	120	101	278
	(ii) Guest worker	121	113	28	262
	(iii) Leave the country	179	126	45	350
	(iv) Not sure	15	4	1	20
	Total	372	363	175	910

(a) What percent of these Tampa, FL voters identify themselves as conservatives?

(b) What percent of these Tampa, FL voters are in favor of the citizenship option?

(c) What percent of these Tampa, FL voters identify themselves as conservatives and are in favor of the citizenship option?

(d) What percent of these Tampa, FL voters who identify themselves as conservatives are also in favor of the citizenship option? What percent of moderates share this view? What percent of liberals share this view?

(e) Do political ideology and views on immigration appear to be independent? Explain your reasoning.

[5] D.J. Graham et al. "Risk of acute myocardial infarction, stroke, heart failure, and death in elderly Medicare patients treated with rosiglitazone or pioglitazone". In: *JAMA* 304.4 (2010), p. 411. ISSN: 0098-7484.

[6] SurveyUSA, News Poll #18927, data collected Jan 27-29, 2012.

2.5 Heart transplant data display. The Stanford University Heart Transplant Study was conducted to determine whether an experimental heart transplant program increased lifespan. Each patient entering the program was officially designated a heart transplant candidate, meaning that they were gravely ill and might benefit from a new heart. Patients were randomly assigned into treatment and control groups. Patients in the treatment group received a transplant, and those in the control group did not. The visualizations below display two different versions of the study results.[7]

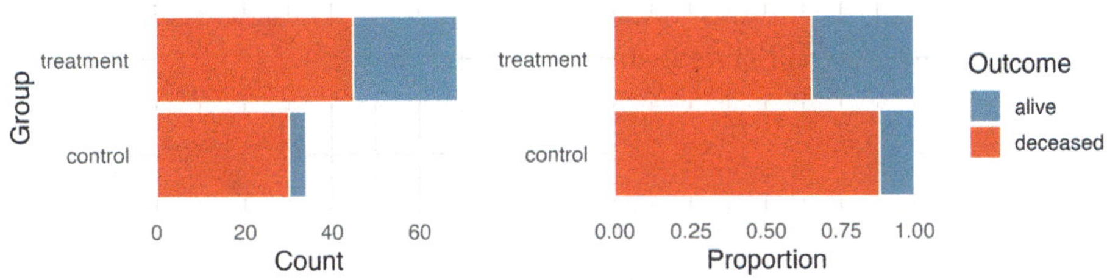

(a) Provide one aspect of the two group comparison that is easier to see from the stacked bar plot (left)?

(b) Provide one aspect of the two group comparison that is easier to see from the standardized bar plot (right)?

(c) For the Heart Transplant Study which of those aspects would be more important to display? That is, which bar plot would be better as a data visualization?

2.6 Shipping holiday gifts data display. A local news survey asked 500 randomly sampled Los Angeles residents which shipping carrier they prefer to use for shipping holiday gifts. The bar plots below show the distribution of responses by age group as well as distribution of responses by shipping method.

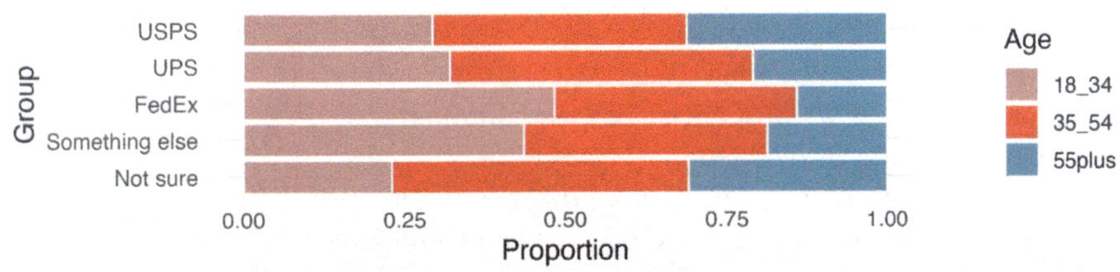

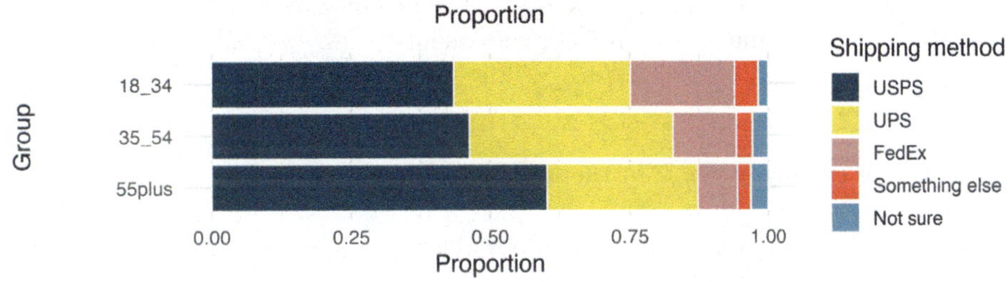

(a) Which graph (top or bottom) would you use to understand the shipping choices of people of different ages? Explain.

(b) Which graph (top or bottom) would you use to understand the age distribution across different types of shipping choices? Explain.

(c) A new shipping company would like to market to people over the age of 55. Who will be their biggest competitor? Explain.

(d) FedEx would like to reach out to grow their market share so as to balance the age demographics of FedEx users. To what age group should FedEx market?

[7] B. Turnbull et al. "Survivorship of Heart Transplant Data". In: *Journal of the American Statistical Association* 69 (1974), pp. 74–80.

2.2 Probability basics

What is the probability of rolling an even number on a die? Of getting 5 heads in row when tossing a coin? Of drawing a Heart or an Ace from a deck of cards? The study of probability is fun and interesting in its own right, but it also forms the foundation for statistical models and inferential procedures, many of which we will investigate in upcoming chapters.

Learning objectives

1. Estimate probabilities using simulations.
2. Calculate probabilities for events and their complements.
3. Justify why two events are mutually exclusive (or disjoint) using joint probability.
4. Calculate probabilities for independent events.

2.2.1 Introductory examples

EXAMPLE 2.6

A "die", the singular of dice, is a cube with six faces numbered 1, 2, 3, 4, 5, and 6. What is the chance of getting 1 when rolling a die?

If the die is fair, then the chance of a 1 is as good as the chance of any other number. Since there are six outcomes, the chance must be 1-in-6 or, equivalently, 1/6.

EXAMPLE 2.7

What is the chance of getting a 1 or 2 in the next roll?

1 and 2 constitute two of the six equally likely possible outcomes, so the chance of getting one of these two outcomes must be 2/6 = 1/3.

EXAMPLE 2.8

What is the chance of getting either 1, 2, 3, 4, 5, or 6 on the next roll?

100%. The outcome must be one of these numbers.

EXAMPLE 2.9

What is the chance of not rolling a 2?

Since the chance of rolling a 2 is 1/6 or $16.\bar{6}\%$, the chance of not rolling a 2 must be $100\% - 16.\bar{6}\% = 83.\bar{3}\%$ or 5/6.

Alternatively, we could have noticed that not rolling a 2 is the same as getting a 1, 3, 4, 5, or 6, which makes up five of the six equally likely outcomes and has probability 5/6.

EXAMPLE 2.10

Consider rolling two dice. If $1/6^{th}$ of the time the first die is a 1 and $1/6^{th}$ of those times the second die is a 1, what is the chance of getting two 1s?

If $16.\bar{6}\%$ of the time the first die is a 1 and $1/6^{th}$ of *those* times the second die is also a 1, then the chance that both dice are 1 is $(1/6) \times (1/6)$ or $1/36$.

2.2.2 Estimating probabilities using simulation

We use probability to build tools to describe and understand apparent randomness. We often frame probability in terms of a **random process** giving rise to an **outcome**. A random process generates results that are determined by chance. An outcome is the result of one trial of a random process.

Roll a die → 1, 2, 3, 4, 5, or 6
Flip a coin → H or T

Rolling a die or flipping a coin is a seemingly random process and each gives rise to an outcome.

PROBABILITY
The **probability** of an outcome is the proportion of times the outcome would occur if we observed the random process an infinite number of times.

Probability is defined as a proportion, and it always takes values between 0 and 1 (inclusively). It may also be displayed as a percentage between 0% and 100%.

An **event** is a collection of outcomes. Probability can be illustrated by rolling a die many times. Consider the event "roll a 1". We can use a **simulation** to model this event and to estimate the likelihood of the event by calculating the relative frequency of the event. The **relative frequency** of an event is the proportion of times the event occurs out of the number of trials. Let $\hat{p}_n$ be the proportion of outcomes that are 1 after the first n rolls. As the number of rolls increases, $\hat{p}_n$ (the relative frequency of rolls) will converge to the probability of rolling a 1, $p = 1/6$. Figure 2.9 shows this convergence for 100,000 die rolls. The tendency of $\hat{p}_n$ to stabilize around p, that is, the tendency of the relative frequency to stabilize around the true probability, is described by the **Law of Large Numbers**.

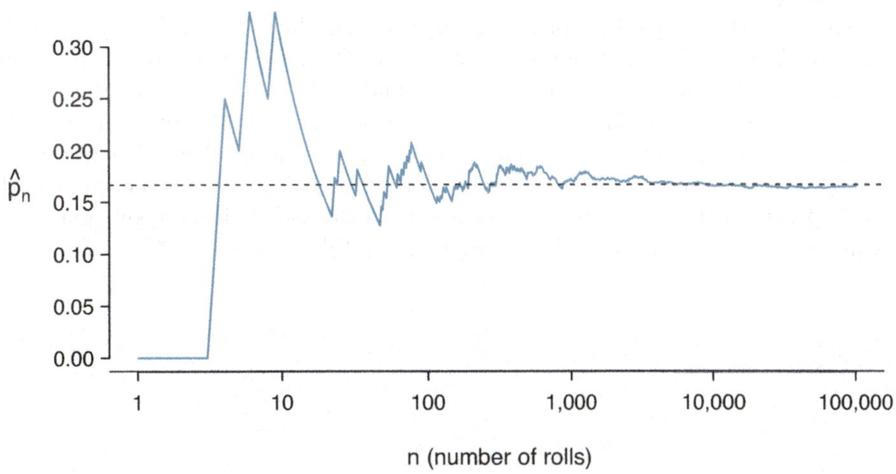

Figure 2.9: The fraction of die rolls that are 1 at each stage in a simulation. The relative frequency tends to get closer to the probability $1/6 \approx 0.167$ as the number of rolls increases.

LAW OF LARGE NUMBERS

The law of large numbers states that for independent trials, as the number of trials increases, the long-run relative frequency of the outcome or event gets closer and closer to a single value.

Occasionally the proportion will veer off from the probability and appear to defy the Law of Large Numbers, as $\hat{p}_n$ does many times in Figure 2.9. However, these deviations become smaller as the number of rolls increases.

Above we write p as the probability of rolling a 1. We can also write this probability as

$$P(\text{rolling a } 1)$$

As we become more comfortable with this notation, we will abbreviate it further. For instance, if it is clear that the process is "rolling a die", we could abbreviate $P(\text{rolling a } 1)$ as $P(1)$.

GUIDED PRACTICE 2.11

G Random processes include rolling a die and flipping a coin. (a) Think of another random process. (b) Describe all the possible outcomes of that process. For instance, rolling a die is a random process with potential outcomes 1, 2, ..., 6. [8]

What we think of as random processes are not necessarily random, but they may just be too difficult to understand exactly. The fourth example in the footnote solution to Guided Practice 2.11 suggests a roommate's behavior is a random process. However, even if a roommate's behavior is not truly random, modeling her behavior as a random process can still be useful.

MODELING A PROCESS AS RANDOM

It can be helpful to model a process as random even if it is not truly random.

2.2.3 Sample space and complement of an event

Rolling a die produces a value in the set {1, 2, 3, 4, 5, 6}. This set of all possible outcomes is called the **sample space** (S) for rolling a die. The probability of the sample space is 1: if we roll a die, it will come up as one of the numbers among 1, 2, 3, 4, 5, or 6. Additionally, the probability of any event is always between 0 and 1, inclusive.

We can say that the probability of rolling an even number when rolling a die is $3/6 = 1/2$, because there are three possible outcomes that are even (2, 4, 6) among the six possible outcomes and each of the outcomes in the sample space is equally likely. More generally, if all outcomes in the sample space are equally likely, then the theoretical probability an event E will occur can be found by dividing the number of outcomes in event E by the total number of outcomes in the sample space. Probability calculations tend to be more straightforward when each possible outcome is equally likely. However, this is often not the case. For example, while someone playing the lottery can either win or not win, the chance of winning is not 50%!

[8] Here are four examples. (i) Whether someone gets sick in the next month or not is an apparently random process with outcomes `sick` and `not`. (ii) We can *generate* a random process by randomly picking a person and measuring that person's height. The outcome of this process will be a positive number. (iii) Whether the stock market goes up or down next week is a seemingly random process with possible outcomes `up`, `down`, and `no_change`. Alternatively, we could have used the percent change in the stock market as a numerical outcome. (iv) Whether your roommate cleans her dishes tonight probably seems like a random process with possible outcomes `cleans_dishes` and `leaves_dishes`.

We often use the sample space to examine the scenario where an event does not occur. Let $D = \{2, 3\}$ represent the event that the outcome of a die roll is 2 or 3. Then the **complement** represents all outcomes in our sample space that are not in D, which is denoted by $D^c = \{1, 4, 5, 6\}$. That is, D^c is the set of all possible outcomes not already included in D. Figure 2.10 shows the relationship between D, D^c, and the sample space S.

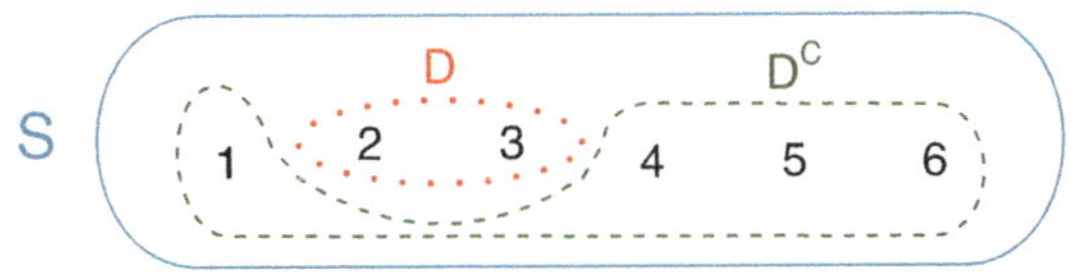

Figure 2.10: Event $D = \{2, 3\}$ and its complement, $D^c = \{1, 4, 5, 6\}$. S represents the sample space, which is the set of all possible events.

GUIDED PRACTICE 2.12

(a) Compute $P(D^c) = P(\text{rolling a } 1, 4, 5, \text{ or } 6)$. (b) What is $P(D) + P(D^c)$?[9]

GUIDED PRACTICE 2.13

Events $A = \{1, 2\}$ and $B = \{4, 6\}$ are shown in Figure 2.11 on page 111. (a) Write out what A^c and B^c represent. (b) Compute $P(A^c)$ and $P(B^c)$. (c) Compute $P(A) + P(A^c)$ and $P(B) + P(B^c)$.[10]

An event A together with its complement A^c comprise the entire sample space. Because of this we can say that $P(A) + P(A^c) = 1$.

COMPLEMENT

The complement of event A is denoted A^c, and A^c represents all outcomes not in A. A and A^c are mathematically related:

$$P(A) + P(A^c) = 1, \quad \text{i.e.} \quad P(A) = 1 - P(A^c)$$

In simple examples, computing A or A^c is feasible in a few steps. However, using the complement can save a lot of time as problems grow in complexity.

GUIDED PRACTICE 2.14

A die is rolled 10 times. (a) What is the complement of getting at least one 6 in 10 rolls of the die? (b) What is the complement of getting at most three 6's in 10 rolls of the die?[11]

[9](a) The outcomes are disjoint and each has probability 1/6, so the total probability is $4/6 = 2/3$. (b) We can also see that $P(D) = \frac{1}{6} + \frac{1}{6} = 1/3$. Because D and D^c are disjoint, $P(D) + P(D^c) = 1$.

[10]Brief solutions: (a) $A^c = \{3, 4, 5, 6\}$ and $B^c = \{1, 2, 3, 5\}$. (b) Noting that each outcome is disjoint, add the individual outcome probabilities to get $P(A^c) = 2/3$ and $P(B^c) = 2/3$. (c) A and A^c are disjoint, and the same is true of B and B^c. Therefore, $P(A) + P(A^c) = 1$ and $P(B) + P(B^c) = 1$.

[11](a) The complement of getting at least one 6 in ten rolls of a die is getting zero 6's in the 10 rolls. (b) The complement of getting at most three 6's in 10 rolls is getting four, five, ..., nine, or ten 6's in 10 rolls.

2.2.4 Disjoint or mutually exclusive outcomes

Two outcomes are called **disjoint** or **mutually exclusive** if they cannot both happen in the same trial. For instance, if we roll a die, the outcomes 1 and 2 are disjoint since they cannot both occur on a single roll. On the other hand, the outcomes 1 and "rolling an odd number" are not disjoint since both occur if the outcome of the roll is a 1. The terms *disjoint* and *mutually exclusive* are equivalent and interchangeable.

Calculating the probability of disjoint outcomes is easy. When rolling a die, the outcomes 1 and 2 are disjoint, and we compute the probability that one of these outcomes will occur by adding their separate probabilities:

$$P(1 \text{ or } 2) = P(1) + P(2) = 1/6 + 1/6 = 1/3$$

What about the probability of rolling a 1, 2, 3, 4, 5, or 6? Here again, all of the outcomes are disjoint so we add the probabilities:

$$\begin{aligned} &P(1 \text{ or } 2 \text{ or } 3 \text{ or } 4 \text{ or } 5 \text{ or } 6) \\ &\quad = P(1) + P(2) + P(3) + P(4) + P(5) + P(6) \\ &\quad = 1/6 + 1/6 + 1/6 + 1/6 + 1/6 + 1/6 = 1. \end{aligned}$$

The Addition Rule guarantees the accuracy of this approach when the outcomes are disjoint.

ADDITION RULE OF DISJOINT OUTCOMES

If A_1 and A_2 represent two disjoint outcomes, then the probability that one of them occurs is given by

$$P(A_1 \text{ or } A_2) = P(A_1) + P(A_2)$$

If there are many disjoint outcomes A_1, ..., A_k, then the probability that one of these outcomes will occur is

$$P(A_1) + P(A_2) + \cdots + P(A_k)$$

GUIDED PRACTICE 2.15

G

We are interested in the probability of rolling a 1, 4, or 5. (a) Explain why the outcomes 1, 4, and 5 are disjoint. (b) Apply the Addition Rule for disjoint outcomes to determine $P(1 \text{ or } 4 \text{ or } 5)$.[12]

GUIDED PRACTICE 2.16

G

In the `email` data set in Chapter 1, the `number` variable described whether no number (labeled `none`), only one or more small numbers (`small`), or whether at least one big number appeared in an email (`big`). Of the 3,921 emails, 549 had no numbers, 2,827 had only one or more small numbers, and 545 had at least one big number. (a) Are the outcomes `none`, `small`, and `big` disjoint? (b) Determine the proportion of emails with value `small` and `big` separately. (c) Use the Addition Rule for disjoint outcomes to compute the probability a randomly selected email from the data set has a number in it, small or big.[13]

[12] (a) The random process is a die roll, and at most one of these outcomes can come up. This means they are disjoint outcomes. (b) $P(1 \text{ or } 4 \text{ or } 5) = P(1) + P(4) + P(5) = \frac{1}{6} + \frac{1}{6} + \frac{1}{6} = \frac{3}{6} = \frac{1}{2}$

[13] (a) Yes. Each email is categorized in only one level of `number`. (b) Small: $\frac{2827}{3921} = 0.721$. Big: $\frac{545}{3921} = 0.139$. (c) $P(\texttt{small} \text{ or } \texttt{big}) = P(\texttt{small}) + P(\texttt{big}) = 0.721 + 0.139 = 0.860$.

Statisticians rarely work with individual outcomes and instead consider *sets* or *collections* of outcomes. Let A represent the event where a die roll results in 1 or 2 and B represent the event that the die roll is a 4 or a 6. We write A as the set of outcomes $\{1, 2\}$ and $B = \{4, 6\}$. These sets are commonly called **events**. Because A and B have no elements in common, they are disjoint events. A and B are represented in Figure 2.11.

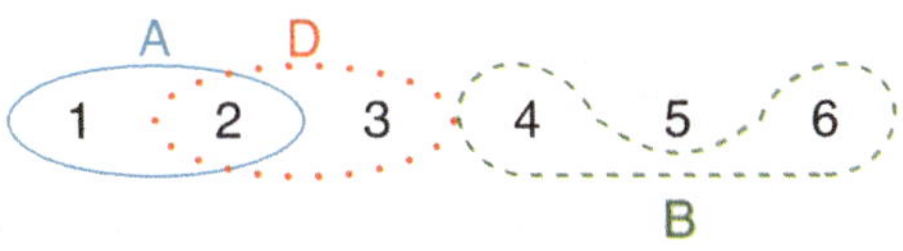

Figure 2.11: Three events, A, B, and D, consist of outcomes from rolling a die. A and B are disjoint since they do not have any outcomes in common.

The Addition Rule applies to both disjoint outcomes and disjoint events. The probability that one of the disjoint events A or B occurs is the sum of the separate probabilities:

$$P(A \text{ or } B) = P(A) + P(B) = 1/3 + 1/3 = 2/3$$

GUIDED PRACTICE 2.17

(a) Verify the probability of event A, $P(A)$, is 1/3 using the Addition Rule. (b) Do the same for event B.[14]

GUIDED PRACTICE 2.18

(a) Using Figure 2.11 as a reference, what outcomes are represented by event D? (b) Are events B and D disjoint? (c) Are events A and D disjoint?[15]

GUIDED PRACTICE 2.19

In Guided Practice 2.18, you confirmed B and D from Figure 2.11 are disjoint. Compute the probability that either event B or event D occurs.[16]

2.2.5 Joint probabilies when events are independent

Just as variables and observations can be independent, random processes can be independent, too. Two processes are **independent** if knowing the outcome of one provides no useful information about the outcome of the other. For instance, flipping a coin and rolling a die are two independent processes – knowing the coin was heads does not help determine the outcome of a die roll. On the other hand, stock prices usually move up or down together, so they are not independent. To define independence mathematically, we will have to wait for the discussion of conditional probability in the next section.

Example 2.10 provides a basic example of two independent processes: rolling two dice. We want to determine the probability that both will be 1. Suppose one of the dice is red and the other white. If the outcome of the red die is a 1, it provides no information about the outcome of the white die. We first encountered this same question in Example 2.10 (page 107), where we calculated the probability using the following reasoning: $1/6^{th}$ of the time the red die is a 1, and $1/6^{th}$ of *those* times the white die will also be 1. This is illustrated in Figure 2.12. Because the rolls are independent, the probabilities of the corresponding outcomes can be multiplied to get the final answer: $(1/6) \times (1/6) = 1/36$. This can be generalized to many independent processes.

[14](a) $P(A) = P(1 \text{ or } 2) = P(1) + P(2) = \frac{1}{6} + \frac{1}{6} = \frac{2}{6} = \frac{1}{3}$. (b) Similarly, $P(B) = 1/3$.

[15](a) Outcomes 2 and 3. (b) Yes, events B and D are disjoint because they share no outcomes. (c) The events A and D share an outcome in common, 2, and so are not disjoint.

[16]Since B and D are disjoint events, use the Addition Rule: $P(B \text{ or } D) = P(B) + P(D) = \frac{1}{3} + \frac{1}{3} = \frac{2}{3}$.

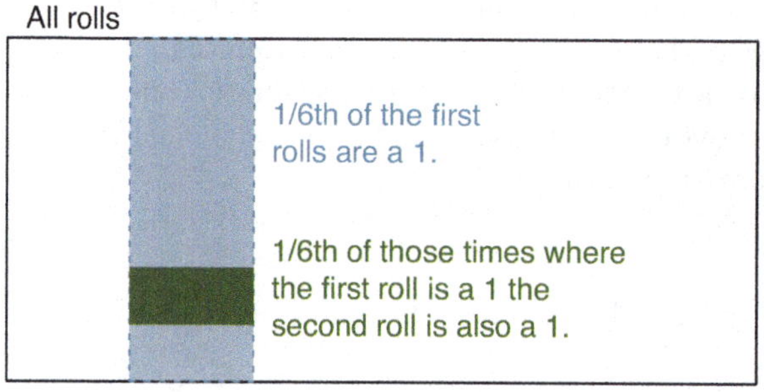

Figure 2.12: $1/6^{th}$ of the time, the first roll is a 1. Then $1/6^{th}$ of *those* times, the second roll will also be a 1.

EXAMPLE 2.20

What if there was also a blue die independent of the other two? What is the probability of rolling the three dice and getting all 1s?

E The same logic applies from Example 2.10. If $1/36^{th}$ of the time the white and red dice are both 1, then $1/6^{th}$ of *those* times the blue die will also be 1, so multiply:

$$\begin{aligned} P(white = 1 \text{ and } red = 1 \text{ and } blue = 1) &= P(white = 1) \times P(red = 1) \times P(blue = 1) \\ &= (1/6) \times (1/6) \times (1/6) = 1/216 \end{aligned}$$

Examples 2.10 and 2.20 illustrate what is called the Multiplication Rule for independent processes.

MULTIPLICATION RULE FOR INDEPENDENT PROCESSES

If A and B represent events from two different and independent processes, then the probability that both A and B occur can be calculated as the product of their separate probabilities:

$$P(A \text{ and } B) = P(A) \times P(B)$$

Similarly, if there are k events A_1, ..., A_k from k independent processes, then the probability they all occur is

$$P(A_1) \times P(A_2) \times \cdots \times P(A_k)$$

GUIDED PRACTICE 2.21

G About 9% of people are left-handed. Suppose 2 people are selected at random from the U.S. population. Because the sample size of 2 is very small relative to the population, it is reasonable to assume these two people are independent. (a) What is the probability that both are left-handed? (b) What is the probability that both are right-handed?[17]

[17](a) The probability the first person is left-handed is 0.09, which is the same for the second person. We apply the Multiplication Rule for independent processes to determine the probability that both will be left-handed: $0.09 \times 0.09 = 0.0081$.

(b) It is reasonable to assume the proportion of people who are ambidextrous (both right- and left-handed) is nearly 0, which results in $P(\text{right-handed}) = 1 - 0.09 = 0.91$. Using the same reasoning as in part (a), the probability that both will be right-handed is $0.91 \times 0.91 = 0.8281$.

GUIDED PRACTICE 2.22

Suppose 5 people are selected at random.[18]

(a) What is the probability that all are right-handed?

(b) What is the probability that all are left-handed?

(c) What is the probability that not all of the people are right-handed?

Suppose the variables `handedness` and `gender` are independent, i.e. knowing someone's `gender` provides no useful information about their `handedness` and vice-versa. Then we can compute whether a randomly selected person is right-handed and female[19] using the Multiplication Rule:

$$\begin{aligned} P(\text{right-handed and female}) &= P(\text{right-handed}) \times P(\text{female}) \\ &= 0.91 \times 0.50 = 0.455 \end{aligned}$$

GUIDED PRACTICE 2.23

Three people are selected at random.[20]

(a) What is the probability that the first person is male and right-handed?

(b) What is the probability that the first two people are male and right-handed?.

(c) What is the probability that the third person is female and left-handed?

(d) What is the probability that the first two people are male and right-handed and the third person is female and left-handed?

Sometimes we wonder if one outcome provides useful information about another outcome. The question we are asking is, are the occurrences of the two events independent? We say that two events A and B are independent if they satisfy Equation (2.21).

EXAMPLE 2.24

If we shuffle up a deck of cards and draw one, is the event that the card is a heart independent of the event that the card is an ace?

The probability the card is a heart is $1/4$ and the probability that it is an ace is $1/13$. The probability the card is the ace of hearts is $1/52$. We check whether Equation 2.21 is satisfied:

$$P(\heartsuit) \times P(\text{ace}) = \frac{1}{4} \times \frac{1}{13} = \frac{1}{52} = P(\heartsuit \text{ and ace})$$

Because the equation holds, the event that the card is a heart and the event that the card is an ace are independent events.

---

[18](a) The abbreviations `RH` and `LH` are used for right-handed and left-handed, respectively. Since each are independent, we apply the Multiplication Rule for independent processes:

$$\begin{aligned} P(\text{all five are RH}) &= P(\text{first} = \text{RH, second} = \text{RH, ..., fifth} = \text{RH}) \\ &= P(\text{first} = \text{RH}) \times P(\text{second} = \text{RH}) \times \cdots \times P(\text{fifth} = \text{RH}) \\ &= 0.91 \times 0.91 \times 0.91 \times 0.91 \times 0.91 = 0.624 \end{aligned}$$

(b) Using the same reasoning as in (a), $0.09 \times 0.09 \times 0.09 \times 0.09 \times 0.09 = 0.0000059$

(c) Use the complement, $P(\text{all five are RH})$, to answer this question:

$$P(\text{not all RH}) = 1 - P(\text{all RH}) = 1 - 0.624 = 0.376$$

[19]The actual proportion of the U.S. population that is `female` is about 50%, and so we use 0.5 for the probability of sampling a woman. However, this probability does differ in other countries.

[20]Brief answers are provided. (a) This can be written in probability notation as $P(\text{a randomly selected person is male and right-handed}) = 0.455$. (b) 0.207. (c) 0.045. (d) 0.0093.

Section summary

- A **random process** generates results that are determined by chance.
- An **outcome** is the result of one trial of a random process.
- An **event** is a collection of outcomes.
- **Simulation** is a way to model random events such that the simulated outcomes closely match real-world outcomes. All possible outcomes are associated with a value to be determined by chance. Record the counts of simulated outcomes and the count total.
- The **probability** of an outcome or event is its long-run relative frequency— that is, its relative frequency over a large number of trials.
- The **relative frequency** of an outcome or event determined from empirical data can be used to estimate the actual, or true, probability of that outcome or event.
- The **law of large numbers** states that for independent trials, as the number of trials increases, the long-run relative frequency of the outcome or event gets closer and closer to a single value.
- The probability of an event is always between 0 and 1, inclusive.
- The **sample space** of a random process is the set of all possible nonoverlapping outcomes. The probability of the sample space is 1.
- If all outcomes in the sample space are equally likely, then the theoretical probability an event E will occur can be found by dividing the number of outcomes in event E by the total number of outcomes in the sample space.
- The probability of the **complement** of an event E, which can written as "not E" or E^c, is equal to 1− P(E), i.e. $\text{P}(E^c) = 1-\text{P}(E)$. This can also be applied as $\text{P}(E) = 1-\text{P}(E^c)$.
- The probability that events A and B both will occur, that is the **joint probability** of A and B, is the probability of the intersection of A and B.
- Two events A and B are **mutually exclusive**, or **disjoint**, if they cannot happen together. In this case, the events do not overlap and $P(A \text{ and } B) = 0$.
- In the *special case* where A and B are **disjoint** events: $P(A \text{ or } B) = P(A) + P(B)$.
- When considering only two events, the probability that one *or* the other happens is equal to the probability that *at least one* of the two events happens.
- To find the probability that *at least one* of several events occurs, use a special case of the rule of **complements**: $P(\text{at least one}) = 1 - P(\text{none})$.
- Two events are **independent** when the occurrence of one does not change the likelihood of the other. Outcomes of coin tosses or rolls of a dice are classic examples of independent events.
- In the *special case* where A and B are **independent**: $P(A \text{ and } B) = P(A) \times P(B)$.

Exercises

2.7 True or false. Determine if the statements below are true or false, and explain your reasoning.

(a) If a fair coin is tossed many times and the last eight tosses are all heads, then the chance that the next toss will be heads is somewhat less than 50%.

(b) Drawing a face card (jack, queen, or king) and drawing a red card from a full deck of playing cards are mutually exclusive events.

(c) Drawing a face card and drawing an ace from a full deck of playing cards are mutually exclusive events.

2.8 Roulette wheel. The game of roulette involves spinning a wheel with 38 slots: 18 red, 18 black, and 2 green. A ball is spun onto the wheel and will eventually land in a slot, where each slot has an equal chance of capturing the ball.

(a) You watch a roulette wheel spin 3 consecutive times and the ball lands on a red slot each time. What is the probability that the ball will land on a red slot on the next spin?

(b) You watch a roulette wheel spin 300 consecutive times and the ball lands on a red slot each time. What is the probability that the ball will land on a red slot on the next spin?

(c) Are you equally confident of your answers to parts (a) and (b)? Why or why not?

Photo by Håkan Dahlström (http://flic.kr/p/93fEzp) CC BY 2.0 license

2.9 Four games, one winner. Below are four versions of the same game. Your archnemesis gets to pick the version of the game, and then you get to choose how many times to flip a coin: 10 times or 100 times. Identify how many coin flips you should choose for each version of the game. It costs $1 to play each game. Explain your reasoning.

(a) If the proportion of heads is larger than 0.60, you win $1.

(b) If the proportion of heads is larger than 0.40, you win $1.

(c) If the proportion of heads is between 0.40 and 0.60, you win $1.

(d) If the proportion of heads is smaller than 0.30, you win $1.

2.10 Backgammon. Backgammon is a board game for two players in which the playing pieces are moved according to the roll of two dice. Players win by removing all of their pieces from the board, so it is usually good to roll high numbers. You are playing backgammon with a friend and you roll two 6s in your first roll and two 6s in your second roll. Your friend rolls two 3s in his first roll and again in his second row. Your friend claims that you are cheating, because rolling double 6s twice in a row is very unlikely. Using probability, show that your rolls were just as likely as his.

2.11 Coin flips. If you flip a fair coin 10 times, what is the probability of

(a) getting all tails?

(b) getting all heads?

(c) getting at least one tails?

2.12 Dice rolls. If you roll a pair of fair dice, what is the probability of

(a) getting a sum of 1?

(b) getting a sum of 5?

(c) getting a sum of 12?

2.3 Conditional probability, intersections, and unions

What is the likelihood that a machine learning algorithm will misclassify a photo as being about fashion if it is not actually about fashion? How does the probability of surviving smallpox given being vaccinated compare to the probability of being vaccinated given having survived smallpox? To answer these questions, we investigate conditional probabilities. We also develop a general formula for "and" probabilities and a general formula for "or" probabilities.

Learning objectives

1. Calculate probabilities using two-way tables and Venn diagrams.
2. Calculate and interpret conditional probabilities.
3. Use a tree diagram along with the conditional probability rule to solve "inverted" conditional probabilities.
4. Calculate the probability of joint events whether or not the events are independent.
5. Calculate the probability of the union of two events, whether or not the events are mutually exclusive.
6. Determine whether two events are independent and whether they are mutually exclusive.

2.3.1 Exploring probabilities with a two-way table

The `photo_classify` data set represents a sample of 1822 photos from a photo sharing website. Data scientists have been working to improve a classifier for whether the photo is about fashion or not, and these 659 photos represent a test for their classifier. Each photo gets two classifications: the first is called `mach_learn` and gives a classification from a machine learning (ML) system of either `pred_fashion` or `pred_not`. Each of these 1822 photos have also been classified carefully by a team of people, which we take to be the source of truth; this variable is called `truth` and takes values `fashion` and `not`. Figure 2.13 summarizes the results.

		truth		
		fashion	not	Total
mach_learn	pred_fashion	197	22	219
	pred_not	112	1491	1603
	Total	309	1513	1822

Figure 2.13: two-way table summarizing the `photo_classify` data set.

EXAMPLE 2.25

If a photo is actually about fashion, what is the chance the ML classifier correctly identified the photo as being about fashion?

E

We can estimate this probability using the data. Of the 309 fashion photos, the ML algorithm correctly classified 197 of the photos:

$$P(\texttt{mach\_learn} \text{ is } \texttt{pred\_fashion} \text{ given } \texttt{truth} \text{ is } \texttt{fashion}) = \frac{197}{309} = 0.638$$

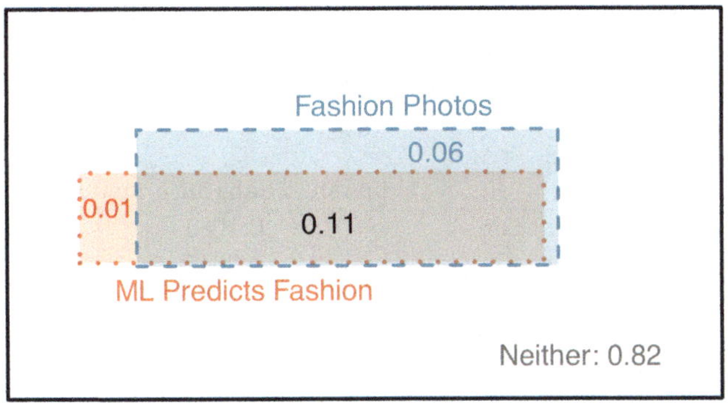

Figure 2.14: A Venn diagram using boxes for the `photo_classify` data set.

EXAMPLE 2.26

We sample a photo from the data set and learn the ML algorithm predicted this photo was not about fashion. What is the probability that it was incorrect and the photo is about fashion?

If the ML classifier suggests a photo is not about fashion, then it comes from the second row in the data set. Of these 1603 photos, 112 were actually about fashion:

$$P(\texttt{truth} \text{ is } \texttt{fashion} \text{ given } \texttt{mach\_learn} \text{ is } \texttt{pred\_not}) = \frac{112}{1603} = 0.070$$

2.3.2 Marginal and joint probabilities

Figure 2.13 includes row and column totals for each variable separately in the `photo_classify` data set. These totals represent **marginal probabilities** for the sample, which are the probabilities based on a single variable without regard to any other variables. For instance, a probability based solely on the `mach_learn` variable is a marginal probability:

$$P(\texttt{mach\_learn} \text{ is } \texttt{pred\_fashion}) = \frac{219}{1822} = 0.12$$

A probability of outcomes for two or more variables or processes is called a **joint probability**:

$$P(\texttt{mach\_learn} \text{ is } \texttt{pred\_fashion} \text{ and } \texttt{truth} \text{ is } \texttt{fashion}) = \frac{197}{1822} = 0.11$$

It is common to substitute a comma for "and" in a joint probability, although using either the word "and" or a comma is acceptable:

$$P(\texttt{mach\_learn} \text{ is } \texttt{pred\_fashion}, \texttt{truth} \text{ is } \texttt{fashion})$$

means the same thing as

$$P(\texttt{mach\_learn} \text{ is } \texttt{pred\_fashion} \text{ and } \texttt{truth} \text{ is } \texttt{fashion})$$

MARGINAL AND JOINT PROBABILITIES

If a probability is based on a single variable, it is a *marginal probability.* The probability of outcomes for two or more variables or processes is called a *joint probability.*

We use **table proportions** to summarize joint probabilities for the `photo_classify` sample. These proportions are computed by dividing each count in Figure 2.13 by the table's total, 1822, to obtain the proportions in Figure 2.15. The joint probability distribution of the `mach_learn` and `truth` variables is shown in Figure 2.16.

	truth: fashion	truth: not	Total
mach_learn: pred_fashion	0.1081	0.0121	0.1202
mach_learn: pred_not	0.0615	0.8183	0.8798
Total	0.1696	0.8304	1.00

Figure 2.15: Probability table summarizing the `photo_classify` data set.

Joint outcome	Probability
`mach_learn` is `pred_fashion` and `truth` is `fashion`	0.1081
`mach_learn` is `pred_fashion` and `truth` is `not`	0.0121
`mach_learn` is `pred_not` and `truth` is `fashion`	0.0615
`mach_learn` is `pred_not` and `truth` is `not`	0.8183
Total	1.0000

Figure 2.16: Joint probability distribution for the `photo_classify` data set.

GUIDED PRACTICE 2.27

G Verify Figure 2.16 represents a probability distribution: events are disjoint, all probabilities are non-negative, and the probabilities sum to 1.[21]

We can compute marginal probabilities using joint probabilities in simple cases. For example, the probability that a randomly selected photo from the data set is about fashion is found by summing the outcomes in which `truth` takes value `fashion`:

$$\begin{aligned} P(\underline{\texttt{truth}\text{ is }\texttt{fashion}}) &= P(\texttt{mach\_learn}\text{ is }\texttt{pred\_fashion}\text{ and }\underline{\texttt{truth}\text{ is }\texttt{fashion}}) \\ &\quad + P(\texttt{mach\_learn}\text{ is }\texttt{pred\_not}\text{ and }\underline{\texttt{truth}\text{ is }\texttt{fashion}}) \\ &= 0.1081 + 0.0615 \\ &= 0.1696 \end{aligned}$$

2.3.3 Defining conditional probability

The ML classifier predicts whether a photo is about fashion, even if it is not perfect. We would like to better understand how to use information from a variable like `mach_learn` to improve our probability estimation of a second variable, which in this example is `truth`.

The probability that a random photo from the data set is about fashion is about 0.17. If we knew the machine learning classifier predicted the photo was about fashion, could we get a better estimate of the probability the photo is actually about fashion? Absolutely. To do so, we limit our view to only those 219 cases where the ML classifier predicted that the photo was about fashion and look at the fraction where the photo was actually about fashion:

$$P(\texttt{truth}\text{ is }\texttt{fashion}\text{ given }\texttt{mach\_learn}\text{ is }\texttt{pred\_fashion}) = \frac{197}{219} = 0.900$$

We call this a **conditional probability** because we computed the probability under a condition: the ML classifier prediction said the photo was about fashion.

[21] Each of the four outcome combination are disjoint, all probabilities are indeed non-negative, and the sum of the probabilities is 0.1081 + 0.0121 + 0.0615 + 0.8183 = 1.00.

There are two parts to a conditional probability, the **outcome of interest** and the **condition**. It is useful to think of the condition as information we know to be true, and this information usually can be described as a known outcome or event. We generally separate the text inside our probability notation into the outcome of interest and the condition with a vertical bar:

$$P(\texttt{truth}\text{ is }\texttt{fashion}\text{ given }\texttt{mach\_learn}\text{ is }\texttt{pred\_fashion})$$
$$= P(\texttt{truth}\text{ is }\texttt{fashion} \mid \texttt{mach\_learn}\text{ is }\texttt{pred\_fashion}) = \frac{197}{219} = 0.900$$

The vertical bar "|" is read as *given*.

In the last equation, we computed the probability a photo was about fashion based on the condition that the ML algorithm predicted it was about fashion as a fraction:

$$P(\texttt{truth}\text{ is }\texttt{fashion} \mid \texttt{mach\_learn}\text{ is }\texttt{pred\_fashion})$$
$$= \frac{\#\text{ cases where }\texttt{truth}\text{ is }\texttt{fashion}\text{ and }\texttt{mach\_learn}\text{ is }\texttt{pred\_fashion}}{\#\text{ cases where }\texttt{mach\_learn}\text{ is }\texttt{pred\_fashion}}$$
$$= \frac{197}{219} = 0.900$$

We considered only those cases that met the condition, `mach_learn` is `pred_fashion`, and then we computed the ratio of those cases that satisfied our outcome of interest, photo was actually about fashion.

Frequently, marginal and joint probabilities are provided instead of count data. For example, disease rates are commonly listed in percentages rather than in a count format. We would like to be able to compute conditional probabilities even when no counts are available, and we use the last equation as a template to understand this technique.

We considered only those cases that satisfied the condition, where the ML algorithm predicted fashion. Of these cases, the conditional probability was the fraction representing the outcome of interest, that the photo was about fashion. Suppose we were provided only the information in Figure 2.15, i.e. only probability data. Then if we took a sample of 1000 photos, we would anticipate about 12.0% or $0.120 \times 1000 = 120$ would be predicted to be about fashion (`mach_learn` is `pred_fashion`). Similarly, we would expect about 10.8% or $0.108 \times 1000 = 108$ to meet both the information criteria and represent our outcome of interest. Then the conditional probability can be computed as

$$P(\texttt{truth}\text{ is }\texttt{fashion} \mid \texttt{mach\_learn}\text{ is }\texttt{pred\_fashion})$$
$$= \frac{\#\ (\texttt{truth}\text{ is }\texttt{fashion}\text{ and }\texttt{mach\_learn}\text{ is }\texttt{pred\_fashion})}{\#\ (\texttt{mach\_learn}\text{ is }\texttt{pred\_fashion})}$$
$$= \frac{108}{120} = \frac{0.108}{0.120} = 0.90$$

Here we are examining exactly the fraction of two probabilities, 0.108 and 0.120, which we can write as

$$P(\texttt{truth}\text{ is }\texttt{fashion}\text{ and }\texttt{mach\_learn}\text{ is }\texttt{pred\_fashion}) \quad\text{and}\quad P(\texttt{mach\_learn}\text{ is }\texttt{pred\_fashion}).$$

The fraction of these probabilities is an example of the general formula for conditional probability.

CONDITIONAL PROBABILITY

The conditional probability of the outcome of interest A given condition B is computed as the following:

$$P(A|B) = \frac{P(A\text{ and }B)}{P(B)}$$

G

GUIDED PRACTICE 2.28

(a) Write out the following statement in conditional probability notation: "*The probability that the ML prediction was correct, if the photo was about fashion*". Here the condition is now based on the photo's `truth` status, not the ML algorithm.

(b) Determine the probability from part (a). Figure 2.15 on page 118 may be helpful.[22]

G

GUIDED PRACTICE 2.29

(a) Determine the probability that the algorithm is incorrect if it is known the photo is about fashion.

(b) Using the answers from part (a) and Guided Practice 2.28(b), compute

$$P(\texttt{mach\_learn} \text{ is } \texttt{pred\_fashion} \mid \texttt{truth} \text{ is } \texttt{fashion}) \\ + P(\texttt{mach\_learn} \text{ is } \texttt{pred\_not} \mid \texttt{truth} \text{ is } \texttt{fashion})$$

(c) Provide an intuitive argument to explain why the sum in (b) is 1.[23]

The `smallpox` data set provides a sample of 6,224 individuals from the year 1721 who were exposed to smallpox in Boston.[24] Doctors at the time believed that inoculation, or vaccination that involved exposing a person to the disease in a controlled form, could reduce the likelihood of death.

Each case represents one person with two variables: `inoculated` and `result`. The variable `inoculated` takes two levels: `yes` or `no`, indicating whether the person was inoculated or not. The variable `result` has outcomes `lived` or `died`. These data are summarized in Tables 2.17 and 2.18.

		inoculated		
		yes	no	Total
result	lived	238	5136	5374
	died	6	844	850
	Total	244	5980	6224

Figure 2.17: Contingency table for the `smallpox` data set.

		inoculated		
		yes	no	Total
result	lived	0.0382	0.8252	0.8634
	died	0.0010	0.1356	0.1366
	Total	0.0392	0.9608	1.0000

Figure 2.18: Table proportions for the `smallpox` data, computed by dividing each count by the table total, 6224.

G

GUIDED PRACTICE 2.30

Write out, in formal notation, the probability a randomly selected person who was not inoculated died from smallpox, and find this probability.[25]

---

[22](a) If the photo is about fashion and the ML algorithm prediction was correct, then the ML algorithm my have a value of `pred_fashion`: $P(\texttt{mach\_learn} \text{ is } \texttt{pred\_fashion} \mid \texttt{truth} \text{ is } \texttt{fashion})$ (b) The equation for conditional probability indicates we should first find $P(\texttt{mach\_learn} \text{ is } \texttt{pred\_fashion} \text{ and } \texttt{truth} \text{ is } \texttt{fashion}) = 0.1081$ and $P(\texttt{truth} \text{ is } \texttt{not}) = 0.1696$. Then the ratio represents the conditional probability: $0.1081/0.1696 = 0.6374$.

[23](a) This probability is $\frac{P(\texttt{mach\_learn is pred\_not, truth is fashion})}{P(\texttt{truth is fashion})} = \frac{0.0615}{0.1696} = 0.3626$. (b) The total equals 1. (c) Under the condition the photo is about fashion, the ML algorithm must have either predicted it was about fashion or predicted it was not about fashion. The complement still works for conditional probabilities, provided the probabilities are conditioned on the same information.

[24]Fenner F. 1988. *Smallpox and Its Eradication (History of International Public Health, No. 6)*. Geneva: World Health Organization. ISBN 92-4-156110-6.

[25]$P(\texttt{result} = \texttt{died} \mid \texttt{not inoculated}) = \frac{P(\texttt{result = died and not inoculated})}{P(\texttt{not inoculated})} = \frac{0.1356}{0.9608} = 0.1411$.

G

GUIDED PRACTICE 2.31

Determine the probability that an inoculated person died from smallpox. How does this result compare with the result of Guided Practice 2.30?[26]

G

GUIDED PRACTICE 2.32

The people of Boston self-selected whether or not to be inoculated. (a) Is this study observational or was this an experiment? (b) Can we draw a causal conclusion using these data? (c) What is a potential confounding variable that might influence whether someone lived or died and also affect whether that person was inoculated?[27]

2.3.4 General multiplication rule for joint probabilities

When finding joint probabilities, we can use the General Multiplication Rule.

GENERAL MULTIPLICATION RULE

If A and B represent two outcomes or events, then to find the joint probability A and B we use:

$$P(A \cap B) = P(A) \times P(B|A)$$

This General Multiplication Rule can also be seen as a rearrangement of the conditional probability rule: $P(B|A) = \frac{P(A \cap B)}{P(A)}$.

E

EXAMPLE 2.33

Consider the `smallpox` data set. Suppose we are given only two pieces of information: 96.08% of residents were not inoculated, and 85.88% of the residents who were not inoculated ended up surviving. How could we compute the probability that a resident was not inoculated and lived?

We will compute our answer using the General Multiplication Rule and then verify it using Figure 2.18. We want to determine

$$P(\texttt{not inoculated} \text{ and } \texttt{lived})$$

and we are given:

$$P(\texttt{not inoculated}) = 0.9608$$
$$P(\texttt{lived} \mid \texttt{not inoculated}) = 0.8588$$

Among the 96.08% of people who were not inoculated, 85.88% survived, so:

$$P(\texttt{not inoculated} \text{ and } \texttt{lived}) = 0.9608 \times 0.8588 = 0.825.$$

This is equivalent to the General Multiplication Rule. We can confirm this probability in Figure 2.18 at the intersection of `no` and `lived` (with a small rounding error).

[26] $P(\texttt{died} \mid \texttt{inoculated}) = \frac{P(\texttt{died and inoculated})}{P(\texttt{inoculated})} = \frac{0.0010}{0.0392} = 0.0255$. The death rate for individuals who were inoculated is only about 1 in 40 while the death rate is about 1 in 7 for those who were not inoculated.

[27] Brief answers: (a) Observational. (b) No, we cannot draw a causal conclusion from this observational study. (However, further research has shown that inoculation is effective at reducing death rates.) (c) Accessibility to the latest and best medical care, so income could be a confounding variable. There are other valid answers for part (c).

G **GUIDED PRACTICE 2.34**

Use $P(\texttt{inoculated}) = 0.0392$ and $P(\texttt{lived} \mid \texttt{inoculated}) = 0.9754$ to determine the probability that a person was both inoculated and lived.[28]

G **GUIDED PRACTICE 2.35**

If 97.54% of the inoculated people lived, what proportion of inoculated people must have died?[29]

2.3.5 Tree diagrams and inverted conditional probabilities

Tree diagrams are a tool to organize outcomes and probabilities around the structure of the data. They are most useful when two or more processes occur in a sequence and each process is conditioned on its predecessors.

The `smallpox` data fit this description. We see the population as split by `inoculation`: `inoculated` and `not inoculated`. Following this split, survival rates were observed for each group. This structure is reflected in the tree diagram shown in Figure 2.19. The first branch for `inoculation` is said to be the **primary** branch while the other branches are **secondary**.

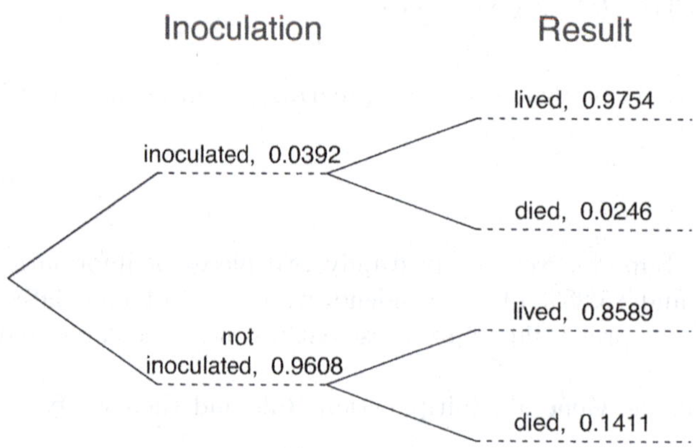

Figure 2.19: A tree diagram of the `smallpox` data set.

Tree diagrams are annotated with probabilities for the primary and secondary branches, as shown in Figure 2.19. This tree diagram splits the smallpox data by `inoculation` into `inoculated` and `not inoculated` with respective probabilities 0.0392 and 0.9608. The secondary branches are conditioned on the first, so we assign conditional probabilities to these branches. For example, the top secondary branch in Figure 2.19 is the probability of `lived` conditioned on `inoculated`.

To calculate a joint probability, we use the General Multiplication Rule. For example:

$$\begin{aligned} P(\texttt{inoculated and lived}) &= P(\texttt{inoculated}) \times P(\texttt{lived} \mid \texttt{inoculated}) \\ &= 0.0392 \times 0.9754 \\ &= 0.0382 \end{aligned}$$

[28]The answer is 0.0382, which can be verified using Figure 2.18.

[29]There were only two possible outcomes: `lived` or `died`. This means that 100% - 97.54% = 2.46% of the people who were inoculated died.

E

EXAMPLE 2.36

What is the probability that a randomly selected person lived?

There are two ways that a person could have lived: be inoculated *and* lived OR not be inoculated *and* lived. To find this probability, we sum the two disjoint probabilities:

$$\begin{aligned} P(\texttt{lived}) &= P(\texttt{inoculated}\ \text{and}\ \texttt{lived}) + P(\texttt{not inoculated}\ \text{and}\ \texttt{lived}) \\ &= 0.0392 \times 0.9754 + 0.9608 \times 0.8589 \\ &= 0.0382 + 0.8252 \\ &= 0.8634 \end{aligned}$$

E

EXAMPLE 2.37

What is the probability that a randomly selected person who was inoculated lived?

This is equivalent to asking the proportion that lived among those who were inoculated. This probability can be written as $P(\texttt{lived} \mid \texttt{inoculated})$ and can be found in the second branch as 0.9754.

Now, instead of asking the probability that a randomly selected person who was inoculated lived, we might want to know the probability that a randomly selected person who lived was inoculated. That is, instead of $P(\texttt{lived} \mid \texttt{inoculated})$, we want to find $P(\texttt{inoculated} \mid \texttt{lived})$. This is more challenging because it cannot be read directly from the tree diagram. However, we can apply the conditional probability rule to find this "inverted" conditional probability.

$$\begin{aligned} P(\texttt{inoculated} \mid \texttt{lived}) &= \frac{P(\texttt{inoculated}\ \text{and}\ \texttt{lived})}{P(\texttt{lived})} \\ &= \frac{0.0392 \times 0.9754}{0.0392 \times 0.9754 + 0.9608 \times 0.8589} \\ &= \frac{0.0382}{0.8634} \\ &= 0.0442 \end{aligned}$$

You might be surprised to see that $P(\texttt{inoculated} \mid \texttt{lived})$ is only 0.0442 while the probability $P(\texttt{lived} \mid \texttt{inoculated})$ is 0.9754. While most people who were inoculated lived, it is not true that most people who lived were inoculated. These two conditional probabilities are very different because the subset `lived` is much larger than the subset `inoculated`.

We can also observe that the events "inoculated" and "lived" are dependent. The conditional probability that someone was inoculated given that they lived (0.0442) is greater than the unconditional probability that someone was inoculated (0.0392).

G

GUIDED PRACTICE 2.38

What is the probability that a random selected person who died was inoculated?[30]

---

[30] This is equivalent to $P(\texttt{inoculated} \mid \texttt{died})$. This conditional probability can be found as $\frac{P(\texttt{inoculated and died})}{P(\texttt{died})} = \frac{0.0392 \times 0.0246}{0.0392 \times 0.0246\ +\ 0.9608 \times 0.1411} = \frac{0.00096}{0.13653} = 0.0070$. Among those who died, only 0.70% were inoculated.

EXAMPLE 2.39

Jose visits campus every Thursday evening. However, some days the parking garage is full, often due to college events. There are academic events on 35% of evenings, sporting events on 20% of evenings, and no events on 45% of evenings. When there is an academic event, the garage fills up about 25% of the time, and it fills up 70% of evenings with sporting events. On evenings when there are no events, it only fills up about 5% of the time. If Jose comes to campus and finds the garage full, what is the probability that there is a sporting event? Use a tree diagram to solve this problem.

The tree diagram, with three primary branches, is shown below.

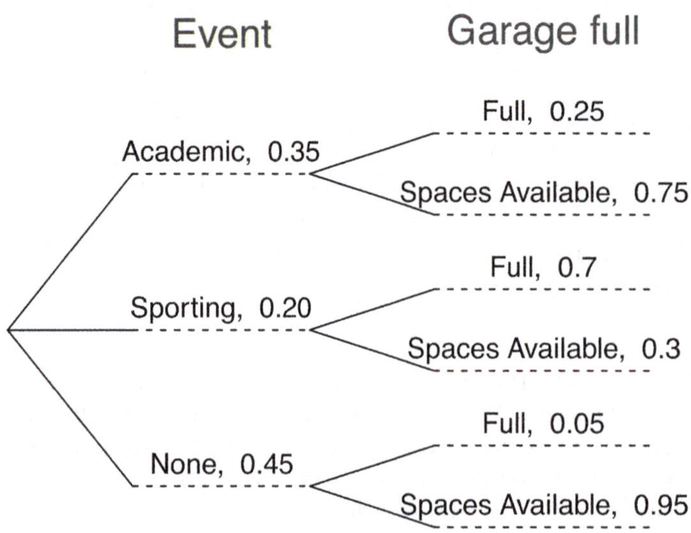

We want to find the probability that there is a sporting event given that the garage is full.

$$\begin{aligned} P(\text{sporting event} \mid \text{garage full}) &= \frac{P(\text{sporting event and garage full})}{P(\text{garage full})} \\ &= \frac{0.20 \times 0.7}{0.35 \times 0.25 + 0.20 \times 0.7 + 0.45 \times 0.05} \\ &= \frac{0.14}{0.0875 + 0.14 + 0.0225} \\ &= 0.56. \end{aligned}$$

If the garage is full, there is a 56% probability that there is a sporting event.

This example offers a way to update our belief about whether there is a sporting event, academic event, or no event going on at the school based on the information that the parking lot is full. This strategy of *updating beliefs* is the foundation of an entire theory of statistics called **Bayesian statistics**. While Bayesian statistics has many applications, we will not have time to cover it in this book.

2.3.6 Sampling without replacement

E

EXAMPLE 2.40

Professors sometimes select a student at random to answer a question. If each student has an equal chance of being selected and there are 15 people in your class, what is the chance that you will get selected for the next question?

If there are 15 people to ask and none are skipping class, then the probability is 1/15, or about 0.067.

E

EXAMPLE 2.41

If the professor asks 3 questions, what is the probability that you will not be selected? Assume that she will not pick the same person twice in a given lecture.

For the first question, she will pick someone else with probability 14/15. When she asks the second question, she only has 14 people who have not yet been asked. Thus, if you were not picked on the first question, the probability you are again not picked is 13/14. Similarly, the probability you are again not picked on the third question is 12/13, and the probability of not being picked for any of the three questions is

$$
\begin{aligned}
&P(\text{not picked in 3 questions}) \\
&\quad = P(\texttt{Q1} = \texttt{not\_picked} \text{ and } \texttt{Q2} = \texttt{not\_picked} \text{ and } \texttt{Q3} = \texttt{not\_picked.}) \\
&\quad = \frac{14}{15} \times \frac{13}{14} \times \frac{12}{13} = 0.80
\end{aligned}
$$

G

GUIDED PRACTICE 2.42

What rule permitted us to multiply the probabilities in Example 2.41?[31]

E

EXAMPLE 2.43

Suppose the professor randomly picks without regard to who she already selected, i.e. students can be picked more than once. What is the probability that you will not be picked for any of the three questions?

Each pick is independent, and the probability of not being picked for any individual question is 14/15. Thus, we can use the Multiplication Rule for independent processes.

$$
\begin{aligned}
&P(\text{not picked in 3 questions}) \\
&\quad = P(\texttt{Q1} = \texttt{not\_picked} \text{ and } \texttt{Q2} = \texttt{not\_picked} \text{ and } \texttt{Q3} = \texttt{not\_picked.}) \\
&\quad = \frac{14}{15} \times \frac{14}{15} \times \frac{14}{15} = 0.813
\end{aligned}
$$

You have a slightly higher chance of not being picked compared to when she picked a new person for each question. However, you now may be picked more than once.

---

[31] We used the General Multiplication Rule to find the product of three probabilities, where the second and third are conditional probabilities:

$$P(\texttt{Q1} = \texttt{not\_picked}) \times P(\texttt{Q2} = \texttt{not\_picked} \mid \texttt{Q1} = \texttt{not\_picked}) \times P(\texttt{Q3} = \texttt{not\_picked} \mid \texttt{Q1} = \texttt{not\_picked} \text{ and } \texttt{Q2} = \texttt{not\_picked})$$

If we sample from a small population **without replacement**, we no longer have independence between our observations. In Example 2.41, the probability of not being picked for the second question was conditioned on the event that you were not picked for the first question. In Example 2.43, the professor sampled her students **with replacement**: she repeatedly sampled the entire class without regard to who she already picked.

GUIDED PRACTICE 2.44

(G) Continuing Examples 2.41 and 2.43, if the professor asks three questions, what is the probability that you will be chosen at least once if she chooses students randomly without replacement (repeats not allowed)? What if she chooses randomly with replacement (repeats allowed)? There are 15 students in the class.[32]

GUIDED PRACTICE 2.45

(G) Now, let's say that you are in a large lecture hall with 150 students. If the professor asks three questions, what is the probability that you will be chosen at least once if she chooses students randomly without replacement? What if she chooses randomly with replacement?[33]

EXAMPLE 2.46

Compare your answers to Guided Practice 2.44 and to Guided Practice 2.45. What do you notice?

(E) The probability of being chosen at least once when the professor randomly samples 3 students out of 15 without replacement versus with replacement differs by 0.003 (0.20–0.187). The probability of being chosen at least once when the professor randomly samples 3 students out of 150 without replacement versus with replacement differs by 0.0001 (0.0200–0.0199). When the population size is larger relative to the sample size, the difference between the conditional and the unconditional probabilities used in the probability calculations is smaller. Therefore, when the population size is large relative to the sample size, the difference between sampling without replacement and sampling with replacement becomes negligible.

When taking a small sample from a much larger population and sampling without replacement, the observations are technically not independent but can be treated *as if* they were independent for calculation purposes. A rule of thumb says that if the sample size is less than 10% of the population size, or equivalently if the population size is at least 10 times greater than the sample size, then the observations can be treated as if they were independent.

SAMPLING WITHOUT REPLACEMENT

When the sample size is only a small fraction of the population (under 10%), observations can be considered independent even when sampling without replacement.

---

[32](a) If you don't get chosen at least once, that means you never get chosen. Therefore, using complements, the probability of being chosen at least once is equal to 1 – the probability of never being chosen.
Without replacement: P(chosen at least once) $= 1 - (14/15)(13/14)(12/13) = 1 - 0.80 = 0.20$.
(b) With replacement: P(chosen at least once) $= 1 - (14/15)^3 = 1 - 0.813 = 0.187$.

[33](a) Without replacement: P(chosen at least once) $= 1 - (149/150)(148/149)(147/148) = 1 - 0.9800 = 0.0200$.
(b) With replacement: P(chosen at least once) $= 1 - (149/150)^3 = 1 - 0.9801 = 0.0199$.

2.3.7 Independence considerations in conditional probability

If two processes are independent, then knowing the outcome of one should provide no information about the other. We can show this is mathematically true using conditional probabilities.

GUIDED PRACTICE 2.47

Let X and Y represent the outcomes of rolling two dice. (a) What is the probability that the first die, X, is 1? (b) What is the probability that both X and Y are 1? (c) Use the formula for conditional probability to compute $P(Y = 1 \mid X = 1)$, then compare this to $P(Y = 1)$.[34]

We can also show that in Guided Practice 2.47(c) the conditioning information has no influence by using the Multiplication Rule for independence processes:

$$\begin{aligned} P(Y = 1 \mid X = 1) &= \frac{P(Y = 1 \text{ and } X = 1)}{P(X = 1)} \\ &= \frac{P(Y = 1) \times P(X = 1)}{P(X = 1)} \\ &= P(Y = 1) \end{aligned}$$

GUIDED PRACTICE 2.48

Ron is watching a roulette table in a casino and notices that the last five outcomes were `black`. He figures that the chances of getting `black` six times in a row is very small (about 1/64) and puts his paycheck on red. What is wrong with his reasoning?[35]

2.3.8 Probabilities when events are not disjoint

Let's consider calculations for two events that are not disjoint in the context of a regular deck of 52 cards, represented in Figure 2.20. If you are unfamiliar with the cards in a regular deck, please see the footnote.[36]

2♣	3♣	4♣	5♣	6♣	7♣	8♣	9♣	10♣	J♣	Q♣	K♣	A♣
2♢	3♢	4♢	5♢	6♢	7♢	8♢	9♢	10♢	J♢	Q♢	K♢	A♢
2♡	3♡	4♡	5♡	6♡	7♡	8♡	9♡	10♡	J♡	Q♡	K♡	A♡
2♠	3♠	4♠	5♠	6♠	7♠	8♠	9♠	10♠	J♠	Q♠	K♠	A♠

Figure 2.20: Representations of the 52 unique cards in a deck.

GUIDED PRACTICE 2.49

(a) What is the probability that a randomly selected card is a diamond? (b) What is the probability that a randomly selected card is a face card?[37]

---

[34] Brief solutions: (a) 1/6. (b) 1/36. (c) $P(Y = 1 \mid X = 1) = \frac{P(Y = 1 \text{ and } X = 1)}{P(X = 1)} = \frac{1/36}{1/6} = 1/6$. This probability is the same as $P(Y = 1) = 1/6$. The probability that $Y = 1$ was unchanged by knowledge about X, which makes sense as X and Y are independent.

[35] He has forgotten that the next roulette spin is independent of the previous spins. Casinos do employ this practice; they post the last several outcomes of many betting games to trick unsuspecting gamblers into believing the odds are in their favor. This is called the **gambler's fallacy**.

[36] The 52 cards are split into four **suits**: ♣ (club), ♢ (diamond), ♡ (heart), ♠ (spade). Each suit has its 13 cards labeled: 2, 3, ..., 10, J (jack), Q (queen), K (king), and A (ace). Thus, each card is a unique combination of a suit and a label, e.g. 4♡ and J♣. The 12 cards represented by the jacks, queens, and kings are called `face cards`. The cards that are ♢ or ♡ are typically colored red while the other two suits are typically colored black.

[37] (a) There are 52 cards and 13 diamonds. If the cards are thoroughly shuffled, each card has an equal chance of being drawn, so the probability that a randomly selected card is a diamond is $P(\diamondsuit) = \frac{13}{52} = 0.250$. (b) Likewise, there are 12 face cards, so $P(\text{face card}) = \frac{12}{52} = \frac{3}{13} = 0.231$.

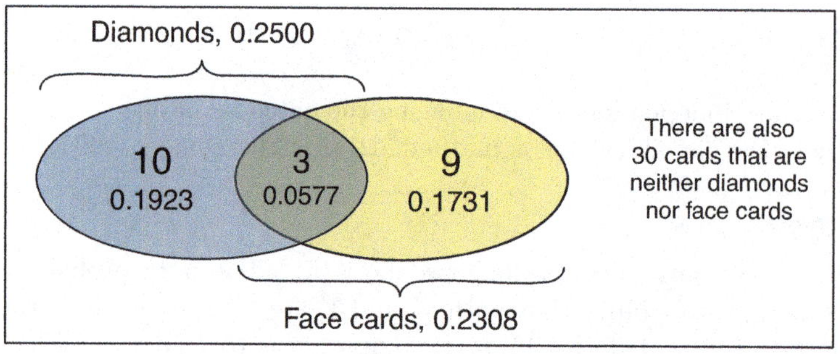

Figure 2.21: A Venn diagram for diamonds and face cards.

Venn diagrams are useful when outcomes can be categorized as "in" or "out" for two or three variables, attributes, or random processes. The Venn diagram in Figure 2.21 uses a oval to represent diamonds and another to represent face cards. If a card is both a diamond and a face card, it falls into the intersection of the ovals. If it is a diamond but not a face card, it will be in part of the left oval that is not in the right oval (and so on). The total number of cards that are diamonds is given by the total number of cards in the diamonds oval: $10 + 3 = 13$. The probabilities are also shown (e.g. $10/52 = 0.1923$).

G

GUIDED PRACTICE 2.50

Using the Venn diagram, verify $P(\text{face card}) = 12/52 = 3/13$.[38]

Let A represent the event that a randomly selected card is a diamond and B represent the event that it is a face card. How do we compute $P(A \text{ or } B)$? Events A and B are not disjoint – the cards $J\diamondsuit$, $Q\diamondsuit$, and $K\diamondsuit$ fall into both categories – so we cannot use the Addition Rule for disjoint events. Instead we use the Venn diagram. We start by adding the probabilities of the two events:

$$P(A) + P(B) = P(\diamondsuit) + P(\text{face card}) = 13/52 + 12/52$$

However, the three cards that are in both events were counted twice, once in each probability. We must correct this double counting:

$$\begin{aligned} P(A \text{ or } B) &= P(\diamondsuit) + P(\text{face card}) \\ &= P(\diamondsuit) + P(\text{face card}) - P(\diamondsuit \text{ and face card}) \\ &= 13/52 + 12/52 - 3/52 \\ &= 22/52 = 11/26 \end{aligned}$$

Equation (2.51) is an example of the **General Addition Rule**.

GENERAL ADDITION RULE

If A and B are any two events, disjoint or not, then the probability that A or B will occur is

$$P(A \text{ or } B) = P(A) + P(B) - P(A \text{ and } B)$$

where $P(A \text{ and } B)$ is the probability that both events occur.

[38] The Venn diagram shows face cards split up into "face card but not $\diamondsuit$" and "face card and $\diamondsuit$". Since these correspond to disjoint events, $P(\text{face card})$ is found by adding the two corresponding probabilities: $\frac{3}{52} + \frac{9}{52} = \frac{12}{52} = \frac{3}{13}$.

SYMBOLIC NOTATION FOR "AND" AND "OR"

The symbol $\cap$ means intersection and is equivalent to "and".
The symbol $\cup$ means union and is equivalent to "or".
It is common to see the General Addition Rule written as

$$P(A \cup B) = P(A) + P(B) - P(A \cap B)$$

"OR" IS INCLUSIVE

When we write, "or" in statistics, we mean "and/or" unless we explicitly state otherwise. Thus, A or B occurs means A, B, or both A and B occur. This is equivalent to at least one of A or B occurring.

GUIDED PRACTICE 2.51

(G) (a) If A and B are disjoint, describe why this implies $P(A \text{ and } B) = 0$. (b) Using part (a), verify that the General Addition Rule simplifies to the simpler Addition Rule for disjoint events if A and B are disjoint.[39]

GUIDED PRACTICE 2.52

(G) In the `email` data set with 3,921 emails, 367 were spam, 2,827 contained some small numbers but no big numbers, and 168 had both characteristics. Create a Venn diagram for this setup.[40]

GUIDED PRACTICE 2.53

(G) (a) Use your Venn diagram from Guided Practice 2.52 to determine the probability a randomly drawn email from the `email` data set is spam and had small numbers (but not big numbers). (b) What is the probability that the email had either of these attributes?[41]

2.3.9 Checking for independent and mutually exclusive events

If A and B are independent events, then the probability of B being true is unchanged if A is true. Mathematically, this is written as

$$P(B|A) = P(B)$$

The General Multiplication Rule states that $P(A \cap B)$ equals $P(A) \times P(B|A)$. If A and B are independent events, we can replace $P(B|A)$ with $P(B)$ and the following multiplication rule applies:

$$P(A \cap B) = P(A) \times P(B)$$

[39] (a) If A and B are disjoint, A and B can never occur simultaneously. (b) If A and B are disjoint, then the last term of Equation (2.51) is 0 (see part (a)) and we are left with the Addition Rule for disjoint events.

[40] Both the counts and corresponding probabilities (e.g. $2659/3921 = 0.678$) are shown. Notice that the number of emails represented in the left oval corresponds to $2659 + 168 = 2827$, and the number represented in the right oval is $168 + 199 = 367$.

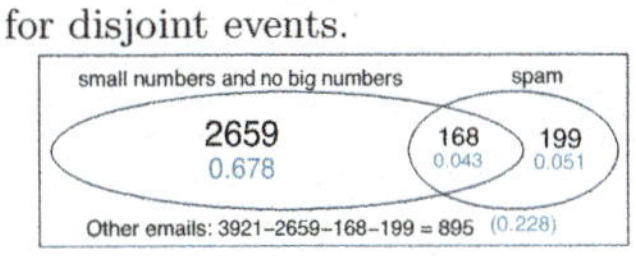

[41] (a) The solution is represented by the intersection of the two ovals: 0.043. (b) This is the sum of the three disjoint probabilities shown in the ovals: $0.678 + 0.043 + 0.051 = 0.772$.

CHECKING WHETHER TWO EVENTS ARE INDEPENDENT

To determine if two events A and B are independent, check whether one of the following equations holds (there is no need to check both equations):

$$P(B|A) = P(B) \qquad\qquad P(A \cap B) = P(A) \times P(B)$$

If the equation that is checked holds true (the left and right sides are equal), then A and B are independent. If the equation does not hold, then A and B are dependent.

EXAMPLE 2.54

Are teenager college attendance and parent college degrees independent or dependent? Use information from Figure 2.22, which shows data from a simulated sample.

We'll check for independence by seeing if the relationship $P(B|A) = P(B)$ holds. If the `teen` and `parents` variables are independent, it must be true that

$$P(\texttt{teen college} \mid \texttt{parent degree}) = P(\texttt{teen college})$$

E Using Figure 2.22, we check whether equality holds in this equation.

$$P(\texttt{teen college} \mid \texttt{parent degree}) \stackrel{?}{=} P(\texttt{teen college})$$

$$\frac{231}{280} \stackrel{?}{=} \frac{445}{792}$$

$$0.83 \neq 0.56$$

Because the sides are not equal, teenager college attendance and parent degree are dependent. We estimate the probability a teenager attended college to be higher if we know that one of the teen's parents has a college degree.

		parents		
		degree	not	Total
teen	college	231	214	445
	not	49	298	347
	Total	280	512	792

Figure 2.22: Contingency table summarizing the `family_college` data set.

GUIDED PRACTICE 2.55

G Also show that teenager college attendance and parent college degree are dependent by showing that $P(A \cap B) \neq P(A) \times P(B)$.[42]

---

[42] We check for equality in the following equation:

$$P(\texttt{teen college} \cap \texttt{parent degree}) \stackrel{?}{=} P(\texttt{teen college}) \times P(\texttt{parent degree})$$

$$\frac{231}{792} \stackrel{?}{=} \frac{445}{792} \times \frac{280}{792}$$

$$0.292 \neq 0.199$$

These terms are not equal, which confirms what we learned in Example 2.54: teenager college attendance and parent college degrees are dependent.

If A and B are mutually exclusive events, then A and B cannot occur at the same time. Mathematically, this is written as

$$P(A \cap B) = 0$$

The General Addition Rule states that $P(A \cup B)$ equals $P(A) + P(B) - P(A \cap B)$. If A and B are mutually exclusive events, we can replace $P(A \text{ and } B)$ with 0 and the following special addition rule applies:

$$P(A \cup B) = P(A) + P(B)$$

CHECKING WHETHER TWO EVENTS ARE MUTUALLY EXCLUSIVE (DISJOINT)

To determine if events A and B are mutually exclusive, check whether one of the following equations holds (there is no need to check both equations):

$$P(A \cap B) = 0 \qquad\qquad P(A \cup B) = P(A) + P(B)$$

If the equation that is checked holds true (the left and right sides are equal), A and B are mutually exclusive. If the equation does not hold, then A and B are not mutually exclusive.

EXAMPLE 2.56

Are teen college attendance and parent college degree mutually exclusive?

Looking in the table, we see that there are 231 instances where both the teenager attended college and parent has a degree, indicating the probability of both events occurring is greater than 0. Since we have found an example where both of these events happen together, these two events are not mutually exclusive. We can write this more formally as:

$$P(\texttt{teen college} \cap \texttt{parent degree}) = \frac{231}{792} \neq 0$$

Since this probability is not zero, teenager college attendance and parent college degree are not mutually exclusive.

MUTUALLY EXCLUSIVE AND INDEPENDENT ARE DIFFERENT

If two events are mutually exclusive, then if one is true, the other cannot be true. This implies the two events are in some way connected, meaning they must be dependent.

If two events are independent, then if one occurs, it is still possible for the other to occur, meaning the events are not mutually exclusive.

DEPENDENT EVENTS NEED NOT BE MUTUALLY EXCLUSIVE.

If two events are dependent, this does not imply that they are mutually exclusive. For example, teenager college attendance and parent college degree are dependent, but these events are not mutually exclusive.

Section summary

- A **conditional probability** can be written as $P(A|B)$ and is read, "Probability of A given B".
- In a conditional probability of the form $P(A|B)$, we are given information about B. In an **unconditional probability** of the form $P(A)$, we are not given any information.
- To find the probability of A given B, use the **conditional probability rule** $P(A|B) = \frac{P(A \cap B)}{P(B)}$.
- To find the probability of A *and* B, that is A intersect B, use the **General Multiplication Rule**: $P(A \cap B) = P(A) \times P(B|A)$. Equivalently, $P(A \cap B)$ can be found using the conditional probability rule by multiplying $P(A|B)$ times $P(B)$.
- Sometimes, the conditional probability $P(B|A)$ may be known, but we want to find the "inverted" probability $P(A|B)$. In this case, draw a tree diagram and apply the conditional probability rule $P(A|B) = \frac{P(A \cap B)}{P(B)}$, where $P(B)$ is computed using multiple pieces of information from the tree diagram.
- Two events are **independent** when the outcome of one has no effect on the outcome of the other. When A and B are independent, $P(B|A) = P(B)$ and $P(A|B) = P(A)$.
- In the *special case* where A and B are **independent**, $P(A \cap B) = P(A) \times P(B)$.
- When sampling **with** replacement, events are independent. When sampling without replacement from a very large population, events may be considered as if they were independent, and the use of conditional or unconditional probabilities will make little difference in the answer when calculating joint probabilities.
- When sampling **without replacement** from a small population, events are dependent and one cannot simply multiply unconditional probabilities to find a joint probability.
- To find the probability of A *or* B, that is A union B, use the **General Addition Rule**: $P(A \text{ or B}) = P(A) + P(B) - P(A \text{ and } B)$, also written as: $P(A \cup B) = P(A) + P(B) - P(A \cap B)$.
- When A and B are **mutually exclusive** (disjoint), $P(A \cap B) = 0$.
- In the special case where A and B are **mutually exclusive** (disjoint), $P(A \cup B) = P(A) + P(B)$.
- If A and B are **mutually exclusive**, they must be **dependent**, because the occurrence of one of them changes the probability that the other occurs to 0.

Exercises

2.13 Joint and conditional probabilities. P(A) = 0.3, P(B) = 0.7

(a) Can you compute P(A and B) if you only know P(A) and P(B)?

(b) Assuming that events A and B arise from independent random processes,

  i. what is P(A and B)?
  ii. what is P(A or B)?
  iii. what is P(A|B)?

(c) If we are given that P(A and B) = 0.1, are the random variables giving rise to events A and B independent?

(d) If we are given that P(A and B) = 0.1, what is P(A|B)?

2.14 PB & J. Suppose 80% of people like peanut butter, 89% like jelly, and 78% like both. Given that a randomly sampled person likes peanut butter, what's the probability that he also likes jelly?

2.15 Global warming. A Pew Research poll asked 1,306 Americans "From what you've read and heard, is there solid evidence that the average temperature on earth has been getting warmer over the past few decades, or not?". The table below shows the distribution of responses by party and ideology, where the counts have been replaced with relative frequencies.[43]

		Response			
		Earth is warming	Not warming	Don't Know Refuse	Total
Party and Ideology	Conservative Republican	0.11	0.20	0.02	0.33
	Mod/Lib Republican	0.06	0.06	0.01	0.13
	Mod/Cons Democrat	0.25	0.07	0.02	0.34
	Liberal Democrat	0.18	0.01	0.01	0.20
	Total	0.60	0.34	0.06	1.00

(a) Are believing that the earth is warming and being a liberal Democrat mutually exclusive?

(b) What is the probability that a randomly chosen respondent believes the earth is warming or is a liberal Democrat?

(c) What is the probability that a randomly chosen respondent believes the earth is warming given that he is a liberal Democrat?

(d) What is the probability that a randomly chosen respondent believes the earth is warming given that he is a conservative Republican?

(e) Does it appear that whether or not a respondent believes the earth is warming is independent of their party and ideology? Explain your reasoning.

(f) What is the probability that a randomly chosen respondent is a moderate/liberal Republican given that he does not believe that the earth is warming?

[43] Pew Research Center, Majority of Republicans No Longer See Evidence of Global Warming, data collected on October 27, 2010.

2.16 Health coverage, relative frequencies. The Behavioral Risk Factor Surveillance System (BRFSS) is an annual telephone survey designed to identify risk factors in the adult population and report emerging health trends. The following table displays the distribution of health status of respondents to this survey (excellent, very good, good, fair, poor) and whether or not they have health insurance.

		Health Status					
		Excellent	Very good	Good	Fair	Poor	Total
Health	No	0.0230	0.0364	0.0427	0.0192	0.0050	0.1262
Coverage	Yes	0.2099	0.3123	0.2410	0.0817	0.0289	0.8738
	Total	0.2329	0.3486	0.2838	0.1009	0.0338	1.0000

(a) Are being in excellent health and having health coverage mutually exclusive?

(b) What is the probability that a randomly chosen individual has excellent health?

(c) What is the probability that a randomly chosen individual has excellent health given that he has health coverage?

(d) What is the probability that a randomly chosen individual has excellent health given that he doesn't have health coverage?

(e) Do having excellent health and having health coverage appear to be independent?

2.17 Swing voters. A Pew Research survey asked 2,373 randomly sampled registered voters their political affiliation (Republican, Democrat, or Independent) and whether or not they identify as swing voters. 35% of respondents identified as Independent, 23% identified as swing voters, and 11% identified as both.[44]

(a) Are being Independent and being a swing voter disjoint, i.e. mutually exclusive?

(b) Draw a Venn diagram summarizing the variables and their associated probabilities.

(c) What percent of voters are Independent but not swing voters?

(d) What percent of voters are Independent or swing voters?

(e) What percent of voters are neither Independent nor swing voters?

(f) Is the event that someone is a swing voter independent of the event that someone is a political Independent?

2.18 Poverty and language. The American Community Survey is an ongoing survey that provides data every year to give communities the current information they need to plan investments and services. The 2010 American Community Survey estimates that 14.6% of Americans live below the poverty line, 20.7% speak a language other than English (foreign language) at home, and 4.2% fall into both categories.[45]

(a) Are living below the poverty line and speaking a foreign language at home disjoint?

(b) Draw a Venn diagram summarizing the variables and their associated probabilities.

(c) What percent of Americans live below the poverty line and only speak English at home?

(d) What percent of Americans live below the poverty line or speak a foreign language at home?

(e) What percent of Americans live above the poverty line and only speak English at home?

(f) Is the event that someone lives below the poverty line independent of the event that the person speaks a foreign language at home?

2.19 Marbles in an urn. Imagine you have an urn containing 5 red, 3 blue, and 2 orange marbles in it.

(a) What is the probability that the first marble you draw is blue?

(b) Suppose you drew a blue marble in the first draw. If drawing with replacement, what is the probability of drawing a blue marble in the second draw?

(c) Suppose you instead drew an orange marble in the first draw. If drawing with replacement, what is the probability of drawing a blue marble in the second draw?

(d) If drawing with replacement, what is the probability of drawing two blue marbles in a row?

(e) When drawing with replacement, are the draws independent? Explain.

[44] Pew Research Center, With Voters Focused on Economy, Obama Lead Narrows, data collected between April 4-15, 2012.

[45] U.S. Census Bureau, 2010 American Community Survey 1-Year Estimates, Characteristics of People by Language Spoken at Home.

2.20 Socks in a drawer. In your sock drawer you have 4 blue, 5 gray, and 3 black socks. Half asleep one morning you grab 2 socks at random and put them on. Find the probability you end up wearing

(a) 2 blue socks
(b) no gray socks
(c) at least 1 black sock
(d) a green sock
(e) matching socks

2.21 Chips in a bag. Imagine you have a bag containing 5 red, 3 blue, and 2 orange chips.

(a) Suppose you draw a chip and it is blue. If drawing without replacement, what is the probability the next is also blue?
(b) Suppose you draw a chip and it is orange, and then you draw a second chip without replacement. What is the probability this second chip is blue?
(c) If drawing without replacement, what is the probability of drawing two blue chips in a row?
(d) When drawing without replacement, are the draws independent? Explain.

2.22 Books on a bookshelf. The table below shows the distribution of books on a bookcase based on whether they are nonfiction or fiction and hardcover or paperback.

		Format		
		Hardcover	Paperback	Total
Type	Fiction	13	59	72
	Nonfiction	15	8	23
	Total	28	67	95

(a) Find the probability of drawing a hardcover book first then a paperback fiction book second when drawing without replacement.
(b) Determine the probability of drawing a fiction book first and then a hardcover book second, when drawing without replacement.
(c) Calculate the probability of the scenario in part (b), except this time complete the calculations under the scenario where the first book is placed back on the bookcase before randomly drawing the second book.
(d) The final answers to parts (b) and (c) are very similar. Explain why this is the case.

2.23 It's never lupus. Lupus is a medical phenomenon where antibodies that are supposed to attack foreign cells to prevent infections instead see plasma proteins as foreign bodies, leading to a high risk of blood clotting. It is believed that 2% of the population suffer from this disease. The test is 98% accurate if a person actually has the disease. The test is 74% accurate if a person does not have the disease. There is a line from the Fox television show *House* that is often used after a patient tests positive for lupus: "It's never lupus." Do you think there is truth to this statement? Use appropriate probabilities to support your answer.

2.24 Predisposition for thrombosis. A genetic test is used to determine if people have a predisposition for *thrombosis*, which is the formation of a blood clot inside a blood vessel that obstructs the flow of blood through the circulatory system. It is believed that 3% of people actually have this predisposition. The genetic test is 99% accurate if a person actually has the predisposition, meaning that the probability of a positive test result when a person actually has the predisposition is 0.99. The test is 98% accurate if a person does not have the predisposition. What is the probability that a randomly selected person who tests positive for the predisposition by the test actually has the predisposition?

2.4 Discrete random variables

We can consider the number of textbooks that a student a particular university will purchase to be a random variable. What does the distribution look like? How can we calculate the expected number of textbooks purchased and the typical variation in that number? If we know the average cost per textbook, how can we calculate the expected *amount* of money a student at this university will spend on textbooks and the typical variation in that amount? In this section, we define and summarize random variables such as these, and we look at some of their properties.

Learning objectives

1. Construct a probability distribution for a discrete random variable.
2. Calculate mean and standard deviation for a discrete random variable
3. Interpret the mean and standard deviation for a discrete random variable.

2.4.1 Probability distributions

The sum of the roll of two dice can vary from roll to roll. We call a variable or process such as this a **random variable**, and we usually represent a random variable with a capital letter such as X, Y, or Z.

RANDOM VARIABLE

A random process or variable with a numerical outcome.

A **probability distribution** for a discrete random variable shows the probability associated with every possible value of the random variable. Figure 2.23 shows the probability distribution for the sum of two dice.

Dice sum	2	3	4	5	6	7	8	9	10	11	12
Probability	$\frac{1}{36}$	$\frac{2}{36}$	$\frac{3}{36}$	$\frac{4}{36}$	$\frac{5}{36}$	$\frac{6}{36}$	$\frac{5}{36}$	$\frac{4}{36}$	$\frac{3}{36}$	$\frac{2}{36}$	$\frac{1}{36}$

Figure 2.23: Probability distribution for the sum of two dice.

In order for a distribution to represent a valid probability distribution, it must satisfy certain rules.

RULES FOR PROBABILITY DISTRIBUTIONS

A probability distribution is a list of the possible outcomes with corresponding probabilities that satisfies three rules:

1. The outcomes listed must be disjoint.
2. Each probability must be between 0 and 1.
3. The probabilities must total 1.

EXAMPLE 2.57

Based on the probability distribution table shown in Figure 2.23, what is the probability of getting a dice sum of greater than 10?

To be greater than 10 means to be 11 or 12. So the probability of the dice sum being greater than 10 is $\frac{2}{36} + \frac{1}{36} = \frac{3}{36}$.

It is common to want to know the probability of getting less than or equal to a certain value. In this case, a **cumulative probability distribution** can be used. Instead of identifying the probability of each value, we record the probability of being less than or equal to each value as shown in the figure below.

Dice sum	2	3	4	5	6	7	8	9	10	11	12
Cumulative probability (≤ dice sum)	$\frac{1}{36}$	$\frac{3}{36}$	$\frac{6}{36}$	$\frac{10}{36}$	$\frac{15}{36}$	$\frac{21}{36}$	$\frac{26}{36}$	$\frac{30}{36}$	$\frac{33}{36}$	$\frac{35}{36}$	$\frac{36}{36}$

Figure 2.24: Cumulative probability distribution for the sum of two dice.

EXAMPLE 2.58

Based on the cumulative probability distribution table shown in Figure 2.24, what is the probability of getting a dice sum less than or equal to 4? Greater than 4?

Because this distribution is a cumulative distribution, we do not need to add up terms. The probability of being less than or equal to 4 is given in the cumulative distribution as $\frac{6}{36}$. To be greater than 4 is the complement of being less than or equal to 4, so the probability of a dice sum greater than 4 can be found as $1 - \frac{6}{36} = \frac{30}{36}$.

Chapter 1 emphasized the importance of plotting data to provide quick summaries. Probability distributions can also be summarized in a histogram or bar plot. The probability distribution for the sum of two dice is shown in Figure 2.23 and its histogram is plotted in Figure 2.25.

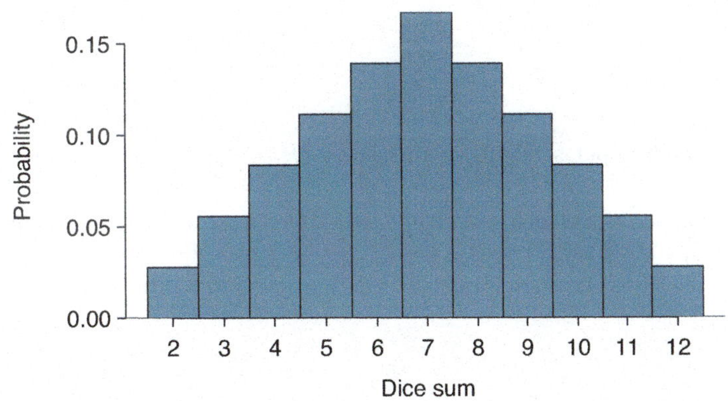

Figure 2.25: A histogram for the probability distribution of the sum of two dice.

In these bar plots, the bar heights represent the probabilities of outcomes. If the outcomes are numerical and discrete, it is usually (visually) convenient to make a histogram, as in the case of the sum of two dice. Another example of plotting the bars at their respective locations is shown in Figure 2.26.

2.4.2 Introduction to expectation

EXAMPLE 2.59

Two books are assigned for a statistics class: a textbook and its corresponding study guide. The university bookstore determined 20% of enrolled students do not buy either book, 55% buy the textbook only, and 25% buy both books, and these percentages are relatively constant from one term to another. If there are 100 students enrolled, how many books should the bookstore expect to sell to this class?

Around 20 students will not buy either book (0 books total), about 55 will buy one book (55 books total), and approximately 25 will buy two books (totaling 50 books for these 25 students). The bookstore should expect to sell about 105 books for this class.

GUIDED PRACTICE 2.60

Would you be surprised if the bookstore sold slightly more or less than 105 books?[46]

EXAMPLE 2.61

The textbook costs \$137 and the study guide \$33. How much revenue should the bookstore expect from this class of 100 students?

About 55 students will just buy a textbook, providing revenue of

$$\$137 \times 55 = \$7,535$$

The roughly 25 students who buy both the textbook and the study guide would pay a total of

$$(\$137 + \$33) \times 25 = \$170 \times 25 = \$4,250$$

Thus, the bookstore should expect to generate about $\$7,535 + \$4,250 = \$11,785$ from these 100 students for this one class. However, there might be some *sampling variability* so the actual amount may differ by a little bit.

EXAMPLE 2.62

What is the average revenue per student for this course?

The expected total revenue is \$11,785, and there are 100 students. Therefore the expected revenue per student is $\$11,785/100 = \$117.85$.

---

[46]If they sell a little more or a little less, this should not be a surprise. Hopefully it is now clear that there is natural variability in observed data. For example, if we would flip a coin 100 times, it will not usually come up heads exactly half the time, but it will probably be close.

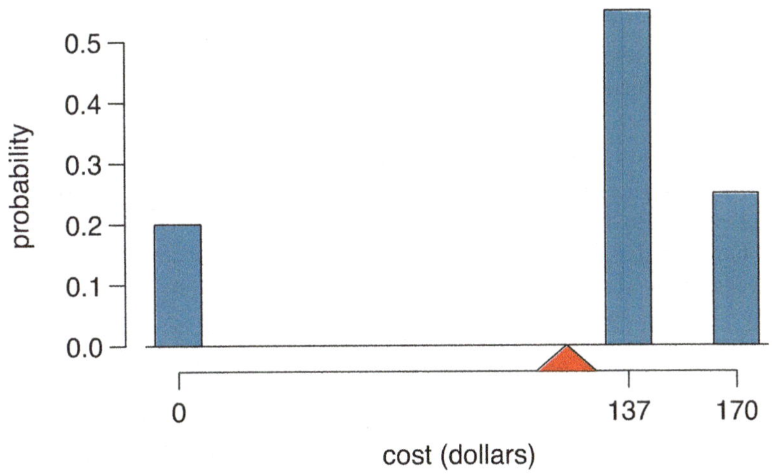

Figure 2.26: Probability distribution for the bookstore's revenue from one student. The triangle represents the average revenue per student.

2.4.3 Expected value

The amount of money a single student will spend on her statistics books is a random variable, and we represent it by X. The possible outcomes of X are labeled with a corresponding lower case letter x and subscripts. For example, we write $x_1 = \$0$, $x_2 = \$137$, and $x_3 = \$170$, which occur with probabilities 0.20, 0.55, and 0.25. The distribution of X is summarized in Figure 2.26 and Figure 2.27.

i	1	2	3	Total
x_i	\$0	\$137	\$170	–
$P(x_i)$	0.20	0.55	0.25	1.00

Figure 2.27: The probability distribution for the random variable X, representing the bookstore's revenue from a single student. We use $P(x_i)$ to represent the probability of x_i.

We computed the average outcome of X as \$117.85 in Example 2.62. We call this average the **expected value** of X, denoted by $E(X)$. The expected value of a random variable is computed by adding each outcome weighted by its probability:

$$\begin{aligned} E(X) &= 0 \cdot P(0) + 137 \cdot P(137) + 170 \cdot P(170) \\ &= 0 \cdot 0.20 + 137 \cdot 0.55 + 170 \cdot 0.25 = 117.85 \end{aligned}$$

EXPECTED VALUE OF A DISCRETE RANDOM VARIABLE

If X takes outcomes x_1, x_2, ..., x_n with probabilities $P(x_1)$, $P(x_2)$, ..., $P(x_n)$, the mean, or expected value, of X is the sum of each outcome multiplied by its corresponding probability:

$$\begin{aligned} \mu_X = E(X) &= \sum_{i=1}^{n} x_i \cdot P(x_i) \\ &= x_1 \cdot P(x_1) + x_2 \cdot P(x_2) + \cdots + x_n \cdot P(x_n) \end{aligned}$$

The expected value for a random variable represents the average outcome. For example, $E(X) = 117.85$ represents the average amount the bookstore expects to make across all students, which we could also write as $\mu = 117.85$. While the bookstore will make more than this on some students and less than this on other students, in the long run, the average will be \$117.85.

INTERPRETING THE MEAN OR EXPECTED VALUE OF A RANDOM VARIABLE

The mean or expected value of a random variable can be interpreted as the long-run average outcome of the random variable.

In physics, the expectation holds the same meaning as the center of gravity. The distribution can be represented by a series of weights at each outcome, and the mean represents the balancing point. This is represented in Figures 2.26 and 2.28. The idea of a center of gravity also expands to continuous probability distributions. Figure 2.29 shows a continuous probability distribution balanced atop a wedge placed at the mean.

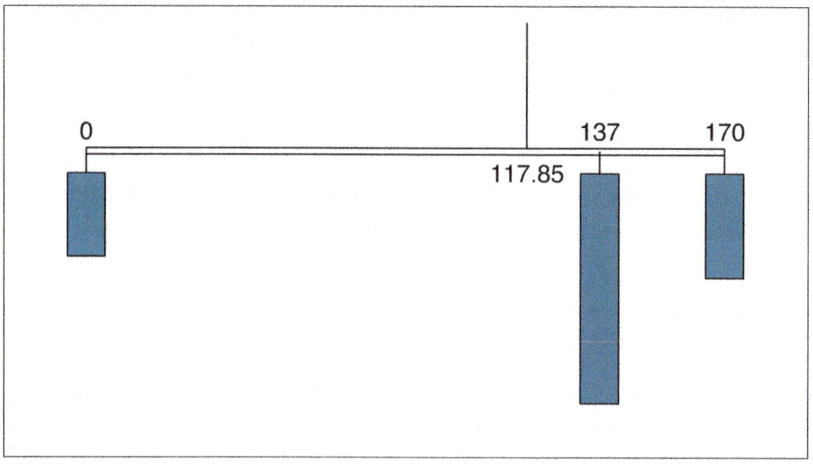

Figure 2.28: A weight system representing the probability distribution for X. The string holds the distribution at the mean to keep the system balanced.

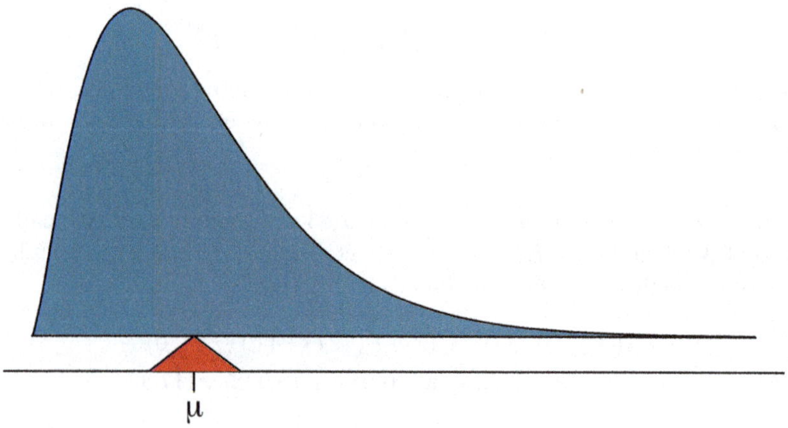

Figure 2.29: A continuous distribution can also be balanced at its mean.

2.4.4 Variability in random variables

Suppose you ran the university bookstore. Besides how much revenue you expect to generate, you might also want to know the volatility (variability) in your revenue.

The variance and standard deviation can be used to describe the variability of a random variable. Section 1.4.2 introduced a method for finding the variance and standard deviation for a data set. We first computed deviations from the mean ($x_i - \mu$), squared those deviations, and took an average to get the variance. In the case of a random variable, we again compute squared deviations. However, we take their sum weighted by their corresponding probabilities, just like we did for the expectation. This weighted sum of squared deviations equals the variance, and we calculate the standard deviation by taking the square root of the variance, just as we did in Section 1.4.2.

VARIANCE AND STANDARD DEVIATION OF A DISCRETE RANDOM VARIABLE

If X takes outcomes x_1, x_2, ..., x_n with probabilities $P(x_1)$, $P(x_2)$, ..., $P(x_n)$ and expected value $\mu_X = E(X)$, then to find the standard deviation of X, we first find the variance and then take its square root.

$$Var(X) = \sigma_X^2 = \sum_{i=1}^{n}(x_i - \mu_X)^2 \cdot P(x_i)$$
$$= (x_1 - \mu_X)^2 \cdot P(x_1) + (x_2 - \mu_X)^2 \cdot P(x_2) + \cdots + (x_n - \mu_X)^2 \cdot P(x_n)$$

$$SD(X) = \sigma_X = \sqrt{\sum_{i=1}^{n}(x_i - \mu_X)^2 \cdot P(x_i)}$$
$$= \sqrt{(x_1 - \mu_X)^2 \cdot P(x_1) + (x_2 - \mu_X)^2 \cdot P(x_2) + \cdots + (x_n - \mu_X)^2 \cdot P(x_n)}$$

EXAMPLE 2.63

Compute the expected value, variance, and standard deviation of X, the revenue of a single statistics student for the bookstore.

It is useful to construct a table that holds computations for each outcome separately, then add up the results.

i	1	2	3	Total
x_i	\$0	\$137	\$170	
$P(x_i)$	0.20	0.55	0.25	
$x_i \cdot P(x_i)$	0	75.35	42.50	117.85

E

Thus, the expected value is μ_X = \$117.85, which we computed earlier. The variance can be constructed using a similar table:

i	1	2	3	Total
x_i	\$0	\$137	\$170	
$P(x_i)$	0.20	0.55	0.25	
$x_i - \mu_X$	-117.85	19.15	52.15	
$(x_i - \mu_X)^2$	13888.62	366.72	2719.62	
$(x_i - \mu_X)^2 \cdot P(x_i)$	2777.7	201.7	679.9	3659.3

The variance of X is $\sigma_X^2 = 3659.3$, which means the standard deviation is $\sigma_X = \sqrt{3659.3} = \$60.49$.

The standard deviation of X, the revenue of a statistics student for the bookstore, is \$60.49. We interpret this by saying that the typical deviation of revenue of a statistics student from the mean revenue of \$117.85 over the long run is \$60.49.

INTERPRETING THE STANDARD DEVIATION OF A RANDOM VARIABLE

The standard deviation of a random variable can be interpreted as the typical deviation of the values of a random variable from the mean value of the random variable over the long-run.

GUIDED PRACTICE 2.64

The bookstore also offers a chemistry textbook for \$159 and a book supplement for \$41. From past experience, they know about 25% of chemistry students just buy the textbook while 60% buy both the textbook and supplement.[47]

(a) What proportion of students don't buy either book? Assume no students buy the supplement without the textbook.

(b) Let Y represent the revenue from a single student. Write out the probability distribution of Y, i.e. a table for each outcome and its associated probability.

(c) Compute the expected revenue from a single chemistry student.

(d) Find the standard deviation to describe the variability associated with the revenue from a single student.

(e) Interpret the mean and the standard deviation in the context of the problem.

[47](a) 100% - 25% - 60% = 15% of students do not buy any books for the class.
(b) is represented by the first three lines in the table below.
(c) is given as the total on the line $y_i \cdot P(y_i)$, which is \$159.75.
(d) is the square-root of the variance listed on in the total on the last line: $\sigma_Y = \sqrt{Var(Y)} = \sqrt{4800} = 69.28$.

i (scenario)	1 (`noBook`)	2 (`textbook`)	3 (`both`)	Total
y_i	0.00	159.00	200.00	
$P(y_i)$	0.15	0.25	0.60	
$y_i \cdot P(y_i)$	0.00	39.75	120.00	$E(Y) = 159.75$
$y_i - \mu_Y$	-159.75	-0.75	40.25	
$(y_i - \mu_Y)^2$	25520.06	0.56	1620.06	
$(y_i - \mu_Y)^2 \cdot P(y_i)$	3828.0	0.1	972.0	$Var(Y) \approx 4800$

(e) In the long run, the average revenue from a chemistry student is \$159.75 and the typical deviation in that revenue from the mean of \$159.75 is \$69.28.

Section summary

- A **random variable** is a variable whose values have numerical outcomes that result from a random phenomenon.
- A **probability distribution** for a discrete random variable shows the probability associated with every possible value of the random variable. The sum of the probabilities over all possible values of a discrete random variable is 1.
- A discrete probability distribution can be determined using the rules of probability or estimated with a simulation.
- A discrete probability distribution can be represented as a graph, table, or function showing the probabilities associated with values of a random variable.
- A cumulative probability distribution can be represented as a table or function and shows the probability of being less than or equal to each value of the discrete random variable.
- A numerical value measuring a characteristic of a probability distribution of a random variable, or a population, is a **parameter**. The value of a parameter is a single, fixed value. The mean and standard deviation are parameters of a probability distribution.
- The **mean** (expected value) and **standard deviation** of a discrete random variable X can be found using the following formulas, where x_i is a value of the random variable and $P(x_i)$ is the probability of that value:

  $$\begin{aligned}
  E(X) = \mu_X &= \sum x_i \cdot P(x_i) \\
  &= x_1 \cdot P(x_1) + x_2 \cdot P(x_2) + \cdots + x_n \cdot P(x_n) \\
  Var(X) = \sigma_X^2 &= \sum (x_i - \mu_X)^2 \cdot P(x_i) \\
  &= (x_1 - \mu_X)^2 \cdot P(x_1) + (x_2 - \mu_X)^2 \cdot P(x_2) + \cdots + (x_n - \mu_X)^2 \cdot P(x_n) \\
  SD(X) = \sigma_X &= \sqrt{\sum (x_i - \mu_X)^2 \cdot P(x_i)} \\
  &= \sqrt{(x_1 - \mu_X)^2 \cdot P(x_1) + (x_2 - \mu_X)^2 \cdot P(x_2) + \cdots + (x_n - \mu_X)^2 \cdot P(x_n)}
  \end{aligned}$$

  We can think of $P(x_i)$ as the *weight*, and each term is weighted its appropriate amount.
- The square of the standard deviation of a random variable is called the variance of the random variable and is denoted as σ_X^2.
- The mean or expected value can be interpreted as the long-run average outcome of the random variable, that is the average after many, many repetitions of the random process. The mean of a probability distribution does not need to be a value in the distribution.
- The standard deviation can be interpreted as the typical deviation of the values of the random variable from the mean value over the long run, that is after many, many repetitions of the random process.
- The parameters mean and standard deviation for the probability distribution of a discrete random variable should be interpreted in the context of a specific population.

Exercises

2.25 Patreon contributions. The distribution of monthly contributions made to a particular Patreon creator is given as follows:

Monthly Contribution	\$3	\$5	\$10	\$25
Proportion	0.50	0.30	0.15	0.05

(a) Compute the average monthly contribution made to this creator.

(b) Compute the standard deviation of the monthly contributions made to this creator.

2.26 Experiment launch impact. An online shopping website develops new features and iterates on its product recommendations on a regular basis. The team also uses an experiment when launching each new feature to assess the impact of the change on the total sales. Below is a summary table of the impacts that the team observes:

Number of Sales From Launch	-200	0	+200	+500
Proportion of Launches	0.1	0.5	0.3	0.1

(a) Compute the average impact for the experiments.

(b) Compute the standard deviation of the impact from these experiments.

2.27 Hearts win. In a new card game, you start with a well-shuffled full deck and draw 3 cards without replacement. If you draw 3 hearts, you win \$50. If you draw 3 black cards, you win \$25. For any other draws, you win nothing.

(a) Create a probability model for the amount you win at this game, and find the expected winnings. Also compute the standard deviation of this distribution.

(b) If the game costs \$5 to play, what would be the expected value and standard deviation of the net profit (or loss)? *(Hint: profit = winnings – cost; $X - 5$)*

(c) If the game costs \$5 to play, should you play this game? Explain.

2.28 Ace of clubs wins. Consider the following card game with a well-shuffled deck of cards. If you draw a red card, you win nothing. If you get a spade, you win \$5. For any club, you win \$10 plus an extra \$20 for the ace of clubs.

(a) Create a probability model for the amount you win at this game. Also, find the expected winnings for a single game and the standard deviation of the winnings.

(b) What is the maximum amount you would be willing to pay to play this game? Explain your reasoning.

2.29 Portfolio return. A portfolio's value increases by 18% during a financial boom and by 9% during normal times. It decreases by 12% during a recession. What is the expected return on this portfolio if each scenario is equally likely?

2.30 Baggage fees. An airline charges the following baggage fees: \$25 for the first bag and \$35 for the second. Suppose 54% of passengers have no checked luggage, 34% have one piece of checked luggage and 12% have two pieces. We suppose a negligible portion of people check more than two bags.

(a) Build a probability model, compute the average revenue per passenger, and compute the corresponding standard deviation.

(b) About how much revenue should the airline expect for a flight of 120 passengers? With what standard deviation? Note any assumptions you make and if you think they are justified.

2.5 Binomial distributions

What is the probability of exactly 50 heads in 100 coin tosses? If 12 people are randomly selected for a jury, what is the probability that more than 9 of them identify as male? Given that the probability of a defective part is 1%, how many defective items would we expect in a random shipment of 200 of those parts? We can model these scenarios and answer these questions using a binomial distribution.

Learning objectives

1. Justify why a random variable is or is not a binomial random variable.
2. Calculate the mean and standard deviation for a binomial distribution.
3. Interpret the mean, standard deviation, and probabilities for a binomial distribution.
4. Estimate probabilities of binomial random variables using data from a simulation.
5. Calculate probabilities for a binomial distribution.

2.5.1 Binary variables

Many health insurance plans in the United States have a deductible, where the insured individual is responsible for costs up to the deductible, and then the costs above the deductible are shared between the individual and insurance company for the remainder of the year.

Suppose a health insurance company found that 70% of the people they insure stay below their deductible in any given year. Each of these people can be thought of as a **trial**. We label a person a **success** if her healthcare costs do not exceed the deductible. We label a person a **failure** if she does exceed her deductible in the year. Because 70% of the individuals will not exceed their deductible, we denote the **probability of a success** as $p = 0.7$. The probability of a failure is $1 - p$, which would be 0.3 for the insurance example.

When an individual trial only has two possible outcomes, often labeled as `success` or `failure`, it is called a **Bernoulli random variable**, or more simply, a **binary variable** or yes/no variable. We chose to label a person who does not exceed her deductible as a "success" and all others as "failures". However, we could just as easily have reversed these labels. The mathematical framework we will build does not depend on which outcome is labeled a success and which a failure, as long as we are consistent.

Binary variables are often denoted as `1` for a success and `0` for a failure. In addition to being convenient in entering data, it is also mathematically handy. Suppose we observe ten trials:

`1 1 1 0 1 0 0 1 1 0`

Then the **sample proportion**, $\hat{p}$, is the sample mean of these observations:

$$\hat{p} = \frac{\text{\# of successes}}{\text{\# of trials}} = \frac{1+1+1+0+1+0+0+1+1+0}{10} = 0.6$$

In general, it is useful to think about a binary random variable as a random process with only two outcomes: a success or failure. Then we build our mathematical framework using the numerical labels `1` and `0` for successes and failures, respectively.

2.5.2 Introducing the binomial formula

Let's imagine ourselves at the insurance agency where 70% of individuals do not exceed their deductible. The probability of success (not exceeding the deductible) is $p = 0.7$.

EXAMPLE 2.65

Suppose the insurance agency is considering a random sample of four individuals they insure. What is the chance exactly one of them will exceed the deductible and the other three will not? Let's call the four people Ariana (A), Brittany (B), Carlton (C), and Damian (D) for convenience.

Let's consider a scenario where one person exceeds the deductible:

(E)

$$\begin{aligned} &P(A = \texttt{exceed},\ B = \texttt{not},\ C = \texttt{not},\ D = \texttt{not}) \\ &\quad = P(A = \texttt{exceed})\ P(B = \texttt{not})\ P(C = \texttt{not})\ P(D = \texttt{not}) \\ &\quad = (0.3)(0.7)(0.7)(0.7) \\ &\quad = (0.7)^3(0.3)^1 \\ &\quad = 0.103 \end{aligned}$$

But there are three other scenarios: Brittany, Carlton, or Damian could have been the one to exceed the deductible. In each of these cases, the probability is again $(0.7)^3(0.3)^1$. These four scenarios exhaust all the possible ways that exactly one of these four people could have exceeded the deductible, so the total probability is $4 \times (0.7)^3(0.3)^1 = 0.412$.

GUIDED PRACTICE 2.66

(G) Verify that the scenario where Brittany is the only one to exceed the deductible has probability $(0.7)^3(0.3)^1$.[48]

In Example 2.65, there were four individuals who could have been the one to exceed the deductible, and each of these four scenarios had the same probability p of 0.7. We use n to represent the number of trials, in this case 4 and we use x to represent the desired number of successes, in this case 1. This leads us to a more general approach to find probabilities involving exactly x successes in n trials, where the probability in each trial of a success is p.

We can write the probability as

$$[\#\text{ of scenarios}] \times P(\text{single scenario})$$

The first component of this equation is the number of ways to arrange the $x = 3$ successes among the $n = 4$ trials. The second component is the probability of any of the four (equally probable) scenarios.

Consider $P(\text{single scenario})$ under the general case of x successes and $n - x$ failures in the n trials. In any such scenario, we apply the Multiplication Rule for independent events:

$$p^x(1-p)^{n-x}$$

This is our general formula for $P(\text{single scenario})$.

---

[48] $P(A = \texttt{not},\ B = \texttt{exceed},\ C = \texttt{not},\ D = \texttt{not}) = (0.7)(0.3)(0.7)(0.7) = (0.7)^3(0.3)^1$.

Secondly, we introduce the **binomial coefficient**, which gives the number of ways to choose x successes in n trials, i.e. arrange x successes and $n - x$ failures:

$$\binom{n}{x} = \frac{n!}{x!(n-x)!}$$

The quantity $\binom{n}{x}$ is read **n choose x**.[49] The exclamation point notation (e.g. $n!$) denotes a **factorial** expression.

$$\begin{aligned} 0! &= 1 \\ 1! &= 1 \\ 2! &= 2 \times 1 = 2 \\ 3! &= 3 \times 2 \times 1 = 6 \\ 4! &= 4 \times 3 \times 2 \times 1 = 24 \\ &\vdots \\ n! &= n \times (n-1) \times \ldots \times 3 \times 2 \times 1 \end{aligned}$$

Using the formula, we can compute the number of ways to choose $x = 3$ successes in $n = 4$ trials:

$$\binom{4}{3} = \frac{4!}{3!(4-3)!} = \frac{4!}{3!1!} = \frac{4 \times 3 \times 2 \times 1}{(3 \times 2 \times 1)(1)} = 4$$

This result is exactly what we found by carefully thinking of each possible scenario in Example 2.65.

Substituting n choose x for the number of scenarios and $p^x(1-p)^{n-x}$ for the single scenario probability yields the **binomial formula**.

BINOMIAL FORMULA

Suppose the probability of a single trial being a success is p. Then the probability of observing exactly x successes in n independent trials is given by:

$$P(X = x) = \binom{n}{x} p^x (1-p)^{n-x} = \frac{n!}{x!(n-x)!} p^x (1-p)^{n-x}$$

where $x = 0, 1, 2, 3 \ldots, n$.

2.5.3 When and how to apply the binomial formula

IS IT BINOMIAL? FOUR CONDITIONS TO CHECK.

Informally, we can say that the binomial formula is used in questions concerned with "how many successes out of n". To be binomial the following four conditions must be met.

(1) Each trial outcome is *binary* (can be classified as a success or failure).
(2) The trials are *independent*.
(3) The number of trials, n, is fixed.
(4) There is the *same* probability of success, p, for each trial.

A useful acronym to remember these conditions is BINS: *b*inary, *i*ndependent, *n* fixed, and *s*ame p.

[49] Other notations for n choose x includes ${}_nC_x$, C_n^x, and $C(n, x)$.

EXAMPLE 2.67

Say we would like to find the probability that 5 of 8 randomly selected individuals from the insurance agency will not exceed the deductible. Can we use the binomial formula?

We are interested in 5 out of 8, which sounds like a "how many successes out of n" scenario. To verify we can use the binomial formula we check the following conditions.

E

1. Each trial outcome is binary (not exceed the insurance deductible or exceed the insurance deductible).
2. The sample is random, but it is not with replacement, so the observations are not entirely independent. However, as we saw in the Section 2.3.6 (Sampling without replacement), when the sample size is very small compared to the population size, we can treat the observations *as if they were with replacement*, since the composition of the population changes very little with each additional person sampled. Since we have a random sample of a very small percent of all individuals from the insurance company, we will consider the independence condition met.
3. The number of trials is fixed ($n = 8$).
4. There is the same probability of a success for each trial. ($p = 0.70$)

SAMPLING WITHOUT REPLACEMENT

When randomly sampling without replacement, if the sample size is small relative to the population size (rule of thumb: sample size less than 1/10 of the population size), we will consider the observations to be independent.

EXAMPLE 2.68

Find the probability that 5 of 8 randomly selected individuals from the insurance agency will not exceed the deductible, i.e. that 3 of them will. Recall that 70% of individuals will not exceed the deductible.

Here a success is not exceeding the deductible and the probability of a success is $p = 0.7$. We want to find the probability of $x = 5$ successes in $n = 8$ trials. We previous verified that this scenario is binomial, so we will use the binomial formula. The probability that 5 of 8 will not exceed the deductible and 3 will exceed the deductible is given by

E

$$\binom{8}{5}(0.7)^5(1-0.7)^{8-5} = \frac{8!}{5!(8-5)!}(0.7)^5(1-0.7)^{8-5}$$
$$= \frac{8!}{5!3!}(0.7)^5(0.3)^3$$

Dealing with the factorial part:

$$\frac{8!}{5!3!} = \frac{8\times 7\times 6\times 5\times 4\times 3\times 2\times 1}{(5\times 4\times 3\times 2\times 1)(3\times 2\times 1)} = \frac{8\times 7\times 6}{3\times 2\times 1} = 56$$

Using $(0.7)^5(0.3)^3 \approx 0.00454$, the final probability is about $56 \times 0.00454 \approx 0.254$.

Evaluating the binomial formula by hand can be tedious. See Section 2.5.7 for ways to evaluate the binomial formula using technology.

COMPUTING BINOMIAL PROBABILITIES

The first step in using the binomial model is to check that the model is appropriate. The second step is to identify n, p, and x. Finally, apply the binomial formula to determine the probability and interpret the results.

EXAMPLE 2.69

Approximately 35% of a population has blood type O+. Suppose four people show up at a hospital and we want to find the probability that exactly one of them has blood type O+. Can we use the binomial formula?

To check if the binomial model is appropriate, we must verify the conditions.

1. Each outcome is binary (blood type O+ or not blood type O+).
2. We will suppose that these 4 people comprise a random sample. This seems reasonable, since one person with a particular blood type showing up at a hospital seems unlikely to affect the chance that other people with that blood type would show up at the hospital. Though this scenario is without replacement, 4 should be less than 1/10th of all people that show up at the hospital. Therefore, we can treat the random sample *as if it were* with replacement and consider the independence condition met.
3. We have a fixed number of trials ($n = 4$).
4. The probability of a success is the same for each trial ($p = 0.35$ if we say a "success" is someone having blood type O+).

EXAMPLE 2.70

Given that 35% of a population has blood type O+, what is the probabilty that in a random sample of 4 people:

(a) none of them have blood type O+?

(b) exactly one will have blood type O+?

(c) no more than one will have blood type O+?

(a) $P(X = 0) = \binom{4}{0}(0.35)^0(0.65)^4 = 1 \times 1 \times 0.65^4 = 0.65^4 = 0.179$
Note that we could have answered this question without the binomial formula, using methods from the previous section.

(b) $P(X = 1) = \binom{4}{1}(0.35)^1(0.65)^3 = 0.384$.

(c) We want to find $P(X \leq 1)$. This can be computed as the sum of parts (a) and (b):
$P(X \leq 1) = P(X = 0) + P(X = 1) = 0.179 + 0.384 = 0.563$.
There is about a 56.3% chance that no more than one of them will have blood type O+.

GUIDED PRACTICE 2.71

What is the probability that at least 3 of 4 people in a random sample will have blood type O+ if 35% of the population has blood type O+?[50]

GUIDED PRACTICE 2.72

The probability that a random smoker will develop a severe lung condition in her lifetime is about 0.3. If you have 4 friends who smoke and you want to find the probability that 1 of them will develop a severe lung condition in her lifetime, can you apply the binomial formula?[51]

---

[50] $P(\text{at least 3 of 4 have blood type O+}) = P(X = 3) + P(X = 4) = \binom{4}{3}(0.35)^3(0.65)^1 + (0.35)^4 = 0.111 + 0.015 = 0.126$

[51] While the binary and fixed n conditions are met, most likely the friends know each other, so the independence assumption is probably not satisfied. For example, acquaintances may have similar smoking habits, or those friends might make a pact to quit together. The same p condition is also not satisfied since this is not a random sample of people.

2.5.4 An example of a binomial distribution

In Guided Practice 2.70, we asked various probability questions regarding the number of people out of 4 with blood type O+. We verified that the scenario was binomial and that each problem could be solved using the binomial formula. Instead of looking at it piecewise, we could describe the entire *distribution* of possible values and their corresponding probabilities. Since there are 4 people, the possible outcomes for the number who will have blood type O+ are: 0, 1, 2, 3, 4. In Figure 2.30, we make a distribution table and a histogram showing the probability of each of these outcomes. Recall that the probability of a randomly sampled person being blood type O+ is about 0.35.

x_i	$P(x_i)$
0	$\binom{4}{0}(0.35)^0(0.65)^4 = 0.179$
1	$\binom{4}{1}(0.35)^1(0.65)^3 = 0.384$
2	$\binom{4}{2}(0.35)^2(0.65)^2 = 0.311$
3	$\binom{4}{3}(0.35)^3(0.65)^1 = 0.111$
4	$\binom{4}{4}(0.35)^4(0.65)^0 = 0.015$

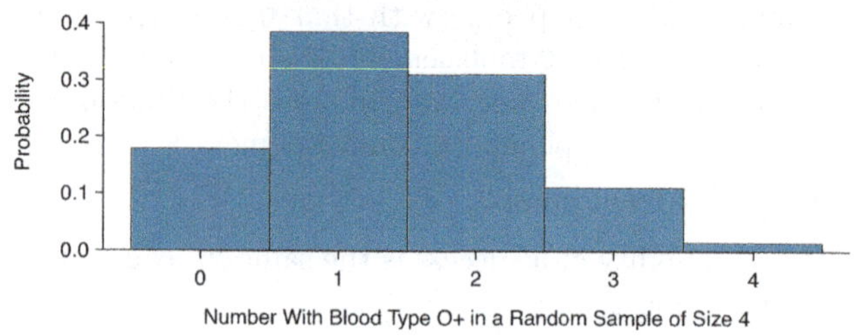

Figure 2.30: Probability distribution for the number with blood type O+ in a random sample of 4 people. This is a binomial distribution. Correcting for rounding error, the probabilities add up to 1, as they must for any probability distribution.

A **binomial distribution** is used to describe the number of successes in a fixed number of independent trials. A binomial distribution has two parameters: n and p, where n is the number of trials and p is the probability of a success. The example in Figure 2.30 shows a binomial distribution with $n = 4$ and $p = 0.35$.

To find the probability that a binomial random variable X has exactly x successes, apply the binomial formula as follows:

$$P(X = x) = \binom{n}{x} p^x (1-p)^{n-x}, \quad \text{where } x = 0, 1, 2, 3 \ldots, n.$$

2.5.5 The mean and standard deviation of a binomial distribution

We can find the mean and standard deviation of the probability distribution in Figure 2.30 using the formulas for the mean and standard deviation of a discrete random variable from Section 2.4. Those formulas require a lot of calculations, so it is fortunate that there is an easier way to compute the mean and standard deviation for a binomial random variable.

MEAN AND STANDARD DEVIATION OF THE BINOMIAL DISTRIBUTION

For a binomial distribution with parameters n and p, where n is the number of trials and p is the probability of a success, the mean and standard deviation of the number of observed successes are

$$\mu_X = np \qquad \sigma_X = \sqrt{np(1-p)}$$

E

EXAMPLE 2.73

The probability that a person has blood type O+ is 0.35. Let X be the number of people with blood type O+ in a random sample of 40 people. Calculate and interpret the mean and standard deviation of X.

In Example 2.69 we confirmed that the binary, independence, n fixed, and same p conditions were met. Therefore we can say that X is Binomial with $n = 40$ and $p = 0.35$.

$$\mu_X = np = 40(0.35) = 14$$

If we were to take many, many random samples of size 40, we would expect to get about 14 people with blood type O+, on average.

$$\sigma_X = \sqrt{np(1-p)} = \sqrt{40(0.35)(0.65)} = 3.0$$

If we were to take many, many random samples of size 40, the number of people in our sample with blood type O+ would typically vary from the mean of 14 people by about 3 people.

The distribution of the number of people with blood type O+ in a random sample of size 40 is shown in Figure 3.2. It is binomial distribution with $n = 40$ and $p = 0.35$, and it has a mean of 14 and a standard deviation of 3.0.

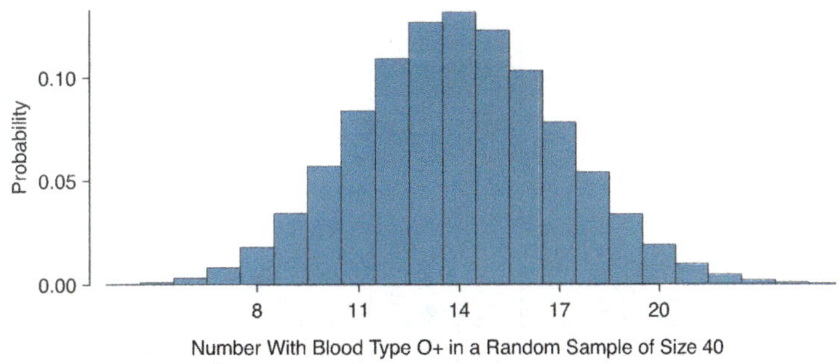

Figure 2.31: Distribution for the number of people with blood type O+ in a random sample of size 40, where $p = 0.35$. The distribution is binomial and is centered on 14 with a standard deviation of 3.

2.5.6 Binomial probabilities for intervals of values

E

EXAMPLE 2.74

Find the probability of getting between 15 and 17 people, inclusive, with blood type O+ in a random sample of 40 people. The probability of a randomly sampled person being blood type O+ is 0.35.

X is the number of people with blood type O+. X has a binomial distribution with $n = 40$ and $p = 0.35$. We want to find $P(15 \le X \le 17)$. Because the binomial distribution is discrete, we can find this by calculating $P(X = 15) + P(X = 16) + P(X = 17)$ as follows:

$$\begin{aligned} P(15 \le X \le 17) &= P(X = 15) + P(X = 16) + P(X = 17) \\ &= \binom{40}{15}(0.35)^{15}(0.65)^{25} + \binom{40}{16}(0.35)^{16}(0.65)^{24} + \binom{40}{17}(0.35)^{17}(0.65)^{23} \\ &= 0.123 + 0.103 + 0.078 \\ &= 0.304 \end{aligned}$$

The probability of getting between 15 and 17 people, inclusive, with blood type O+ in a random sample of 40 people is 0.304, or 30.4%.

EXAMPLE 2.75

Find the probability that at least 18 people have blood type O+ in a random sample of 40 people. The probability of a randomly sampled person being blood type O+ is 0.35.

X is the number of people with blood type O+. Here, we want to find the probability that X ≥18.

$P(X \geq 18) = P(X = 18) + P(X = 19) + P(X = 20) + \ldots P(X = 40)$.

Evaluating this with the binomial formula would involve applying the binomial formula 23 times! Fortunately, we can use technology to help us. We first identify the distribution and its parameters. Here X is Binomial with $n = 50$ and $p = 0.35$. We sometimes write this as X is Binomial($n = 40$, $p = 0.35$). Then we can use a technology option from Section 2.5.7 to find that $P(X \geq 18) = 0.124$. There is about a 12.4% probability of at least 18 people with blood type O+ in a random sample of size 40.

When using technology to find binomial probabilities, make sure to identify the distribution being used and its parameters, then write a clear probability statement communicating what is being evaluated. A graph, such as the one shown below in Figure 2.32, helps to visualize the distribution and the individual probabilities used in the calculation.

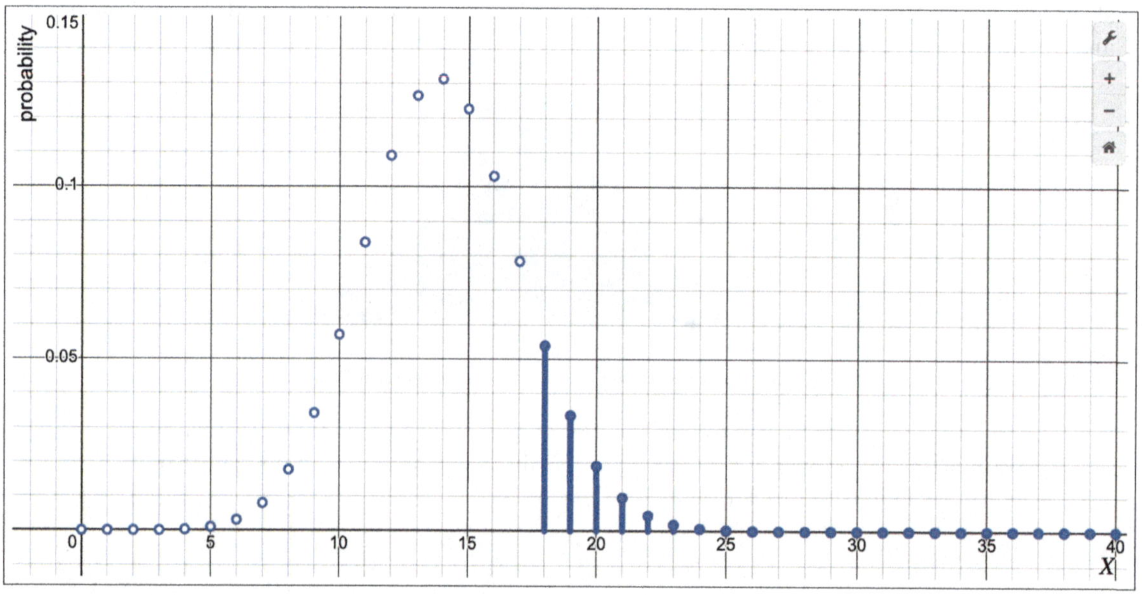

Figure 2.32: A Desmos graph illustrating a Binomial distribution with $n = 40$ and $p = 0.35$. Values greater than or equal to 18 are highlighted.

GUIDED PRACTICE 2.76

What is the probability that at most 12 people in a random sample of 40 will have blood type O+ if 35% of the population has blood type O+? [52]

[52] X is the number of people with blood type O+. X is Binomial($n = 40$, $p = 0.35$). $P(X \leq 12) = 0.314$. There is a 31.4% probability that at most 12 people in a random sample of size 40 will have blood type O+.

2.5.7 Technology: binomial probabilities

A spinner has four equally likely options: red, green, blue, yellow. If you spin the spinner 6 times, what is the probability you get:

a) Exactly 2 red?

b) Less than 2 red?

c) At least 3 red?

This scenario describes a **binomial distribution** with $n = 6$ and $p = 0.25$.

Desmos: Use `binomialdist(n, p)`, replacing `n` and `p` with appropriate values.

1. Type `binomialdist(6, 0.25)`.
2. Click the triangle next to `Cumulative Probability`.
3. Select `Inner`, `Outer`, `Left`, or `Right` as illustrated below.
4. Enter the appropriate boundary value(s) for x as illustrated below.
5. Click the magnifying glass to Zoom Fit the graphing window.

a) P(Exactly 2 red): $P(x = 2)$.

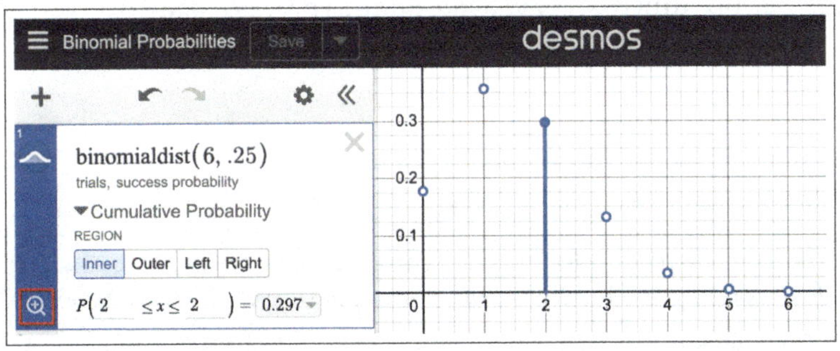

b) P(Less than 2 red): $P(x < 2) = P(x \leq 1)$.

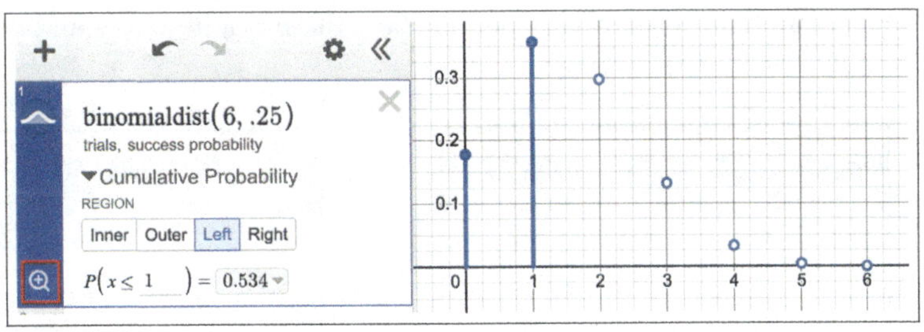

c) P(At least 3 red): $P(x \geq 3)$.

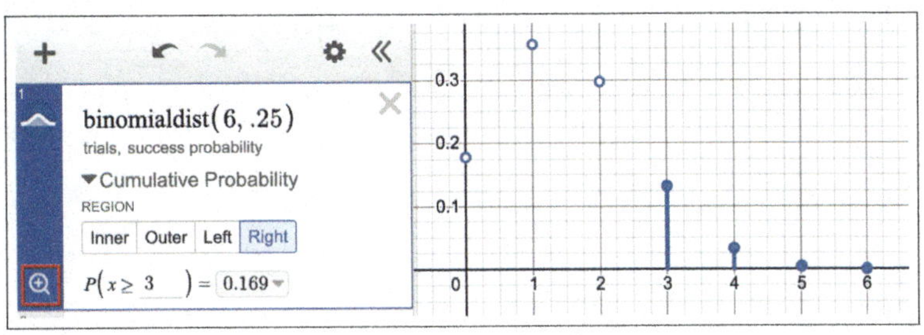

R: Given a binomial distribution with $n = 6$ and $p = 0.25$, calculate the following probabilities.

a) P(X = 2). Use `dbinom(x, size, prob)` for the probability of exactly `x`.
Note that "`size =`" and "`prob =`" can be omitted, but the labels are often included for clarity.
`> dbinom(2, size = 6, prob = 0.25)` or `> dbinom(2, 6, 0.25)`
`[1] 0.2966309`

b) P(X ≤ 1). Use `pbinom(q, size, prob)` for the probability of ≤ `q` as shown.

```
> pbinom(1, size = 6, prob = 0.25)
[1] 0.5339355
```

c) P(X ≥ 3).

```
> 1 - pbinom(2, size = 6, prob = 0.25)
[1] 0.1694336
```

or

```
> pbinom(2, size = 6, prob = 0.25, lower.tail = FALSE)
[1] 0.1694336
```

Calculator: NumWorks calculator instructions and example output are included. For the TI-83/84 and Casio calculators, general instructions are provided, and worked example videos can be accessed via the icon or at **openintro.org/ti** and **openintro.org/casio**.

NUMWORKS: BINOMIAL CALCULATIONS

Use `OK` or `EXE` to make a selection.

1. From the home screen, select `Distributions` then select `Binomial`. If a list of distributions does not appear, hit the ← button (next to `OK`) as many times as needed.
2. Enter the values of `n` and `p` then choose `Next`. If the screen shows a graph rather than asking for `n` and `p`, hit the ← button.
3. Hit the left arrow to highlight the graph. Hit the down arrow to choose whether you want left, inner, right, or exactly, then hit `OK`. Hit the right arrow and enter the boundary value(s) followed by `EXE`.

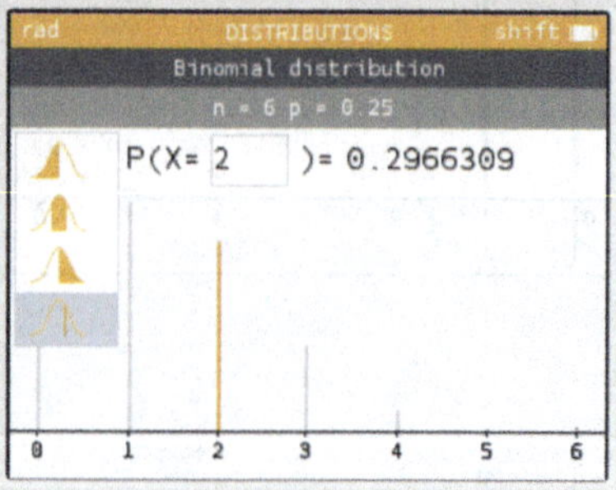

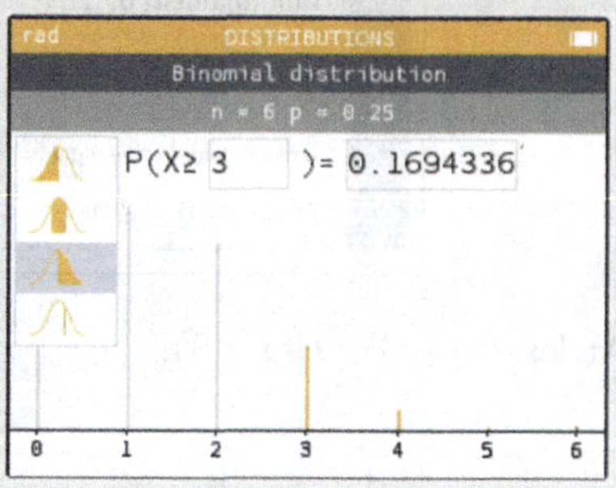

TI-84: COMPUTING THE BINOMIAL FORMULA, $P(X = x) = \binom{n}{x}p^x(1-p)^{n-x}$

Use `2ND VARS`, `binompdf` to evaluate the probability of *exactly* x occurrences out of n independent trials of an event with probability p.

1. Select `2ND VARS` (i.e. `DISTR`)
2. Choose `A:binompdf` (use the down arrow to scroll down).
3. Let `trials` be n.
4. Let `p` be p
5. Let `x value` be x.
6. Select `Paste` and hit `ENTER`.

TI-83: Do step 1, choose `0:binompdf`, then enter n, p, and x separated by commas: `binompdf(n, p, x)`. Then hit `ENTER`.

TI-84: COMPUTING $P(X \le x) = \binom{n}{0}p^0(1-p)^{n-0} + \ldots + \binom{n}{x}p^x(1-p)^{n-x}$

Use `2ND VARS`, `binomcdf` to evaluate the cumulative probability of *at most* x occurrences out of n independent trials of an event with probability p.

1. Select `2ND VARS` (i.e. `DISTR`)
2. Choose `B:binomcdf` (use the down arrow).
3. Let `trials` be n.
4. Let `p` be p
5. Let `x value` be x.
6. Select `Paste` and hit `ENTER`.

TI-83: Do steps 1-2, then enter the values for n, p, and x separated by commas as follows: `binomcdf(n, p, x)`. Then hit `ENTER`.

CASIO FX-9750GII: BINOMIAL CALCULATIONS

1. Navigate to `STAT` (`MENU`, then hit `2`).
2. Select `DIST` (`F5`), and then `BINM` (`F5`).
3. Choose whether to calculate the binomial distribution for a specific number of successes, $P(X = k)$, or for a range $P(X \le k)$ of values (0 successes, 1 success, ..., x successes).
   - For a specific number of successes, choose `Bpd` (`F1`).
   - To consider the range 0, 1, ..., x successes, choose `Bcd(F1)`.
4. If needed, set `Data` to `Variable` (`Var` option, which is `F2`).
5. Enter the value for `x` (x), `Numtrial` (n), and `p` (probability of a success).
6. Hit `EXE`.

Section summary

- A **binomial random variable**, X, is a discrete random variable that counts the number of successes in n repeated *independent* trials that have only two possible outcomes (success or failure), with a fixed probability of success p and probability of failure $1 - p$ for each trial.
- If X follows a **binomial distribution** with parameters n and p, then:
  - The mean is given by $\mu_X = np$. (*center*)
  - The standard deviation is given by $\sigma_X = \sqrt{np(1-p)}$. (*spread*)
- The mean of a binomial distribution tells us about how many successes we expect, on average, across many, many samples of size n from the same population.
- The standard deviation of a binomial distribution tells us the typical variation of values from the mean across many, many samples of size n from the same population.
- The mean and standard deviation should always be interpreted in context.
- A probability distribution can be constructed using the rules of probability or estimated with a simulation.
- The probability that a binomial random variable, X, has exactly x successes among n independent trials, when the probability of success is p, is calculated as:

$$P(X = x) = \binom{n}{x} p^x (1-p)^{n-x}, \text{ where } x = 0, 1, 2, 3 \ldots, n.$$

Exercises

2.31 Exploring combinations. A coin is tossed 5 times. How many sequences / combinations of Heads/Tails are there that have:

(a) Exactly 1 Tail?
(b) Exactly 4 Tails?
(c) Exactly 3 Tails?
(d) At least 3 Tails?

2.32 Political affiliation. Suppose that in a large population, 51% identify as Democrat. A researcher takes a random sample of 3 people.

(a) Use the binomial model to calculate the probability that two of them identify as Democrat.
(b) Write out all possible orderings of 3 people, 2 of whom identify as Democrat. Use these scenarios to calculate the same probability from part (a) but using the Addition Rule for disjoint events. Confirm that your answers from parts (a) and (b) match.
(c) If we wanted to calculate the probability that a random sample of 8 people will have 3 that identify as Democrat, briefly describe why the approach from part (b) would be more tedious than the approach from part (a).

2.33 Underage drinking, Part I. Data collected by the Substance Abuse and Mental Health Services Administration (SAMSHA) suggests that 69.7% of 18-20 year olds consumed alcoholic beverages in any given year.[53]

(a) Suppose a random sample of ten 18-20 year olds is taken. Is the use of the binomial distribution appropriate for calculating the probability that exactly six consumed alcoholic beverages? Explain.
(b) Calculate the probability that exactly 6 out of 10 randomly sampled 18- 20 year olds consumed an alcoholic drink.
(c) What is the probability that exactly four out of ten 18-20 year olds have *not* consumed an alcoholic beverage?
(d) What is the probability that at most 2 out of 5 randomly sampled 18-20 year olds have consumed alcoholic beverages?
(e) What is the probability that at least 1 out of 5 randomly sampled 18-20 year olds have consumed alcoholic beverages?

2.34 Chickenpox, Part I. Boston Children's Hospital estimates that 90% of Americans have had chickenpox by the time they reach adulthood.[54]

(a) Suppose we take a random sample of 100 American adults. Is the use of the binomial distribution appropriate for calculating the probability that exactly 97 out of 100 randomly sampled American adults had chickenpox during childhood? Explain.
(b) Calculate the probability that exactly 97 out of 100 randomly sampled American adults had chickenpox during childhood.
(c) What is the probability that exactly 3 out of a new sample of 100 American adults have *not* had chickenpox in their childhood?
(d) What is the probability that at least 1 out of 10 randomly sampled American adults have had chickenpox?
(e) What is the probability that at most 3 out of 10 randomly sampled American adults have *not* had chickenpox?

[53]SAMHSA, Office of Applied Studies, National Survey on Drug Use and Health, 2007 and 2008.
[54]Boston Children's Hospital, Chickenpox summary page, referenced April 29, 2021.

2.35 Game of dreidel. A dreidel is a four-sided spinning top with the Hebrew letters *nun*, *gimel*, *hei*, and *shin*, one on each side. Each side is equally likely to come up in a single spin of the dreidel. Suppose you spin a dreidel three times. Calculate the probability of getting

(a) at least one *nun*?

(b) exactly 2 *nuns*?

(c) exactly 1 *hei*?

(d) at most 2 *gimels*?

Photo by Staccabees, cropped (http://flic.kr/p/7gLZTf) CC BY 2.0 license

2.36 Sickle cell anemia. Sickle cell anemia is a genetic blood disorder where red blood cells lose their flexibility and assume an abnormal, rigid, "sickle" shape, which results in a risk of various complications. If both parents are carriers of the disease, then a child has a 25% chance of having the disease, 50% chance of being a carrier, and 25% chance of neither having the disease nor being a carrier. If two parents who are carriers of the disease have 3 children, what is the probability that

(a) two will have the disease?

(b) none will have the disease?

(c) at least one will neither have the disease nor be a carrier?

(d) the first child with the disease will the be 3rd child?

2.37 Underage drinking, Part II. We learned in Exercise 2.33 that about 70% of 18-20 year olds consumed alcoholic beverages in any given year. We now consider a random sample of fifty 18-20 year olds.

(a) How many people would you expect to have consumed alcoholic beverages? And with what standard deviation?

(b) Would you be surprised if there were 45 or more people who have consumed alcoholic beverages?

(c) What is the probability that 45 or more people in this sample have consumed alcoholic beverages? How does this probability relate to your answer to part (b)?

2.38 Chickenpox, Part II. We learned in Exercise 2.34 that about 90% of American adults had chickenpox before adulthood. We now consider a random sample of 120 American adults.

(a) How many people in this sample would you expect to have had chickenpox in their childhood? And with what standard deviation?

(b) Would you be surprised if there were 105 people who have had chickenpox in their childhood?

(c) What is the probability that 105 or fewer people in this sample have had chickenpox in their childhood? How does this probability relate to your answer to part (b)?

2.6 Normal distributions

What proportion of adults have systolic blood pressure above 140? If the average weight of a piece of carry-on luggage is 11 pounds, what is the probability that 100 random carry on pieces will weigh more than 1200 pounds? If 55% of a population supports a certain candidate, what is the probability that she will have less than 50% support in a random sample of size 200?

There is one distribution that can help us answer all of these questions. Can you guess what it is? That's right – it's the normal distribution.

Learning objectives

1. Describe the standard normal distribution.
2. Use Z-scores and the standard normal model to approximate a distribution where appropriate.
3. Calculate the proportion of values in a specified interval for a normal distribution.
4. Find boundary values of an interval associated with a given proportion under a normal distribution.
5. Calculate the mean or standard deviation of a normal distribution given the value of a percentile.
6. Compare measures of relative position for distributions.
7. Estimate the proportion of values in a specified interval of a normal distribution using the empirical rule.

2.6.1 Normal distribution model

Among all the distributions we see in practice, one is overwhelmingly the most common. The symmetric, unimodal, bell curve is ubiquitous throughout statistics. Indeed it is so common, that people often know it as the **normal curve** or **normal distribution**.[55] A normal curve is shown in Figure 2.33.

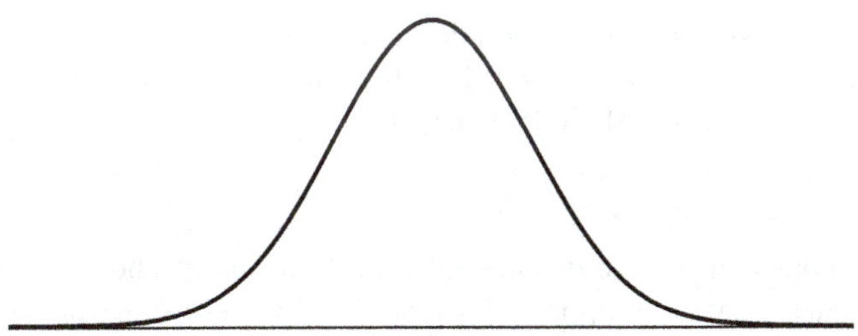

Figure 2.33: A normal curve.

[55] It is also introduced as the Gaussian distribution after Frederic Gauss, the first person to formalize its mathematical expression.

The **normal distribution** always describes a symmetric, unimodal, bell-shaped curve. However, these curves can look different depending on the details of the model. Specifically, the normal distribution model can be adjusted using two parameters: mean and standard deviation. As you can probably guess, changing the mean shifts the bell curve to the left or right, while changing the standard deviation stretches or constricts the curve.

Figure 2.34 shows the normal distribution with mean 0 and standard deviation 1 in the left panel and the normal distributions with mean 19 and standard deviation 4 in the right panel. Figure 2.35 shows these distributions on the same axis.

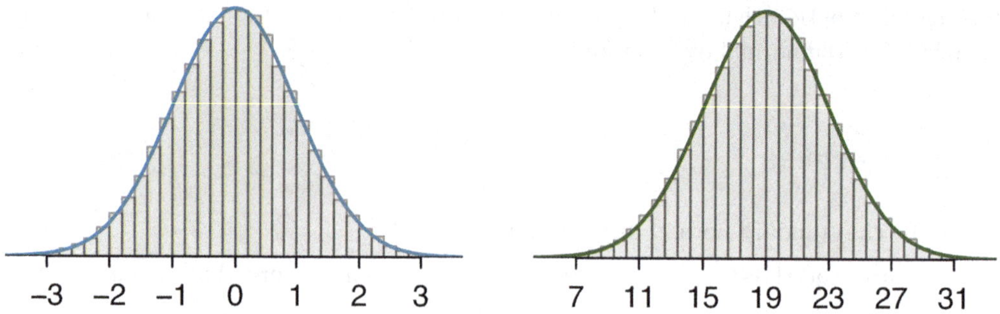

Figure 2.34: Both curves represent a normal distribution. However, they differ in their center and spread.

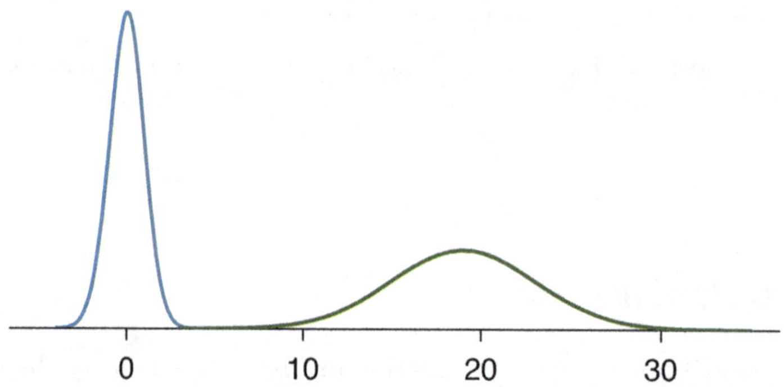

Figure 2.35: The normal distributions shown in Figure 2.34 but plotted together and on the same scale.

Because the mean and standard deviation describe a normal distribution exactly, they are called the distribution's **parameters**. The normal distribution with mean $\mu = 0$ and standard deviation $\sigma = 1$ is called the **standard normal distribution**..

NORMAL DISTRIBUTION FACTS

Many variables are nearly normal, but none are exactly normal. The normal distribution, while never perfect, provides very close approximations for a variety of scenarios. We will use it to model data as well as probability distributions.

2.6.2 Using the normal distribution to approximate empirical distributions

We often want to put data onto a standardized scale, which can make comparisons more reasonable.

EXAMPLE 2.77

Figure 2.36 shows the mean and standard deviation for total scores on the SAT and ACT. The distribution of SAT and ACT scores are both nearly normal. Suppose Ann scored 1300 on her SAT and Tom scored 24 on his ACT. Who performed better?

As we saw in section 1.4.7, we can use Z-scores to compare observations from different distributions. Using Ann's SAT score, 1300 , along with the SAT mean and SD, we can find Ann's Z-score.

E

$$Z_{\text{Ann}} = \frac{x_{\text{Ann}} - \mu_{\text{SAT}}}{\sigma_{\text{SAT}}} = \frac{1300 - 1100}{200} = 1$$

Similarly, using Tom's ACT score, 24, along with the ACT mean and SD we can find his Z-score.

$$Z_{Tom} = \frac{x_{\text{Tom}} - \mu_{\text{ACT}}}{\sigma_{\text{ACT}}} = \frac{24 - 21}{6} = 0.5$$

Because Ann's score was 1 standard deviation above the mean, while Tom's score was 0.5 standard deviations above the mean, we can say that Ann did better than Tom.

	SAT	ACT
Mean	1100	21
SD	200	6

Figure 2.36: Mean and standard deviation for the SAT and ACT.

Assuming that both the SAT and ACT distributions are nearly normally distributed, what percent of test takers scored lower than Ann? What percent scored lower than Tom? To answer these question exactly, we would need all of the data. However, if we use the information that SAT and ACT distributions are nearly normal, we can estimate these percents. Figure 2.37 shows these distributions modeled with a normal curve. If we can find the percent of the normal curve that is to the left of Ann's score, we could use that percent as our estimate of the percent of the data points that are smaller than Ann's score. We call this process *normal approximation*. The steps are:

1. First verify that the distribution can be reasonably modeled with a normal distribution.
2. Convert value or values of interest to Z-scores.
3. Find the relevant area/percent under the standard normal curve.

We use the area/percent that we find from the normal curve as our *estimate* of the desired percent.

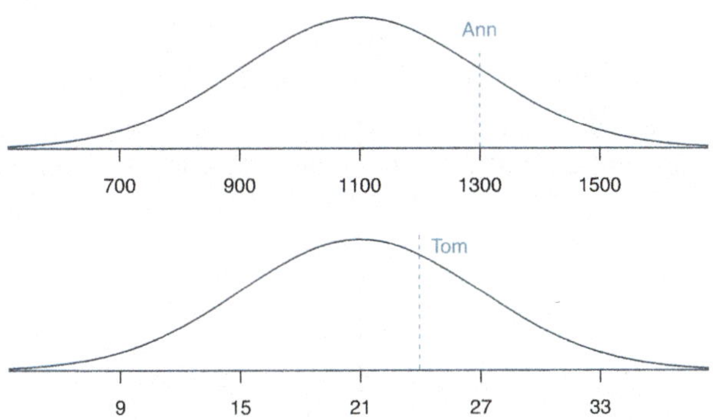

Figure 2.37: Ann's and Tom's scores shown with the distributions of SAT and ACT scores.

There are many techniques for finding areas under a normal distribution. The most common approach is to use technology. In practice, statisticians use statistical software, such as R, Python, or SAS. In classrooms, it is common to use Desmos or a handheld graphing calculator such as NumWorks, TI, or Casio. Instructions for finding areas of a normal distribution using these R, Desmos, and these handheld graphing calculators are provided in Section 2.6.4.

Another option for finding tail areas is to use what's called a **probability table**; these are occasionally used in classrooms but rarely in practice. Appendix C.2 contains such a table and a guide for how to use it.

2.6.3 Normal probability examples

Combined SAT scores are approximated well by a normal model with mean 1100 and standard deviation 200.

EXAMPLE 2.78

How many standard units above average is a score of 1190 on the SAT? What is the probability that a randomly selected SAT taker scores at least 1190 on the SAT?

The probability that a randomly selected SAT taker scores at least 1190 on the SAT is equivalent to the proportion of all SAT takers that score at least 1190 on the SAT. First, always draw and label a picture of the normal distribution. (Drawings need not be exact to be useful.) We are interested in the probability that a randomly selected score will be above 1190, so we shade this upper tail:

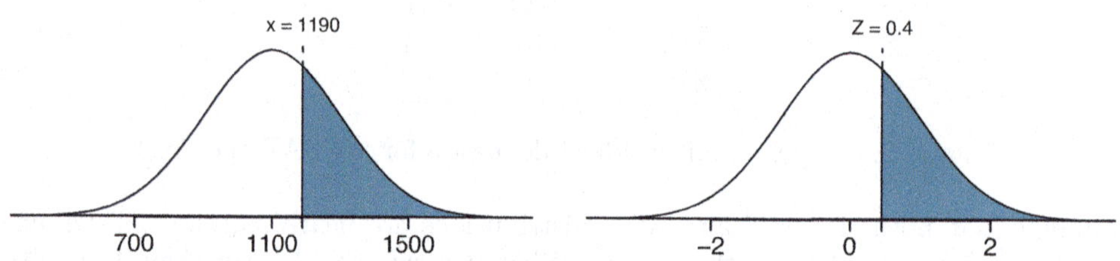

The labels on the distributions correspond to the mean and to values 2 standard deviations above and below the mean.

The number of standard units 1190 is above average on the SAT corresponds to the Z-score of 1190. With $\mu = 1100$, $\sigma = 200$, the Z-score that corresponds to $x = 1190$ is computed as

$$Z = \frac{x - \mu}{\sigma} = \frac{1190 - 1100}{200} = \frac{90}{200} = 0.45$$

To estimate the probability that a randomly selected SAT taker scores at least 1190 on the SAT, we can find the area under the standard normal curve to right of $Z = 0.45$. Using technology, we find $P(Z \geq 0.45) = 0.326$. The probability that a randomly selected score is at least 1190 on the SAT is 0.326.

ALWAYS DRAW A PICTURE FIRST, AND DO THE CALCULATIONS SECOND

For any normal probability situation, *always* draw and label the normal curve and shade the area of interest first. The picture will provide an estimate of the probability.

E

EXAMPLE 2.79

Mika earned a 1030 on her SAT. What is her percentile?

First, a picture is needed. Mika's percentile is the proportion of people who got less than or equal to 1030.

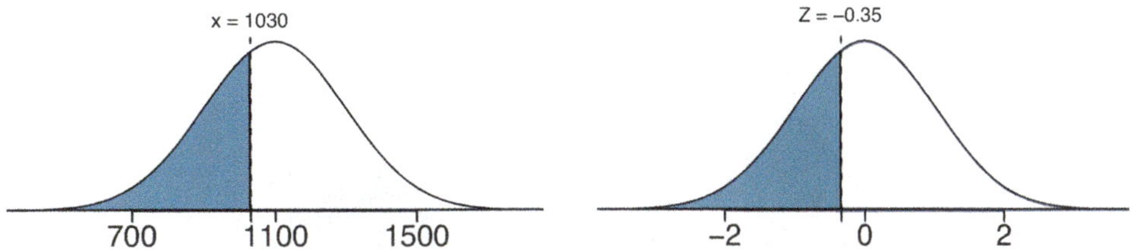

Identifying the mean $\mu = 1100$, the standard deviation $\sigma = 200$, and the cutoff for the tail area $x = 1030$ makes it easy to compute the Z-score:

$$Z = \frac{x - \mu}{\sigma} = \frac{1030 - 1100}{200} = -0.35$$

Using technology, we determine that $P(Z \le -0.35) = 0.363$. Mika is at the 36th percentile.

In Example 2.79, we standardized the value $x = 1030$ and used technology to find the area to the left of $Z = -0.35$ under the standard normal distribution ($\mu = 0$, $\sigma = 1$). Instead of this traditional method, we can also use technology to find the area to the left of 1030 under the normal distribution with $\mu = 1100$ and $\sigma = 200$. When using this approach, full communication should include the following:

- Draw a graph and shade the desired boundary region as we did in the upper left graph.
- Identify the distribution and its parameters. X is SAT score. X is Normal with $\mu = 1100$ and $\sigma = 200$. We sometimes write this as X is Normal($\mu = 1100$, $\sigma = 200$).
- Write a probability statement that corresponds to the shaded area: $P(X \le 1030) = 0.363$.
- Answer the question in context: Mika is at the 36th percentile.

G

GUIDED PRACTICE 2.80

Use the results of Example 2.79 to compute the proportion of SAT takers who did better than Mika. Also draw a new picture.[56]

The last several problems have focused on finding the probability or percentile for a particular observation. It is also possible to identify the value corresponding to a particular percentile.

[56]If about 36% of SAT takers got less than or equal to Mika, then about 64% must have done better than her.

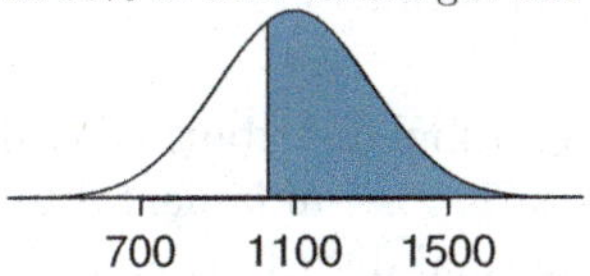

EXAMPLE 2.81

Carlos believes he can get into his preferred college if he scores in the 80th percentile or better on the SAT. How many standard deviations above the average SAT score will Carlos have to be and what score does this correspond to?

Here, we are given a percentile rather than an SAT score, so we work backwards. As always, first draw the picture. We shade the lower 80% of the area under the curve.

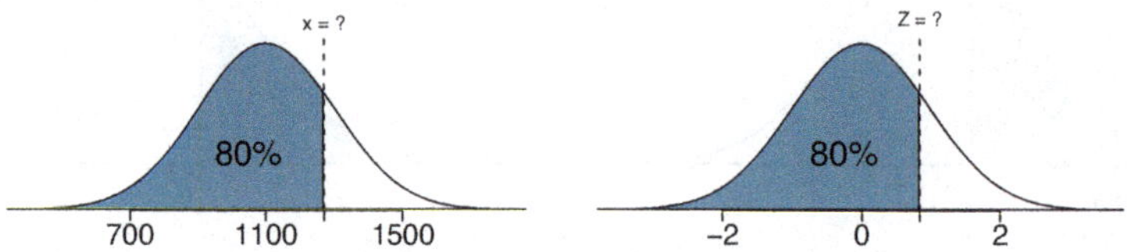

First, we find the Z-score associated with the 80th percentile. Using technology, we find that $P(Z \le 0.8416) = 0.80$. In any normal distribution, a value with a Z-score of 0.8416 will be at the 80th percentile. Once we have the Z-score, we work backwards to find the x value that corresponds to the 80th percentile using the Z-score formula.

$$\begin{aligned} Z &= \frac{x - \mu}{\sigma} \\ 0.8416 &= \frac{x - 1100}{200} \\ 0.8416 \times 200 + 1100 &= x \\ x &= 1268.32 \end{aligned}$$

Carlos will have to be 0.8416 standard deviations above the mean SAT score to be at the 80th percentile, and he will need a score of about 1268.

To find the score needed to be at the 80th percentile, we can also work with the non-standardized normal distribution with $\mu = 1100$ and $\sigma = 200$. As we did in Example 2.79, we communicate our work as follows.

- First draw the distribution, shading the desired region as we did in the upper left graph.
- Identify the distribution and its parameters. X is SAT score. X is Normal($\mu = 1100$, $\sigma = 200$).
- Write a probability statement that corresponds to the shaded area. $P(X \le 1268) = 0.80$.
- Answer the question in context. Carlos should aim for a score of 1268 to be at the 80th percentile.

GUIDED PRACTICE 2.82

Erica scored at the 20th percentile on the SAT. What was her SAT score?[57]

While it is convenient to be able to directly find areas under a normal distribution with arbitrary parameters, understanding Z-scores and the concept of standardization is fundamental to statistics. Consider the following example, where working with a standard normal distribution and Z-scores is necessary to solve the problem.

---

[57] First, draw a picture, shading the lower 20% of the normal distribution.

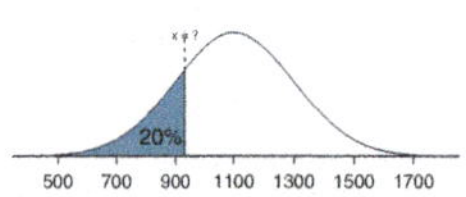

Using Normal($\mu = 0$, $\sigma = 1$), we find that the Z-score with 20% of the area to the left of it is −0.8416: $P(Z \le -0.8416) = 0.20$. Next, set up the Z-score formula and solve for x: $-0.8416 = \frac{x-1100}{200} \to x = 931.7$. Or, using X is Normal($\mu = 1100$, $\sigma = 200$), we find $P(x \le 931.7) = 0.20$. Erica scored approximately 932. Notice that the 20th percentile and 80th percentile values are the same distance away from the mean of 1100.

E

EXAMPLE 2.83

Assume that systolic blood pressure for adults is approximately normally distributed with a mean of 125 mmHg. If 10 percent of adults have systolic blood pressure above 138, what is the standard deviation of the distribution?

Here we are looking for the standard deviation, not a particular x-value. We know that the distribution is approximately normal and that the x-value of 138 corresponds to the 90th percentile, so we can find the Z-score for that x-value. Then we can use the Z-score formula to solve for the standard deviation.

Using technology and the Normal($\mu = 0$, $\sigma = 1$) distribution, we find that $P(Z \geq 1.282) = 0.10$. We then plug $Z = 1.282$ into the Z-score formula and solve for σ.

$$Z = \frac{x - \mu}{\sigma}$$

$$1.282 = \frac{138 - 125}{\sigma}$$

$$\sigma = \frac{138 - 125}{1.282}$$

$$\sigma = 10.1$$

The standard deviation of systolic blood pressure is estimated as 10.1 mmHg.

IF THE DATA ARE NOT NEARLY NORMAL, DON'T USE THE NORMAL APPROXIMATION

Before using the normal approximation method, verify that the data or distribution is approximately normal. If it is not, the normal approximation will give incorrect results. Also remember that all answers based on normal approximations are in fact approximations and are not exact.

Finally, we should observe that it is possible for a normal random variable to fall 4, 5, or even more standard deviations from the mean. The probability of being further than 4 standard deviations from the mean is about 1-in-15,000. For 5 and 6 standard deviations, it is about 1-in-2 million and 1-in-500 million, respectively. However, while the tails of the normal distribution extend infinitely in either direction, our data sets are finite and normal approximation in the extreme tails is unlikely to be very accurate, even for bell-shaped data sets.

2.6.4 Technology: normal probabilities and boundary values

Given a standard normal distribution ($\mu = 0$, $\sigma = 1$),

(i) find the following probabilities:

(a) Probability of a value between −2 and 2.

(b) Probability of a value less than −1.5.

(ii) find the following boundary values:

(a) The value that corresponds to the 40th percentile.

(b) The value $z^\star$ such that 95% of the distribution lies between $-z^\star$ and $z^\star$.

*With the standard normal distribution ($\mu = 0$, $\sigma = 1$), the values of the distribution correspond to Z-scores. If a problem involves a normal distribution with a different mean and standard deviation (e.g. battery life), we can use the mean and standard deviation of that distribution in place of 0 and 1.

Desmos: Use `normaldist(mean, stdev)`, replacing `mean` and `stdev` with appropriate values.

1. Type `normaldist(0,1)`.
2. Click the triangle next to `Cumulative Probability`.
3. Select `Inner`, `Outer`, `Left`, or `Right` as illustrated below.
4. (i) To find a probability, choose `Area` and enter the boundary value(s) for x.
   (ii) To find a boundary value, choose `Bounds` and enter the area as a decimal to the right of the = sign.
5. Click the magnifying glass to Zoom Fit the graphing window.

(i) Finding probabilities/areas.

(a) $P(-2 \le x \le 2)$.

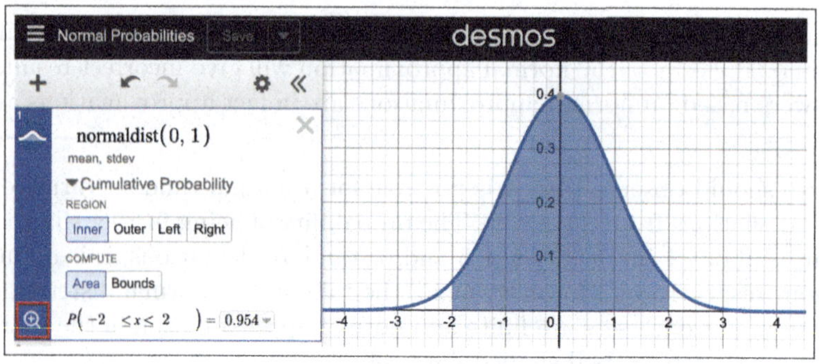

(b) $P(x < -1.5)$. Note: for a continuous distribution $P(x < -1.5) = P(x \le -1.5)$.

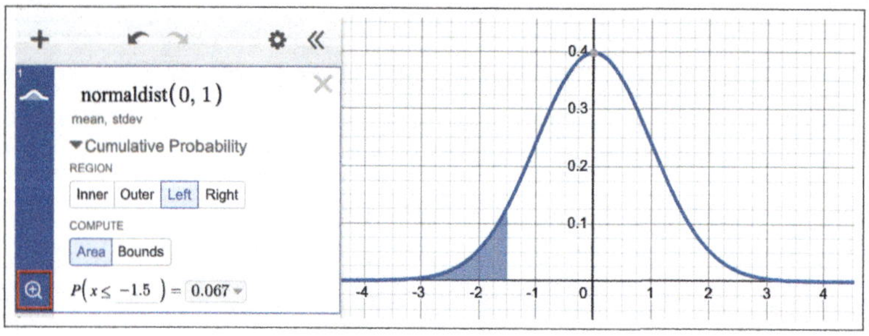

(ii) Finding boundary values.

(a) The value that corresponds to the 40th percentile. $P(x \le ?) = 0.40$.

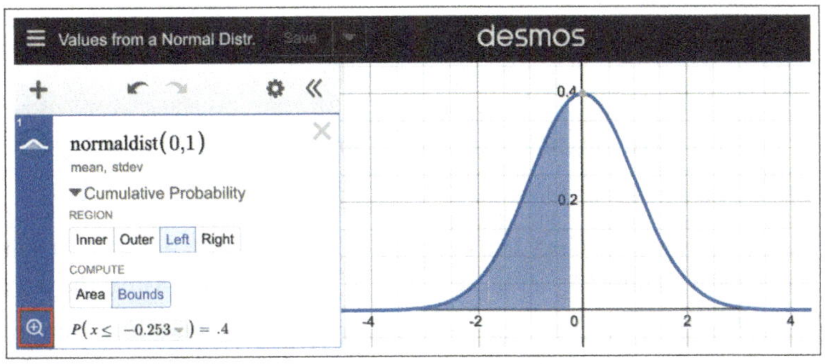

(b) The value $z^\star$ such that 95% of the distribution lies between $-z^\star$ and $z^\star$. $P(-z^\star \le x \le z^\star) = 0.95$.

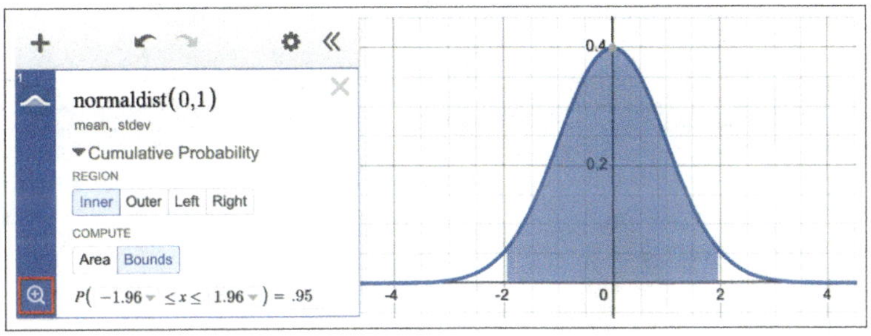

R: Normal probabilities and boundary values for the standard normal distribution.

(i) Finding probabilities/areas. Use the `pnorm(x, mean, stdev)` function, which gives the area to the *left* of the `x` value entered, unless specified otherwise. Note that "`mean =`" and "`sd =`" can be omitted, but the labels are often included for clarity.

(a) $P(-2 \le X \le 2)$.

```
> pnorm(2, 0, 1) - pnorm(-2, 0, 1)
[1] 0.9544997
```

(b) $P(X \le 1.5)$.

```
> pnorm(-1.5, mean = 0, sd = 1)
[1] 0.0668072
```

(ii) Finding boundary values. Use the `qnorm(p, mean, stdev)` function. `qnorm()` returns the value that has the entered probability `p` to the *left* of it, unless specified otherwise.

(a) Find the value that corresponds to the 40th percentile.

```
> qnorm(0.40, 0, 1)
[1] -0.2533471
```

(b) Find the value $z^\star$ such that 95% of the distribution lies between $-z^\star$ and $z^\star$. This implies that 2.5% in the upper (and lower) tail.

```
> qnorm(0.025, mean = 0, sd = 1, lower.tail = FALSE)
[1] 1.959964
```

Calculator: NumWorks calculator instructions and example output are included. For the TI-83/84 and Casio calculators, general instructions are provided, and worked example videos can be accessed via the icon or at **openintro.org/ti** and **openintro.org/casio**.

NUMWORKS: NORMAL PROBABILITIES AND BOUNDARY VALUES

Use `OK` or `EXE` to make a selection.

1. From the home screen, select `Distributions` then select `Normal`. If a list of distributions does not show up, hit the ↩ button (next to `OK`) as many times as needed.
2. Enter μ and σ then choose `Next`. Note: if the screen shows a graph, hit the ↩ button.
3. Hit the left arrow to highlight the graph. Hit the down arrow to choose whether you want left, inner, or right, then hit `OK`. Hit the right arrow to enter the desired value(s).

   (i) For a probability, enter the boundary value(s), then hit `EXE`.

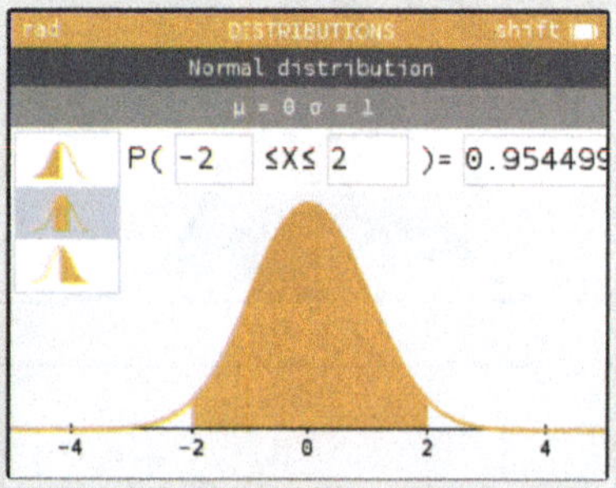

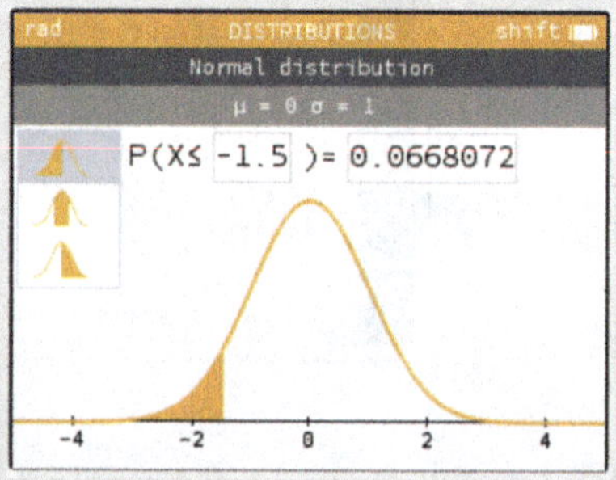

   (ii) For a boundary value, enter the desired area as a decimal to the right of the = sign, then hit `EXE`.

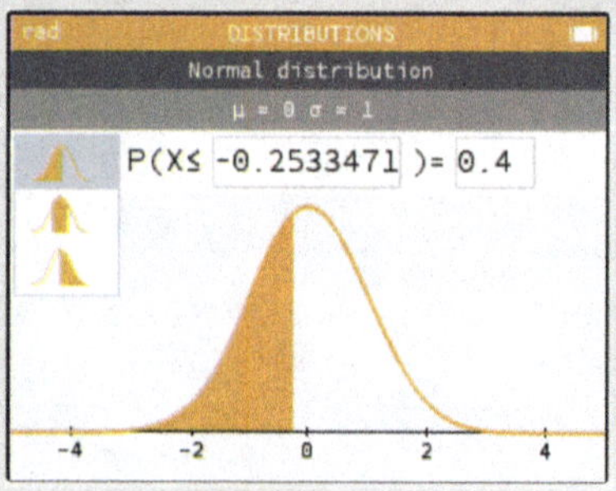

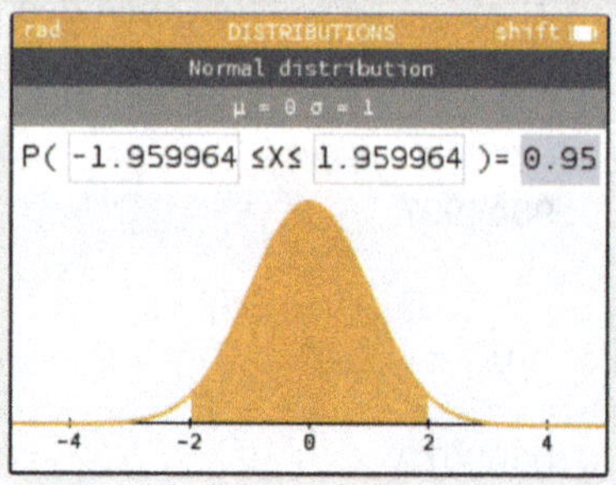

TI-84: NORMAL PROBABILITIES

1. Choose `2ND VARS` (i.e. `DISTR`).
2. Choose `2:normalcdf`.
3. Enter the `lower` (left) value and the `upper` (right) value.
   - If finding just a lower tail area, set `lower` to `-∞` (−1E99).
   - If finding just an upper tail area, set `upper` to `∞` (1E99).
4. Enter μ and σ.
5. Down arrow, choose `Paste`, and hit `ENTER`.

TI-83: Do steps 1-2, then enter the lower bound, upper bound, μ, and σ separated by commas as follows: `normalcdf(lower, upper, μ, σ)`. Then hit `ENTER`.

TI-84: BOUNDARY VALUES FOR A NORMAL DISTRIBUTION

Use `2ND VARS`, `invNorm` to find the X-value that corresponds to a given percentile.

1. Choose `2ND VARS` (i.e. `DISTR`).
2. Choose `3:invNorm`.
3. Let `Area` be the desired percent as a decimal.
4. Enter the appropriate values for μ and σ. If you want a Z-score enter μ as `0` and σ as `1`.
5. Let `Tail` be LEFT if entering a percentile. For a value with a certain percent above it, choose RIGHT. For a value with a certain percent between ± that value, choose CENTER.
6. Down arrow, choose `Paste`, and hit `ENTER`.

TI-83: Do steps 1-2, then enter the percentile as a decimal, μ, and σ separated by commas as follows: `invNorm(area to left, μ, σ)`. Then hit `ENTER`.

CASIO FX-9750GII: NORMAL PROBABILITIES

1. Navigate to `STAT` (`MENU`, then hit `2`).
2. Select `DIST` (`F5`), then `NORM` (`F1`), and then `Ncd` (`F2`).
3. If needed, set `Data` to `Variable` (`Var` option, which is `F2`).
4. Enter the `Lower` Z-score and the `Upper` Z-score. Set σ to `1` and μ to `0`.
   - If finding just a lower tail area, set `Lower` to `-5`.
   - For an upper tail area, set `Upper` to `5`.
5. Hit `EXE`, which will return the area probability (`p`) along with the Z-scores for the lower and upper bounds.

CASIO FX-9750GII: BOUNDARY VALUES FOR A NORMAL DISTRIBUTION

1. Navigate to `STAT` (`MENU`, then hit `2`).
2. Select `DIST` (`F5`), then `NORM` (`F1`), and then `InvN` (`F3`).
3. If needed, set `Data` to `Variable` (`Var` option, which is `F2`).
4. Decide which tail area to use (`Tail`), the tail area (`Area`), and then enter the σ and μ values.
5. Hit `EXE`.

2.6.5 68-95-99.7 rule

Here, we present a useful rule of thumb for the probability of falling within 1, 2, and 3 standard deviations of the mean in the normal distribution. The 68-95-99.7 rule, also known as the *empirical rule*, will be useful in a wide range of practical settings, especially when trying to make a quick estimate without technology or a Z-table.

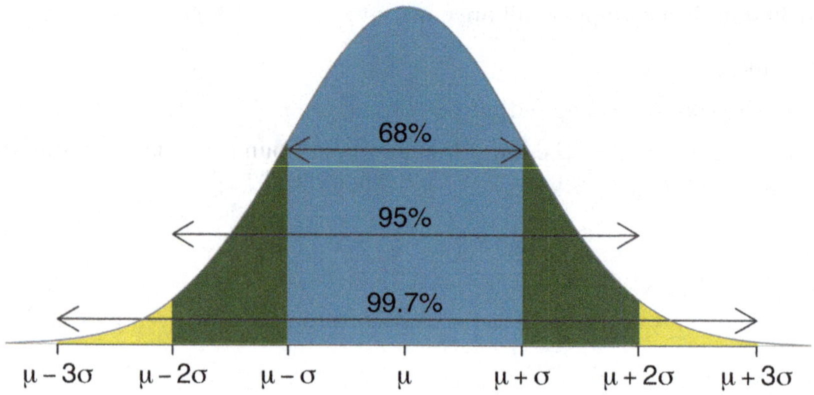

Figure 2.38: Probabilities for falling within 1, 2, and 3 standard deviations of the mean in a normal distribution.

This implies that for an approximately normal distribution, the area that falls between $Z = -1$ and $Z = 1$ is about 68% and the area between $Z = -2$ and $Z = 2$ is about 95%.

It is possible for a normal random variable to fall 4, 5, or even more standard deviations from the mean. However, these occurrences are very rare if the data are nearly normal. The probability of being further than 4 standard deviations from the mean is about 1-in-15,000. For 5 and 6 standard deviations, it is about 1-in-2 million and 1-in-500 million, respectively.

GUIDED PRACTICE 2.84

G

SAT scores closely follow the normal model with mean $\mu = 1100$ and standard deviation $\sigma = 200$. (a) About what percent of test takers score 700 to 1500? (b) About what percent score between 1100 and 1500?[58]

[58] (a) 700 and 1500 represent two standard deviations above and below the mean, which means about 95% of test takers score between 700 and 1500. (b) Because 1100 is the mean and the normal model is symmetric, half of the test takers from part (a) score 1100 to 1500. So about $\frac{95\%}{2} = 47.5\%$ score between 1100 and 1500.

Section summary

- A **Z-score** represents the number of standard deviations a value in a data set is above or below the mean. To calculate a Z-score use: $Z = \frac{x - \text{mean}}{SD}$.
- A **continuous random variable** is a variable that can take on any value within a specified domain. Every interval within the domain has a probability associated with it. The total probability or area under a continuous distribution is 1.
- A **normal distribution** can be described as a continuous, unimodal, bell-shaped, and symmetric curve. Normal distributions are the most commonly used distribution in Statistics. Many continuous random variables and some large data sets have an approximately normal distribution, but none are exactly normal.
- The normal distribution, or the normal curve, is identified by two parameters, the mean μ and the standard deviation σ. The smaller the standard deviation, the taller and more concentrated the normal curve is around its mean. The larger the standard deviation, the shorter and less concentrated the normal curve is around its mean.
- A **standard normal distribution** is a normal distribution with mean $\mu = 0$ and standard deviation $\sigma = 1$.
- For a normal distribution, approximately 68% of observations are within 1 standard deviation of the mean, approximately 95% of observations are within 2 standard deviations of the mean, and approximately 99.7% of observations are within 3 standard deviations of the mean. This is called the **empirical rule**, or the 68–95–99.7 rule. The empirical rule is useful for estimating the area of a region under an approximately normal distribution.
- If the distribution of a random variable is approximately normal, the probability that the random variable takes on values within a particular interval of the random variable is determined by the area under the normal curve within that interval. The total probability or area under the normal curve is 1.
- The boundaries of an interval associated with a given area in a normal distribution can be determined using technology or using z-scores and a standard normal table.
- Two type of normal approximation problems involve (1) finding a probability or percent, which requires finding an area under a normal distribution given one ore more boundary values and (2) finding an x-value that bounds a given area or percentile under the normal distribution. For both types of problems, first draw a normal distribution and shade the area of interest. Then, identify the distribution and its parameters, write the relevant probability statement, and answer the question in context.
- Percentiles and proportions may be used to compare relative positions of individual values within a normal distribution or between normal distributions.

Exercises

2.39 Area under the curve, Part I. What percent of a standard normal distribution $N(\mu = 0, \sigma = 1)$ is found in each region? Be sure to draw a graph.

(a) $Z < -1.35$ (b) $Z > 1.48$ (c) $-0.4 < Z < 1.5$ (d) $|Z| > 2$

2.40 Area under the curve, Part II. What percent of a standard normal distribution $N(\mu = 0, \sigma = 1)$ is found in each region? Be sure to draw a graph.

(a) $Z > -1.13$ (b) $Z < 0.18$ (c) $Z > 8$ (d) $|Z| < 0.5$

2.41 GRE scores, Part I. Sophia who took the Graduate Record Examination (GRE) scored 160 on the Verbal Reasoning section and 157 on the Quantitative Reasoning section. The mean score for Verbal Reasoning section for all test takers was 151 with a standard deviation of 7, and the mean score for the Quantitative Reasoning was 153 with a standard deviation of 7.67. Suppose that both distributions are nearly normal.

(a) What is Sophia's Z-score on the Verbal Reasoning section? On the Quantitative Reasoning section? Draw a standard normal distribution curve and mark these two Z-scores.
(b) What do these Z-scores tell you?
(c) Relative to others, which section did she do better on?
(d) Find her percentile scores for the two exams.
(e) What percent of the test takers did better than her on the Verbal Reasoning section? On the Quantitative Reasoning section?
(f) Explain why simply comparing raw scores from the two sections could lead to an incorrect conclusion as to which section a student did better on.
(g) If the distributions of the scores on these exams are not nearly normal, would your answers to parts (b) - (e) change? Explain your reasoning.

2.42 Triathlon times, Part I. In triathlons, it is common for racers to be placed into age and gender groups. Friends Leo and Mary both completed the Hermosa Beach Triathlon, where Leo competed in the *Men, Ages 30 - 34* group while Mary competed in the *Women, Ages 25 - 29* group. Leo completed the race in 1:22:28 (4948 seconds), while Mary completed the race in 1:31:53 (5513 seconds). Obviously Leo finished faster, but they are curious about how they did within their respective groups. Can you help them? Here is some information on the performance of their groups:

- The finishing times of the *Men, Ages 30 - 34* group has a mean of 4313 seconds with a standard deviation of 583 seconds.
- The finishing times of the *Women, Ages 25 - 29* group has a mean of 5261 seconds with a standard deviation of 807 seconds.
- The distributions of finishing times for both groups are approximately Normal.

Remember: a better performance corresponds to a faster finish.

(a) What are the Z-scores for Leo's and Mary's finishing times? What do these Z-scores tell you?
(b) Did Leo or Mary rank better in their respective groups? Explain your reasoning.
(c) What percent of the triathletes did Leo finish faster than in his group?
(d) What percent of the triathletes did Mary finish faster than in her group?
(e) If the distributions of finishing times are not nearly normal, would your answers to parts (a) - (d) change? Explain your reasoning.

2.43 GRE scores, Part II. In Exercise 2.41 we saw two distributions for GRE scores: $N(\mu = 151, \sigma = 7)$ for the verbal part of the exam and $N(\mu = 153, \sigma = 7.67)$ for the quantitative part. Use this information to compute each of the following:

(a) The score of a student who scored in the 80th percentile on the Quantitative Reasoning section.
(b) The score of a student who scored worse than 70% of the test takers in the Verbal Reasoning section.

2.44 Triathlon times, Part II. In Exercise 2.42 we saw two distributions for triathlon times: $N(\mu = 4313, \sigma = 583)$ for *Men, Ages 30 - 34* and $N(\mu = 5261, \sigma = 807)$ for the *Women, Ages 25 - 29* group. Times are listed in seconds. Use this information to compute each of the following:

(a) The cutoff time for the fastest 5% of athletes in the men's group, i.e. those who took the shortest 5% of time to finish.

(b) The cutoff time for the slowest 10% of athletes in the women's group.

2.45 LA weather, Part I. The average daily high temperature in June in LA is 77°F with a standard deviation of 5°F. Suppose that the temperatures in June closely follow a normal distribution.

(a) What is the probability of observing an 83°F temperature or higher in LA during a randomly chosen day in June?

(b) How cool are the coldest 10% of the days (days with lowest high temperature) during June in LA?

2.46 CAPM. The Capital Asset Pricing Model (CAPM) is a financial model that assumes returns on a portfolio are normally distributed. Suppose a portfolio has an average annual return of 14.7% (i.e. an average gain of 14.7%) with a standard deviation of 33%. A return of 0% means the value of the portfolio doesn't change, a negative return means that the portfolio loses money, and a positive return means that the portfolio gains money.

(a) What percent of years does this portfolio lose money, i.e. have a return less than 0%?

(b) What is the cutoff for the highest 15% of annual returns with this portfolio?

2.47 Scores on stats final, Part I. Below are final exam scores of 20 Introductory Statistics students.

1	2	3	4	5	6	7	8	9	10	11	12	13	14	15	16	17	18	19	20
57,	66,	69,	71,	72,	73,	74,	77,	78,	78,	79,	79,	81,	81,	82,	83,	83,	88,	89,	94

The mean score is 77.7 points. with a standard deviation of 8.44 points. Use this information to determine if the scores approximately follow the 68-95-99.7% Rule.

2.48 Heights of female college students, Part I. Below are heights of 25 female college students.

1	2	3	4	5	6	7	8	9	10	11	12	13	14	15	16	17	18	19	20	21	22	23	24	25
54,	55,	56,	56,	57,	58,	58,	59,	60,	60,	60,	61,	61,	62,	62,	63,	63,	63,	64,	65,	65,	67,	67,	69,	73

The mean height is 61.52 inches with a standard deviation of 4.58 inches. Use this information to determine if the heights approximately follow the 68-95-99.7% Rule.

2.7 Sampling distributions and the CLT

How unlikely is it to get a sample proportion a certain distance from the population proportion based on a random sample of a certain size? In an experiment, how large does an observed difference need to be for it to provide convincing evidence that one treatment is more effective than another treatment? Formally answering these question requires the tools that we will encounter in the next two chapters. However, we take a first look at them here using simulation methods.

Learning objectives

1. Explain the concept of a sampling distribution and a randomization distribution.
2. Describe sampling distributions with simulations.
3. Explain the central limit theorem and its importance.

2.7.1 Visualizing a sampling distribution through simulation

Suppose the proportion of American adults who support the expansion of solar energy is 0.88.[59]. We consider this our parameter of interest and label it p. If we were to take a poll of 1000 American adults on this topic, the estimate would not be perfect, but how close might we expect the sample proportion in the poll would be to 88%? We want to understand, how does the *sample proportion* behave when the *true population proportion is 0.88.* Let's find out! We can simulate responses we would get from a simple random sample of 1000 American adults, which is only possible because we know the actual support expanding solar energy to be 0.88. Here's how we might go about constructing such a simulation:

1. There were about 250 million American adults in 2018. On 250 million pieces of paper, write "support" on 88% of them and "not" on the other 12%.

2. Mix up the pieces of paper and pull out 1000 pieces to represent our sample of 1000 American adults.

3. Compute the fraction of the sample that say "support".

Any volunteers to conduct this simulation? Probably not. While this physical simulation is totally impractical, we can simulate it thousands, even millions, of times using computer code. We've written a short computer simulation and run it 10,000 times. The results are show in Figure 2.39 in case you are curious what the computer code looks like. We use $\hat{p}$ to represent a sample proportion. In this simulation, the sample proportion was $\hat{p}_1 = 0.894$. We know the population proportion for the simulation was $p = 0.88$, so we know the estimate had an error of $0.894 - 0.88 = +0.014$.

One simulation isn't enough to get a great sense of the distribution of estimates we might expect in the simulation, so we should run more simulations. In a second simulation, we get $\hat{p}_2 = 0.885$, which has an error of +0.005. In another, $\hat{p}_3 = 0.878$ for an error of -0.002. And in another, an estimate of $\hat{p}_4 = 0.859$ with an error of -0.021. With the help of a computer, we've run the simulation 10,000 times and created a histogram of the results from all 10,000 simulations in Figure 2.40. This graph of many, many values of the sample statistic $\hat{p}$ approximates what is called the *sampling distribution* of the statistic. The **sampling distribution** of a sample statistic is the distribution of values of the statistic for all possible random samples of a given size from a given population.

[59]We haven't actually conducted a census to measure this value perfectly. However, a very large sample has suggested the actual level of support is about 88%.

```
# 1. Create a set of 250 million entries, where 88% of them are "support"
#    and 12% are "not".
pop_size <- 250000000
possible_entries <- c(rep("support", 0.88 * pop_size), rep("not", 0.12 * pop_size))

# 2. Sample 1000 entries without replacement.
sampled_entries <- sample(possible_entries, size = 1000)

# 3. Compute p-hat: count the number that are "support", then divide by
#    the sample size.
sum(sampled_entries == "support") / 1000
```

Figure 2.39: This is code for a single $\hat{p}$ simulation using the statistical software called R. Each line that starts with # is a **code comment**, which is used to describe in regular language what the code is doing. We've provided software labs in R at openintro.org/book/statlabs for anyone interested in learning more.

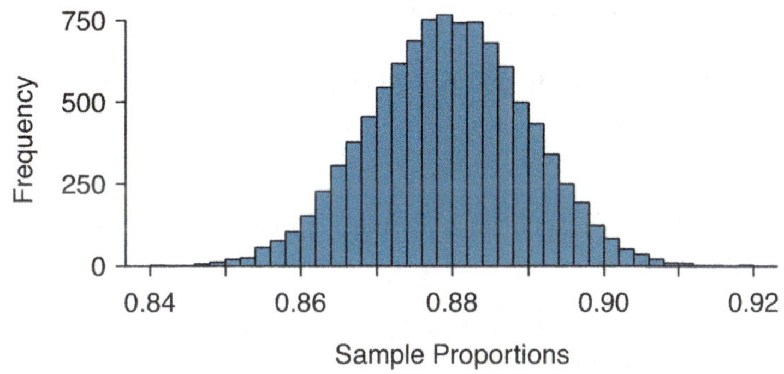

Figure 2.40: A histogram of 10,000 sample proportions sampled from a population where the population proportion is 0.88 and the sample size is $n = 1000$.

THE SAMPLING DISTRIBUTION OF A SAMPLE STATISTIC

The sampling distribution of a sample statistic is the distribution of values of the statistic for all possible random samples of a given size from a given population.

It is useful to think of sample statistics as coming from the theoretical sampling distribution.

EXAMPLE 2.85

Based on the simulated sampling distribution in Figure 2.40 above, what would be your estimate for the typical variability among sample proportions for random samples of size n=1000 from this population?

The typical variability is given by the standard deviation. Because the distribution appears approximately normal, we can use the empirical rule. About 68% of the sample proportions lie between 0.87 and 0.89 and about 95% of the sample proportions lie between 0.86 and 0.90. Therefore, we estimate the standard deviation of the sample proportions as 0.01.

GUIDED PRACTICE 2.86

Assuming the true population proportion is 0.88 as it was in our simulation, would you be surprised to get a sample proportion as large as 0.92 in a random sample of size $n = 1000$? Explain your reasoning.[60]

[60]Yes, this would be surprising, because when the true population proportion is 0.88, we almost never got a value as large as 0.92 among our simulated sample proportions.

2.7.2 Randomization distributions

We consider a study on a new malaria vaccine called PfSPZ. In this study, volunteer patients were randomized into one of two experiment groups: 14 patients received an experimental vaccine or 6 patients received a placebo vaccine. Nineteen weeks later, all 20 patients were exposed to a drug-sensitive malaria parasite strain; the motivation of using a drug-sensitive strain of parasite here is for ethical considerations, allowing any infections to be treated effectively. The results are summarized in Figure 2.41, where 9 of the 14 treatment patients remained free of signs of infection while all of the 6 patients in the control group patients showed some baseline signs of infection.

		outcome		
		infection	no infection	Total
`treatment`	vaccine	5	9	14
	placebo	6	0	6
	Total	11	9	20

Figure 2.41: Summary results for the malaria vaccine experiment.

However, the sample is very small, and it is unclear whether the difference provides *convincing evidence* that the vaccine is effective.

EXAMPLE 2.87

How do the proportions that developed an infection compare between those in the vaccine group and those in the placebo group?

In this study, a smaller proportion of patients who received the vaccine showed signs of an infection: $\frac{5}{14}$ = 35.7% for those that received the vaccine versus $\frac{6}{6}$= 100% for those that received the placebo.

The observed infection rates (35.7% for the treatment group versus 100% for the placebo group) suggest the vaccine may be effective at preventing infection. However, we cannot be sure if the observed difference represents the vaccine's efficacy or is just from random chance. Generally there is a little bit of fluctuation in sample data, and we wouldn't expect the sample proportions to be *exactly* equal, even if the truth was that the vaccine did not work. Additionally, with such small samples, perhaps it's common to observe such large differences when we randomly split a group due to chance alone!

How much variation would we expect between the infection rates of the two groups *if the vaccine had no effect*? To answer this, we implement a simulation, where we will pretend we know that the malaria vaccine being tested does *not* work. Ultimately, we want to understand if the large difference we observed is common or uncommon in these simulations. If it is common, then maybe the difference we observed was purely due to chance. If it is very uncommon, then the claim that the vaccine was helpful seems more plausible.

Figure 2.41 shows that in this overall study, 11 patients developed infections and 9 did not. For our simulation, we will suppose the vaccine does not work and that we are able to *rewind* back to when the researchers randomized the patients in the study. If we happened to randomize the patients differently, we may get a different result in this hypothetical world where the vaccine doesn't influence the infection. Let's perform another *randomization* to treatment groups using a simulation.

To simulate this scenario, we take 20 notecards to represent the 20 patients, where we write down "infection" on 11 cards and "no infection" on 9 cards. In this hypothetical world, we believe each patient that got an infection was going to get it regardless of which group they were in, so let's see what happens if we randomly assign the patients to the treatment and control groups again. We thoroughly shuffle the notecards and deal 14 into a `vaccine` pile and 6 into a `placebo` pile. Finally, we tabulate the results, which are shown in Figure 2.42.

		outcome		
		infection	no infection	Total
treatment	vaccine	7	7	14
(simulated)	placebo	4	2	6
	Total	11	9	20

Figure 2.42: Simulation results, assuming the vaccine does not work and that any difference in infection rates is due to chance.

GUIDED PRACTICE 2.88

What is the difference in infection rates between the two simulated groups in Figure 2.42? How does this compare to the observed 64.3% difference (35.7% – 100%) in the actual data?[61]

We computed one possible difference under the assumption that the vaccine does not work in Guided Practice 2.88, which represents one difference due to chance. While in this first simulation, we physically dealt out notecards to represent the patients, it is more efficient to perform this simulation using a computer. Repeating the simulation on a computer, we get another difference due to chance:

$$\frac{2}{6} - \frac{9}{14} = -0.310$$

And another:

$$\frac{3}{6} - \frac{8}{14} = -0.071$$

And so on until we repeat the simulation enough times that we have a good idea of what represents the *distribution of differences from chance alone*. Figure 2.43 shows a stacked dot plot of the differences found from 100 simulations, where each dot represents a simulated difference between the infection rates (control rate minus treatment rate).

This distribution represents a randomization distribution, which is analogous to a sampling distribution in the context of experiments. A **randomization distribution** of a statistic, for example the difference in infection rates between two treatment groups, is the distribution of a statistic generated by a simulation that repeatedly reassigns response values to treatment groups and recalculates the resulting statistic. The resulting distribution of the statistic approximates the sampling distribution of the statistic and provides a range of expected values assuming no difference in the treatments.

The distribution of these simulated differences is centered around 0. We simulated these differences assuming that the vaccine did not work, and under this condition, we expect the difference to be near zero with random fluctuation. The fluctuation is pretty large here because the treatment group sizes are so small in this study.

EXAMPLE 2.89

Given the results of the simulation shown in Figure 2.43, about how often would you expect to observe a result as large as 64.3% if the vaccine does not work?

Because a result this large happened 2 times out the 100 simulations, we would expect such a large value only 2% of the time if the vaccine does not work.

Based on our simulation, we might be led to believe that the vaccine works, because assuming that it doesn't work, there is only about a 2% chance of getting a difference as big as we got in our study. In the next chapter, we will see how to calculate this conditional probability, known as the p-value, using a normal model when certain conditions are met.

[61] $4/6 - 7/14 = 0.167$ or about 16.7% in favor of the vaccine. This difference due to chance is much smaller than the difference observed in the actual groups.

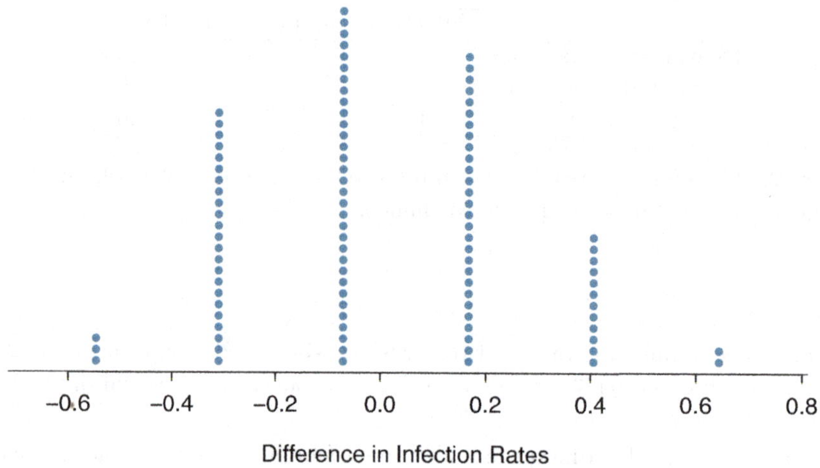

Figure 2.43: A stacked dot plot of differences from 100 simulated values produced under the assumption that the vaccine does not work.

2.7.3 The Central Limit Theorem

We can characterize the sampling distribution in Figure 2.40 as symmetric and bell-shaped, and it *resembles a normal distribution.* The randomization distribution shown in Figure 2.43 also looks somewhat normal in shape. This is not an accident; it is the result of a general principle called the **Central Limit Theorem.** The Central Limit Theorem tells us that the sampling distribution of certain statistics, specifically the sample proportion and sample mean, become more normal in shape as the sample size increases.

CENTRAL LIMIT THEOREM (CLT)

For any population with a fixed mean and standard deviation, the shape of the sampling distribution of the proportion and of the mean of a random sample becomes more normal as the sample size n gets larger.

The Central Limit Theorem is fundamental to statistics and will form the foundation for working with sampling distributions and inference procedures encountered in the next two chapters. In any sample or experiment, we would like to know if the result we observe is significant or if the result is within the realm of expected variation. Sampling distributions and randomization distributions help us visualize and understand the likelihood of getting a result as extreme as we got, assuming the distribution is centered on a certain value. The Central Limit Theorem will enable us to calculate this likelihood using a normal model.

Section summary

- A **sampling distribution** of a statistic is the distribution of values of the statistic for all possible random samples of a given size from a given population.
- The **sampling distribution** of a statistic can be simulated by repeatedly generating a large number of random samples from the population assuming known value(s) for the parameter(s). The value of the statistic is determined and recorded for each sample. The resulting distribution of the sample statistic values approximates the sampling distribution of the statistic.
- A **randomization distribution** is the distribution of a statistic generated by simulation from repeatedly randomly reallocating, or reassigning, the response values to treatment groups. The value of the statistic is determined and recorded for each reallocation, or reassignment. The resulting distribution of the statistic values approximates the sampling distribution of the statistic.
- The **Central Limit Theorem** (CLT) states that the sampling distribution of a sample proportion and a sample mean become more normal in shape as the sample size increases.

Exercises

2.49 Side effects of Avandia. Rosiglitazone is the active ingredient in the controversial type 2 diabetes medicine Avandia and has been linked to an increased risk of serious cardiovascular problems such as stroke, heart failure, and death. A common alternative treatment is pioglitazone, the active ingredient in a diabetes medicine called Actos. In a nationwide retrospective observational study of 227,571 Medicare beneficiaries aged 65 years or older, it was found that 2,593 of the 67,593 patients using rosiglitazone and 5,386 of the 159,978 using pioglitazone had serious cardiovascular problems. These data are summarized in the contingency table below.[62]

		Cardiovascular problems		
		Yes	No	Total
Treatment	Rosiglitazone	2,593	65,000	67,593
	Pioglitazone	5,386	154,592	159,978
	Total	7,979	219,592	227,571

(a) Determine if each of the following statements is true or false. If false, explain why. *Be careful:* The reasoning may be wrong even if the statement's conclusion is correct. In such cases, the statement should be considered false.

i. Since more patients on pioglitazone had cardiovascular problems (5,386 vs. 2,593), we can conclude that the rate of cardiovascular problems for those on a pioglitazone treatment is higher.

ii. The data suggest that diabetic patients who are taking rosiglitazone are more likely to have cardiovascular problems since the rate of incidence was (2,593 / 67,593 = 0.038) 3.8% for patients on this treatment, while it was only (5,386 / 159,978 = 0.034) 3.4% for patients on pioglitazone.

iii. The fact that the rate of incidence is higher for the rosiglitazone group proves that rosiglitazone causes serious cardiovascular problems.

iv. Based on the information provided so far, we cannot tell if the difference between the rates of incidences is due to a relationship between the two variables or due to chance.

(b) What proportion of all patients had cardiovascular problems?

(c) If the type of treatment and having cardiovascular problems were independent, about how many patients in the rosiglitazone group would we expect to have had cardiovascular problems?

(d) We can investigate the relationship between outcome and treatment in this study using a randomization technique. While in reality we would carry out the simulations required for randomization using statistical software, suppose we actually simulate using index cards. In order to simulate from the independence model, which states that the outcomes were independent of the treatment, we write whether or not each patient had a cardiovascular problem on cards, shuffled all the cards together, then deal them into two groups of size 67,593 and 159,978. We repeat this simulation 1,000 times and each time record the number of people in the rosiglitazone group who had cardiovascular problems. Use the relative frequency histogram of these counts to answer (i)-(iii).

i. What are the claims being tested?

ii. Compared to the number calculated in part (c), which would provide more support for the alternative hypothesis, *more* or *fewer* patients with cardiovascular problems in the rosiglitazone group?

iii. What do the simulation results suggest about the relationship between taking rosiglitazone and having cardiovascular problems in diabetic patients?

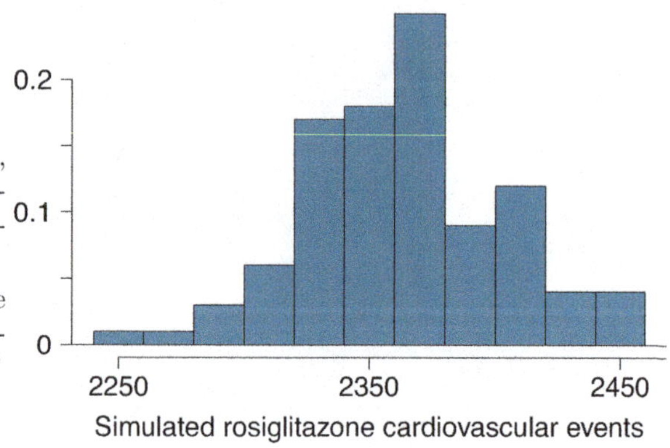

[62]D.J. Graham et al. "Risk of acute myocardial infarction, stroke, heart failure, and death in elderly Medicare patients treated with rosiglitazone or pioglitazone". In: *JAMA* 304.4 (2010), p. 411. ISSN: 0098-7484.

2.50 Heart transplants. The Stanford University Heart Transplant Study was conducted to determine whether an experimental heart transplant program increased lifespan. Each patient entering the program was designated an official heart transplant candidate, meaning that he was gravely ill and would most likely benefit from a new heart. Some patients got a transplant and some did not. The variable `transplant` indicates which group the patients were in; patients in the treatment group got a transplant and those in the control group did not. Of the 34 patients in the control group, 30 died. Of the 69 people in the treatment group, 45 died. Another variable called `survived` was used to indicate whether or not the patient was alive at the end of the study.[63]

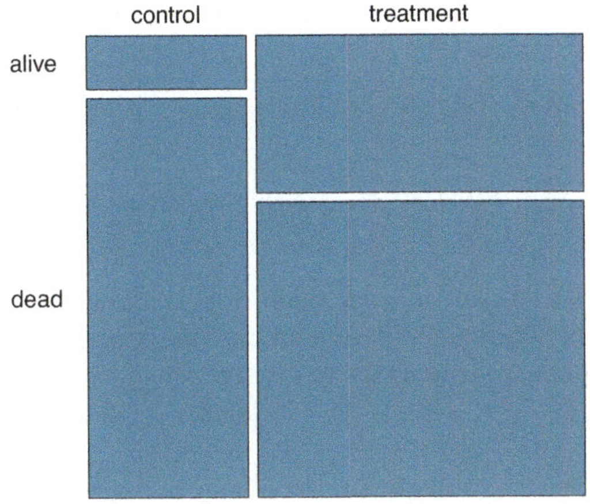

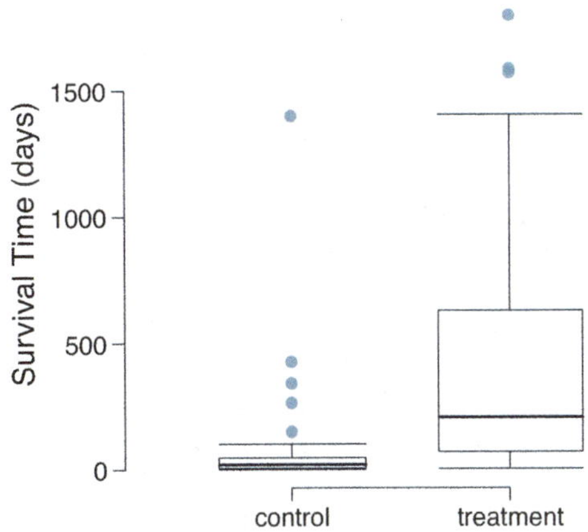

(a) Based on the mosaic plot, is survival independent of whether or not the patient got a transplant? Explain your reasoning.

(b) What do the box plots below suggest about the efficacy (effectiveness) of the heart transplant treatment.

(c) What proportion of patients in the treatment group and what proportion of patients in the control group died?

(d) One approach for investigating whether or not the treatment is effective is to use a randomization technique.

   i. What are the claims being tested?

   ii. The paragraph below describes the set up for such approach, if we were to do it without using statistical software. Fill in the blanks with a number or phrase, whichever is appropriate.

   We write *alive* on ____________ cards representing patients who were alive at the end of the study, and *dead* on ____________ cards representing patients who were not. Then, we shuffle these cards and split them into two groups: one group of size ____________ representing treatment, and another group of size ____________ representing control. We calculate the difference between the proportion of *dead* cards in the treatment and control groups (treatment - control) and record this value. We repeat this 100 times to build a distribution centered at ____________. Lastly, we calculate the fraction of simulations where the simulated differences in proportions are ____________. If this fraction is low, we conclude that it is unlikely to have observed such an outcome by chance and that the null hypothesis should be rejected in favor of the alternative.

   iii. What do the simulation results shown below suggest about the effectiveness of the transplant program?

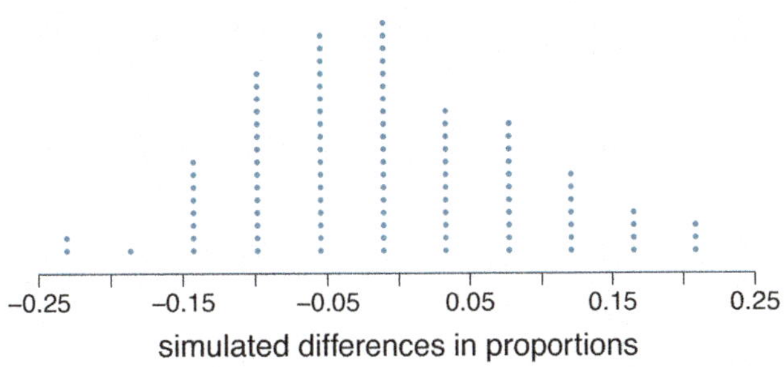

[63]B. Turnbull et al. "Survivorship of Heart Transplant Data". In: *Journal of the American Statistical Association* 69 (1974), pp. 74–80.

Chapter highlights

This chapter focused on understanding likelihood and chance variation, first by solving individual probability questions and then by investigating probability distributions.

The main probability techniques covered in this chapter are as follows:

- The **Conditional Probability Rule**.
- The **General Multiplication Rule** for **and** (intersection) probabilities, along with the special case when events are **independent**.
- The **General Addition Rule** for **or** (union) probabilities, along with the special case when events are **mutually exclusive**.

Fundamental to all of these problems is understanding when events are independent and when they are mutually exclusive. Two events are **independent** when the outcome of one does not affect the outcome of the other, i.e. $P(A \mid B) = P(A)$. Two events are **mutually exclusive** when they cannot both happen together, i.e. $P(A \cap B) = 0$.

Moving from solving individual probability questions to studying probability distributions helps us better understand chance processes and quantify expected chance variation.

- For a **discrete probability distribution**, the **sum** of the probabilities must equal 1.
- As with any distribution, one can calculate the mean and standard deviation of a probability distribution. In the context of a probability distribution, the **mean** and **standard deviation** describe the average and the typical deviation of values from the average, respectively, in the long run, or after many, many repetitions of the chance process.
- A probability distribution can be summarized by its **center** (mean, median), **spread** (SD, IQR), and **shape** (right skewed, left skewed, approximately symmetric).
- The **binomial distribution** is a discrete probability distribution which provides a model for the number of successes in n independent trials.
- The **normal distribution** is a continuous probability distribution which can be used to model various empirical and probability distributions.
- The **Central Limit Theorem** is a powerful theorem thats tells us that as the sample size n increases, the sampling distribution of a sample proportion and a sample mean become more normal in shape. For large enough n, we can model sampling distributions using a normal distribution.

The study of probability is useful for measuring uncertainty and assessing risk. In addition, probability serves as the foundation for inference, providing a framework for evaluating when an outcome falls outside of the range of what would be expected by chance alone.

Chapter exercises

2.51 Grade distributions. Each row in the table below is a proposed grade distribution for a class. Identify each as a valid or invalid probability distribution, and explain your reasoning.

	Grades				
	A	B	C	D	F
(a)	0.3	0.3	0.3	0.2	0.1
(b)	0	0	1	0	0
(c)	0.3	0.3	0.3	0	0
(d)	0.3	0.5	0.2	0.1	-0.1
(e)	0.2	0.4	0.2	0.1	0.1
(f)	0	-0.1	1.1	0	0

2.52 Health coverage, frequencies. The Behavioral Risk Factor Surveillance System (BRFSS) is an annual telephone survey designed to identify risk factors in the adult population and report emerging health trends. The following table summarizes two variables for the respondents: health status and health coverage, which describes whether each respondent had health insurance.[64]

		Health Status					
		Excellent	Very good	Good	Fair	Poor	Total
Health	No	459	727	854	385	99	2,524
Coverage	Yes	4,198	6,245	4,821	1,634	578	17,476
	Total	4,657	6,972	5,675	2,019	677	20,000

(a) If we draw one individual at random, what is the probability that the respondent has excellent health and doesn't have health coverage?

(b) If we draw one individual at random, what is the probability that the respondent has excellent health or doesn't have health coverage?

2.53 HIV in Eswatini. Eswatini (formerly named in English as Swaziland) has the highest HIV prevalence in the world: 25.9% of this country's population is infected with HIV.[65] The ELISA test is one of the first and most accurate tests for HIV. For those who carry HIV, the ELISA test is 99.7% accurate. For those who do not carry HIV, the test is 92.6% accurate. If an individual from Eswatini has tested positive, what is the probability that he carries HIV?

2.54 Twins. About 30% of human twins are identical, and the rest are fraternal. Identical twins are necessarily the same sex – half are males and the other half are females. One-quarter of fraternal twins are both male, one-quarter both female, and one-half are mixes: one male, one female. You have just become a parent of twins and are told they are both girls. Given this information, what is the probability that they are identical?

2.55 Disjoint vs. independent. In parts (a) and (b), identify whether the events are disjoint, independent, or neither (events cannot be both disjoint and independent).

(a) You and a randomly selected student from your class both earn A's in this course.

(b) You and your class study partner both earn A's in this course.

(c) If two events can occur at the same time, must they be dependent?

2.56 Guessing on an exam. In a multiple choice exam, there are 5 questions and 4 choices for each question (a, b, c, d). Nancy has not studied for the exam at all and decides to randomly guess the answers. What is the probability that:

(a) the first question she gets right is the 5th question?

(b) she gets all of the questions right?

(c) she gets at least one question right?

[64] Office of Surveillance, Epidemiology, and Laboratory Services Behavioral Risk Factor Surveillance System, BRFSS 2010 Survey Data.

[65] Source: CIA Factbook, Country Comparison: HIV/AIDS - Adult Prevalence Rate.

2.57 Educational attainment. The `family_college` data set contains a sample of 792 cases with two variables, `teen` and `parents`, and is summarized below.[66] The `teen` variable is either `college` or `not`, where the `college` label means the teen went to college immediately after high school. The `parents` variable takes the value `degree` if at least one parent of the teenager completed a college degree.

		parents		
		degree	not	Total
teen	college	231	214	445
	not	49	298	347
	Total	280	512	792

Table 2.44: Contingency table summarizing the `family_college` data set.

(a) For a randomly selected case, what is the probability that a parent completed a college degree?

(b) For a randomly selected case, what is the probability that the teen went to college immediately after high school?

(c) For a randomly selected case, what is the probability that a parent completed a college degree *and* teen went to college immediately after high school?

(d) Is P(a parent completed college degree *and* teen went to college immediately after high school) = P(parent completed college degree) x P(teen went to college immediately after high school)? Explain why this is or is not the case.

2.58 Speeding on the I-5, Part I. The distribution of passenger vehicle speeds traveling on the Interstate 5 Freeway (I-5) in California is nearly normal with a mean of 72.6 miles/hour and a standard deviation of 4.78 miles/hour.[67]

(a) What percent of passenger vehicles travel slower than 80 miles/hour?

(b) What percent of passenger vehicles travel between 60 and 80 miles/hour?

(c) How fast do the fastest 5% of passenger vehicles travel?

(d) The speed limit on this stretch of the I-5 is 70 miles/hour. Approximate what percentage of the passenger vehicles travel above the speed limit on this stretch of the I-5.

2.59 College smokers. At a university, 13% of students smoke.

(a) Calculate the expected number of smokers in a random sample of 100 students from this university.

(b) The university gym opens at 9 am on Saturday mornings. One Saturday morning at 8:55 am there are 27 students outside the gym waiting for it to open. Should you use the same approach from part (a) to calculate the expected number of smokers among these 27 students?

2.60 Roulette winnings. In the game of roulette, a wheel is spun and you place bets on where it will stop. One popular bet is that it will stop on a red slot; such a bet has an 18/38 chance of winning. If it stops on red, you double the money you bet. If not, you lose the money you bet. Suppose you play 3 times, each time with a $1 bet. Let Y represent the total amount won or lost. Write a probability model for Y.

2.61 Multiple choice quiz. In a multiple choice quiz there are 5 questions and 4 choices for each question (a, b, c, d). Robin has not studied for the quiz at all, and decides to randomly guess the answers. What is the probability that

(a) the first question she gets right is the 3rd question?

(b) she gets exactly 3 or exactly 4 questions right?

(c) she gets the majority of the questions right?

[66] A simulated data set based on real population summaries at nces.ed.gov/pubs2001/2001126.pdf.

[67] S. Johnson and D. Murray. "Empirical Analysis of Truck and Automobile Speeds on Rural Interstates: Impact of Posted Speed Limits". In: *Transportation Research Board 89th Annual Meeting*. 2010.

2.62 Birth weight. In a large study of birth weight of newborns, the weights of 23,419 newborn boys were recorded. The distribution of weights was approximately normal with a mean of 7.44 lbs (3376 grams) and a standard deviation of 1.33 lbs (603 grams). The government classifies a newborn as having low birth weight if the weight is less than 5.5 pounds.[68]

(a) What percent of these newborns had a low birth weight?

(b) Approximately what percent of these babies weighed greater than 10 pounds?

(c) Approximately *how many* of these newborns weighed greater than 10 pounds?

(d) How much would a newborn have to weigh in order to be at the 90th percentile among this group?

2.63 Auto insurance premiums. Suppose a newspaper article states that the distribution of auto insurance premiums for residents of California is approximately normal with a mean of $1,650. The article also states that 25% of California residents pay more than $1,800.

(a) What is the Z-score that corresponds to the top 25% (or the 75th percentile) of the standard normal distribution?

(b) What is the mean insurance cost? What is the cutoff for the 75th percentile?

(c) Identify the standard deviation of insurance premiums in California.

2.64 Heights of 10 year olds, Part I. Heights of 10 year olds, regardless of gender, closely follow a normal distribution with mean 55 inches and standard deviation 6 inches.

(a) What is the probability that a randomly chosen 10 year old is shorter than 48 inches?

(b) What is the probability that a randomly chosen 10 year old is between 60 and 65 inches?

(c) If the tallest 10% of the class is considered "very tall", what is the height cutoff for "very tall"?

2.65 Student outfits. In a classroom with 24 students, 7 students are wearing jeans, 4 are wearing shorts, 8 are wearing skirts, and the rest are wearing leggings. If we randomly select 3 students without replacement, what is the probability that one of the selected students is wearing leggings and the other two are wearing jeans? Note that these are mutually exclusive clothing options.

2.66 Find the SD. Cholesterol levels for women aged 20 to 34 follow an approximately normal distribution with mean 185 milligrams per deciliter (mg/dl). Women with cholesterol levels above 220 mg/dl are considered to have high cholesterol and about 18.5% of women fall into this category. What is the standard deviation of the distribution of cholesterol levels for women aged 20 to 34?

[68] www.biomedcentral.com/1471-2393/8/5

Chapter 3

Inference for categorical data: proportions

Statistical inference is primarily concerned with understanding and quantifying the uncertainty of estimates of parameters. While the equations and details change depending on the setting, the foundations for inference are the same throughout all of statistics.

We start with a familiar topic: the idea of using a sample proportion to estimate a population proportion. Next, we create what's called a *confidence interval*, which is a range of plausible values for the true population value. Then, we introduce a *hypothesis testing framework*, which allows us to use data to formally evaluate claims about the population, such as whether a survey provides strong evidence against a claim that a candidate has the support of a majority of the voting population.

After developing inference techniques for a population proportion, we apply these same techniques to analyze the difference of two population proportions. Lastly, we develop a hypothesis test for categorical variables arranged in two-way tables; while we will use a different distribution in this context, the core ideas of hypothesis testing remain the same.

For videos, slides, and other resources, please visit
www.openintro.org/ahss

3.1 Point estimators

Companies such as Gallup and Pew Research frequently conduct polls as a way to understand the state of public opinion or knowledge on many topics, including politics, scientific understanding, brand recognition, and more. What quantities can these polls use to estimate the opinion or knowledge of the broader population? What does it mean for their estimate to be biased, and how can they attempt to minimize such bias?

Learning objectives

1. Describe the purpose and use of a point estimator.
2. Justify why an estimator is or is not unbiased.
3. Calculate estimates for a population parameter.
4. Recognize that point estimators have variability.

3.1.1 Introducing point estimators

Suppose we want to estimate the approval rating of the governor of a particular state among adult residents of that state. We generally estimate unknown quantities such as this by collecting a sample. Let's say that in a sample of 500 adult residents of that state, 225 of the sampled individuals report that they approve of the governor. We calculate the sample proportion as $\frac{225}{500} = 0.45$ and we consider 45% to be a point estimate of the approval rating we might see if we collected responses from adults in the entire state.

This entire-population response proportion is generally referred to as the **parameter** of interest, and when the parameter is a proportion, we denote it with the letter p. An estimate calculated from a sample is called a sample **statistic**. The sample proportion, denoted $\hat{p}$, is an example of a statistic. Unless we collect responses from every individual in the population, the parameter remains unknown, and we use the sample statistic as our **point estimator**, or just **estimator**, for the parameter. For example, the sample proportion $\hat{p}$ is a point estimator for the population proportion p.

The strategy of using a sample statistic to estimate a parameter is quite common, and it's a strategy that we can apply to other statistics besides a proportion. For instance, if we want to estimate the average difference in product prices for two websites, we might take a random sample of products available on both sites, check the prices on each, and then compute the average of the differences in price; this strategy certainly wouldn't give us a perfect measurement of the average difference in price for all the products, but it would give us a point estimate.

EXAMPLE 3.1

Consider the summary statistics for the number of characters in 50 randomly selected emails. These values are summarized below.

$\bar{x}$	11,160
median	6,890
s_x	13,130

E

What quantity should we use as a point estimator for the **population mean** μ. What is the point estimate for the population mean based on this sample?

We use the sample mean $\bar{x}$ as our point estimator for the population mean μ. Based on this sample, the point estimate for μ is $\bar{x}$ =11,160.

GUIDED PRACTICE 3.2

Ⓖ Using the email data, what quantity should we use as a point estimator for the population standard deviation σ? What is the point estimate for σ based on this sample?[1]

3.1.2 Biased and unbiased estimators

When estimating a parameter such as the approval rating of a governor, we aim to use an *unbiased* estimator. **Bias** describes a systematic tendency to overestimate or underestimate the true population parameter. For instance, if we took a political poll but our sample didn't include a roughly representative distribution of the political parties, the sample would likely skew in a particular direction and be biased. Taking a truly random sample from the entire population of interest avoids sampling bias. However, as we saw in Chapter 1, even with a random sample, nonresponse bias and various types of response bias can still be present.

EXAMPLE 3.3

Ⓔ Consider a student poll designed to estimate support for a new college stadium. Say that a random sample of students is taken and asked the question: *Do you support your school by supporting funding for the new stadium?* Do expect to get a biased or an unbiased estimate? If biased, do you expect to overestimate or underestimate the true proportion of students that support a new college stadium?

Because the wording of the questions is leading, we would expect a biased estimate. In this case our estimate would likely *overestimate* the true parameter, as the wording of the question invites a positive response.

When estimating a population parameter, an estimator is **unbiased** if, on average, the value of the estimator does not underestimate or overestimate the population parameter. When an estimator is unbiased, the distribution of values of the estimator for all random samples of a given size, known as the sampling distribution, will be *centered* on the true value.

EXAMPLE 3.4

Ⓔ Figure 3.1 shows the sampling distributions associated with four different estimators of a parameter, whose value is indicated. Which of the estimators is biased? Which of the estimators is unbiased? Also, order the estimators from lowest variability to highest variability).

The estimators in (a) and (c) are biased, because those sampling distributions are not centered on the parameter (true value). The estimators in (b) and (d) are centered on the parameter, so those estimators are unbiased. From lowest to greatest variability, we have (a), (b), (c), (d).

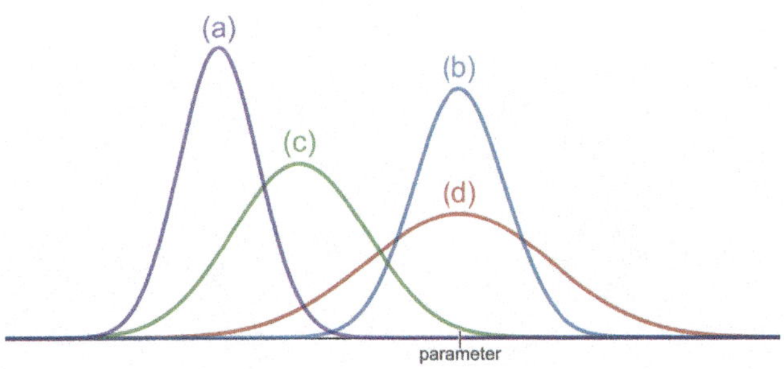

Figure 3.1: The sampling distribution of four different estimators of a parameter. This is a screenshot from the interactive Desmos Activity: Bias and Variability in Sampling Distributions. Find it at openintro.org/ahss/desmos.

---

[1] Intuitively we would use the sample standard deviation s as a point estimator for σ. Based on this sample, the point estimate for σ is s =13,130.

Section summary

- In this section we laid the groundwork for our study of **inference**. Inference involves using sample statistics as point estimators to estimate or justify claims about unknown population parameters.

- A sample statistic is a point estimator of the corresponding population parameter and can be thought of as an estimate of the population parameter. For example, the sample mean $\bar{x}$ is a point estimator for an unknown population mean μ, and the sample proportion $\hat{p}$ is a point estimator for an unknown population proportion p.

- It is helpful to imagine point estimates as being drawn from the sampling distribution of the statistic or point estimator.

- A **point estimator**, or just estimator, is **unbiased** if, on average, the values of the estimator do not underestimate or overestimate the population parameter. For an unbiased estimator, the sampling distribution of the estimator (i.e., the distribution of values of the estimator for all random samples of the same size from the same population) is *centered* on the population parameter.

Exercises

3.1 Identify the parameter, Part I. For each of the following situations, state whether the parameter of interest is a mean or a proportion. It may be helpful to examine whether individual responses are numerical or categorical.

(a) In a survey, one hundred college students are asked how many hours per week they spend on the Internet.

(b) In a survey, one hundred college students are asked: "What percentage of the time you spend on the Internet is part of your course work?"

(c) In a survey, one hundred college students are asked whether or not they cited information from Wikipedia in their papers.

(d) In a survey, one hundred college students are asked what percentage of their total weekly spending is on alcoholic beverages.

(e) In a sample of one hundred recent college graduates, it is found that 85 percent expect to get a job within one year of their graduation date.

3.2 Identify the parameter, Part II. For each of the following situations, state whether the parameter of interest is a mean or a proportion.

(a) A poll shows that 64% of Americans personally worry a great deal about federal spending and the budget deficit.

(b) A survey reports that local TV news has shown a 17% increase in revenue within a two year period while newspaper revenues decreased by 6.4% during this time period.

(c) In a survey, high school and college students are asked whether or not they use geolocation services on their smart phones.

(d) In a survey, smart phone users are asked whether or not they use a web-based taxi service.

(e) In a survey, smart phone users are asked how many times they used a web-based taxi service over the last year.

3.3 Quality control. As part of a quality control process for computer chips, an engineer at a factory samples 212 chips during a week of production to test the current rate of chips with severe defects. She finds that 27 of the chips are defective.

(a) What is the population of interest for the data set?

(b) What parameter is being estimated?

(c) What is the value of the point estimator (sample statistic) for the parameter?

(d) Briefly describe a method that would lead to a biased estimator. Also, say whether you think the estimator will underestimate or overestimate the parameter.

(e) How could you get an unbiased estimator of the parameter?

3.4 Unexpected expense. In a sample of 765 adults in the United States, 322 say they could not cover a $400 unexpected expense without borrowing money or going into debt.

(a) What population of interest for the data set?

(b) What parameter is being estimated?

(c) What is the value of the point estimator (sample statistic) for the parameter?

(d) Briefly describe a method that would lead to a biased estimator. Also, say whether you think the estimator will underestimate or overestimate the parameter.

(e) How could you get an unbiased estimator of the parameter?

3.2 Sampling distribution of a sample proportion

Given a fair coin, what is the probability that in 200 tosses you would get greater than 52% tails just by random variation? In a particular state, 48% support a controversial measure. When estimating the percent through polling, what is the probability that a random sample of size 200 will mistakenly estimate the percent support to be greater than 50%? In general, how far do we expect our sample proportion to be from the population proportion? In this section, we consider the sampling distribution of the sample proportion in order to answer questions such as these.

Learning objectives

1. Interpret and apply the concept of a sampling distribution in the context of a sample proportion.
2. Calculate the mean and standard deviation of a sampling distribution of a sample proportion.
3. Justify whether conditions for independence are met when sampling from a population.
4. Determine whether or not the shape of the sampling distribution of a sample proportion is approximately normal.
5. Interpret the mean, standard deviation, and probabilities for a sampling distribution of a sample proportion.

3.2.1 Visualizing a sampling distribution of a sample proportion

In a large population, the proportion of people with blood type O+ is 0.35. If we take a random sample of size 40 from this population, how far off do we expect our sample proportion to be from the true proportion? To answer this question we wish to understand and visualize the sampling distribution of the proportion of people with blood type O+ in a random sample of size 40. Recall that the **sampling distribution** of a sample statistic is the distribution of values of the statistic for all random samples of a given size from a given population.

THE SAMPLING DISTRIBUTION OF A SAMPLE PROPORTION

The sampling distribution of a sample proportion $\hat{p}$ is the distribution of $\hat{p}$ values for all random samples of a given size from a given population.

In Section 2.5, we saw that the distribution for the *number* of people with blood type O+ in a random sample of size 40 follows a binomial distribution with $n = 40$ and $p = 0.35$. What do we expect the sampling distribution for the *proportion* of people with blood type O+ in a random sample of size 40 to look like?

In this scenario, we can actually calculate the probability of each of the possible sample proportions and draw the exact sampling distribution. The right panel of Figure 3.2 shows the distribution for the *proportion* of people with blood type O+ in a random sample of size 40, given a population proportion $p = 0.35$, and the left panel of Figure 3.2 shows the distribution for the *number* of people with blood type O+ in a random sample of size 40, given a population proportion $p = 0.35$.

We can see that the distribution for number with blood type O+ and the distribution for proportion with blood type O+ look the same, but with a change of scale. Instead of showing counts along the horizontal axis, the graph of the sampling distribution of the sample proportion shows proportions. To convert from a count to a proportion, we divide the count (number of successes) by the sample size, $n = 40$. For example, 8 becomes $8/40 = 0.20$ and 14 becomes $14/40 = 0.35$.

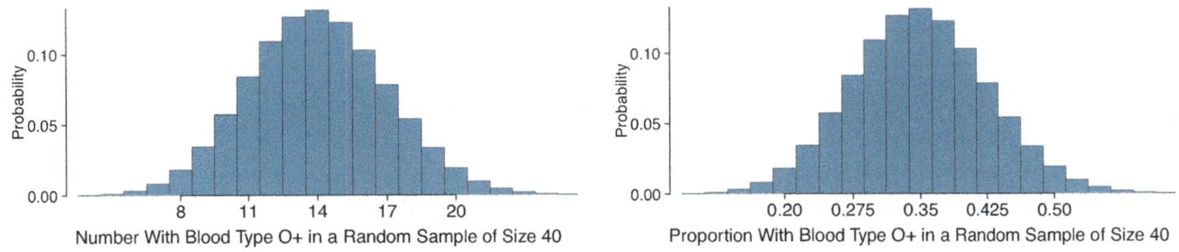

Figure 3.2: Two distributions where $p = 0.35$ and $n = 40$: the binomial distribution for the *number* with blood type O+ and the sampling distribution of the *proportion* with blood type O+.

To better understand what a sampling distribution of $\hat{p}$ represents, we can also conduct a simulation just as we did in Section 2.7.1. Here we will simulate drawing a sample of size 40 from a population where $p = 0.35$. We calculate the number of successes in the sample and the proportion of successes in the sample. We repeat this 300 times. The results are graphed in Figure 3.3.

Figure 3.3: 300 simulated values for the number of successes (left) and the proportion of successes (right) in a random sample of size 40 from a population with p=0.35.

E

EXAMPLE 3.5

What does the graph in the right panel of Figure 3.3 represent? What does each dot represent?

The graph represents an approximation of the sampling distribution of $\hat{p}$. While the theoretical sampling distribution of $\hat{p}$ is the distribution of $\hat{p}$ values for *all* random samples of size 40 from this population, this simulated distribution represents the distribution of $\hat{p}$ values for 300 random samples of size 40 from this population. Each dot represents the value of a sample proportion $\hat{p}$ calculated from *one* random sample of size 40.

G

GUIDED PRACTICE 3.6

Why do the graphs in Figure 3.3 look different than the graphs in Figure 3.2?[2]

[2]The graphs in Figure 3.3 include only 300 simulated values, whereas the graphs in Figure 3.2 show the theoretical sampling distributions. If we used a much larger number of trials in our simulation, we would expect the simulation graphs to look more like the theoretical sampling distributions.

3.2.2 The mean and standard deviation of $\hat{p}$

Looking at Figure 3.2, we see that the sampling distribution for the proportion of people with blood type O+ in a random sample of size 40 is centered on 0.35 (the population proportion) and has a standard deviation of approximately 0.075.

If we were to look at more sampling distributions for proportions, we'd eventually find that there are some patterns that depend on n and p, and these patterns can be characterized mathematically. As one might expect, the distribution of $\hat{p}$ is centered on the true proportion p, assuming the sample is random and the estimator is unbiased. When the observations are independent, the standard deviation of $\hat{p}$ is given by: $\sigma_{\hat{p}} = \sqrt{\frac{p(1-p)}{n}}$.

MEAN AND STANDARD DEVIATION OF A SAMPLE PROPORTION

The mean and standard deviation of the sampling distribution of a sample proportion describe the center and spread of the distribution of sample proportions $\hat{p}$ for all random samples of size n from a population with size N. Let p represent the population proportion. We find the mean and standard deviation of the sampling distribution of $\hat{p}$ as follows:

$$\mu_{\hat{p}} = p$$

$$\sigma_{\hat{p}} = \sqrt{\frac{p(1-p)}{n}} \quad \text{when } n < 0.10(N) \text{ if sampling without replacement}$$

When sampling without replacement, independence of observations is technically not met, and the standard deviation formula will overestimate the true standard deviation. However, as we saw in Examples 2.46 and 2.67, if the sample size is small compared to the population size, then the observations can be treated *as if* they were independent. In this case, the standard deviation formula will provide a good estimate. A rule of thumb is that if the sample size is less than 10% of the population size, or equivalently, the population size is at least 10 times larger than the sample size, then the standard deviation formula will provide a good estimate and can be safely used.

The standard deviation of a sample proportion, $\sigma_{\hat{p}}$, tells us about the "typical" deviation in the sample proportions from the true population proportion. In analyses, we think of the standard deviation of a sample proportion as describing the uncertainty associated with the estimate $\hat{p}$. That is, $\sigma_{\hat{p}}$ can be thought of as a way to quantify the typical error in our sample estimate $\hat{p}$ of the true proportion p. Understanding the variability of statistics such as $\hat{p}$ is a central component in the study of statistics.

E

EXAMPLE 3.7

If the proportion of people in the county with blood type O+ is really 35%, find and interpret the mean and standard deviation of the sample proportion for a random sample of size 400.

The mean of the sample proportion $\mu_{\hat{p}}$ is calculated as follows:

$$\mu_{\hat{p}} = p = 0.35.$$

For all random samples of size 400 from this population, the sample proportions with blood type O+ will have a mean of 0.35. In other words, if we took many, many random samples of size 400 from this population and calculated $\hat{p}$ for each sample, the average of all the $\hat{p}$ values would be about 0.35.

We will assume that the sample size of 400 is less than 10% of the population size, i.e. that the number of people in the county is at least 4000. The standard deviation of $\hat{p}$ is calculated as follows:

$$\sigma_{\hat{p}} = \sqrt{\frac{p(1-p)}{n}} = \sqrt{\frac{0.35(0.65)}{400}} = 0.024.$$

For all random samples of size 400 from this population, the sample proportions with blood type O+ would typically vary from the population proportion of 0.35 by about 0.024.

E

EXAMPLE 3.8

Would you be surprised in you took a random sample of size 400 from this population and 37% of the sample had blood type O+? That is, would you be surprised to get a $\hat{p}$ value of 0.37?

The value 0.37 is less than one standard deviation from the mean, so it would not be surprising to have 37% with blood type O+ in a random sample of size 400 from this population.

G

GUIDED PRACTICE 3.9

Would you be surprised in you took a random sample of size 400 from this population and 27% of the sample had blood type O+? That is, would you be surprised to get a $\hat{p}$ value of 0.27?[3]

G

GUIDED PRACTICE 3.10

If instead of taking a random sample of size 400 from this population, you took a random sample of size 100, how would the mean and standard deviation of the sample proportion change?[4]

---

[3]0.27 is more than three standard deviations below the mean, so it would be surprising to have only 27% with blood type O+ in a random sample of size 400 from this population.

[4]The mean is just p so the mean would not change. The standard deviation would be larger because the sample size is smaller. The standard deviation gets smaller by a factor of $1/\sqrt{n}$, so the standard deviation would be twice as large (not four times as large).

3.2.3 The Central Limit Theorem revisited

The sampling distribution for the sample proportion in Figure 3.2 looks an awful lot like a normal distribution. This is not surprising and is a result of the Central Limit Theorem, which was introduced in Section 2.7.

CENTRAL LIMIT THEOREM FOR SAMPLE PROPORTIONS

When observations are independent and the sample size is sufficiently large, the sample proportion $\hat{p}$ will tend to follow a normal distribution.

In order for the Central Limit Theorem to hold, the sample size is typically considered sufficiently large when $np \geq 10$ and $n(1-p) \geq 10$, where n is the sample size and p is the population proportion. This is called the **large counts condition**, or the success-failure condition.

The Central Limit Theorem is incredibly important, and it provides a foundation for much of statistics. As we begin applying the Central Limit Theorem in the context of a sample proportion, be mindful of the two technical conditions: the observations must be independent, and the sample size must be sufficiently large such that the expected number of successes, np, and the expected number of failures, $n(1-p)$, are both at least 10.

HOW TO VERIFY SAMPLE OBSERVATIONS ARE INDEPENDENT

If the observations are from a random process such as tossing a coin, then they are independent.

If the observations are from a random sample with replacement, then they are independent.

If the observations are from a simple random sample (without replacement), we can treat them as independent if the sample size is less than 10% of the population size.

If a sample is from a seemingly random process, e.g. an occasional error on an assembly line, checking independence is more difficult. In this case, use your best judgement.

When the sample exceeds 10% of the population size, the methods we discuss tend to overestimate the sampling error slightly versus what we would get using more advanced methods.[5]

An interesting question to answer is, *what happens when* $np < 10$ *or* $n(1-p) < 10$? We can simulate drawing samples of different sizes where, say, the true proportion is $p = 0.25$. Here's a sample of size 10:

no, no, yes, yes, no, no, no, no, no, no

In this sample, we observe a sample proportion of yeses of $\hat{p} = \frac{2}{10} = 0.2$. We can simulate many such proportions to understand the sampling distribution of $\hat{p}$ when $n = 10$ and $p = 0.25$, which we've plotted in Figure 3.4 alongside a normal distribution with the same mean and variability. These distributions have a number of important differences.

	Unimodal?	Smooth?	Symmetric?
Normal: $N(0.25, 0.14)$	Yes	Yes	Yes
$n = 10$, $p = 0.25$	Yes	*No*	*No*

Notice that the large counts condition was not satisfied when $n = 10$ and $p = 0.25$:

$$np = 10 \times 0.25 = 2.5 \qquad n(1-p) = 10 \times 0.75 = 7.5$$

This single sampling distribution does not show that the large counts condition is the perfect guideline, but we have found that the guideline did correctly identify that a normal distribution might not be appropriate.

[5]For example, we could use what's called the **finite population correction factor**: if the sample is of size n and the population size is N, then we can multiply the typical standard deviation formula by $\sqrt{\frac{N-n}{N-1}}$ to obtain a smaller, more precise estimate of the actual standard deviation. When $n < 0.1 \times N$, this correction factor is relatively close to 1.

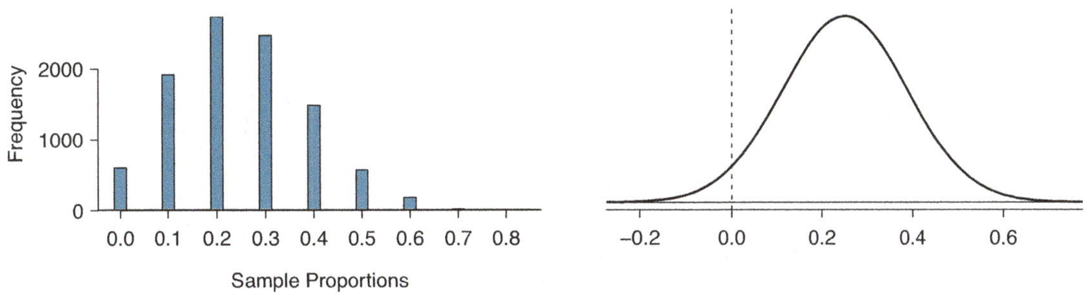

Figure 3.4: Left: simulations of $\hat{p}$ when the sample size is $n = 10$ and the population proportion is $p = 0.25$. Right: a normal distribution with the same mean (0.25) and standard deviation (0.137).

We can complete several additional simulations, shown in Figures 3.5 and 3.6, and we can see some trends:

1. When either np or $n(1-p)$ is small, the distribution is more **discrete**, i.e. *not continuous.*
2. When np or $n(1-p)$ is smaller than 10, the skew in the distribution is more noteworthy.
3. The larger both np *and* $n(1-p)$, the more normal the distribution. This may be a little harder to see for the larger sample size in these plots as the variability also becomes much smaller.
4. When np and $n(1-p)$ are both very large, the distribution's discreteness is hardly evident, and the distribution looks much more like a normal distribution.

So far we've only focused on the skew and discreteness of the distributions. We haven't considered how the mean and standard deviation of the distributions change. Take a moment to look back at the graphs, and pay attention to three things:

1. The centers of the distribution are always at the population proportion, p, that was used to generate the simulation. Because the sampling distribution for $\hat{p}$ is always centered at the population parameter p, it means the sample proportion $\hat{p}$ is **unbiased** when the data are independent and drawn from such a population.
2. For a particular population proportion p, the variability in the sampling distribution decreases as the sample size n becomes larger. This will likely align with your intuition: an estimate based on a larger sample size will tend to be more accurate.
3. For a particular sample size, the variability will be largest when $p = 0.5$. The differences may be a little subtle, so take a close look. This reflects the role of the proportion p in the standard deviation formula: $SD = \sqrt{\frac{p(1-p)}{n}}$. The standard deviation is largest when $p = 0.5$.

At no point will the distribution of $\hat{p}$ look *perfectly* normal, since $\hat{p}$ will always be take discrete values (x/n). It is always a matter of degree, and we will use the standard large counts condition with minimums of 10 for np and $n(1-p)$ as our guideline within this book.

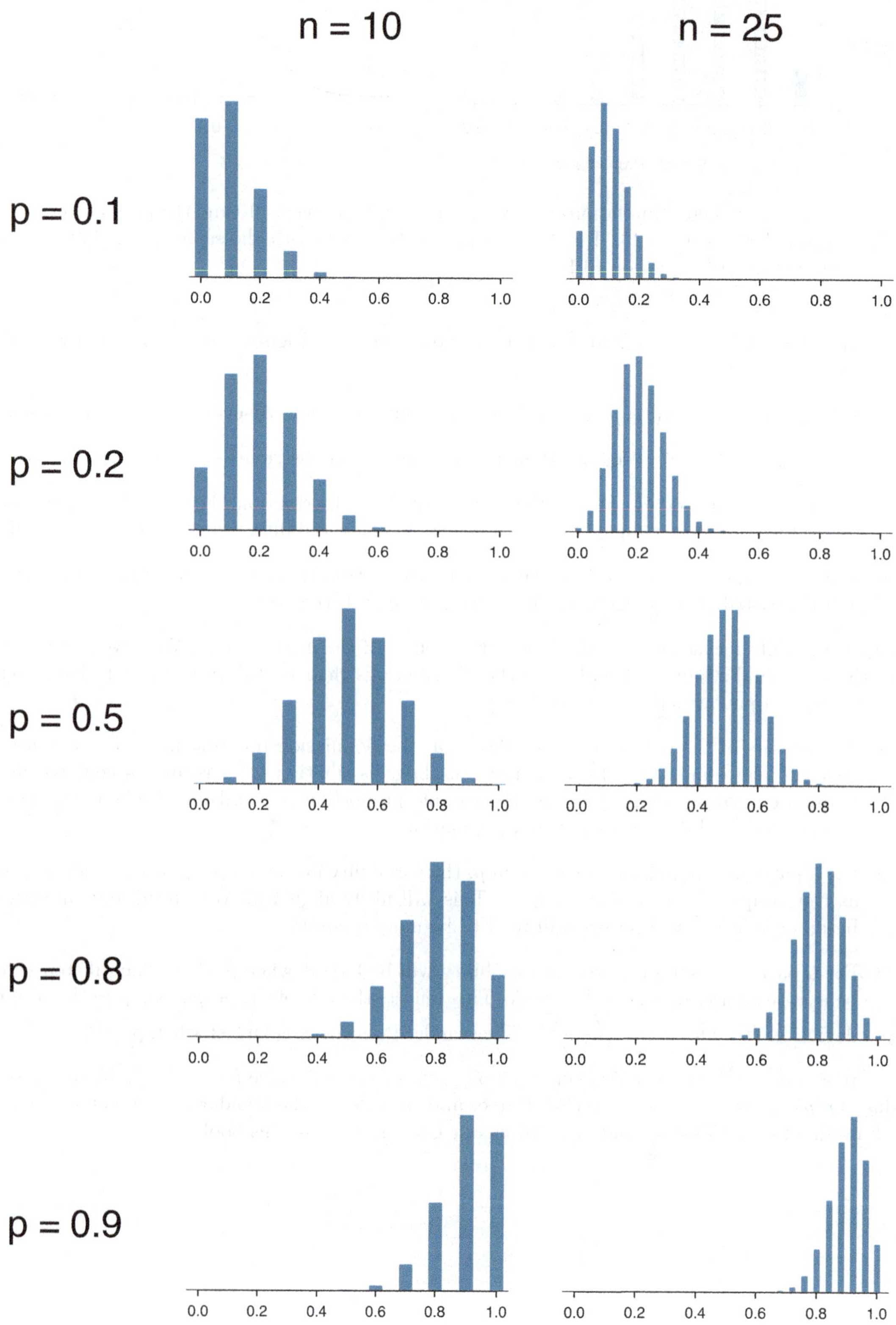

Figure 3.5: Sampling distributions for several scenarios of p and n.
Rows: $p = 0.10$, $p = 0.20$, $p = 0.50$, $p = 0.80$, and $p = 0.90$.
Columns: $n = 10$ and $n = 25$.

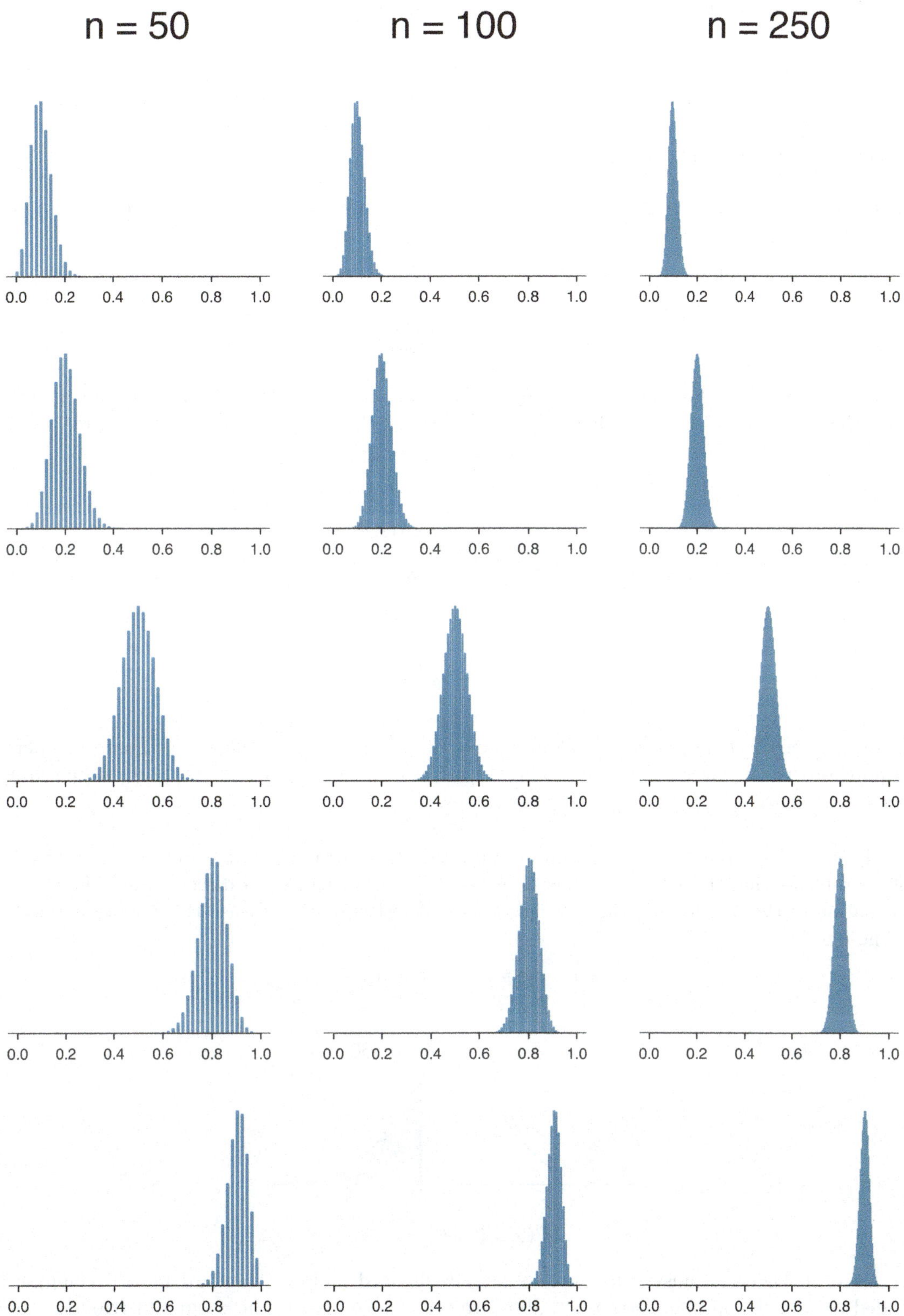

Figure 3.6: Sampling distributions for several scenarios of p and n.
Rows: $p = 0.10$, $p = 0.20$, $p = 0.50$, $p = 0.80$, and $p = 0.90$.
Columns: $n = 50$, $n = 100$, and $n = 250$.

3.2.4 Using a normal model for the sampling distribution of $\hat{p}$

We can now answer a question posed at the beginning of this section.

EXAMPLE 3.11

Assume that in a particular state, 48% support a controversial measure. When estimating the percent through polling, what is the probability that a random sample of size 200 will mistakenly estimate the percent support to be at least 50%?

$\hat{p}$ is the proportion in a random sample of size 200 that support the controversial measure, and the mean of $\hat{p}$ is: $\mu_{\hat{p}} = p = 0.48$.

Because 200 < 10% of all people in the state, we can calculate the standard deviation of $\hat{p}$ as follows:

$$\sigma_{\hat{p}} = \sqrt{\frac{p(1-p)}{n}} = \sqrt{\frac{0.48(1-0.48)}{200}} = 0.035.$$

Because $200(0.48) \geq 10$ and $200(0.52) \geq 10$, $\hat{p}$ is approximately Normal. Using a technology option from Section 2.6.4 and the fact that $\hat{p}$ is approximately Normal($\mu = 0.48$, $\sigma = 0.035$), we find that $P(\hat{p} \geq 0.50) = 0.286$.

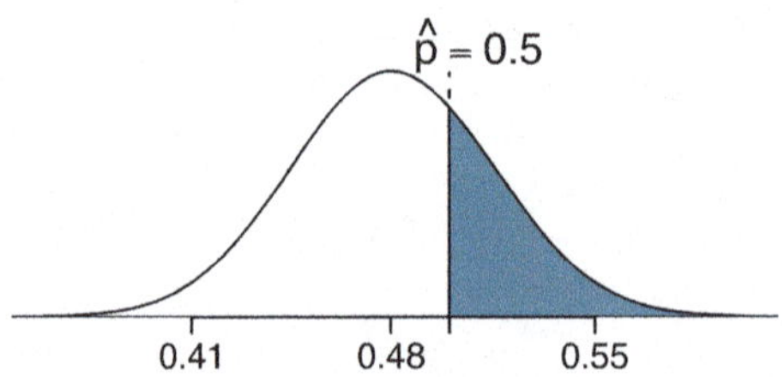

There is a 28.6% chance that a random sample of size 200 will mistakenly estimate the percent support for the controversial measure to be at least 50%, assuming the true proportion who support it is 48%.

It is common to calculate a Z-score and use the standard normal distribution ($\mu = 0$, $\sigma = 1$) to solve problems similar to the one above. In Section 3.4, we will see a direct parallel between the calculation of the Z-score and the calculation of a Z test statistic. To use the Z-score method, we do the following:

$$Z = \frac{\hat{p} - \mu_{\hat{p}}}{\sigma_{\hat{p}}} = \frac{\hat{p} - p}{\sqrt{\frac{p(1-p)}{n}}} = \frac{0.50 - 0.48}{\sqrt{\frac{0.48(1-0.48)}{200}}} = 0.566$$

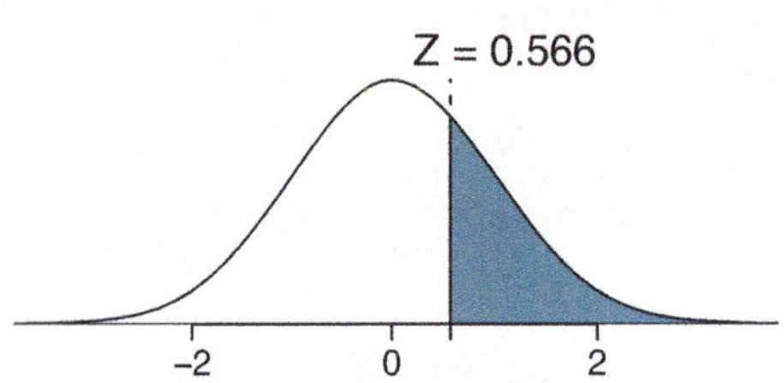

$$P(Z \geq 0.566) = 0.286.$$

We arrive at the same answer because we simply changed units to standard units – compare the shaded area in the normal distribution above to the shaded area in the normal distribution shown in Example 3.11.

GUIDED PRACTICE 3.12

In the example above, the probability of mistakenly thinking that a majority support the measure is quite high (more than 1 in 4 chance). What could be done differently to lower this probability?[6]

Section summary

- The symbol $\hat{p}$ denotes a sample proportion. $\hat{p}$ for any particular sample is a number. However, $\hat{p}$ can vary from sample to sample. The **sampling distribution of a sample proportion** $\hat{p}$ is the distribution of values of $\hat{p}$ for all random samples of size n from a given population.
- When the observations can be considered independent, such as from a *random* sample:
  - The **mean** of the sampling distribution of a sample proportion $\hat{p}$ is given by: $\mu_{\hat{p}} = p$, where p is the population proportion.
  - The **standard deviation** of the sampling distribution of a sample proportion $\hat{p}$ is: $\sigma_{\hat{p}} = \sqrt{\frac{p(1-p)}{n}}$. When sampling without replacement, the sample size n should be than 10% of the population size N, i.e. $n < 0.10(N)$, in order for the standard deviation formula to be used.
  - The **shape** of the sampling distribution of a sample proportion $\hat{p}$ is approximately normal when the sample size is large enough that the expected number of success, np, and the expected number of failures, $n(1-p)$, are both at least 10, i.e. $np \geq 10$ and $n(1-p) \geq 10$. This is sometimes called the large counts condition or the success-failure condition.
- $\mu_{\hat{p}}$, the mean of $\hat{p}$, describes the average of the sample proportions among all random samples of size n from a given population.
- $\sigma_{\hat{p}}$, the standard deviation of $\hat{p}$, measures how far the sample proportions typically vary from the population proportion p for all random samples of size n from a given population.
- To use a normal model to find probabilities involving a sample proportion, first verify that the conditions for independence are met and that the large counts condition is met. Identify the distribution and its parameters, write the relevant probability statement, and answer the question in context.
- The mean, standard deviation, and probabilities for a sampling distribution of a sample proportion should be interpreted in the context of a specific population.

[6]To lower this tail probability, we would want to take a larger sample, which would reduce the standard devation of the sampling distribution.

Exercises

3.5 Distribution of $\hat{p}$. Suppose the true population proportion were $p = 0.95$. The figure below shows what the distribution of a sample proportion looks like when the sample size is $n = 20$, $n = 100$, and $n = 500$. (a) What does each point (observation) in each of the samples represent? (b) Describe the distribution of the sample proportion, $\hat{p}$. How does the distribution of the sample proportion change as n becomes larger?

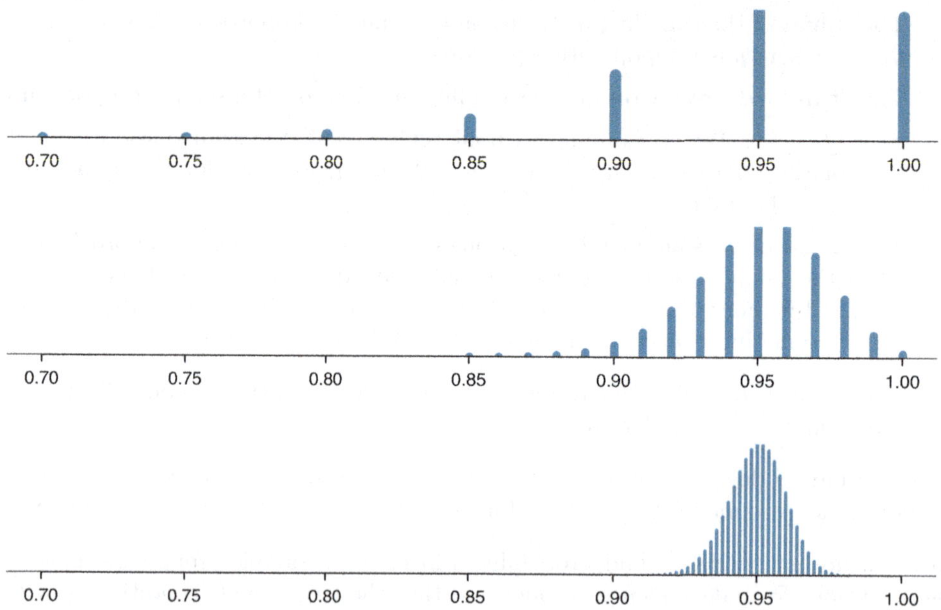

3.6 Distribution of $\hat{p}$. Suppose the true population proportion were $p = 0.5$. The figure below shows what the distribution of a sample proportion looks like when the sample size is $n = 20$, $n = 100$, and $n = 500$. What does each point (observation) in each of the samples represent? Describe how the distribution of the sample proportion, $\hat{p}$, changes as n becomes larger.

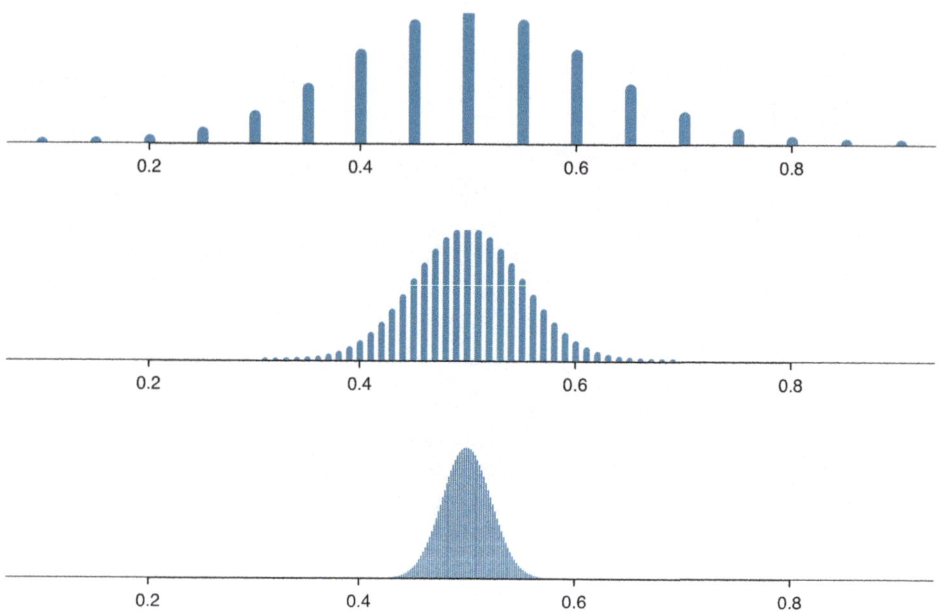

3.7 Vegetarian college students. Suppose that 8% of college students are vegetarians. Determine if the following statements are true or false, and explain your reasoning.

(a) The distribution of the sample proportions of vegetarians in random samples of size 60 is approximately normal since $n \geq 30$.

(b) The distribution of the sample proportions of vegetarian college students in random samples of size 50 is right skewed.

(c) A random sample of 125 college students where 12% are vegetarians would be considered unusual.

(d) A random sample of 250 college students where 12% are vegetarians would be considered unusual.

(e) The standard deviation of the sample proportion would be reduced by one-half if we increased the sample size from 125 to 250.

3.8 Young Americans, Part I. About 77% of young adults think they can achieve the American dream. Determine if the following statements are true or false, and explain your reasoning.[7]

(a) The distribution of sample proportions of young Americans who think they can achieve the American dream in random samples of size 20 is left skewed.

(b) The distribution of sample proportions of young Americans who think they can achieve the American dream in random samples of size 40 is approximately normal since $n \geq 30$.

(c) A random sample of 60 young Americans where 85% think they can achieve the American dream would be considered unusual.

(d) A random sample of 120 young Americans where 85% think they can achieve the American dream would be considered unusual.

3.9 Distribution of $\hat{p}$. Suppose the true population proportion were $p = 0.5$ and a researcher takes a simple random sample of size $n = 50$.

(a) Find and interpret the standard deviation of the sample proportion $\hat{p}$.

(b) Calculate the probability that the sample proportion will be larger than 0.55 for a random sample of size 50.

3.10 Distribution of $\hat{p}$. Suppose the true population proportion were $p = 0.6$ and a researcher takes a simple random sample of size $n = 50$.

(a) Find and interpret the standard deviation of the sample proportion $\hat{p}$.

(b) Calculate the probability that the sample proportion will be larger than 0.65 for a random sample of size 50.

3.11 Nearsighted children. It is believed that nearsightedness affects about 8% of all children. We are interested in finding the probability that fewer than 12 out of 200 randomly sampled children will be nearsighted.

(a) Estimate this probability using the normal approximation to the binomial distribution.

(b) Estimate this probability using the distribution of the sample proportion.

(c) How do your answers from parts (a) and (b) compare?

3.12 Social network use. The Pew Research Center estimates that as of January 2014, 89% of 18-29 year olds in the United States use social networking sites.[8] Calculate the probability that at least 95% of 500 randomly sampled 18-29 year olds use social networking sites.

3.13 CLT for proportions, Part 1. Define the term "sampling distribution" of the sample proportion, and describe how the shape, center, and spread of the sampling distribution change as the sample size increases when $p = 0.1$.

3.14 CLT for proportions, Part 2. Define the term "sampling distribution" of the sample proportion, and describe how the shape, center, and spread of the sampling distribution change as the sample size increases when $p = 0.8$.

[7] A. Vaughn. "Poll finds young adults optimistic, but not about money". In: *Los Angeles Times* (2011).

[8] Pew Research Center, Washington, D.C. Social Networking Fact Sheet, accessed on May 9, 2015.

3.3 Confidence intervals for a population proportion

What percent of adults in the US approve of the way the Supreme Court is handling its job? Do greater than half of adults in the US believe in intelligent life on other planets? When polling companies report point estimates, they usually also report a margin of error and an interval estimate. In this section we discuss margin of error and how to construct and interpret what is called a confidence interval for a population proportion.

Learning objectives

1. Identify and set up an appropriate confidence interval procedure for estimating a population proportion p.
2. Justify the appropriateness of constructing a confidence interval for a population proportion by verifying that the conditions are met.
3. Calculate a confidence interval for a population proportion.
4. Calculate the standard error and margin of error of a sample proportion, and estimate a sample size from the margin of error.
5. Interpret a confidence interval for a population proportion in context.
6. Justify a claim about a population proportion based on an appropriate confidence interval.
7. Identify the relationships among sample size, confidence interval width, confidence level, and margin of error.
8. Recognize that margin of error calculations only measure sampling error, and that other types of errors such as bias may be present.

3.3.1 Introducing confidence intervals and margin of error

In Section 3.1, we saw that a point estimate provides a single estimate for a parameter. However, a point estimate isn't perfect and we do not expect it to hit the parameter exactly. To increase our confidence, we provide a *range* of plausible values, called an interval estimate or **confidence interval**.

CONFIDENCE INTERVAL

A confidence interval is an interval estimate and provides a range of plausible values for a parameter based on sample data.

A point estimate is our best estimate for the value of the parameter, so it makes sense to build the confidence interval around that value. How can we quantify the expected variability or error in a point estimate? We use a quantity called the **standard error** of the estimate. The standard error of an estimate, as we will see later in this section, is closely related to the standard deviation of an estimate.

EXAMPLE 3.13

How many standard errors should we extend above and below the point estimate if we want to be 95% confident of capturing the true value?

E

First, we observe that the area under the standard normal distribution between -1.96 and 1.96 is 95%. When conditions for a normal model are met, the point estimate we observe will be within 1.96 standard deviations of the true value about 95% of the time. Thus, if we want to be 95% confident of capturing the true value, we should go 1.96 standard errors on either side of the point estimate.

CONSTRUCTING A 95% CONFIDENCE INTERVAL USING A NORMAL MODEL

When the sampling distribution of a point estimate can reasonably be modeled as normal, a 95% confidence interval for the unknown parameter can be constructed as:

$$\text{point estimate} \pm 1.96 \times SE \text{ of estimate}$$

We can be **95% confident** that this interval captures the true value.

EXAMPLE 3.14

In Section 3.1, we considered a point estimate for the proportion in the population that approve of a particular governor. The point estimate was 45% based on a sample size of 500. The standard error of this estimate is $SE = 0.02$. Assuming that conditions for a normal model are met, construct a 95% confidence interval.

E

$$\begin{aligned}\text{point estimate} &\pm 1.96 \times SE \text{ of estimate}\\ 0.45 &\pm 1.96 \times 0.02\\ 0.45 &\pm 0.039\\ &(0.411, 0.489)\end{aligned}$$

Based on our calculations, we can be 95% confident that the interval (0.411, 0.489) contains the true proportion of the population who approve of the governor. In general, we can be 95% confident that a 95% confidence interval captures the true population parameter. However, confidence intervals are imperfect. About 1-in-20 (5%) properly constructed 95% confidence intervals will fail to capture the parameter of interest. Figure 3.7 shows 25 confidence intervals for a proportion that were constructed from simulations where the true proportion was $p = 0.3$. However, 1 of these 25 confidence intervals happened not to include the true value.

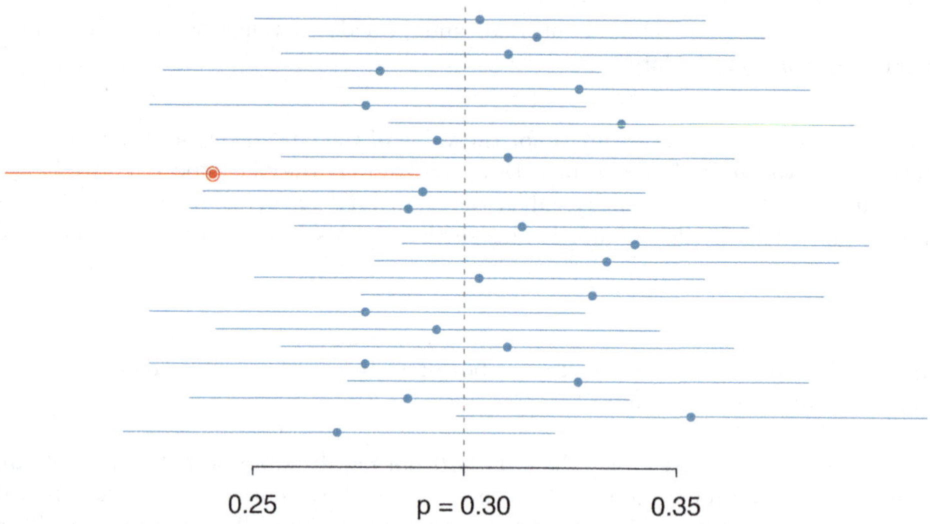

Figure 3.7: Twenty-five samples of size $n = 300$ were simulated when $p = 0.30$. For each sample, a confidence interval was created to try to capture the true proportion p. However, 1 of these 25 intervals did not capture $p = 0.30$. View this simulation in an interactive Desmos calculator here, or find it at openintro.org/ahss/desmos.

Using this simulation, we can better understand what we mean by 95% confidence. In the long run, about 95% of the 95% confidence intervals, based on random samples of the same size from the same population, will capture the true value. This is why we can say that we are 95% confident that an individual interval captures the true value.

GUIDED PRACTICE 3.15

G In Figure 3.7, one interval does not contain the true proportion, $p = 0.3$. Does this imply that there was a problem with the simulations?[9]

Confidence intervals are also often reported as:

$$\text{point estimate} \pm \text{margin of error}$$

In Example 3.14 the 95% confidence interval was calculated as 0.45 ± 0.039. For this interval, the point estimate is 0.45 and the margin of error of the estimate is 0.039. This tells us that we can be 95% confident that our estimate is within 0.039 of the true proportion in the population who approve of the governor.

In Figure 3.7, we see the point estimates represented by the dot in the middle of each interval. For each confidence interval, the ± margin of error is represented by the line extending to the right of the point estimate and to the left of the point estimate. Numerically, the margin of error corresponds to the distance between the point estimate and the lower or upper bound of a confidence interval, and thus is half of the total width of the interval.

[9]No. Just as some observations occur more than 1.96 standard deviations from the mean, some point estimates will be more than 1.96 standard errors from the parameter. A confidence interval only provides a plausible range of values for a parameter. While we might say other values are implausible based on the data, this does not mean they are impossible.

MARGIN OF ERROR

When using a normal model, the margin of error of a 95% confidence interval is given by

$$1.96 \times SE \text{ of estimate}$$

The margin of error of a 95% confidence interval tells us that we can be 95% confident that our point estimate is within that margin of error of the true value.

GUIDED PRACTICE 3.16

For a 95% confidence interval given by: (0.035, 0.145), find the point estimate and the margin of error of the estimate.[10]

3.3.2 Changing the confidence level

A researcher chooses a confidence level, and then calculates the margin of error of the estimate at that confidence level. Suppose we want to construct a confidence interval with a confidence level somewhat greater than 95%: perhaps we would like a confidence level of 99%.

EXAMPLE 3.17

Other things being equal, would a 99% confidence interval have a larger margin or error or a smaller margin of error than a 95% confidence interval?

All other things being equal, a 99% confidence interval will have a larger margin of error than a 95% confidence interval; to be more confident we will need a wider interval and a wider interval corresponds to a larger margin of error.

The 95% confidence interval structure provides guidance in how to make intervals with new confidence levels. Below is a general 95% confidence interval for a point estimate that comes from a nearly normal distribution:

$$\text{point estimate} \pm 1.96 \times SE \text{ of estimate}$$

There are three components to this interval: the point estimate, "1.96", and the standard error of the estimate. The choice of $1.96 \times SE$ was based on capturing 95% of the distribution since the estimate is within 1.96 standard deviations of the true value about 95% of the time. The choice of 1.96 corresponds to a 95% confidence level.

GUIDED PRACTICE 3.18

If X is a normally distributed random variable, how often will X be within 2.58 standard deviations of the mean?[11]

[10]The point estimate is the middle value of the interval and can be calculated as: $(0.035 + 0.145)/2 = 0.09$. The margin of error is *half* of the total width of the interval and can be calculated as: $\frac{0.145-0.035}{2} = 0.055$, or as $0.145 - 0.9 = 0.055$

[11]This is equivalent to asking how often the Z-score will be larger than -2.58 but less than 2.58. (For a picture, see Figure 3.8.) There is ≈ 0.99 probability that a normally distributed random variable X will be within 2.58 standard deviations of the mean.

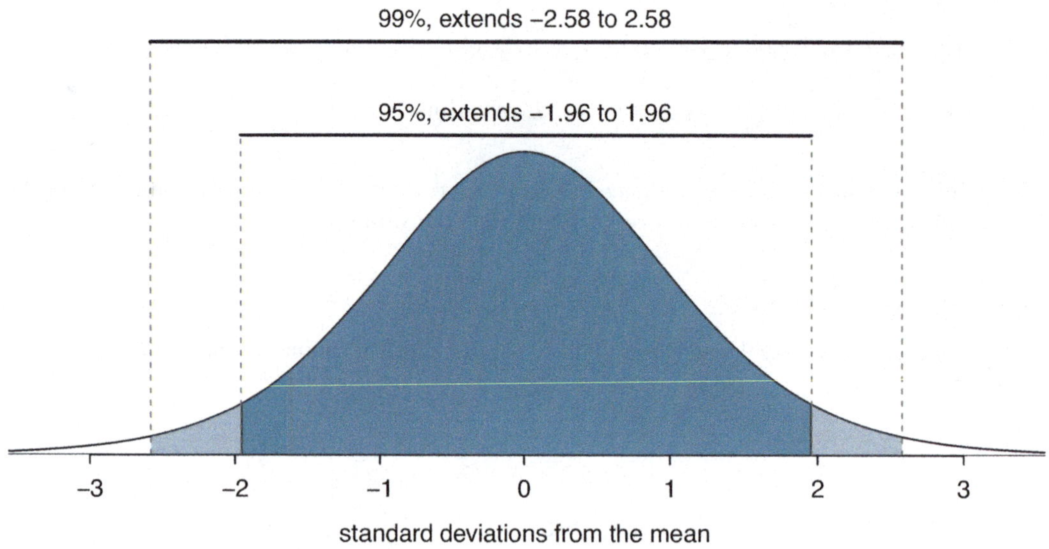

Figure 3.8: The area between $-z^\star$ and $z^\star$ increases as $|z^\star|$ becomes larger. If the confidence level is 99%, we choose $z^\star$ such that 99% of the normal distribution is between $-z^\star$ and $z^\star$, which corresponds to 0.5% in the lower tail and 0.5% in the upper tail: $z^\star = 2.58$.

Guided Practice 3.18 highlights that 99% of the time a normal random variable will be within 2.58 standard deviations of its mean. To create a 99% confidence interval, change 1.96 in the 95% confidence interval formula to be 2.58. Thus, the formula for a 99% confidence interval is

$$\text{point estimate} \ \pm\ 2.58 \times SE \text{ of estimate}$$

In the interval above, the value 2.58 is called the **critical value**. When the critical value is determined based on a normal model, we label the critical value $z^\star$. Figure 3.8 provides a picture of how to identify $z^\star$ based on a confidence level. We use the $z^\star$ value such that the area under the standard normal distribution between $-z^\star$ and $z^\star$ corresponds to our confidence level of C%.

CONFIDENCE INTERVAL FOR ANY CONFIDENCE LEVEL

If the point estimate follows a normal model, then a C% confidence interval for the population parameter is

$$\text{point estimate} \ \pm\ z^\star \times SE \text{ of estimate}$$

$z^\star$ is the value such that the area under the standard normal curve between $-z^\star$ and $z^\star$ is C%.

Finding the value of $z^\star$ that corresponds to a particular confidence level can be accomplished by using a new table, called the t-table. For now, what is noteworthy about this table is that the bottom row corresponds to confidence levels. The numbers inside the table are the critical values, but which row should we use? When finding $z^\star$, we use the t-table at row ∞. The reason for this will be explained in the next chapter.

Another way to find the value of $z^\star$ for a C% confidence interval is to use technology to find the boundary value such that C% of the standard normal distribution falls between $-z^\star$ and $z^\star$. We encountered problems such as this in the Normal distribution section. See problem (ii)(b) on page 167 for a technology refresher.

	one tail	0.100	0.050	0.025	0.010	0.005
df	1	3.078	6.314	12.71	31.82	63.66
	2	1.886	2.920	4.303	6.965	9.925
	3	1.638	2.353	3.182	4.541	5.841
	⋮	⋮	⋮	⋮	⋮	
	1000	1.282	1.646	1.962	2.330	2.581
	∞	1.282	1.645	1.960	2.326	2.576
Confidence level C		80%	90%	95%	98%	99%

Figure 3.9: An abbreviated look at the t-table. The columns correspond to confidence levels. Row ∞ corresponds to the normal distribution.

GUIDED PRACTICE 3.19

Find the $z^\star$ critical value for an 80% confidence interval.[12]

We now see how to determine the margin error for a any confidence level, not just a 95% confidence level.

MARGIN OF ERROR

When using a normal model, the margin of error for a C% confidence interval is given by

$$z^\star \times SE \text{ of estimate}$$

$z^\star$ is the value such that the area under the standard normal curve between $-z^\star$ and $z^\star$ is C%. The margin of error for a C% confidence interval tells us that we can be C% confident that our point estimate is within that margin of error of the true value.

Choosing a higher confidence level, for example 95% instead of 90%, increases the margin of error, as it requires a larger $z^\star$ to capture the desired percent between $-z^\star$ and $z^\star$. We can also think about this as: to be more confident, we need to cast a wider net. For a given sample, increasing the confidence level will result in the following:

i. The critical value will increase.

ii. The margin of error will increase.

iii. The width of the confidence interval will increase.

The normal approximation is crucial to the precision of these confidence intervals. In this chapter, we provide detailed discussions about when a normal model can safely be applied to a variety of situations. When a normal model is not a good fit, we will use alternate distributions that better characterize the sampling distribution.

[12] Using technology or using row ∞ on the t-table, we find that the $z^\star$ critical value that corresponds to an 80% confidence level is 1.282.

3.3.3 Verifying conditions for a confidence interval for a proportion

When the sampling distribution of a sample proportion, $\hat{p}$, is approximately normal, we can estimate a population proportion using confidence intervals based on a normal distribution. We call these intervals "Z-intervals" for short. We check that $\hat{p}$ can be modeled using a normal distribution by assessing the independence assumption and verifying that the large counts condition is met.

Independence. Observations can be considered independent when the data are collected from a *random process*, such as tossing a coin, or from a *random sample*. When sampling without replacement from a finite population, the observations can be considered independent when sampling less than 10% of the population.[13]

Large counts. In Section 3.2, we applied a version of the large counts condition. Here, because p is unknown, we check that the number of observed success and failures in the sample is at least 10, that is, that $n\hat{p} \geq 10$ and $n(1-\hat{p}) \geq 10$.

When these conditions are met, we can use what is called a **one-sample Z-interval for p**, where p is a population proportion.

3.3.4 Carrying out a one-sample Z-interval for p

The Gallup organization began measuring the public's view of the Supreme Court's job performance in 2000, and has measured it every year since then with the question: "Do you approve or disapprove of the way the Supreme Court is handling its job?". In 2025, the Gallup poll randomly sampled 1,033 adults in the US and found that 42% of them approved. We know that 42% is just a point estimate. What range of values are reasonable or plausible estimates for the percent of the population that approved of the job the Supreme Court is doing? We can use the confidence interval procedure introduced in the previous section to answer this question, but first we must clearly identify the parameter we're trying to estimate and be sure that a Z-interval will be appropriate. The following examples walk through the various steps for carrying out a confidence interval procedure using the Gallup poll data.

EXAMPLE 3.20

Identify the population of interest and the parameter of interest for the Gallup poll about the US Supreme Court.

Gallup sampled from US adults, therefore the population of interest, and the population to which we can make an inference, is US adults. We know the percent of the sample who said they approve of the job the Supreme Court is doing. However, we do not know what percent of the population would approve. The parameter of interest, which is unknown, is the percent of *all* US adults that approve of the job the Supreme Court is doing. This is the quantity that we seek to estimate with the confidence interval.

[13]When sampling without replacement and sampling greater than 10% of the population, a modified standard error formula should be used.

EXAMPLE 3.21

Can the sample proportion $\hat{p}$ be modeled using a normal distribution?

In order to construct a Z-interval, the sample statistic must be able to be modeled using a normal distribution. Gallup took a random sample of adults in the US. The sample is random and the sample size is much less than 10% of the population size, so the first condition (the independence condition) is satisfied. We must also test the second condition (the large counts condition) to ensure that the sample size is large enough for the Central Limit Theorem to apply. The large counts condition is met when np and $n(1-p)$ are at least 10. Because p is always unknown when constructing a confidence interval for p, we use the sample proportion $\hat{p}$ to check this condition. Here we have:

$$n\hat{p} = 1033(0.42) = 434 \text{ (“successes”)}$$
$$n(1-\hat{p}) = 1033(1-0.42) = 599 \text{ (“failures”)}$$

The second condition is satisfied since 434 and 599 are both at least 10. With the two conditions satisfied, we can model the sample proportion $\hat{p}$ using a normal model and we can construct a Z-interval.

EXAMPLE 3.22

Calculate the point estimate and the SE of the estimate.

The point estimate for the unknown parameter p (the proportion of all US adults that approve of the job the Supreme Court is doing) is the sample proportion. The point estimate here is $\hat{p} = 0.42$.

We now want to find the SE of the estimate, which is the SE of $\hat{p}$. In Section 3.2, we learned that the formula for the standard deviation of $\hat{p}$ is

$$\sigma_{\hat{p}} = \sqrt{\frac{p(1-p)}{n}}$$

Because p is unknown, we use the sample proportion $\hat{p}$ as our best estimate for p, and we call the estimate of the standard deviation of $\hat{p}$ the standard error (SE) of $\hat{p}$. The SE of $\hat{p}$ is calculated as

$$SE_{\hat{p}} = \sqrt{\frac{\hat{p}(1-\hat{p})}{n}}$$

Here $\hat{p} = 0.42$ and n =1,033, so the SE of the sample proportion is:

$$SE_{\hat{p}} = \sqrt{\frac{0.42(1-0.42)}{1033}} = 0.015$$

The notation $s_{\hat{p}}$ is sometimes used to refer to the standard error of a sample proportion. However, for clarity, we will use $SE_{\hat{p}}$ or SE of $\hat{p}$ in this book.

EXAMPLE 3.23

Interpret the standard error of 0.015 calculated above.

The standard error of 0.015 tells us that for random samples of size $n = 1,033$ from this population, the typical amount that a sample proportion varies from the true proportion of all US adults that approve of the job to Supreme Court is doing is 1.5%.

EXAMPLE 3.24

Construct a 90% confidence interval for p, the proportion of all US adults that approve of the job the Supreme Court is doing.

Recall that the general form of a confidence interval is:

$$\text{point estimate} \pm \text{critical value} \times SE \text{ of estimate}$$

We have already found the point estimate and the SE of the estimate. Because we previously verified that $\hat{p}$ can be modeled using a normal distribution, the critical value is a $z^\star$ value. The $z^\star$ value can be found using technology (see the Technology: normal probabilities and boundary values problem (ii)(b) on page 167) or using the t-table on page 468 at row ∞. For a confidence level of 90%, $z^\star$=1.645. We can now construct the 90% confidence interval as follows.

$$\text{point estimate} \pm z^\star \times SE \text{ of estimate}$$

$$0.42 \pm 1.645\sqrt{\frac{0.42(1-0.42)}{1033}}$$

$$(0.395,\ 0.445)$$

EXAMPLE 3.25

Interpret the calculated confidence interval in context.

Here, we are trying to estimate the true proportion of US adults who approve of the job the Supreme Court is doing, so that context is an important component of our interpretation. A correct interpretation is: We are 90% confident that the interval (0.395, 0.445) contains the true proportion of US adults who approve of the job the Supreme Court is doing.

EXAMPLE 3.26

The calculated confidence interval may or may not provide evidence to justify a claim. Based on this interval, is there evidence to justify a claim that less than half of US adults approve of the job the Supreme Court is doing?

The 90% confidence interval (0.395, 0.445) provides an interval of plausible values for the parameter. The interval does not contain 0.50 or values higher than 0.50, therefore those can be considered implausible, based on the sample. Because the *entire* interval is below 0.50, we do have evidence, at the 90% confidence level, that less than half of US adults (at the time of this poll) approve of the job the Supreme Court is doing.

GUIDED PRACTICE 3.27

Using a 90% confidence level, calculate the margin of error for the estimate in Example 3.24. Interpret this quantity in context.[14]

---

[14]Using a 90% confidence level, the margin of error is $1.645 \times \sqrt{\frac{0.42(1-0.42)}{1033}} = 0.025$. This can also be calculated by finding half of the width of the 90% confidence interval (0.395, 0.445): $(0.445 - 0.395)/2 = 0.025$. We are 90% confident that our estimate is within 0.025 of the true proportion of US adults who approve of the job the Supreme Court is doing.

EXAMPLE 3.28

All other things being equal, when estimating a population proportion, what would we have to do to the sample size in order to halve the margin of error (decrease it by a factor of 2)?

(E) To decrease the error, we would need to increase the sample size. We note that $\sqrt{n}$ is in the denominator of the SE formula, so we would have to *quadruple* the sample size in order to decrease the SE by a factor of 2. For a confidence interval for a population proportion with a given confidence level, the margin of error as well as the width of the confidence interval is approximately proportional to $\frac{1}{\sqrt{n}}$.

3.3.5 Interpreting confidence levels and intervals revisited

What do we really mean when we say we are 95% confident an interval contains the true value? As we saw in Figure 3.7, the 95% confidence interval *method* has a 95% probability of producing an interval that will capture the population parameter. A correct interpretation of a 95% confidence *level* is that in repeated random sampling with the same sample size, approximately 95% of confidence intervals calculated will capture the population proportion.

INTERPRETING THE CONFIDENCE LEVEL

The correct way to interpret a C% confidence level is:

In repeated random sampling with the same sample size, approximately C% of confidence intervals calculated will capture the population parameter.

While 95% of all 95% confidence intervals should contain the population parameter, each individual interval either does or does not. This is why we cannot say that there is a 95% probability that our calculated interval contains the true value.[15] Applying the language of probability to a fixed interval or to a fixed parameter is one of the most common errors when interpreting confidence intervals.

INTERPRETING A CONFIDENCE INTERVAL

The correct way to interpret a particular confidence interval is:

We are C% *confident* that the interval (___, ___) contains the population parameter.

It is also correct to describe the parameter first. For example, we could say we are C% confident that the true proportion of US adults who approve of the job the Supreme Court is doing is between ___ and ___. Referencing the interval first is sometimes seen as preferable as it more clearly attributes the variability to the interval, not to the fixed parameter. However, both interpretations are valid. The three important elements when interpreting a particular confidence interval are: use the word "confident" not "probability", describe the parameter in the context of the problem, and provide the interval or interval endpoints.

Another especially important consideration of confidence intervals is that they only try to capture the *population parameter*. Our intervals say nothing about the confidence of capturing individual observations, a proportion of the observations, or point estimates. Confidence intervals only attempt to capture population parameters.

[15]To see that this interpretation is incorrect, imagine taking two random samples and constructing two 95% confidence intervals for an unknown proportion. If these intervals are disjoint, can we say that there is a 95%+95%=190% chance that the first or the second interval captures the true value?

3.3.6 A four-step framework for confidence interval procedures

Having worked through the examples in Section 3.3.4, we see that a complete confidence interval procedure involves multiple steps. We will find it useful to have a framework for summarizing and remembering the relevant steps. Throughout the textbook we will use the following four-step framework for confidence interval procedures.

- **Identify**: Identify the appropriate interval procedure, the parameter, and the confidence level.
- **Check**: Check that the conditions for the interval procedure are met.
- **Calculate**: Calculate the confidence interval and record it in interval form.

$$\text{point estimate} \ \pm \ \text{critical value} \times SE \text{ of estimate}$$

- **Conclude**: Interpret the interval and, if applicable, draw a conclusion based on whether the interval is entirely above, is entirely below, or contains the value of interest.

We will apply this four-step framework to confidence intervals for a proportion as follows.

CONSTRUCTING A CONFIDENCE INTERVAL FOR A PROPORTION

To carry out a complete confidence interval procedure to estimate a single population proportion,

Identify: Identify the interval procedure, parameter, and confidence level.

Use a **one-sample Z-interval for a population proportion p**. Define the (unknown) population proportion p in words, referencing the population of interest. Choose a confidence level (C%).

Check: Check conditions for constructing a confidence interval using a normal distribution.

1. Independence: Data come from a random sample or random process. When sampling without replacement, check that sample size is less than 10% of the population size.
2. Large counts: $n\hat{p} \geq 10$ and $n(1 - \hat{p}) \geq 10$.

Calculate: Calculate the confidence interval and record it in interval form.

point estimate $\pm$ $z^\star \times SE$ of estimate

point estimate: $\hat{p}$, the sample proportion

SE of estimate: $\sqrt{\frac{\hat{p}(1-\hat{p})}{n}}$

$z^\star$: use technology or a t-table at row ∞ and confidence level C%

(____, ____)

Conclude: Interpret the interval and, if applicable, draw a conclusion in context.

We are C% confident that the interval (____, ____) contains the true *proportion* of [...]. If applicable, draw a conclusion based on whether the interval is entirely above, is entirely below, or contains the value of interest.

E

EXAMPLE 3.29

A Marist Poll reports: "Many Americans (68%) think there is intelligent life on other planets." The results were based on a random sample of 1,033 adults in the US. Does this poll provide evidence at the 95% confidence level that greater than half of all US adults think there is intelligent life on other planets? Carry out a confidence interval procedure to answer this question. Use the four-step framework to organize your work.

Identify: Because the parameter to be estimated is a single proportion, we will use a one-sample Z-interval for a population proportion p. Here, the parameter p is the true proportion of US adults that think there is intelligent life on other planets. We will estimate this at the 95% confidence level.

Check: We must check that a Z-interval is appropriate. The problem states that the data come from a random sample, and since the population is adults in the US, the population size is much more than 10 times larger than the sample size of 1,033. Next we must check the large counts condition. Here, we have that $1033(.68) \geq 10$ and $1033(1-0.68) \geq 10$. The nearly normal sampling distribution conditions are met, so we can proceed with a one-sample Z-interval for p.

Calculate: We will calculate the interval:

$$\text{point estimate} \ \pm \ z^\star \times SE \text{ of estimate}$$

The point estimate is the sample proportion: $\hat{p} = 0.68$.

$$SE \text{ of } \hat{p} = \sqrt{\frac{\hat{p}(1-\hat{p})}{n}} = \sqrt{\frac{0.68(1-0.68)}{1033}} = 0.015.$$

For a 95% confidence level, $z^\star = 1.96$, which can be found using technology or using the t-table at row ∞. The 95% confidence interval is given by:

$$0.68 \pm 1.96 \times \sqrt{\frac{0.68(1-0.68)}{1033}}$$
$$0.68 \pm 1.96 \times 0.015$$
$$(0.651, \ 0.709)$$

Conclude: We are 95% confident that the interval (0.651, 0.709) contains the true *proportion* of US adults that think there is intelligent life on other planets. Because the entire interval is above 0.5 we have evidence that greater than half of all US adults think there is intelligent life on other planets.

G

GUIDED PRACTICE 3.30

True or False: There is a 95% probability that between 65.1% and 70.9% of US adults think that there is intelligent life on other planets.[16]

---

[16]False. The true percent of US adults that think there is intelligent life on other planets either falls in that interval or it doesn't. A correct interpretation of the confidence level would be that if we were to repeat this process over and over, about 95% of the 95% confidence intervals constructed would contain the true value.

3.3.7 Choosing a sample size when estimating a proportion

Planning a sample size before collecting data is important. If researchers collect too little data, the standard error of the point estimate may be so large that the estimate is not very useful. On the other hand, collecting data in some contexts is time-consuming and expensive, so researchers don't want to waste resources on collecting more data than they need.

When considering the sample size, we want to put an upper bound on the margin of error. Recall that the margin of error is calculated as: critical value × *SE* of estimate. For a sample proportion, the margin of error is given by:

$$z^\star \times \sqrt{\frac{\hat{p}(1-\hat{p})}{n}}$$

EXAMPLE 3.31

Suppose we are conducting a university survey to determine whether students support a \$200 per year increase in fees to pay for a new football stadium. Find the smallest sample size n so that the margin of error of the point estimate $\hat{p}$ will be no larger than 0.04 when using a 95% confidence level.

We want the margin of error to be less than or equal to 0.04, so we want:

$$z^\star \times \sqrt{\frac{\hat{p}(1-\hat{p})}{n}} \le 0.04$$

E We know that $z^\star$ for a 95% confidence interval is 1.96. There are still two unknown quantities in the inequality: n and $\hat{p}$. We haven't taken a sample yet, so we do not have a value for $\hat{p}$. If we have an estimate of what we are expecting for $\hat{p}$, perhaps from a similar survey, we could use that value. If we have no such estimate, we use 0.50 as a conservative estimate. It turns out that the margin of error is largest when $\hat{p}$ is 0.5, so we typically use this *worst case estimate* of $\hat{p} = 0.5$ if no other estimate is available.

$$\begin{aligned}
1.96 \times \sqrt{\frac{0.5(1-0.5)}{n}} &\le 0.04 \\
1.96^2 \times \frac{0.5(1-0.5)}{n} &\le 0.04^2 \\
1.96^2 \times \frac{0.5(1-0.5)}{0.04^2} &\le n \\
600.25 &\le n \\
n &= 601
\end{aligned}$$

The sample size must be an integer and we round up because n must be greater than or equal to 600.25. We need at least 601 participants to ensure the sample proportion is within 0.04 of the true proportion with 95% confidence, so 601 is the smallest sample size that will suffice.

In sample size computations for a proportion, if we have a reliable estimate of the proportion, we should use it. If not, we use the conservative estimate of 0.5.

E

EXAMPLE 3.32

A recent estimate of Congress' approval rating was 17%. If another poll were taken, what minimum sample size does this estimate suggest should be used to have a margin of error no greater than 0.04 with 95% confidence?

We complete the same computations as before, except now we use 0.17 instead of 0.5 for the proportion:

$$\begin{aligned} 1.96 \times \sqrt{\frac{0.17(1-0.17)}{n}} &\le 0.04 \\ 1.96^2 \times \frac{0.17(1-0.17)}{n} &\le 0.04^2 \\ 1.96^2 \times \frac{0.17(1-0.17)}{0.04^2} &\le n \\ 338.8 &\le n \\ n &= 339 \end{aligned}$$

If the true proportion is 0.17, then 339 is the minimum sample size that will ensure a margin of error no greater than 0.04 with 95% confidence.

IDENTIFY A SAMPLE SIZE FOR A PARTICULAR MARGIN OF ERROR

When estimating a single proportion at a given confidence level, we find the minimum sample size n to have no greater than a certain margin of error MOE as follows:

$$z^\star \times \sqrt{\frac{\hat{p}(1-\hat{p})}{n}} \le MOE$$

$$(z^\star)^2 \times \frac{\hat{p}(1-\hat{p})}{(MOE)^2} \le n$$

where $z^\star$ depends on the confidence level. If no rough expected value for $\hat{p}$ exists, use $\hat{p} = 0.5$.

G

GUIDED PRACTICE 3.33

A manager is about to oversee the mass production of a new tire model in her factory, and she would like to estimate the proportion of these tires that will be rejected through quality control. The quality control team has previously found that about 6.2% of tires fail inspection.

(a) How many tires should the manager examine to estimate the failure rate of the new tire model to within 2% with a 90% confidence level?[17]

(b) What if the estimate of p is 1.7% rather than 6.2%?[18]

3.3.8 Technology: the one-sample Z-interval for p

Section 3.4.8 demonstrates how to calculate the one-sample Z-interval for p and the one-sample Z-test for p (introduced in the next section) using Desmos, R, and the NumWorks, TI-83/84 and Casio calculators.

[17]The $z^\star$ corresponding to a 90% confidence level is 1.645. Since we have an estimate for p of 6.2%, we use it. So we have: $1.645 \times \sqrt{\frac{0.062(1-0.062)}{n}} \le 0.02$. Rearranging for n gives: $n \ge 393.4$, so she should use $n = 394$.

[18]Substituting 0.017 for p gives an n of 114. We can note that in this case $n \times p = 114 \times 0.017 = 1.9 < 10$. Since the large counts condition is not met, the use of $z^\star = 1.645$ based on a normal model is not appropriate. We would need additional methods than what we've covered so far to get a good estimate for the minimum sample size in this scenario.

Section summary

- A confidence interval is an interval estimate for a population parameter based on a sample statistic. The appropriate confidence interval procedure to estimate a single population proportion p is a **one-sample Z-interval for a population proportion p.** The parameter p should be identified in context.
- A one-sample Z-interval for a population proportion requires the following conditions be met:
  1. Independence: The data should come from a random sample or random process. When sampling without replacement, check that the sample size n is less than 10% of the population size.
  2. Large counts : $n\hat{p} \geq 10$ and $n(1-\hat{p}) \geq 10$.
- The general form for a C% confidence interval is:

  point estimate ± margin of error, or
  point estimate ± critical value × SE of estimate.
- A one-sample Z-interval for a population proportion p can be written as:

$$\hat{p} \pm z^{\star}\sqrt{\frac{\hat{p}(1-\hat{p})}{n}}$$

  where $z^{\star}$ denotes the critical value, such that C% of the standard normal distribution is enclosed between $-z^{\star}$ and $+z^{\star}$. C% represents the confidence level, e.g. 95%.
- The **standard error** (SE) of an estimate/statistic is an estimate of the standard deviation of the sampling distribution of the statistic. The SE quantifies the typical amount that a statistic will vary from the value of the corresponding population parameter. SE of $\hat{p} = \sqrt{\frac{\hat{p}(1-\hat{p})}{n}}$.
- The **margin of error** of $\hat{p}$ is: $z^{\star}\sqrt{\frac{\hat{p}(1-\hat{p})}{n}}$, and is half of the width of the confidence interval.
- To find the **minimum sample size** needed to estimate a proportion with a given confidence level and no greater than a certain margin of error MOE , set up an inequality of the form:

$$z^{\star}\sqrt{\frac{\hat{p}(1-\hat{p})}{n}} \leq MOE, \text{ which can be rewritten as: } (z^{\star})^2\frac{\hat{p}(1-\hat{p})}{(MOE)^2} \leq n$$

  $z^{\star}$ depends on the desired confidence level. Unless an approximate proportion is available, use $\hat{p} = 0.5$. Solve for the sample size n. The final answer for n should be an *integer*.
- Because the confidence interval is based on a sample, the point estimate has associated error and the confidence interval may or may not contain the true value of the population proportion.
- The interpretation of a C% **confidence level** is that in repeated random sampling with the same sample size, approximately C% of confidence intervals calculated will capture the population proportion.
- We say we are C% confident that a *particular* interval (_, _) contains the population proportion.
- A confidence interval provides a range of plausible values for a parameter and can be used as evidence to justify a claim about a population proportion. At a particular confidence level, values are considered plausible if they are inside the confidence interval and values are considered implausible if they are outside the confidence interval.
- For a given sample, increasing the confidence level will result in a larger critical value, a larger margin of error, and a wider confidence interval.
- Increasing the sample size n decreases the standard error of $\hat{p}$ and, when all other things remain the same, decreases the width of a confidence interval for p. The width of the interval is approximately proportional to $\frac{1}{\sqrt{n}}$.

Exercises

3.15 Chronic illness, Part I. In 2013, the Pew Research Foundation reported that "45% of U.S. adults report that they live with one or more chronic conditions".[19] However, this value was based on a sample, so it may not be a perfect estimate for the population parameter of interest on its own. The study reported a standard error of about 1.2%, and a normal model may reasonably be used in this setting. Create a 95% confidence interval for the proportion of U.S. adults who live with one or more chronic conditions. Also interpret the confidence interval in the context of the study.

3.16 Twitter users and news, Part I. A poll conducted in 2013 found that 52% of U.S. adult Twitter users get at least some news on Twitter.[20]. The standard error for this estimate was 2.4%, and a normal distribution may be used to model the sample proportion. Construct a 99% confidence interval for the fraction of U.S. adult Twitter users who get some news on Twitter, and interpret the confidence interval in context.

3.17 Waiting at an ER, Part I. A hospital administrator hoping to improve wait times decides to estimate the average emergency room waiting time at her hospital. She collects a simple random sample of 64 patients and determines the time (in minutes) between when they checked in to the ER until they were first seen by a doctor. A 95% confidence interval based on this sample is (128 minutes, 147 minutes), which is based on the normal model for the mean. Determine whether the following statements are true or false, and explain your reasoning.

(a) We are 95% confident that the average waiting time of these 64 emergency room patients is between 128 and 147 minutes.

(b) We are 95% confident that the average waiting time of all patients at this hospital's emergency room is between 128 and 147 minutes.

(c) 95% of random samples have a sample mean between 128 and 147 minutes.

(d) A 99% confidence interval would be narrower than the 95% confidence interval since we need to be more sure of our estimate.

(e) The margin of error is 9.5 and the sample mean is 137.5.

(f) In order to decrease the margin of error of a 95% confidence interval to half of what it is now, we would need to double the sample size.

3.18 Mental health. The General Social Survey asked the question: "For how many days during the past 30 days was your mental health, which includes stress, depression, and problems with emotions, not good?" Based on responses from 1,151 US residents, the survey reported a 95% confidence interval of 3.40 to 4.24 days in 2010.

(a) Interpret this interval in context of the data.

(b) What does "95% confident" mean? Explain in the context of the application.

(c) Suppose the researchers think a 99% confidence level would be more appropriate for this interval. Will this new interval be smaller or wider than the 95% confidence interval?

(d) If a new survey were to be done with 500 Americans, do you think the standard error of the estimate be larger, smaller, or about the same.

3.19 Cyberbullying rates. Teens were surveyed about cyberbullying, and 54% to 64% reported experiencing cyberbullying (95% confidence interval).[21] Answer the following questions based on this interval.

(a) A newspaper claims that a majority of teens have experienced cyberbullying. Is this claim supported by the confidence interval? Explain your reasoning.

(b) A researcher conjectured that 70% of teens have experienced cyberbullying. Is this claim supported by the confidence interval? Explain your reasoning.

(c) Without actually calculating the interval, determine if the claim of the researcher from part (b) would be supported based on a 90% confidence interval?

---

[19] Pew Research Center, Washington, D.C. The Diagnosis Difference, November 26, 2013.

[20] Pew Research Center, Washington, D.C. Twitter News Consumers: Young, Mobile and Educated, November 4, 2013.

[21] Pew Research Center, A Majority of Teens Have Experienced Some Form of Cyberbullying. September 27, 2018.

3.20 Waiting at an ER, Part II. Exercise 3.17 provides a 95% confidence interval for the mean waiting time at an emergency room (ER) of (128 minutes, 147 minutes). Answer the following questions based on this interval.

(a) A local newspaper claims that the average waiting time at this ER exceeds 3 hours. Is this claim supported by the confidence interval? Explain your reasoning.

(b) The Dean of Medicine at this hospital claims the average wait time is 2.2 hours. Is this claim supported by the confidence interval? Explain your reasoning.

(c) Without actually calculating the interval, determine if the claim of the Dean from part (b) would be supported based on a 99% confidence interval?

3.21 Orange tabbies. Suppose that 90% of orange tabby cats are male. Determine if the following statements are true or false, and explain your reasoning.

(a) The distribution of sample proportions of random samples of size 30 is left skewed.

(b) Using a sample size that is 4 times as large will reduce the standard error of the sample proportion by one-half.

(c) The distribution of sample proportions of random samples of size 140 is approximately normal.

(d) The distribution of sample proportions of random samples of size 280 is approximately normal.

3.22 Young Americans, Part II. About 25% of young Americans have delayed starting a family due to the continued economic slump. Determine if the following statements are true or false, and explain your reasoning.[22]

(a) The distribution of sample proportions of young Americans who have delayed starting a family due to the continued economic slump in random samples of size 12 is right skewed.

(b) In order for the distribution of sample proportions of young Americans who have delayed starting a family due to the continued economic slump to be approximately normal, we need random samples where the sample size is at least 40.

(c) A random sample of 50 young Americans where 20% have delayed starting a family due to the continued economic slump would be considered unusual.

(d) A random sample of 150 young Americans where 20% have delayed starting a family due to the continued economic slump would be considered unusual.

(e) Tripling the sample size will reduce the standard error of the sample proportion by one-third.

3.23 Gender equality. The General Social Survey asked a random sample of 1,390 Americans the following question: "On the whole, do you think it should or should not be the government's responsibility to promote equality between men and women?" 82% of the respondents said it "should be". At a 95% confidence level, this sample has 2% margin of error. Based on this information, determine if the following statements are true or false, and explain your reasoning.[23]

(a) We are 95% confident that between 80% and 84% of Americans in this sample think it's the government's responsibility to promote equality between men and women.

(b) We are 95% confident that between 80% and 84% of all Americans think it's the government's responsibility to promote equality between men and women.

(c) If we considered many random samples of 1,390 Americans, and we calculated 95% confidence intervals for each, 95% of these intervals would include the true population proportion of Americans who think it's the government's responsibility to promote equality between men and women.

(d) In order to decrease the margin of error to 1%, we would need to quadruple (multiply by 4) the sample size.

(e) Based on this confidence interval, there is sufficient evidence to conclude that a majority of Americans think it's the government's responsibility to promote equality between men and women.

3.24 Elderly drivers. The Marist Poll published a report stating that 66% of adults nationally think licensed drivers should be required to retake their road test once they reach 65 years of age. It was also reported that interviews were conducted on a random sample of 1,018 American adults, and that the margin of error was 3% using a 95% confidence level.[24]

(a) Verify the margin of error reported by The Marist Poll.

(b) Based on a 95% confidence interval, does the poll provide convincing evidence that *more than* two thirds of the population think that licensed drivers should be required to retake their road test once they turn 65?

---

[22] Demos.org. "The State of Young America: The Poll". In: (2011).

[23] National Opinion Research Center, General Social Survey, 2018.

[24] Marist Poll, Road Rules: Re-Testing Drivers at Age 65?, March 4, 2011.

3.25 Fireworks on July 4th. A local news outlet reported that 56% of 600 randomly sampled Kansas residents planned to set off fireworks on July 4th. Determine the margin of error for the 56% point estimate using a 95% confidence level.[25]

3.26 Life rating in Greece. Greece has faced a severe economic crisis since the end of 2009. A Gallup poll surveyed 1,000 randomly sampled Greeks in 2011 and found that 25% of them said they would rate their lives poorly enough to be considered "suffering".[26]

(a) Describe the population parameter of interest. What is the value of the point estimate of this parameter?

(b) Check if the conditions required for constructing a confidence interval based on these data are met.

(c) Construct a 95% confidence interval for the proportion of Greeks who are "suffering".

(d) Without doing any calculations, describe what would happen to the confidence interval if we decided to use a higher confidence level.

(e) Without doing any calculations, describe what would happen to the confidence interval if we used a larger sample.

3.27 Study abroad. A survey on 1,509 high school seniors who took the SAT and who completed an optional web survey shows that 55% of high school seniors are fairly certain that they will participate in a study abroad program in college.[27]

(a) Is this sample a representative sample from the population of all high school seniors in the US? Explain your reasoning.

(b) Let's suppose the conditions for inference are met. Even if your answer to part (a) indicated that this approach would not be reliable, this analysis may still be interesting to carry out (though not report). Construct a 90% confidence interval for the proportion of high school seniors (of those who took the SAT) who are fairly certain they will participate in a study abroad program in college, and interpret this interval in context.

(c) What does "90% confidence" mean?

(d) Based on this interval, would it be appropriate to claim that the majority of high school seniors are fairly certain that they will participate in a study abroad program in college?

3.28 Legalization of marijuana, Part I. The General Social Survey asked a random sample of 1,578 US residents: "Do you think the use of marijuana should be made legal, or not?" 61% of the respondents said it should be made legal.[28]

(a) Is 61% a sample statistic or a population parameter? Explain.

(b) Construct a 95% confidence interval for the proportion of US residents who think marijuana should be made legal, and interpret it in the context of the data.

(c) A critic points out that this 95% confidence interval is only accurate if the statistic follows a normal distribution, or if the normal model is a good approximation. Is this true for these data? Explain.

(d) A news piece on this survey's findings states, "Majority of Americans think marijuana should be legalized." Based on your confidence interval, is this news piece's statement justified?

3.29 National Health Plan, Part II. Exercise 3.35 presents the results of a poll evaluating support for a generic "National Health Plan" in the US in 2019, reporting that 55% of Independents are supportive. If we want to estimate the percent of Independents who are supportive this year to within 1% with 90% confidence, what would be an appropriate sample size?

3.30 Legalize Marijuana, Part II. As discussed in Exercise 3.28, the General Social Survey reported a sample where about 61% of US residents thought marijuana should be made legal. If we wanted to limit the margin of error of a 95% confidence interval to 2%, about how many Americans would we need to survey?

---

[25]Survey USA, News Poll #19333, data collected on June 27, 2012.

[26]Gallup World, More Than One in 10 "Suffering" Worldwide, data collected throughout 2011.

[27]studentPOLL, College-Bound Students' Interests in Study Abroad and Other International Learning Activities, January 2008.

[28]National Opinion Research Center, General Social Survey, 2018.

3.4 Hypothesis testing for a population proportion

A consultant carries out a random sample and finds support for a particular candidate for office to be 52%. Is this convincing evidence that the candidate's support is really above 50% in the overall population? If so, how much evidence is there for this claim? In this section, we will set up a framework for answering questions such as this and will look at the different types of decision errors that researchers can make when drawing conclusions based on data.

Learning objectives

1. Identify an appropriate testing method for a population proportion p and describe the parameter in context.
2. Identify the null and alternative hypotheses for a population proportion.
3. Justify the appropriateness of a hypothesis test for a population proportion using a normal distribution by verifying conditions are met.
4. Interpret the p-value of a hypothesis test for a population proportion.
5. Calculate an appropriate test statistic and p-value for testing a hypothesis about a population proportion.
6. Justify a claim about the population proportion based on the results of the hypothesis test.
7. Identify Type I and Type II errors.
8. Find the probability of Type I and Type II errors.
9. Identify the factors that affect the probability of errors in hypothesis testing.
10. Interpret Type I and Type II errors.

3.4.1 Case study: medical consultant

People providing an organ for donation sometimes seek the help of a special medical consultant. These consultants assist the patient in all aspects of the surgery, with the goal of reducing the possibility of complications during the medical procedure and recovery. Patients might choose a consultant based in part on the historical complication rate of the consultant's clients.

One consultant tried to attract patients by noting the overall complication rate for liver donor surgeries in the US is about 10%, but her clients have had only 9 complications in the 142 liver donor surgeries she has facilitated. She claims this is strong evidence that her work meaningfully contributes to reducing complications (and therefore she should be hired!).

E

EXAMPLE 3.34

We will let p represent the true complication rate for liver donors working with this consultant. Calculate the best estimate for p using the data. Label the point estimate as $\hat{p}$.

The sample proportion for the complication rate is 9 complications divided by the 142 surgeries the consultant has worked on: $\hat{p} = 9/142 = 0.063$.

EXAMPLE 3.35

Is it possible to prove that the consultant's work reduces complications?

No. The claim implies that there is a causal connection, but the data are observational. For example, maybe patients who can afford a medical consultant can afford better medical care, which can also lead to a lower complication rate.

EXAMPLE 3.36

While it is not possible to assess the causal claim, it is still possible to ask whether the low complication rate of $\hat{p} = 0.063$ provides evidence that the consultant's true complication rate is different than the US complication rate. Why might we be tempted to immediately conclude that the consultant's true complication rate is different than the US complication rate? Can we draw this conclusion?

Her sample complication rate is $\hat{p} = 0.063$, which is 0.037 lower than the US complication rate of 10%. However, we cannot yet be sure if the observed difference represents a real difference or is just the result of random variation. We wouldn't expect the sample proportion to be *exactly* 0.10, even if the truth was that her real complication rate was 0.10.

3.4.2 Setting up the null and alternative hypothesis

We can set up two competing hypotheses about the consultant's true complication rate. The first is call the **null hypothesis** and represents either a skeptical perspective or a perspective of no difference. The second is called the **alternative hypothesis** (or alternate hypothesis) and represents a new perspective such as the possibility that there has been a change or that there is a treatment effect in an experiment.

NULL AND ALTERNATIVE HYPOTHESES

The **null hypothesis** is abbreviated H_0. It represents a skeptical perspective and is often a claim of no change or no difference.

The **alternative hypothesis** is abbreviated H_A. It is the claim researchers hope to prove or find evidence for, and it often asserts that there has been a change or an effect.

Our job as data scientists is to play the skeptic: before we buy into the alternative hypothesis, we need to see strong supporting evidence.

EXAMPLE 3.37

Identify the null and alternative claim regarding the consultant's complication rate.

H_0: The true complication rate for the consultant's clients is equal to 10%.

H_A: The true complication rate for the consultant's clients is not equal to 10%.

Often it is convenient to write the null and alternative hypothesis in abbreviated mathematical or numerical terms. To do so, we first identify the parameter of interest. The parameter in a hypothesis test is a true but unknown value regarding the population of interest. When the parameter is a proportion, as in Example 3.37, we label it p. For this example, we would define p and write the hypotheses as:

p: the true complication rate for the consultant's clients

H_0: $p = 0.10$

H_A: $p \neq 0.10$

The true complication rate for this consultant's clients is unknown, but the null hypothesis is that it equals 0.10, the overall proportion of complications. This hypothesized value is called the **null value**.

NULL VALUE OF A HYPOTHESIS TEST

The **null value** is the value hypothesized for the parameter in H_0, and it is sometimes represented with a subscript 0, e.g. p_0 (just like H_0).

The null claim is always framed as an equality: it tells us what quantity we should use for the parameter when carrying out calculations for the hypothesis test. There are three choices for the alternative hypothesis, depending upon whether the researcher is trying to prove that the value of the parameter is greater than, less than, or not equal to the null value.

ALWAYS WRITE THE NULL HYPOTHESIS AS AN EQUALITY

We will find it most useful if we always list the null hypothesis as an equality (e.g. $p = 0.1$) while the alternative always uses an inequality (e.g. $p \neq 0.1$, $p > 0.1$, or $p < 0.1$).

These hypotheses are part of what is called a **hypothesis test**. A hypothesis test is a statistical technique used to evaluate competing claims based on data. Often times, the null hypothesis takes a stance of *no difference* or *no effect*. If the null hypothesis and the data notably disagree, then we will reject the null hypothesis in favor of the alternative hypothesis.

GUIDED PRACTICE 3.38

G According to the 2020 US Census, 6.6% of residents in the state of Alaska were under 5 years old.[29] A researcher plans to take a random sample of residents from Alaska to test whether or not this is still the case. Identify the parameter of interest and write the hypotheses that the researcher should test in both plain and statistical language.[30]

When the alternative claim uses a $\neq$, we call the test a **two-sided** test, because either extreme provides evidence against H_0. When the alternative claim uses a $<$ or a $>$, we call it a **one-sided** test.

ONE-SIDED AND TWO-SIDED TESTS

If the researchers are only interested in showing an increase or a decrease, but not both, use a one-sided test. If the researchers would be interested in any difference from the null value – an increase or decrease – then the test should be two-sided.

EXAMPLE 3.39

E Let's re-examine hypothesis test for the consultant's complication rate from Example 3.37. The hypotheses were H_0: $p = 0.10$ versus H_A: $p \neq 0.10$. We knew that her sample complication rate was 0.063, which was lower than the US complication rate of 0.10. Why did we conduct a two-sided hypothesis test for this setting?

The setting was framed in the context of the consultant being helpful, but what if the consultant actually performed worse than the US complication rate? Would we care? More than ever! Since we care about a finding in either direction, we should run a two-sided test.

[30] `www.census.gov/library/visualizations/interactive/exploring-age-groups-in-the-2020-census.html`

[30] p: the *current* proportion of residents in Alaska that are under 5 years old.
H_0: $p = 0.066$; The proportion is *unchanged* from 2020.
H_A: $p \neq 0.066$; The proportion has changed from 2020. It could have increased or decreased.

ONE-SIDED HYPOTHESES ARE ALLOWED ONLY *BEFORE* SEEING DATA

After observing data, it is tempting to turn a two-sided test into a one-sided test. Avoid this temptation. Hypotheses must be set up *before* observing the data. If they are not, the test must be two-sided.

3.4.3 Evaluating the hypotheses with a p-value

EXAMPLE 3.40

E

If the null claim is true, what proportion would we expect to have had a complication?

If the null claim is true, we would expect about 10% of the patients to have a complication.

The consultant's complication rate for her 142 clients was $9/142 = 0.063$. What is the probability that a sample of size 142 would produce a complication rate this far from the expected rate of 0.10, assuming H_0 were true? We call this probability the **p-value**. When conditions are met, the p-value can be estimated using a a normal model as shown in Figure 3.10. When a normal model is not appropriate the p-value can be estimated using a simulation technique as described in Section 3.4.4 on page 228.

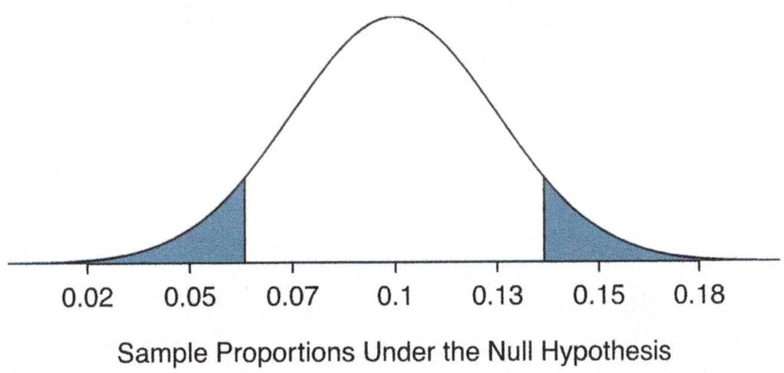

Figure 3.10: The shaded area represents the p-value. We observed $\hat{p} = 0.063$. Any observations smaller than this are at least as extreme relative to the null value, $p_0 = 0.10$, and so the lower tail is shaded. However, since this is a two-sided test, values above 0.137 are also at least as extreme as 0.063 (relative to 0.10), and so they are also shaded. The tail areas together represent the p-value.

When working with proportions, we can say that the p-value is the probability of getting a sample proportion as far from or farther from the null proportion in the direction of H_A if the null hypothesis is true. In general, we calculate and interpret a p-value under one of three scenarios.

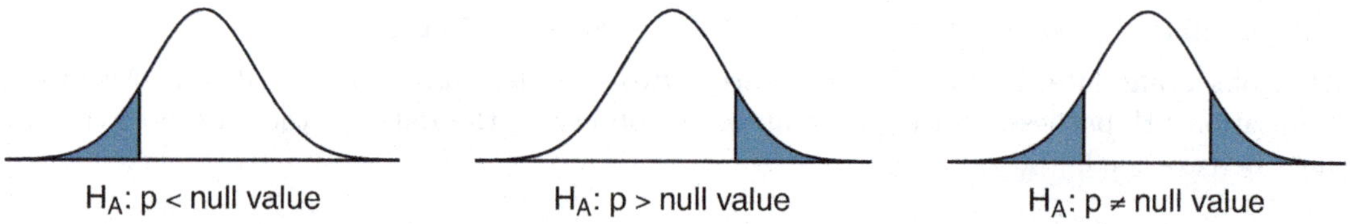

Figure 3.11: When the alternative hypothesis takes the form $p <$ null value, the p-value is represented by the lower tail. When it takes the form $p >$ null value, the p-value is represented by the upper tail. When using $p \neq$ null value, then the p-value is represented by both tails.

FINDING AND INTERPRETING THE P-VALUE

We find and interpret the **p-value** according to the nature of the alternative hypothesis.

H_A: **parameter > null value.** The p-value corresponds to the area in the *upper* tail and is probability of getting a test statistic as large or larger than the observed test statistic if the null hypothesis is true.

H_A: **parameter < null value.** The p-value corresponds to the area in the *lower* tail and is the probability of observing a test statistic as small or smaller than the observed test statistic if the null hypothesis is true.

H_A: **parameter ≠ null value.** The p-value corresponds to the area in *both* tails and is the probability of observing a test statistic as extreme or more extreme than the observed test statistic if the null hypothesis is true.

More generally, we can say that the p-value is the probability of getting a test statistic as extreme or more extreme than the observed test statistic in the direction of H_A if the null hypothesis is true.

When the p-value is small, i.e. less than or equal to a previously set threshold, we say the results are **statistically significant**. This means the data provide such strong evidence against H_0 that we reject the null hypothesis in favor of the alternative hypothesis. The threshold is called the **significance level** and is represented by α (the Greek letter *alpha*). The significance level is typically set to $\alpha = 0.05$, but it can vary depending on the field or the application.

STATISTICAL SIGNIFICANCE

If the p-value is less than or equal to the significance level α (usually 0.05), we say that the result is **statistically significant**. We reject H_0, and we have strong evidence favoring H_A.

If the p-value is greater than the significance level α, we say that the result is not statistically significant. We do not reject H_0, and we do not have sufficient evidence for H_A.

Recall that the null claim is the claim of no difference. If we reject H_0, we are asserting that there is strong evidence of a real difference. If we do not reject H_0, we are saying that the null claim is not unreasonable based on the data analyzed, but we are not saying that the null claim has been proven.

GUIDED PRACTICE 3.41

In our consultant's complication rate example the p-value is 0.1754. This is larger than the significance level 0.05, so we do not reject the null hypothesis. Explain what this means in the context of the problem using plain language.[31]

[31] The data do not provide evidence that the consultant's complication rate is significantly lower or higher than the US complication rate of 10%.

EXAMPLE 3.42

In the previous exercise, we did not reject H_0. This means that we did not disprove the null claim. Is this equivalent to proving the null claim is true?

No. We did not prove that the consultant's complication rate is *exactly* equal to 10%. Recall that the test of hypothesis starts by *assuming the null claim is true.* That is, the test proceeds as an argument by contradiction. *If the null claim is true*, there is a 0.1754 chance of seeing sample data as divergent from 10% as we saw in our sample. Because 0.1754 is large, it is within the realm of chance error, and we cannot say the null hypothesis is unreasonable.[32]

DOUBLE NEGATIVES CAN SOMETIMES BE USED IN STATISTICS

In many statistical explanations, we use double negatives. For instance, we might say that the null hypothesis is *not implausible* or we *failed to reject* the null hypothesis. Double negatives are used to communicate that while we are not rejecting a position, we are also not saying that we know it to be true.

EXAMPLE 3.43

Does the conclusion in Guided Practice 3.41 ensure that there is no real association between the surgical consultant's work and the risk of complications? Explain.

No. It is possible that the consultant's work is associated with a lower or higher risk of complications. If this was the case, the sample may have been too small to reliable detect this effect.

EXAMPLE 3.44

An experiment was conducted where study participants were randomly divided into two groups. Both were given the opportunity to purchase a DVD, but one half was reminded that the money, if not spent on the DVD, could be used for other purchases in the future, while the other half was not. The half that was reminded that the money could be used on other purchases was 20% less likely to continue with a DVD purchase. We determined that such a large difference would only occur about 1-in-150 times if the reminder actually had no influence on student decision-making. What is the p-value in this study? Was the result statistically significant?

The p-value was 0.006 (about 1/150). Since the p-value is less than 0.05, the data provide statistically significant evidence that US college students were actually influenced by the reminder.

WHAT'S SO SPECIAL ABOUT 0.05?

We often use a threshold of 0.05 to determine whether a result is statistically significant. But why 0.05? Maybe we should use a bigger number, or maybe a smaller number. If you're a little puzzled, that probably means you're reading with a critical eye – good job! We've made a video to help clarify *why 0.05*:

www.openintro.org/why05

Sometimes it's a good idea to deviate from the standard. We'll discuss when to choose a threshold different than 0.05 in Section 3.4.7.

Statistical inference is the practice of making decisions and conclusions from data in the context of uncertainty. Just as a confidence interval may occasionally fail to contain the true value of the parameter, a test of hypothesis may occasionally lead us to an incorrect conclusion. While a given data set may not always lead us to a correct conclusion, statistical inference gives us tools to control and evaluate how often these errors occur.

---

[32]The p-value is a conditional probability. It is P(getting data at least as divergent from the null value as we observed | H_0 is true). It is NOT P(H_0 is true | we got data this divergent from the null value).

3.4.4 Calculating the p-value by simulation

When conditions for the applying a normal model are met, we use a normal model to find the p-value of a test of hypothesis. A more general approach, though, for calculating p-values when a normal model does not apply is to use what is known as **simulation**, the idea of which was introduced in Section 2.7. While performing this procedure is outside of the scope of the course, we provide an example here that will help us better understand the concept of a p-value.

We simulate 142 new patients to see what result might happen if the complication rate really is 0.10. To do this, we could use a deck of cards. Take one red card, nine black cards, and mix them up. If the cards are well-shuffled, drawing the top card is one way of simulating the chance a patient has a complication if the true rate is 0.10: if the card is red, we say the patient had a complication, and if it is black then we say they did not have a complication. If we repeat this process 142 times and compute the proportion of simulated patients with complications, $\hat{p}_{sim}$, then this simulated proportion is exactly a draw from the null distribution.

There were 12 simulated cases with a complication and 130 simulated cases without a complication: $\hat{p}_{sim} = 12/142 = 0.085$.

One simulation isn't enough to get a sense of the null distribution, so we repeated the simulation 10,000 times using a computer. Figure 3.12 shows the randomization distribution from these 10,000 simulations. The simulated proportions that are less than or equal to $\hat{p} = 0.063$ are shaded. There were 0.0877 simulated sample proportions with $\hat{p}_{sim} \le 0.063$, which represents a fraction 0.0877 of our simulations:

$$\text{left tail} = \frac{\text{Number of observed simulations with } \hat{p}_{sim} \le \ 0.063}{10000} = \frac{877}{10000} = 0.0877$$

However, this is not our p-value! Remember that we are conducting a two-sided test, so we should double the one-tail area to get the p-value:[33]

$$\text{p-value} = 2 \times \text{left tail} = 2 \times 0.0877 = 0.1754$$

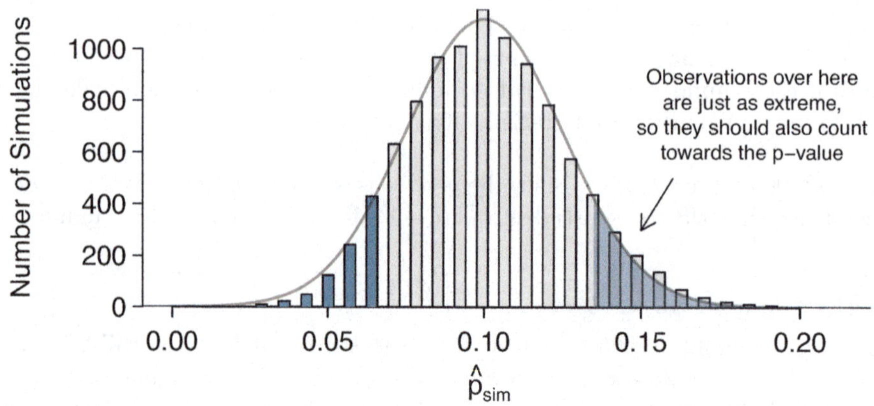

Figure 3.12: The null distribution or randomization distribution for $\hat{p}$, created from 10,000 simulated studies. The left tail contains 8.77% of the simulations. For a two-sided test, we double the tail area to get the p-value. This doubling accounts for the observations we might have observed in the upper tail, which are also at least as extreme (relative to 0.10) as what we observed, $\hat{p} = 0.063$.

[33]This doubling approach is preferred even when the distribution isn't symmetric, as in this case.

3.4.5 Checking conditions and carrying out a test for a proportion

Now that we have discussed the basic logic and framework for hypothesis testing, we consider the conditions and calculations necessary for a hypothesis test for a proportion.

EXAMPLE 3.45

Deborah Toohey is running for Congress. A large campaign donor will choose to support Toohey if they are confident that she has a majority of support from the district's electorate and so is more likely to win. A research team collects a random sample of 500 likely voters in the district and estimates Toohey's support to be 52%. Identify the parameter of interest and the hypotheses to be tested. What value should we use as the null value, p_0?

(E) The parameter of interest is p: the true proportion of support for Deborah Toohey among likely voters in the district. The alternative hypothesis, the one that bears the burden of proof, argues that Toohey has more than 50% support. Therefore, H_A will be one-sided and the null value will be $p_0 = 0.5$. So we have:

H_0: $p = 0.5$

H_A: $p > 0.5$.

Note that the hypotheses are about a population parameter. The hypotheses are never about the sample.

To calculate the p-value for this test, we will first calculate a test statistic, which takes the general form:

$$\text{test statistic} = \frac{\text{point estimate} - \text{null value}}{SE \text{ of estimate}}$$

When conditions for a normal model are met, the test statistic is called a Z-statistic and the test is called a Z-test. The conditions for a normal model for a **one-sample Z-test for p**, a population proportion, are:

Independence. Data should be collected using a random sample or process. If sampling without replacement, the sample size must be less than 10% of the population size, i.e. $n < 0.10(N)$.

Large counts. The expected number of success and expected number of failures, assuming H_0 is true must be at least 10, i.e. $np_0 \geq 10$ and $n(1 - p_0) \geq 10$.

These conditions will look familiar: they are very similar to the ones we used for the confidence interval for a proportion. The only difference is that for the confidence interval we use the sample proportion for the large counts check, whereas for the hypothesis test we use the null proportion for the large counts check.

EXAMPLE 3.46

Check whether conditions for using a normal model are met in the Deborah Toohey example.

(E) First, we observe that the problem states that a random sample was chosen. We will assume that the size of the electorate in Toohey's district is more than 10 times the size of the sample, that is will assume that the size of the electorate in her district is greater than $10 \times 500 = 5,000$. Next, we check the large counts condition. Because we assume that $p = p_0$ for the calculations of the hypothesis test, we use the hypothesized value p_0 rather than the sample value $\hat{p}$ when verifying the large counts condition.

$$\begin{aligned} np_0 \geq 10 \quad &\rightarrow \quad 500(0.5) \geq 10 \\ n(1 - p_0) \geq 10 \quad &\rightarrow \quad 500(1 - 0.5) \geq 10 \end{aligned}$$

The conditions for a normal model are met.

CONFIDENCE INTERVALS VERSUS HYPOTHESIS TESTS FOR A SINGLE PROPORTION

one-sample Z-interval for p:

Check: $n\hat{p} \geq 10$ and $n(1-\hat{p}) \geq 10$ Use: $SE_{\hat{p}} = \sqrt{\frac{\hat{p}(1-\hat{p})}{n}}$

one-sample Z-test for p:

Check: $np_0 \geq 10$ and $n(1-p_0) \geq 10$ Use: $SE_{\hat{p}} = \sqrt{\frac{p_0(1-p_0)}{n}}$

With the conditions met, we can calculate the test statistic.

$$\text{test statistic} = \frac{\text{point estimate} - \text{null value}}{SE \text{ of estimate}}$$

- The test statistic here is a Z-statistic.
- The point estimate is the sample proportion $\hat{p}$ (same as for a confidence interval).
- The null value is p_0.
- We compute the test statistic assuming the null hypothesis is true, that is that $p = p_0$. Because we have a hypothesized value p_0, we use it in place of $\hat{p}$ for the standard error calculation as follows: $\sqrt{\frac{p_0(1-p_0)}{n}}$. This is essentially $\sigma_{\hat{p}}$, with p_0 substituted in for p. For a one-sample Z-test, we compute the test statistic as follows:

$$Z = \frac{\hat{p} - p_0}{\sqrt{\frac{p_0(1-p_0)}{n}}}$$

EXAMPLE 3.47

(Continues previous example). A large campaign donor will choose to support Toohey for Congress if they are confident that she has a majority of support from the district's electorate. A research team collects a random sample of 500 likely voters in the district and finds that 52% of 500 likely voters who were sampled support Toohey. Does this provide convincing evidence, at the 5% significance level, that Tahooey has more than 50% support among all likely voters in the district?

We will use a one-sample Z-test for p, where
p: the true proportion of support for Deborah Toohey among likely voters in the district.

(E) H_0: $p = 0.5$. Toohey's support is 50%.

H_A: $p > 0.5$. Toohey's support is higher than 50%.

We will use a significance level of $\alpha = 0.05$ for the test. The test statistic can be computed as:

$$Z = \frac{\hat{p} - p_0}{\sqrt{\frac{p_0(1-p_0)}{n}}} = \frac{0.52 - 0.5}{\sqrt{\frac{0.5(0.5)}{500}}} = \frac{0.52 - 0.5}{0.022} = 0.89$$

Because the alternative hypothesis uses a greater than sign (>), this is an upper-tail test. We find the area under the standard normal distribution to the *right* of $Z = 0.89$. Figure 3.13 shows the p-value as the shaded region.

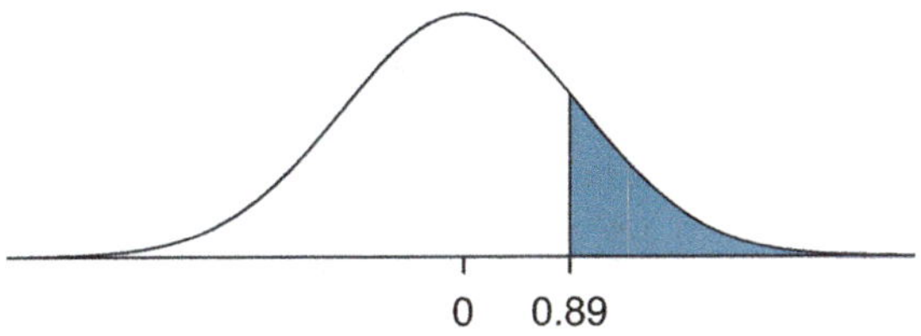

Figure 3.13: Sampling distribution of the sample proportion if the null hypothesis is true for Example 3.47. The p-value for the test with H_A: $p > 0.5$ is shaded.

The p-value, which is the area under the standard normal distribution to the right of Z = 0.89, equals 0.19. This p-value of 0.19 is greater than $\alpha = 0.05$, so we do not reject H_0. That is, we do not have sufficient evidence to support the claim that Tahooey has more than 50% support within the district. The campaign donor will not support her.

EXAMPLE 3.48

Based on the result above, do we have evidence that Toohey's support equals 50%?

No. In a hypothesis test we look for degrees of evidence *against* the null hypothesis. We cannot ever prove the null hypothesis directly. The value 0.5 is reasonable, but many other values are reasonable as well. There are many values that would not get rejected by this test.

EXAMPLE 3.49

Interpret the p-value of 0.19 in the context of this problem

There is a 19% chance of getting a test statistic as large or larger than 0.89 assuming that the true proportion who support Toohey is 0.50 (i.e. assuming the null hypothesis is true). Equivalently, we could say: there is a 19% chance of getting a sample proportion as large or larger than 0.52 assuming that the true proportion who support Toohey is 0.50.

3.4.6 A four-step framework for hypothesis testing procedures

When carrying out a formal hypothesis testing procedure, we will find it useful to use the same four-step framework introduced previously for confidence interval procedures. The general framework is as follows:

- **Identify**: Identify the appropriate test procedure, parameter, significance level, and hypotheses.
- **Check**: Check that the conditions for the test procedure are met.
- **Calculate**: Calculate the test statistic and the p-value.

$$\text{test statistic} = \frac{\text{point estimate} - \text{null value}}{SE \text{ of estimate}}$$

- **Conclude**: Compare the p-value to the significance level to determine whether to reject H_0 or not reject H_0. Draw a conclusion in the context of H_A.

We will apply this framework to hypothesis testing for a single proportion as follows.

HYPOTHESIS TESTING FOR A PROPORTION

To carry out a complete hypothesis test to evaluate a claim about a population proportion,

Identify: Identify the test procedure, parameter, significance level, and hypotheses.

Use a **one-sample Z-test for a population proportion p**. Define the (unknown) population proportion p in words, referencing the population of interest. Choose a significance level (α) and test the following hypotheses.

H_0: $p = p_0$

H_A: $p \neq p_0$; $p > p_0$; or $p < p_0$ (p_0 is the null or hypothesized proportion)

Check: Check conditions for the test statistic to be nearly normal, assuming H_0 is true.

1. Independence: Data come from a random sample or random process. When sampling without replacement, check that sample size is less than 10% of the population size.
2. Large counts: $np_0 \geq 10$ and $n(1 - p_0) \geq 10$

Calculate: Calculate the Z-statistic and p-value.

$$Z = \frac{\text{point estimate} - \text{null value}}{SE \text{ of estimate}}$$

point estimate: $\hat{p}$, the sample proportion

null value: p_0

SE of estimate: $\sqrt{\frac{p_0(1-p_0)}{n}}$ (use p_0 because we are assuming assuming $p = p_0$)

p-value = (based on the Z-statistic and the direction of H_A)

Conclude: Compare the p-value to α, and draw a conclusion in context.

If the p-value is $\leq \alpha$, reject H_0; there is sufficient evidence that [H_A in context].

If the p-value is $> \alpha$, do not reject H_0; there is not sufficient evidence that [H_A in context].

EXAMPLE 3.50

A certain convenience store currently stocks RC Cola (among other colas). The manager is considering replacing RC Cola with Shasta Cola, but only if there is sufficient evidence that customers like Shasta Cola better. To test this, the manager randomly chooses 30 customers from the store and offers each one a free can of either RC Cola or Shasta Cola and records which they choose. Out of the 30 customers, 20 choose Shasta Cola. At the 5% significance level is there sufficient evidence that the customers at this convenience store prefer Shasta Cola to RC Cola?

Identify: Because the hypotheses are about a single proportion, we choose the one-sample Z-test for a population proportion p. Here, p is the proportion of all customers of this convenience store that prefer Shasta Cola to RC Cola. We will test the following hypotheses at the $\alpha = 0.05$ significance level.

H_0: $p = 0.5$
H_A: $p > 0.5$

Check: We must check the independence and large counts conditions.
- We will need to assume that the sample can be considered a random sample of all customers at this convenience store.
- The sample size of 30 should be less than 10% of all customers at this convenience store.
- $30(0.5) = 15 \geq 10$ and $30(1 - 0.5) = 15 \geq 10$, so conditions for a normal model are met.

E

Calculate: First we calculate the Z-statistic: $Z = \frac{\text{point estimate} - \text{null value}}{SE \text{ of estimate}}$

The point estimate is the sample proportion: $\hat{p} = \frac{20}{30} = 0.667$, and the null value is: $p_0 = 0.5$.

The SE of $\hat{p}$, assuming H_0 is true, is: $\sqrt{\frac{p_0(1-p_0)}{n}} = \sqrt{\frac{0.5(1-0.5)}{30}} = 0.0913$.

$$Z = \frac{0.667 - 0.5}{\sqrt{\frac{0.5(1-0.5)}{30}}} = 1.83$$

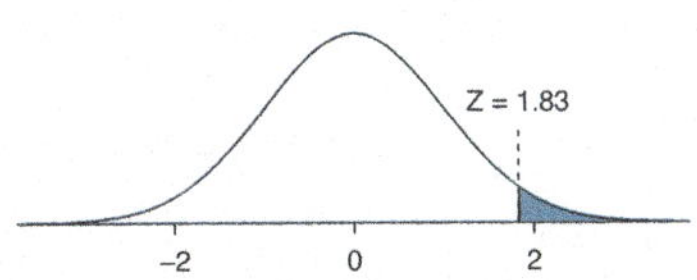

The p-value = 0.034, which corresponds to the area under the standard normal distribution with $Z \geq 1.83$.

Conclude: The p-value = $0.034 < 0.05$, so we reject H_0; there is sufficient evidence that customers of this convenience store prefer Shasta Cola to RC Cola.

G **GUIDED PRACTICE 3.51**

In context, interpret the p-value of 0.034 from the previous example.[34]

G **GUIDED PRACTICE 3.52**

If we had used a significance level of $\alpha = 0.01$, would our conclusion be the same or would it be different?[35]

---

[34]There is a 3.4% chance of getting a test statistic as big or bigger than 1.83 assuming the null hypothesis is true, i.e. that the proportion of all customers at this convenience store that prefer Shasta Cola to RC Cola is 50%.

[35]Because our p-value = $0.034 > 0.010$, we would not reject H_0 and we would not have sufficient evidence to conclude that customers of this convenience store prefer Shasta Cola to RC Cola. With this stricter significance level, our p-value does not provide enough evidence for H_A and we come to a different conclusion. The choice of significance level is important and should be decided before seeing the data.

3.4.7 Decision errors and power

The hypothesis testing framework is a very general tool, and we often use it without a second thought. If a person makes a somewhat unbelievable claim, we are initially skeptical. However, if there is sufficient evidence that supports the claim, we set aside our skepticism. The hallmarks of hypothesis testing are also found in the US court system.

EXAMPLE 3.53

A US court considers two possible claims about a defendant: she is either innocent or guilty. If we set these claims up in a hypothesis framework, which would be the null hypothesis and which the alternative?

(E)

The jury considers whether the evidence is so convincing (strong) that there is evidence beyond a reasonable doubt of the person's guilt. That is, the starting assumption (null hypothesis) is that the person is innocent until evidence is presented that convinces the jury that the person is guilty (alternative hypothesis). In statistics, our evidence comes in the form of data, and we use the significance level to decide what is beyond a reasonable doubt.

Jurors examine the evidence to see whether it convincingly shows a defendant is guilty. Notice that a jury finds a defendant either guilty or not guilty. They either reject the null claim or they do not reject the null claim. They never prove the null claim, that is, they never find the defendant innocent. If a jury finds a defendant *not guilty*, this does not necessarily mean the jury is confident in the person's innocence. They are simply not convinced of the alternative that the person is guilty.

This is also the case with hypothesis testing: *even if we fail to reject the null hypothesis, we typically do not accept the null hypothesis as truth.* Failing to find strong evidence for the alternative hypothesis is not equivalent to providing evidence that the null hypothesis is true.

Hypothesis tests are not flawless. Just think of the court system: innocent people are sometimes wrongly convicted and the guilty sometimes walk free. Similarly, data can point to the wrong conclusion. However, what distinguishes statistical hypothesis tests from a court system is that our framework allows us to quantify and control how often the data lead us to the incorrect conclusion.

There are two competing hypotheses: the null and the alternative. In a hypothesis test, we make a statement about which one might be true, but we might choose incorrectly. There are four possible scenarios in a hypothesis test, which are summarized in Figure 3.14.

		Test conclusion	
		do not reject H_0	reject H_0 in favor of H_A
Truth	H_0 true	correct conclusion	Type I error
	H_A true	Type II error	correct conclusion

Figure 3.14: Four different scenarios for hypothesis tests.

TYPE I AND TYPE II ERRORS

A **Type I error** is rejecting H_0 when H_0 is actually true. When we reject the null hypothesis, it is possible that we make a Type I error.

A **Type II error** is failing to reject H_0 when H_A is actually true. When we do not reject the null hypothesis, it is possible that we make a Type II error.

EXAMPLE 3.54

In a US court, the defendant is either innocent (H_0) or guilty (H_A). What does a Type I error represent in this context? What does a Type II error represent? Figure 3.14 may be useful.

(E)

If the court makes a Type I error, this means the defendant is innocent (H_0 true) but wrongly convicted. A Type II error means the court failed to reject H_0 (i.e. failed to convict the person) when they were in fact guilty (H_A true).

E

EXAMPLE 3.55

How could we reduce the Type I error rate in US courts? What influence would this have on the Type II error rate?

To lower the Type I error rate, we might raise our standard for conviction from "beyond a reasonable doubt" to "beyond a conceivable doubt" so fewer people would be wrongly convicted. However, this would also make it more difficult to convict the people who are actually guilty, so we would make more Type II errors.

G

GUIDED PRACTICE 3.56

How could we reduce the Type II error rate in US courts? What influence would this have on the Type I error rate?[36]

G

GUIDED PRACTICE 3.57

A group of women bring a class action lawsuit that claims discrimination in promotion rates. What would a Type I error represent in this context?[37]

These examples provide an important lesson: if we reduce how often we make one type of error, we generally make more of the other type.

If H_0 is true, what is the probability that we will incorrectly reject it? In hypothesis testing, we perform calculations under the premise that H_0 is true, and we reject H_0 if the p-value is less than or equal to the significance level α. That is, α *is* the probability of making a Type I error. The choice of what to make α is not arbitrary. It depends on the gravity of the consequences of a Type I error.

RELATIONSHIP BETWEEN TYPE I AND TYPE II ERRORS

The probability of a Type I error is called α and corresponds to the significance level of a test. If we make α smaller (P(Type I error) smaller), P(Type II error) gets larger; if we make α larger (P(Type I error) larger), P(Type II error) gets smaller .

E

EXAMPLE 3.58

If making a Type I error is especially dangerous or especially costly, should we choose a smaller significance level or a higher significance level?

Under this scenario, we want to be very cautious about rejecting the null hypothesis, so we demand very strong evidence before we are willing to reject the null hypothesis. Therefore, we want a smaller significance level, maybe $\alpha = 0.01$.

E

EXAMPLE 3.59

If making a Type II error is especially dangerous or especially costly, should we choose a smaller significance level or a higher significance level?

We should choose a higher significance level (e.g. 0.10). Here we want to be cautious about failing to reject H_0 when the null is actually false.

---

[36]To lower the Type II error rate, we want to convict more guilty people. We could lower the standards for conviction from "beyond a reasonable doubt" to "beyond a little doubt". Lowering the bar for guilt will also result in more wrongful convictions, raising the Type I error rate.

[37]We must first identify which is the null hypothesis and which is the alternative. The alternative hypothesis is the one that bears the burden of proof, so the null hypothesis is that there was no discrimination and the alternative hypothesis is that there was discrimination. Making a Type I error in this context would mean we concluded that women were discriminated against when in fact there was no discrimination. Notice that this does *not* necessarily mean something was wrong with the data or that we made a computational mistake. Sometimes data simply point us to the wrong conclusion, which is why scientific studies are often repeated to check initial findings.

SIGNIFICANCE LEVELS SHOULD REFLECT CONSEQUENCES OF ERRORS

The significance level selected for a test should reflect the real-world consequences associated with making a Type I or Type II error. If a Type I error is very dangerous, make α smaller.

The power of a test is an important concept in hypothesis testing. The **power** of a test is the probability that a hypothesis test will correctly reject a false null hypothesis. Stated another way, the power of a test is the probability of correctly detecting an effect of a particular size when it is present. It is common for researchers to perform a power analysis to ensure their study collects enough data to detect the effects they anticipate finding. As you might imagine, if the effect they care about is small or subtle, the researchers will need to collect a large sample size in order to have a good chance of detecting the effect if it is real. However, if the effect they are interested in is large, they do not need to collect as much data.

THE POWER OF A TEST AND THE PROBABILITY OF A TYPE II ERROR ARE COMPLEMENTS

P(Type II error) = 1 − power; power = 1 − P(Type II error)

The Type II error rate and the magnitude of the error for a point estimate are controlled by the sample size. As the sample size n goes up, the Type II error rate goes down, and power goes up. Real differences from the null value, even large ones, may be difficult to detect with small samples. However, if we take a very large sample, we might find a statistically significant difference but the size of the difference might be so small that it is of no practical value. The size of the effect and the standard error also affect the probability of a Type II error. We can summarize this as follows.

FACTORS THAT AFFECT THE PROBABILITY OF A TYPE II ERROR AND POWER

The probability of a Type II error decreases and the power increases when any one of the following occurs, provided the others do not change:

i. Sample size(s) increase.
ii. Standard error decreases.
iii. True parameter value is farther from the null hypothesis.
iv. Significance level α of a test increases.

The role of a data scientist in conducting a study often includes planning the size of the study. The data scientist might first consult experts or scientific literature to learn what would be the smallest meaningful difference from the null value. She also would obtain some reasonable estimate for the standard deviation. With these important pieces of information, she would choose a sufficiently large sample size so that the power for the meaningful difference is perhaps 80% or 90%. While larger sample sizes may still be used, she might advise against using them in some cases, especially in sensitive areas of research, such as a high-risk clinical trial where a new drug is being tested.

When a result is statistically significant at the $\alpha = 0.05$ level, we have evidence that the result is real. However, when there is no difference or effect, we can expect that 5% of the time the test conclusion will lead to a Type I error and incorrectly reject the null hypothesis. Therefore we must beware of what is called p-hacking, in which researchers may test many, many hypotheses and then publish the ones that come out statistically significant. As we noted, we can expect 5% of the results to be significant when the null hypothesis is true even if there is no difference or effect.[38]

[38] The problem is even greater than p-hacking. In what has been called the "reproducibility crisis", researchers have failed to reproduce a large proportion of results that were found significant and were published in scientific journals. This problem highlights the importance of research that reproduces earlier work. Ideally, one would be familiar with multiple studies on a topic, rather than taking the word of a single study.

3.4.8 Technology: the one-sample Z-interval and Z-test for p

Evaluate the 95% confidence interval from Example 3.29 to estimate the proportion of US adults who think there is intelligent life on other planets. Also find the test statistic and p-value for a test to see if there is evidence that the true proportion differs from 0.70. The sample percent was 68% and the sample size was 1,033. Conditions were verified to be met.

Desmos: Use the `zproptest(x, n)` function as explained below.

1. Click + in the upper left, then choose `inference`.
2. Choose `z-test for proportions` in the pop-up window.

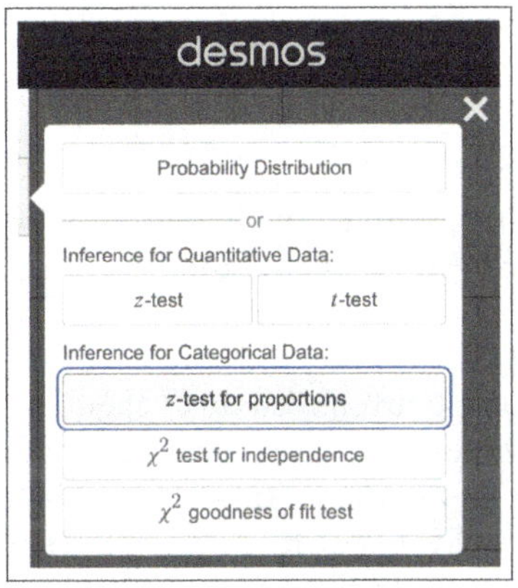

3. Enter `successes` (x) and `sample size` (n). Here, successes = $0.68 * 1,033 = 702.44$, and we need to round this to an integer. So we enter 702 for successes and 1033 for sample size. Click `Create Test`.

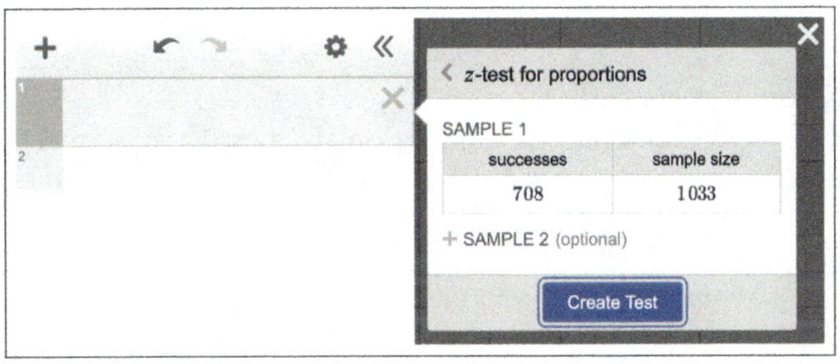

* You can type `zproptest(702, 1033)` in place of steps 1-3 above.

4. Click the triangle next to `Confidence Interval` and input the desired `Confidence level`. Here we use 0.95, which is entered by default. Click the ⋮ to the right of the confidence interval to see more information. Hover over the dot in the middle of the confidence interval to see the point estimate.
5. Click the triangle next to `Significance Test`. Enter the hypothesized value for `p` and select `Tails` to be `Left`, `Right` or `Both` depending on the direction of the alternative hypothesis. Here the hypothesized value of p is 0.70 and H_A uses a ≠, so we select Tails to be Both.

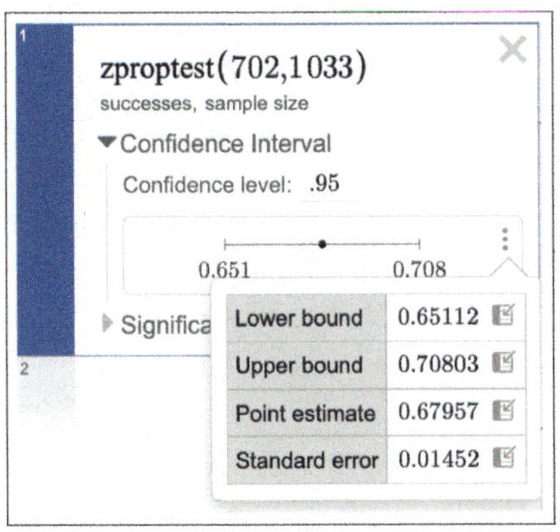

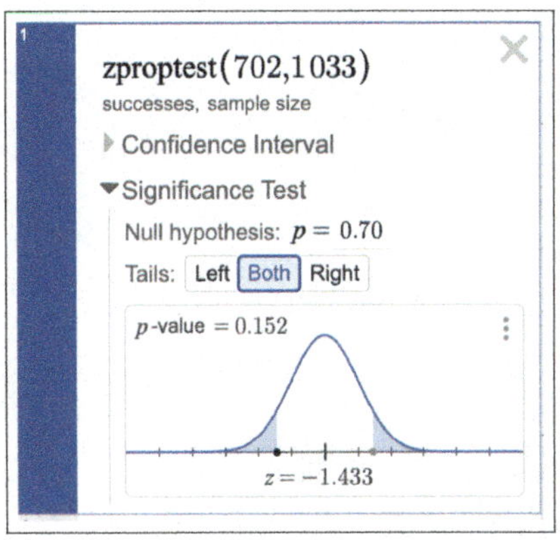

R: 1-sample Z-interval/test for p

CONFIDENCE INTERVAL.

```
> prop.test(x = 702, n = 1033, correct = FALSE, conf.level = 0.95)[39]
1-sample proportions test without continuity correction
data:  702 out of 1033, null probability 0.5
X-squared = 133.24, df = 1, p-value < 0.00000000000000022
alternative hypothesis:  true p is not equal to 0.5
95 percent confidence interval:
 0.6504973 0.7073202
sample estimates:
p
0.6795741
```

HYPOTHESIS TEST.
Make sure to specify `p`, the hypothesized or null proportion.
`alternative` can be `"two.sided"`, `"greater"`, or `"less"`.

```
> prop.test(x = 702, n = 1033, p = 0.70, correct = FALSE, alternative = "two.sided")
1-sample proportions test without continuity correction
data:  702 out of 1033, null probability 0.7
X-squared = 2.0523, df = 1, p-value = 0.152
alternative hypothesis:  true p is not equal to 0.7
95 percent confidence interval:
0.6504973 0.7073202
sample estimates:
p
0.6795741
```

This test returns X-squared instead of Z.
Z = $+\sqrt{\texttt{X-squared}}$ or $-\sqrt{\texttt{X-squared}}$, depending on if `(sample p - null p)` is +. or −.
Here (0.6504973 − 0.70) is negative, so Z = $-\sqrt{2.0523}$.

```
> Z = -sqrt(2.0523)
> Z
[1] -1.432585
```

A note on default values:

- If `p` is omitted a default null value of 0.5 is used.
- If `correct = FALSE` is omitted, what is called the "continuity correction" will be applied.
- If `alternative` is omitted, a default value of `"two.sided"` is used.
- If `conf.level` is omitted, a default value of 0.95 is used.

[39]R uses a different formula for calculating the one proportion interval/test than what is presented in this textbook. Results may differ slightly from Desmos and handheld calculators.

Calculator: NumWorks calculator instructions and example output are included. For the TI-83/84 and Casio calculators, general instructions are provided, and worked example videos can be accessed via the icon or at **openintro.org/ti** and **openintro.org/casio**.

NUMWORKS: 1-PROPORTION Z-INTERVAL

Use `OK` or `EXE` to make a selection.

1. From the home screen, select `Inference`, then `Intervals`, then `One proportion`. If these options do not appear, click the ← button in the upper right as many times as needed.
2. Enter the values of `x`, `n`, and `Confidence level`. Hit the down arrow and choose `Next`.
3. Note the quantities returned. Click the down arrow and choose `Next`.
4. In addition to seeing the confidence interval displayed in two ways, you can press the up and down arrows to quickly change confidence level and see the resulting interval and margin of error.

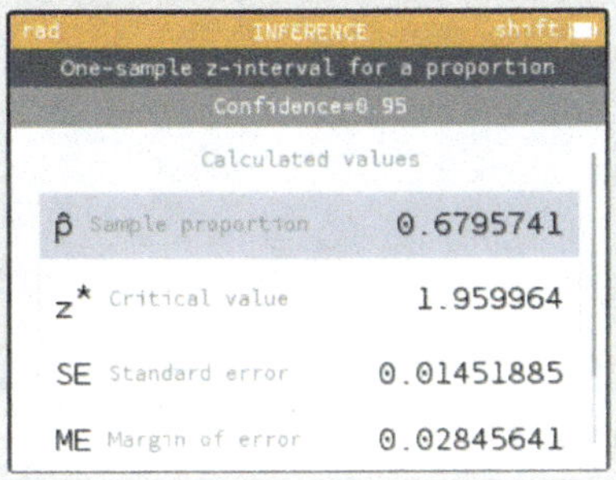

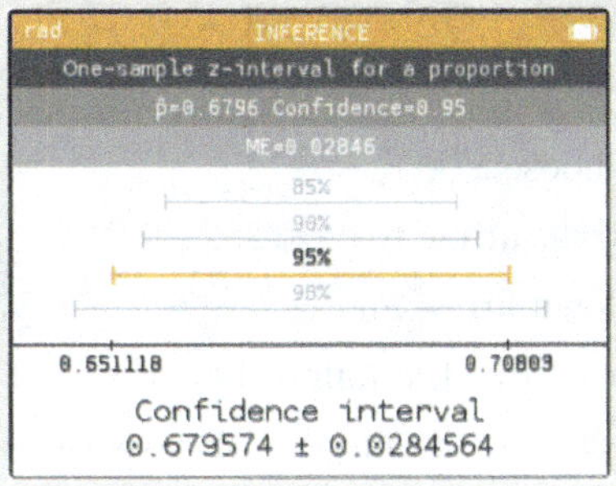

NUMWORKS: 1-PROPORTION Z-TEST

Use `OK` or `EXE` to make a selection.

1. From the home screen, select `Inference`, then `Tests`, then `One proportion`. If these options do not appear, click the ← button in the upper right as many times as needed.
2. Enter the value of the hypothesized proportion for the Null hypothesis. Press the down arrow. Press `OK` and choose `<`, `≠` , or `>` for the Alternative hypothesis. Press the down arrow and choose `Next`.
3. Enter the values of `x`, `n`, and α. Hit the down arrow and choose `Next`.
4. Note the quantities returned. Click the down arrow and choose `Next`.
5. On this screen, the p-value and alpha are shaded on the normal distribution and can be visually compared.

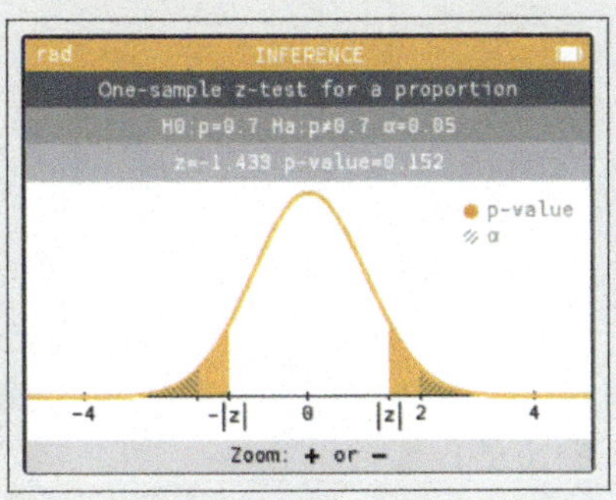

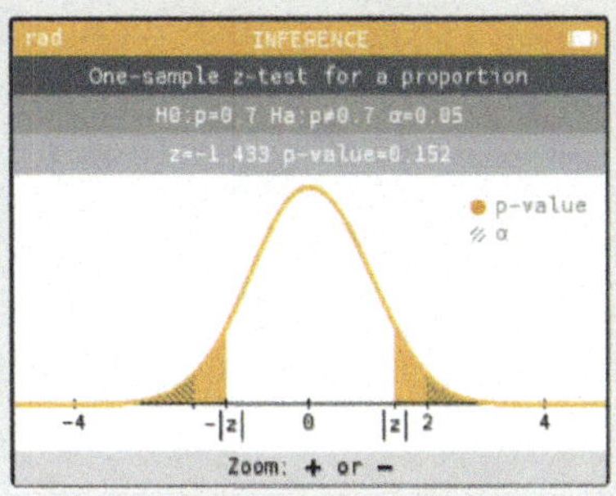

TI-83/84: 1-PROPORTION Z-INTERVAL

Use `STAT`, `TESTS`, `1-PropZInt`.

1. Choose `STAT`.
2. Right arrow to `TESTS`.
3. Down arrow and choose `A:1-PropZInt`.
4. Let `x` be the *number* of yeses (must be an integer).
5. Let `n` be the sample size.
6. Let `C-Level` be the desired confidence level.
7. Choose `Calculate` and hit `ENTER`, which returns

   (__,__) the confidence interval
   $\hat{p}$ the sample proportion
   n the sample size

TI-83/84: 1-PROPORTION Z-TEST

Use `STAT`, `TESTS`, `1-PropZTest`.

1. Choose `STAT`.
2. Right arrow to `TESTS`.
3. Down arrow and choose `5:1-PropZTest`.
4. Let p_0 be the null or hypothesized value of p.
5. Let `x` be the *number* of yeses (must be an integer).
6. Let `n` be the sample size.
7. Choose ≠, <, or > to correspond to H_A.
8. Choose `Calculate` or `Draw` and hit `ENTER`. `Draw` shows the Z-statistic and p-value as well as a graph of the normal curve with p-value shaded. `Calculate` returns:

   z Z-statistic
   p p-value
   $\hat{p}$ the sample proportion
   n the sample size

CASIO FX-9750GII: 1-PROPORTION Z-INTERVAL

1. Navigate to `STAT` (`MENU` button, then hit the `2` button or select `STAT`).
2. Choose the `INTR` option (`F4` button).
3. Choose the `Z` option (`F1` button).
4. Choose the `1-P` option (`F3` button).
5. Specify the interval details:
   - Confidence level of interest for `C-Level`.
   - Enter the number of successes, `x`.
   - Enter the sample size, `n`.
6. Hit the `EXE` button, which returns

`Left`, `Right`	ends of the confidence interval
$\hat{p}$	sample proportion
`n`	sample size

CASIO FX-9750GII: 1-PROPORTION Z-TEST

The steps closely match those of the 1-proportion confidence interval.

1. Navigate to `STAT` (`MENU` button, then hit the `2` button or select `STAT`).
2. Choose the `TEST` option (`F3` button).
3. Choose the `Z` option (`F1` button).
4. Choose the `1-P` option (`F3` button).
5. Specify the test details:
   - Specify the sidedness of the test using the `F1`, `F2`, and `F3` keys.
   - Enter the null value, `p0`.
   - Enter the number of successes, `x`.
   - Enter the sample size, `n`.
6. Hit the `EXE` button, which returns

`z`	Z-statistic
`p`	p-value
$\hat{p}$	the sample proportion
`n`	the sample size

Section summary

- A **hypothesis test** is a statistical inference procedure used to evaluate competing claims based on data in order to make a decision.
- The appropriate hypothesis test for a population proportion p is a **one-sample Z-test for a population proportion p**. The parameter p should be identified in context.
- The competing claims are called **hypotheses**.
  - The **null hypothesis** is abbreviated H_0 and is a statement about a parameter that is assumed to be correct unless there is convincing statistical evidence suggesting otherwise. It is the status quo condition.
  - The **alternative hypothesis** is abbreviated H_A and is the claim or belief about a parameter for which evidence is being collected. A researcher's claim or belief about the population parameter is represented by the alternative hypothesis.
- The null hypotheses for a one-sample Z-test for a population proportion p is:

  H_0: $p = p_0$, where p_0 is the null hypothesized value for the population mean.

  The null hypothesis is sometimes written using a $\le$ or a $\ge$, but we will always use the equality.
- The alternative hypothesis may be one-sided ($<$ or $>$) or two-sided ($\ne$).

  H_A: $p < p_0$. The p-value will correspond to a lower tail.

  H_A: $p > p_0$. The p-value will correspond to an upper tail.

  H_A: $p \ne p_0$. The p-value will correspond to both tails.
- We set a **significance level**, denoted α, that determines the probability of rejecting the null hypothesis given that it is true. The most common significance level is $\alpha = 0.05$. If we want to require more evidence to reject the null hypothesis, we use a smaller α.
- The **logic of a hypothesis test**: In a hypothesis test, we begin by *assuming that the null hypothesis is true.* Then, we calculate how unlikely it would be to get a sample value as extreme as we actually got in our sample, in the direction of H_A, assuming that the null value is correct. If this likelihood is too small (below our α threshold), it casts doubt on the null hypothesis and provides evidence for the alternative hypothesis.
- The one-sample Z-test for a population proportion requires the following **conditions** be met:

  1. Independence: The data come from a random sample or random process. When sampling without replacement, check that the sample size is less than 10% of the population size ($n < 0.10(N)$).
  2. Large counts: the expected number of successes and failures assuming the null hypothesis is true is at least 10. $np_0 \ge 10$ and $n(1 - p_0) \ge 10$.
- A **test statistic** has the form: test statistic $= \frac{\text{point estimate} - \text{null value}}{SE \text{ of estimate}}$.

  The test statistic for a one-sample Z-test for p is: $Z = \frac{\hat{p}-p_0}{\sqrt{\frac{p_0(1-p_0)}{n}}}$.
  The SE is calculated under the assumption that H_0 is true and uses p_0.
- The **p-value** for a one-sample Z-test for p is the probability of obtaining a Z-statistic as small or smaller, as large or larger, or as extreme or more extreme than the Z-statistic that was observed, depending on whether the direction of the alternate hypothesis is $<$, $>$, or $\ne$, assuming the null hypothesis is true (i.e. that the population proportion really equals p_0).
- Small p-values indicate that the observed value of the test statistic would be unusual if the null hypothesis were true and therefore provide evidence for the alternative hypothesis. The lower the p-value, the more convincing the statistical evidence for the alternative hypothesis.

- p-values that are not small indicate that the observed value of the test statistic would not be unusual if the null hypothesis were true and therefore do not provide convincing statistical evidence for the alternative hypothesis, nor do they provide evidence that the null hypothesis is true.
- The conclusion or decision of a hypothesis test is based on whether the p-value is smaller or larger than the preset significance level α.
  - When the p-value $\leq \alpha$, we reject H_0 and have convincing statistical evidence for H_A. We say the results are *statistically significant* at the α level.
  - When the p-value $> \alpha$, we fail to reject H_0 and we do not have convincing statistical evidence for H_A. We say the results are not statistically significant at the α level.
- A hypothesis test can lead to rejecting or not rejecting the null hypothesis but can never lead to concluding or proving that the null hypothesis is true. Lack of statistical evidence for the alternative hypothesis is not the same as evidence for the null hypothesis.
- The results of a hypothesis test for a population proportion can serve as the statistical reasoning to support the answer to an investigative question about the population that was sampled. The conclusion of a hypothesis test should be stated in terms of the alternative hypothesis and in context using non-causal language.
- **Decision errors**. In a hypothesis test, there are two types of decision errors that could be made. These are called Type I and Type II errors.
  - A **Type I error** is rejecting H_0, when H_0 is actually true. We commit a Type I error if we call a result significant (find enough evidence for H_A) when there is *no* real difference or effect.
  - A **Type II error** is not rejecting H_0, when H_A is actually true. We commit a Type II error if we call a result not significant (do not find enough evidence for H_A) when there *is* a real difference or effect.
- The **power** of a hypothesis test is the probability that a hypothesis test will correctly reject the false null hypothesis. Stated another way, the power of a test is the probability of correctly detecting an effect of a particular size when it is present.
- Probabilities of Type I and Type II errors:
  - The probability of making a Type I error is defined as the significance level α. For a given study and hypothesis test, the probability of making a Type I error is typically set to a small value (e.g., 0.01, 0.05, 0.10) prior to collecting the data.
  - The probability of making a Type II error is 1 – power.
- For a given study and hypothesis test, the probability of a Type II error should ideally be small, and thus, the power will be large (e.g., P(Type II error) = 0.20 and power = 0.80).
- The probability of a Type II error decreases and the power increases when any one of the following occurs, provided the others do not change:
  i. Sample size(s) increase
  ii. Standard error decreases.
  iii. True parameter value is farther from the null hypothesis.
  iv. Significance level α of a test increases.
- In some studies, making a Type I error may have more serious consequences than making a Type II error. In other studies, making a Type II error may have more serious consequences than making a Type I error. The consequences of each error should be considered prior to conducting the study.
- Because the significance level, α, is the probability of making a Type I error, the consequences of a Type I error influence decisions about a significance level.
- Because sample size influences the probability of making a Type II error, the consequences of a Type II error influence decisions about how large the sample size should be.

Exercises

3.31 Identify hypotheses, Part I. Write the null and alternative hypotheses in words and then symbols for each of the following situations.

(a) A tutoring company would like to understand if most students tend to improve their grades (or not) after they use their services. They sample 200 of the students who used their service in the past year and ask them if their grades have improved or declined from the previous year.

(b) Employers at a firm are worried about the effect of March Madness, a basketball championship held each spring in the US, on employee productivity. They estimate that on a regular business day employees spend on average 15 minutes of company time checking personal email, making personal phone calls, etc. They also collect data on how much company time employees spend on such non-business activities during March Madness. They want to determine if these data provide convincing evidence that employee productivity changed during March Madness.

3.32 Identify hypotheses, Part II. Write the null and alternative hypotheses in words and using symbols for each of the following situations.

(a) Since 2008, chain restaurants in California have been required to display calorie counts of each menu item. Prior to menus displaying calorie counts, the average calorie intake of diners at a restaurant was 1100 calories. After calorie counts started to be displayed on menus, a nutritionist collected data on the number of calories consumed at this restaurant from a random sample of diners. Do these data provide convincing evidence of a difference in the average calorie intake of a diners at this restaurant?

(b) The state of Wisconsin would like to understand the fraction of its adult residents that consumed alcohol in the last year, specifically if the rate is different from the national rate of 70%. To help them answer this question, they conduct a random sample of 852 residents and ask them about their alcohol consumption.

3.33 Online communication. A study suggests that 60% of college student spend 10 or more hours per week communicating with others online. You believe that this is incorrect and decide to collect your own sample for a hypothesis test. You randomly sample 160 students from your dorm and find that 70% spent 10 or more hours a week communicating with others online. A friend of yours, who offers to help you with the hypothesis test, comes up with the following set of hypotheses. Indicate any errors you see.

$$H_0 : \hat{p} < 0.6$$
$$H_A : \hat{p} > 0.7$$

3.34 Married at 25. A study suggests that the 25% of 25 year olds have gotten married. You believe that this is incorrect and decide to collect your own sample for a hypothesis test. From a random sample of 25 year olds in census data with size 776, you find that 24% of them are married. A friend of yours offers to help you with setting up the hypothesis test and comes up with the following hypotheses. Indicate any errors you see.

$$H_0 : \hat{p} = 0.24$$
$$H_A : \hat{p} \neq 0.24$$

3.35 National Health Plan, Part I. A *Kaiser Family Foundation* poll for a random sample of US adults in 2019 found that 79% of Democrats, 55% of Independents, and 24% of Republicans supported a generic "National Health Plan". There were 347 Democrats, 298 Republicans, and 617 Independents surveyed.[40]

(a) A political pundit on TV claims that a majority of Independents support a National Health Plan. Do these data provide strong evidence to support this type of statement? Remember to Identify, Check, Calculate and Conclude.

(b) Would you expect a confidence interval for the proportion of Independents who oppose the public option plan to include 0.5? Explain.

[40] Kaiser Family Foundation, The Public On Next Steps For The ACA And Proposals To Expand Coverage, data collected between Jan 9-14, 2019.

3.36 Is college worth it? Part I. Among a simple random sample of 331 American adults who do not have a four-year college degree and are not currently enrolled in school, 48% said they decided not to go to college because they could not afford school.[41]

(a) A newspaper article states that only a minority of the Americans who decide not to go to college do so because they cannot afford it and uses the point estimate from this survey as evidence. Conduct a hypothesis test to determine if these data provide strong evidence supporting this statement.

(b) Would you expect a confidence interval for the proportion of American adults who decide not to go to college because they cannot afford it to include 0.5? Explain.

3.37 Testing for Fibromyalgia. A patient named Diana was diagnosed with Fibromyalgia, a long-term syndrome of body pain, and was prescribed anti-depressants. Being the skeptic that she is, Diana didn't initially believe that anti-depressants would help her symptoms. However after a couple months of being on the medication she decides that the anti-depressants are working, because she feels like her symptoms are in fact getting better.

(a) Write the hypotheses in words for Diana's skeptical position when she started taking the anti-depressants.

(b) What is a Type 1 Error in this context?

(c) What is a Type 2 Error in this context?

3.38 Testing for food safety. A food safety inspector is called upon to investigate a restaurant with a few customer reports of poor sanitation practices. The food safety inspector uses a hypothesis testing framework to evaluate whether regulations are not being met. If he decides the restaurant is in gross violation, its license to serve food will be revoked.

(a) Write the hypotheses in words.

(b) What is a Type 1 Error in this context?

(c) What is a Type 2 Error in this context?

(d) Which error is more problematic for the restaurant owner? Why?

(e) Which error is more problematic for the diners? Why?

(f) As a diner, would you prefer that the food safety inspector requires strong evidence or very strong evidence of health concerns before revoking a restaurant's license? Explain your reasoning.

3.39 Which is higher? In each part below, there is a value of interest and two scenarios (I and II). For each part, report if the value of interest is larger under scenario I, scenario II, or whether the value is equal under the scenarios.

(a) The standard error of $\hat{p}$ when (I) $n = 125$ or (II) $n = 500$.

(b) The margin of error of a confidence interval when the confidence level is (I) 90% or (II) 80%.

3.40 Which is higher? In each part below, there is a value of interest and two scenarios (I and II). For each part, report if the value of interest is larger under scenario I, scenario II, or whether the value is equal under the scenarios.

(a) The p-value for a Z-statistic of 2.5 calculated based on a (I) sample with $n = 500$ or based on a (II) sample with $n = 1000$.

(b) The probability of making a Type 2 Error when the alternative hypothesis is true and the significance level is (I) 0.05 or (II) 0.10.

[41] Pew Research Center Publications, Is College Worth It?, data collected between March 15-29, 2011.

3.5 Sampling distribution for $\hat{p}_1 - \hat{p}_2$

If two treatments work equally well, how much variation in their success rates would we expect due to random variation? How different are the voting rates between two adjacent counties? Often, researchers are interested in comparing the proportion of individuals with a certain characteristic between two treatment groups or populations. In this section, we lay the groundwork for these types of questions by investigating the sampling distribution for a difference in sample proportions.

Learning objectives

1. Calculate the mean and standard deviation of a sampling distribution for a difference in sample proportions.
2. Justify whether the independence condition is satisfied when considering properties of the sampling distribution for a difference in sample proportions.
3. Determine whether or not the shape of the sampling distribution for a difference in sample proportions is approximately normal.
4. Interpret the mean, standard deviation, and probabilities for the sampling distribution for the difference between two sample proportions.

3.5.1 Visualizing a distribution for a difference in sample proportions

In Section 2.7.2, we looked at a randomization distribution for the difference in proportion that would contract malaria if given the PfSPZ vaccine versus given a placebo. Understanding the expected variation in the difference helps us determine whether an observed difference should be considered surprising or not.

Imagine that the owners of a new specialty grocery store are deciding which of two equally-sized counties, which we'll call County 1 and County 2, they should open their store in. They hire a market researcher to help them collect data to inform their decision. Unknown to the researcher, 45% of the people County 1 would be interested in the new specialty grocery store, while 35% of the people in County 2 would be interested. The researcher uses random digit dialing to select a random sample of adults from each county and asks the each person whether they would be interested in the new specialty grocery store.

Let $\hat{p}_1$ be the proportion of adults in a random sample of size 50 from County 1 who answer yes, they are interested, and let $\hat{p}_2$ be the proportion of adults in a random sample of size 60 from County 2 who answer yes, they are interested. How large of a difference do we expect in the sample proportions: $\hat{p}_1 - \hat{p}_2$? What is the likelihood that this difference is negative, meaning that the proportion in the sample from County 2 is larger than from County 1, even though in the populations it is not?

First, we want to visualize the sampling distribution of the difference: $\hat{p}_1 - \hat{p}_2$. To do this we run a simulation. We randomly select 50 people from County 1 where $p_1 = 0.45$ and calculate the sample proportion $\hat{p}_1$ and we randomly select 60 people from County 2 where $p_2 = 0.35$ and calculate the sample proportion $\hat{p}_2$, then we calculate the difference $\hat{p}_1 - \hat{p}_2$. We do this 300 times, giving us 300 values for $\hat{p}_1 - \hat{p}_2$. These 300 sample differences are graphed in Figure 3.15.

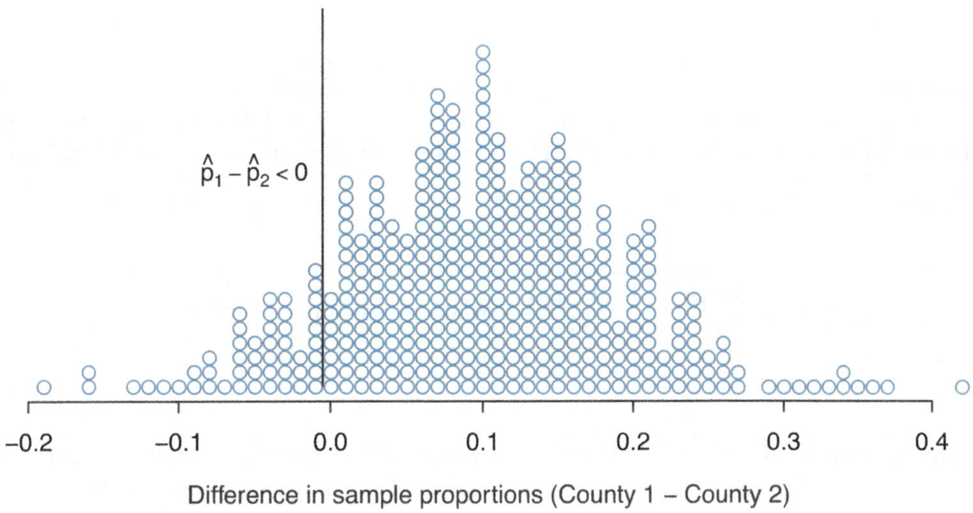

Figure 3.15: 300 simulated differences in sample proportions

The distribution of sample differences in Figure 3.15 seems to be centered on 0.10. This makes sense because the mean of possible sample differences should be centered on the difference in the true proportions of 0.45 and 0.35. Each dot in Figure 3.15 represents a difference in sample proportions. We can count that 49 of the 300 dots have a value less than 0 so, based on the simulation, we estimate that there is a $\frac{49}{300}$, or about a 16% chance that the sample proportion from County 2 will be higher, misleading the market researcher.

3.5.2 The mean and standard deviation for $\hat{p}_1 - \hat{p}_2$

We saw that the mean of $\hat{p}_1 - \hat{p}_2$ for our grocery store example is centered on 0.10, which is $p_1 - p_2$. In general, for two random variables X and Y, the mean of the difference is the difference of the individual means:

$$\mu_{X-Y} = \mu_X - \mu_Y.$$

To calculate the mean of a sampling distribution of $\hat{p}_1 - \hat{p}_2$, we apply this property to find that:

$$\mu_{\hat{p}_1-\hat{p}_2} = \mu_{\hat{p}_1} - \mu_{\hat{p}_2} = p_1 - p_2.$$

The situation is a little more complex when looking at the variability of the difference in X and Y. Here we're going to require a condition be met, specifically that X and Y are independent random variables. When that independence condition is met, then the following formula for the standard deviation of the difference, $X - Y$, holds:

$$\sigma_{X-Y} = \sqrt{\sigma_X^2 + \sigma_Y^2}.$$

The two samples are random and independent of each other so, assuming that the sample sizes are less then 10% of the county population sizes, we can apply this property as follows:

$$\sigma_{\hat{p}_1-\hat{p}_2} = \sqrt{(\sigma_{\hat{p}_1})^2 + (\sigma_{\hat{p}_2})^2} = \sqrt{\frac{p_1(1-p_1)}{n_1} + \frac{p_2(1-p_2)}{n_2}}.$$

MEAN AND STANDARD DEVIATION OF A DIFFERENCE IN SAMPLE PROPORTIONS

The mean and standard deviation of the sampling distribution for a difference in sample proportions describe the center and spread of the distribution of differences $\hat{p}_1 - \hat{p}_2$ for all random samples of size n_1 and n_2 from the given populations. Given population proportions p_1 and p_2, population sizes N_1 and N_2, and independent random samples of size n_1 and n_2, we have the following:

$$\mu_{\hat{p}_1-\hat{p}_2} = p_1 - p_2$$

$$\sigma_{\hat{p}_1-\hat{p}_2} = \sqrt{\frac{p_1(1-p_1)}{n_1} + \frac{p_2(1-p_2)}{n_2}} \quad \text{when } n_1 < 0.10(N_1) \text{ and } n_2 < 0.10(N_2)$$

EXAMPLE 3.60

In our grocery store example, 45% of County 1 is interested in the new specialty grocery store, and 35% of County 2 is interested. Calculate the mean and standard deviation for the difference in sample proportions if the market researcher takes a random sample of size 60 from County 1 and a random sample of size 50 from County 2.

E

$$\mu_{\hat{p}_1-\hat{p}_2} = p_1 - p_2 = 0.45 - 0.35 = 0.10.$$

$$\sigma_{\hat{p}_1-\hat{p}_2} = \sqrt{\frac{p_1(1-p_1)}{n_1} + \frac{p_2(1-p_2)}{n_2}} = \sqrt{\frac{0.45(1-0.45)}{50} + \frac{0.35(1-0.35)}{60}} = 0.093.$$

EXAMPLE 3.61

Interpret the mean and standard deviation calculated in Example 3.60.

E

For all random samples of 50 people from County 1 and 60 people from County 2, the difference in sample proportions of people with an interest in the new specialty grocery store (County 1 – County 2) will have a mean of 0.10 and will typically vary by 0.093 from the mean of 0.10.

3.5.3 Using a normal model for the sampling distribution of $\hat{p}_1 - \hat{p}_2$

We used the simulation shown in Figure 3.15 to estimate a probability involving the sampling distribution of $\hat{p}_1 - \hat{p}_2$, the difference in sample proportions who have an interest in the new specialty store. We would like a method for estimating such probabilities that does not depend on a simulation. Fortunately, when certain conditions are met, the distribution of $\hat{p}_1 - \hat{p}_2$ can be modeled with a normal distribution and we can use a normal approximation to estimate probabilities based on this sampling distribution.

When looking at the sum or difference of two random variables, if each variable is nearly normal, then the sum and difference are also nearly normal random variables. This property will be very useful, as it says that when each sample proportion has a nearly normal distribution, then the difference in sample proportions will also be nearly normal. The sampling distribution for $\hat{p}_1 - \hat{p}_2$ can be modeled with a normal distribution when the following two conditions are met:

Independence. The observations within and between groups should be independent. The independence condition is satisfied if the data is collected from 2 independent random samples, where each sample size is less than 10% of the population size if done without replacement. We also consider the independence condition satisfied if the data is collected from an experiment with two randomly assigned treatments (in this case the 10% condition is not relevant and does not need to be checked).

Large counts. In the two-sample case, the number of expected successes and failures should be at least 10 for *both* groups. We must check that the following four inequalities are satisfied: $n_1p_1 \geq 10$, $n_1(1 - p_1) \geq 10$, $n_2p_2 \geq 10$, and $n_2(1 - p_2) \geq 10$.

EXAMPLE 3.62

Let's return to our original question about sampling from County 1 and County 2. In County 1, it is true that 45% of the people have an interest in the new specialty grocery store . In County 2, it is true that 35% of the people have an interest in the new specialty grocery store. If we take a random sample of size 50 from County 1 and a random sample of size 60 from County 2, what is the probability that we will get a *higher* proportion with an interest in the store in sample 2 than in sample 1?

Here $p_1 = 0.45$, $n_1 = 50$, $p_2 = 0.35$, and $n_2 = 60$. Define $\hat{p}_1 - \hat{p}_2$ as follows:

$\hat{p}_1 - \hat{p}_2$: difference in sample proportions (County 1 –County 2) with an interest in the new specialty grocery store.

We want to find $P(\hat{p}_1 - \hat{p}_2 < 0)$, which is equivalent to $P(\hat{p}_1 < \hat{p}_2)$. We have already found:

$$\mu_{\hat{p}_1-\hat{p}_2} = 0.45 - 0.35 = 0.10$$

$$\sigma_{\hat{p}_1-\hat{p}_2} = \sqrt{\frac{0.45(1-0.45)}{50} + \frac{0.35(1-0.35)}{60}} = 0.093$$

Our random samples are independent and we are assuming sample sizes are less than 10% of the county populations. We now check the large counts condition for each group to determine if the distribution of $\hat{p}_1 - \hat{p}_2$ can be modeled by a normal distribution.

$$n_1p_1 = 50 \times 0.45 = 22.5 \geq 10 \qquad n_1(1-p_1) = 50 \times (1-0.45) = 27.5 \geq 10$$

$$n_2p_2 = 60 \times 0.35 = 21 \geq 10 \qquad n_2(1-p_2) = 60 \times (1-0.35) = 39 \geq 10$$

All four inequalities are satisfied, so we can use a normal model for the difference in sample proportions. Given that $\hat{p}_1 - \hat{p}_2$ is approximately Normal($\mu = 0.10$, $\sigma = 0.093$), we can use technology to find that $P(\hat{p}_1 - \hat{p}_2 < 0) = 0.141$.

Alternately, we can find the Z-score that corresponds to the boundary value of 0 and use the standard normal distribution. In Section 3.7 we will see a parallel between the calculation of this Z-score and the calculation of a Z test statistic for a test for a difference of population proportion.

$$Z = \frac{x-\mu}{\sigma} = \frac{0-(0.45-0.35)}{\sqrt{\frac{0.45(1-0.45)}{50} + \frac{0.35(1-0.35)}{60}}} = \frac{0-0.10}{0.093} = -1.075$$

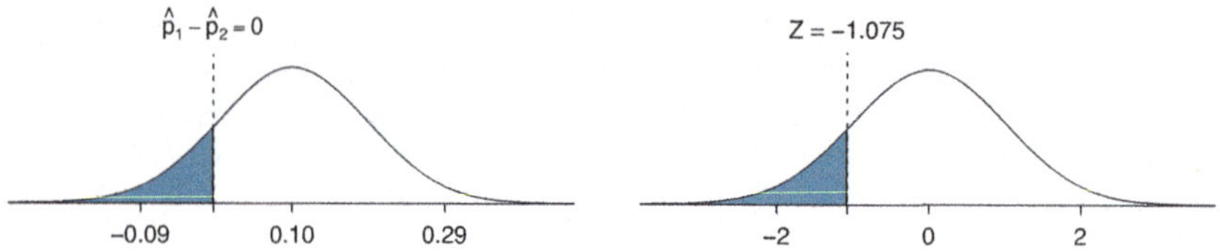

Using technology and the Normal($\mu = 0$, $\sigma = 1$) distribution, we find $P(Z < -1.075) = 0.141$.

Even though County 1 has a higher proportion of people with an interest in the new specialty grocery store, with these small sample sizes, there is still about a 14.1% chance that the sample from County 2 has a higher proportion with an interest in the new specialty grocery store than the sample from County 1.

EXAMPLE 3.63

What could the market researcher have done to lower the probability of mistakenly thinking that a higher proportion in County 2 are interested in the new specialty grocery store?

If the sample sizes were larger, the sampling distribution of the difference in sample proportions would the same mean but smaller spread, resulting in a lower probability of $\hat{p}_1 - \hat{p}_2 < 0$.

Section summary

- $\hat{p}_1 - \hat{p}_2$ represents a difference in sample proportions and can take on different values for different samples. For two independent populations, the **sampling distribution for a difference in sample proportions**, $\hat{p}_1 - \hat{p}_2$, is the distribution of $\hat{p}_1 - \hat{p}_2$ values for all random samples of size n_1 and n_2 from the given populations.

- When the observations can be treated as independent, such as from two independent random samples or two randomly assigned treatments:

  - The **mean** of the sampling distribution of $\hat{p}_1 - \hat{p}_2$ is given by: $\mu_{\hat{p}_1 - \hat{p}_2} = p_1 - p_2$, where p_1 and p_2 are population proportions.
  - The **standard deviation** of the sampling distribution of $\hat{p}_1 - \hat{p}_2$ is given by: $\sigma_{\hat{p}_1 - \hat{p}_2} = \sqrt{\frac{p_1(1-p_1)}{n_1} + \frac{p_2(1-p_2)}{n_2}}$. If sampling is done without replacement, both sample sizes should be less than 10% of the size of their corresponding populations, i.e. $n_1 < 0.10(N_1)$ and $n_2 < 0.10(N_2)$, in order for this standard deviation formula to used. If the data come from an experiment with two randomly assigned treatments, the 10% condition does not need to be checked.
  - The **shape** of the sampling distribution for a difference between sample proportions, $\hat{p}_1 - \hat{p}_2$, will be approximately normal when both sample sizes are large enough to satisfy the large counts condition: $n_1 p_1 \geq 10$, $n_1(1 - p_1) \geq 10$, $n_2 p_2 \geq 10$, and $n_2(1 - p_2) \geq 10$.

- $\mu_{\hat{p}_1 - \hat{p}_2}$, the mean of $\hat{p}_1 - \hat{p}_2$, describes the average of values of $\hat{p}_1 - \hat{p}_2$ among all random samples of size n_1 and n_2 from the given populations.

- $\sigma_{\hat{p}_1 - \hat{p}_2}$, the standard deviation of $\hat{p}_1 - \hat{p}_2$, describes the typical variation in values of $\hat{p}_1 - \hat{p}_2$ from $p_1 - p_2$ for all random samples of size n_1 and n_2 from the given populations.

- To use a normal model to find probabilities involving a difference in sample proportions, first verify that the conditions for independence are met and that the large counts condition is met for both groups. Identify the distribution and its parameters, write the relevant probability statement, and answer the question in context.

- The mean, standard deviation, and probabilities for the sampling distribution for a difference between two sample proportions should be interpreted within the context of two specific populations.

Exercises

3.41 Difference of proportions, Part 1. The fraction of workers who are considered "supercommuters", because they commute more than 90 minutes to get to work, varies by state. Suppose the following were the exact values for Nebraska and New York:

State	Proportion Supercommuters
Nebraska	0.01
New York	0.06

Now suppose that we plan a study to survey 1000 people from each state, and we will compute the sample proportions $\hat{p}_{NE}$ for Nebraska and $\hat{p}_{NY}$ for New York.

(a) What is the associated mean and standard deviation of $\hat{p}_{NE}$?

(b) What is the associated mean and standard deviation of $\hat{p}_{NY}$?

(c) Calculate and interpret the mean and standard deviation associated with the difference in sample proportions for the two groups, $\hat{p}_{NY} - \hat{p}_{NE}$.

(d) How are the standard deviations from parts (a), (b), and (c) related?

3.42 Difference of proportions, Part 2. The fraction of workers who are considered "supercommuters", because they commute more than 90 minutes to get to work, varies by state. Suppose the following were the exact values for Nebraska and New York:

State	Proportion Supercommuters
Nebraska	0.01
New York	0.06

Now suppose that we plan a study to survey 1000 people from each state, and we will compute the sample proportions $\hat{p}_{NE}$ for Nebraska and $\hat{p}_{NY}$ for New York.

(a) What distribution is associated with the difference $\hat{p}_{NY} - \hat{p}_{NE}$? Justify your answer.

(b) Determine the probability that $\hat{p}_{NY} - \hat{p}_{NE}$ will be larger than 0.055.

(c) Determine the probability that $\hat{p}_{NY} - \hat{p}_{NE}$ will be smaller than 0.04.

3.6 Confidence intervals for $p_1 - p_2$

How much more effective is a blood thinner than a placebo for those who undergo CPR for a heart attack? Researchers often wish to estimate how large a difference there is between two treatments or populations. In this section we develop a framework for confidence intervals for a difference in proportions by combining what we learned about confidence intervals for a single proportion with our understanding of the sampling distribution for a difference in proportions. While some of the details change, the general confidence interval framework remains the same.

Learning objectives

1. Identify and set up an appropriate confidence interval procedure for estimating the difference in population proportions $p_1 - p_2$.
2. Justify the appropriateness of constructing a confidence interval for a difference between two population proportions by verifying conditions.
3. Calculate an appropriate confidence interval for a difference in population proportions.
4. Calculate the standard error and margin of error for a confidence interval for a difference in population proportions.
5. Interpret a confidence interval in context for a difference in population proportions.
6. Justify a claim about the difference in population proportions based on an appropriate confidence interval.

3.6.1 Conditions for a confidence interval for a difference of proportions

We consider an experiment for patients who underwent CPR for a heart attack and were subsequently admitted to a hospital. These patients were randomly divided into a treatment group where they received a blood thinner or the control group where they did not receive a blood thinner. The outcome variable of interest was whether the patients survived for at least 24 hours. The results are shown in Figure 3.16.

	Survived	Died	Total
Treatment	14	26	40
Control	11	39	50
Total	25	65	90

Figure 3.16: Results for the CPR study. Patients in the treatment group were given a blood thinner, and patients in the control group were not.

Here, the parameter of interest is a difference in population proportions, specifically, the difference in the proportion of similar patients that would survive for at least 24 hours if in the treatment group versus if in the control group. Let:

$$p_1 : \text{proportion who would survive in treatment group, and}$$
$$p_2 : \text{proportion who would survive in control group}$$

Then the parameter of interest is $p_1 - p_2$, which is the difference in proportions (treatment - control) that would survive. In order to use a Z-interval to estimate this difference, we must see if the point estimate, $\hat{p}_1 - \hat{p}_2$, follows a normal distribution. Because the patients were randomly assigned to one of the two groups and one heart attack patient is unlikely to influence the next that was in the study, the observations are considered independent, both within the groups and between the groups (since there is no sampling, there is no need to check the 10% condition). Next, the large counts condition should be verified for each group. Here, we do not know the true proportions, so we must use the sample proportions along with the sample sizes to check the condition.

$$n_1\hat{p}_1 \geq 10 \qquad n_1(1-\hat{p}_1) \geq 10 \qquad n_2\hat{p}_2 \geq 10 \qquad n_2(1-\hat{p}_2) \geq 10$$
$$40 \times \frac{14}{40} \geq 10 \qquad 40 \times (1 - \frac{14}{40}) \geq 10 \qquad 50 \times \frac{11}{50} \geq 10 \qquad 50 \times (1 - \frac{11}{50}) \geq 10$$

Because all conditions are met, the normal model can be used for a difference in survival rates, and we can apply a **two-sample Z-interval for $p_1 - p_2$**.

The conditions for a two-sample Z-interval for a difference in proportions are the same as the conditions for a two-sample Z-test for a difference in proportions with one modification: instead of using the pooled proportion to check the large counts condition, we use the sample proportions $\hat{p}_1$ and $\hat{p}_2$. For the two-sample Z-interval for $p_1 - p_2$, the following conditions should be met:

Independence. The data is collected from 2 independent random samples, where each sample size is less than 10% of the population size if done without replacement, or the data is collected from an experiment with two randomly assigned treatments (in this case the 10% condition is not relevant and does not need to be checked).

Large counts. We check the large counts condition for each sample separately using the observed sample proportions: $n_1\hat{p}_1 \geq 10$, $n_1(1-\hat{p}_1) \geq 10$, $n_2\hat{p}_2 \geq 10$, and $n_2(1-\hat{p}_2) \geq 10$.

3.6.2 Calculating a confidence interval for a difference in proportions

To calculate a confidence interval for a difference in proportions, we apply the same confidence interval structure from the single-proportion context but with a different point estimate and SE.

The point estimate for the difference in population proportions is:

$$\hat{p}_1 - \hat{p}_2 = \frac{14}{40} - \frac{11}{50} = 0.35 - 0.22 = 0.13$$

Because the point estimate is a difference in sample proportions, we now computer the standard error for a difference in sample proportions. We compute this in the same way that we compute the standard deviation for a difference in sample proportions – the only difference is that we use the sample proportions in place of the population proportions:

$$SE_{\hat{p}_1 - \hat{p}_2} = \sqrt{\frac{\hat{p}_1(1-\hat{p}_1)}{n_1} + \frac{\hat{p}_2(1-\hat{p}_2)}{n_2}} = \sqrt{\frac{0.35(1-0.35)}{40} + \frac{0.22(1-0.22)}{50}} = 0.095$$

The standard error tells us about the typical error when using the difference in sample proportions as an estimate for the difference in population proportions. Sometimes the error will be higher and sometimes it will be lower, but this gives us a sense of how large we expect it to be on average.

EXAMPLE 3.64

Estimate the true difference in survival rate with 90% confidence.

For a 90% confidence level, we use $z^\star = 1.645$. The 90% confidence interval is calculated as:

$$\begin{aligned}
&\text{point estimate} \pm z^\star \times SE \text{ of estimate} \\
&(0.35 - 0.22) \pm 1.65 \times \sqrt{\frac{0.35(1-0.35)}{40} + \frac{0.22(1-0.22)}{50}} \\
&0.13 \pm 1.65 \times 0.095 \\
&0.13 \pm 0.157 \\
&(-0.027,\ 0.287)
\end{aligned}$$

In this example, we see that the standard error (SE) is 0.095 and the margin of error for the 90% confidence interval is 0.157. Unlike the standard error, the margin of error depends on the *confidence level.* The margin of error of 0.157 in this problem tells us that we can be 90% confident that our estimate is within 0.157 of the true difference (treatment – control) in the proportions that would survive at least 24 hours.

The general form of a two-sample Z-interval for $p_1 - p_2$ is given by:

$$(\hat{p}_1 - \hat{p}_2) \pm z^\star \times \sqrt{\frac{\hat{p}_1(1-\hat{p}_1)}{n_1} + \frac{\hat{p}_2(1-\hat{p}_2)}{n_2}}$$

where $\hat{p}_1 - \hat{p}_2$ is the point estimate and $z^\star \times \sqrt{\frac{\hat{p}_1(1-\hat{p}_1)}{n_1} + \frac{\hat{p}_2(1-\hat{p}_2)}{n_2}}$ is the margin of error.

3.6.3 Interpreting and applying a confidence interval for a difference of proportions

In this problem, we calculated the 90% two-sample Z-interval as: (−0.027, 0.287). The interpretation of this interval is analogous to the interpretation of a one-sample Z-interval, instead here we are estimating a difference in population proportions rather than a single population proportion. We can interpret the interval as follows: We are 90% confident that the interval (−0.027, 0.287) contains the difference in the true proportions of patients like the ones in this study that would survive at least 24 hours after receiving blood thinner versus not (treatment − control). That is, we are 90% confident that, for patients like those in the study, the survival rate for those who would receive blood thinners (treatment) is 2.7% *lower* to +28.7% *higher* than the survival rate for those who would not receive the blood thinner (control).

EXAMPLE 3.65

Based on this confidence interval of (−0.027, 0.287), do we have evidence at the 90% confidence level that blood thinners help or harm heart attack patients who have been admitted after they have undergone CPR?

Recall that we took the difference as (treatment − control). If the entire interval were greater than 0, we would have evidence that the true difference is positive, meaning that we would have evidence that the survival rate for those who would be in the treatment group would be *higher*. This would mean that we have evidence that the blood thinners would help patients such as these.

On the other hand, if the entire interval was less than 0, we would have evidence that the true difference is negative, meaning that we would have evidence that the survival rate for those who would be in the treatment group would be *lower*. This would mean that we have evidence that the blood thinners would hurt patients such as these.

What can we say based on the interval we calculated? Because the interval contains both negative and positive values, we do not have evidence and we cannot say with confidence whether blood thinners would harm or help heart attack patients who have been admitted after they have undergone CPR.

In general, when using a confidence interval for a difference, we justify claims about the true difference based on whether the interval is entirely above 0 (evidence the first is greater), entirely below 0 (evidence the first is less), or contains 0 (not enough evidence of a real difference).

3.6.4 Technology: the two-sample Z-interval for $p_1 - p_2$

Section 3.7.4 demonstrates how to calculate the two-proportion Z-interval and the two-proportion Z-test (introduced in the next section) using Desmos, R, and the NumWorks, TI-83/84 and Casio calculator.

3.6.5 Summary and worked example

CONSTRUCTING A CONFIDENCE INTERVAL FOR A DIFFERENCE IN PROPORTIONS

To carry out a complete confidence interval procedure to estimate the difference in population proportions,

Identify: Identify the interval procedure, parameter, and confidence level.

Use a **two-sample Z-interval for a difference in population proportions $p_1 - p_2$**. Define the difference in population proportions $p_1 - p_2$ in words, referencing the populations of interest. Choose a confidence level (C%).

Check: Check conditions for constructing a confidence interval using a normal distribution.

1. Independence: Data come from 2 independent random samples or from a randomized experiment with two treatments. When sampling without replacement, check that the sample size is less than 10% of the population size for both samples.
2. Large counts: $n_1\hat{p}_1 \geq 10$, $n_1(1-\hat{p}_1) \geq 10$, $n_2\hat{p}_2 \geq 10$, and $n_2(1-\hat{p}_2) \geq 10$.

Calculate: Calculate the confidence interval and record it in interval form.

point estimate $\pm$ $z^\star \times SE$ of estimate

point estimate: $\hat{p}_1 - \hat{p}_2$, the difference in sample proportions

SE of estimate: $\sqrt{\frac{\hat{p}_1(1-\hat{p}_1)}{n_1} + \frac{\hat{p}_2(1-\hat{p}_2)}{n_2}}$

$z^\star$: use technology or a t-table at row ∞ and confidence level C%

(____, ____)

Conclude: Interpret the interval and, if applicable, draw a conclusion in context.

We are C% confident that the interval (____, ____) contains the *difference* (specify order) in the true proportions that [...]. If applicable, draw a conclusion based on whether the interval is entirely above, is entirely below, or contains the value 0.

EXAMPLE 3.66

A remote control car company is considering a new manufacturer for wheel gears. The new manufacturer would be more expensive but their higher quality gears are more reliable, resulting in happier customers and fewer warranty claims. However, management must be convinced that the more expensive gears are worth the conversion before they approve the switch. The quality control engineer collects a sample of gears from each supplier, examining 1000 gears from each company, and finds that 879 gears pass inspection from the current supplier and 958 pass inspection from the prospective supplier. Using these data, construct a 95% confidence interval for a difference in the proportion from each supplier that would pass inspection. Use the four-step framework described above to organize your work.

Identify: Because the parameter to be estimated is a difference in proportions, we will use a two-sample Z-interval for a difference in population proportions $p_1 - p_2$. We can define p_1 and p_2 separately as:

p_1: true proportion that would pass inspection from current supplier

p_2: true proportion that would pass inspection from prospective supplier

Alternately, we can define the difference $p_1 - p_2$ as follows:

$p_1 - p_2$: the difference (current – prospective) in the true proportions that would pass inspection between current and prospective supplier

We will estimate the difference at the 95% confidence level.

E **Check:** The samples are independent, but not necessarily random, so to proceed we must assume the gears are all independent. For this sample we will suppose this assumption is reasonable, but the engineer would be more knowledgeable as to whether this assumption is appropriate. We will also assume that the 1000 gears represents less than 10% of the total gears from each supplier. Next, we verify the minimum sample size conditions:

$$1000 \times \frac{879}{1000} \geq 10 \qquad 1000 \times \frac{121}{1000} \geq 10 \qquad 1000 \times \frac{958}{1000} \geq 10 \qquad 1000 \times \frac{42}{1000} \geq 10$$

The large counts condition is met for both samples.

Calculate: We will calculate the interval:

$$\text{point estimate} \ \pm \ z^\star \times SE \text{ of estimate}$$

The point estimate is the difference in sample proportions: $\hat{p}_1 - \hat{p}_2 = 0.879 - 0.958 = -0.079$.

$$SE \text{ of } \hat{p}_1 - \hat{p}_2 = \sqrt{\frac{\hat{p}_1(1-\hat{p}_1)}{n_1} + \frac{\hat{p}_2(1-\hat{p}_2)}{n_2}} = \sqrt{\frac{0.879(1-0.879)}{1000} + \frac{0.958(1-0.958)}{1000}} = 0.0121.$$

The 95% confidence interval is given by:

$$(0.879 - 0.958) \pm 1.96 \times \sqrt{\frac{0.879(1-0.879)}{1000} + \frac{0.958(1-0.958)}{1000}}$$
$$-0.079 \pm 1.96 \times 0.0121$$
$$(-0.103, \ -0.055)$$

Conclude: We are 95% confident that the interval (−0.103, −0.055) contains the difference (current – prospective) in the true proportions that would pass inspection between the current and prospective supplier, meaning that we are 95% confident that the current supplier would have between a 10.3% and 5.5% *lower* rate of passing inspection. Because the entire interval is below zero, the data provide sufficient evidence that the prospective gears pass inspection more often than the current gears. The remote control car company should go with the new manufacturer.

Section summary

- Based on the sample data, a confidence interval can be calculated to estimate the difference between two population proportions. The appropriate confidence interval procedure is a **two-sample Z-interval for a difference in population proportions $p_1 - p_2$**. The parameters p_1 and p_2 should be identified in context.

- A two-sample Z-interval for a difference in population proportion requires the following conditions be met:

  1. Independence: The data come from two independent random samples, each with sample size < 10% of its corresponding population size if sampling without replacement OR the data come from an experiment with two randomly assigned treatments.
  2. Large counts: the number of successes and failures for both samples is at least 10, that is, $n_1\hat{p}_1 \geq 10$, $n_1(1-\hat{p}_1) \geq 10$, $n_2\hat{p}_2 \geq 10$, and $n_2(1-\hat{p}_2) \geq 10$.

- The general form for a C% confidence interval is:

$$\text{point estimate} \pm \text{margin of error, or}$$
$$\text{point estimate} \pm \text{critical value} \times SE \text{ of estimate.}$$

- A two-sample Z-interval for a difference of populations proportion $p_1 - p_2$ can be written as:

$$(\hat{p}_1 - \hat{p}_2) \pm z^\star\sqrt{\frac{\hat{p}_1(1-\hat{p}_1)}{n_1} + \frac{\hat{p}_2(1-\hat{p}_2)}{n_2}}$$

  where $z^\star$ is the critical value for the middle C% of a standard normal distribution.

- The SE of $\hat{p}_1 - \hat{p}_2$ is: $\sqrt{\frac{\hat{p}_1(1-\hat{p}_1)}{n_1} + \frac{\hat{p}_2(1-\hat{p}_2)}{n_2}}$.

- The margin of error of $\hat{p}_1 - \hat{p}_2$ is: $z^\star\sqrt{\frac{\hat{p}_1(1-\hat{p}_1)}{n_1} + \frac{\hat{p}_2(1-\hat{p}_2)}{n_2}}$.

- Because the confidence interval is based on a samples, the point estimate has associated error and the confidence interval may or may not contain the true value of the population proportion.

- The interpretation of the confidence level C% is that in repeated random sampling with the same sample size from the same populations, approximately C% of confidence intervals created will capture the true difference between the population proportions.

- We say we are C% confident that a particular interval (___, ___) contains the difference in population proportions. The parameter and the populations should be stated in the context of the study.

- A confidence interval provides a range of plausible values for a parameter and can be used as evidence to justify a claim about a difference in population proportions. For example, if the interval contains 0, there is not sufficient evidence to conclude that there is a difference between the population proportions. If the interval does not contain 0, there is sufficient evidence to conclude that there is a difference between the two population proportions.

Exercises

3.43 Social experiment, Part I. A "social experiment" conducted by a TV program questioned what people do when they see a very obviously bruised woman getting picked on by her boyfriend. On two different occasions at the same restaurant, the same couple was depicted. In one scenario the woman was dressed "provocatively" and in the other scenario the woman was dressed "conservatively". The table below shows how many restaurant diners were present under each scenario, and whether or not they intervened.

		Scenario		
		Provocative	Conservative	Total
Intervene	Yes	5	15	20
	No	15	10	25
	Total	20	25	45

Explain why the sampling distribution of the difference between the proportions of interventions under provocative and conservative scenarios does not follow an approximately normal distribution.

3.44 Heart transplant success. The Stanford University Heart Transplant Study was conducted to determine whether an experimental heart transplant program increased lifespan. Each patient entering the program was officially designated a heart transplant candidate, meaning that he was gravely ill and might benefit from a new heart. Patients were randomly assigned into treatment and control groups. Patients in the treatment group received a transplant, and those in the control group did not. The table below displays how many patients survived and died in each group.[42]

	control	treatment
survived	4	24
died	30	45

Suppose we are interested in estimating the difference in survival rate between the control and treatment groups using a confidence interval. Explain why we cannot construct such an interval using the normal approximation. What might go wrong if we constructed the confidence interval despite this problem?

3.45 Gender and color preference. A study asked 1,924 male and 3,666 female undergraduate college students their favorite color. A 95% confidence interval for the difference between the proportions of males and females whose favorite color is black ($p_{male} - p_{female}$) was calculated to be (0.02, 0.06). Based on this information, determine if the following statements about undergraduate college students are true or false, and explain your reasoning for each statement you identify as false.[43]

(a) We are 95% confident that the true proportion of males whose favorite color is black is 2% lower to 6% higher than the true proportion of females whose favorite color is black.

(b) We are 95% confident that the true proportion of males whose favorite color is black is 2% to 6% higher than the true proportion of females whose favorite color is black.

(c) 95% of random samples will produce 95% confidence intervals that include the true difference between the population proportions of males and females whose favorite color is black.

(d) We can conclude that there is a significant difference between the proportions of males and females whose favorite color is black and that the difference between the two sample proportions is too large to plausibly be due to chance.

(e) The 95% confidence interval for ($p_{female} - p_{male}$) cannot be calculated with only the information given in this exercise.

3.46 An apple a day keeps the doctor away. A physical education teacher at a high school wanting to increase awareness on issues of nutrition and health asked her students at the beginning of the semester whether they believed the expression "an apple a day keeps the doctor away", and 40% of the students responded yes. Throughout the semester she started each class with a brief discussion of a study highlighting positive effects of eating more fruits and vegetables. She conducted the same apple-a-day survey at the end of the semester, and this time 60% of the students responded yes. Can she used a two-proportion method from this section for this analysis? Explain your reasoning.

[42] B. Turnbull et al. "Survivorship of Heart Transplant Data". In: *Journal of the American Statistical Association* 69 (1974), pp. 74–80.

[43] L Ellis and C Ficek. "Color preferences according to gender and sexual orientation". In: *Personality and Individual Differences* 31.8 (2001), pp. 1375–1379.

3.47 National Health Plan, Part III. Exercise 3.35 presents the results of a poll evaluating support for a generically branded "National Health Plan" in the United States. 79% of 347 Democrats and 55% of 617 Independents support a National Health Plan.

(a) Calculate a 95% confidence interval for the difference between the proportion of Democrats and Independents who support a National Health Plan ($p_D - p_I$), and interpret it in this context. We have already checked conditions for you.

(b) True or false: If we had picked a random Democrat and a random Independent at the time of this poll, it is more likely that the Democrat would support the National Health Plan than the Independent.

3.48 Sleep deprivation, CA vs. OR, Part I. According to a report on sleep deprivation by the Centers for Disease Control and Prevention, the proportion of California residents who reported insufficient rest or sleep during each of the preceding 30 days is 8.0%, while this proportion is 8.8% for Oregon residents. These data are based on simple random samples of 11,545 California and 4,691 Oregon residents. Calculate a 95% confidence interval to estimate the difference between the proportions of Californians and Oregonians who are sleep deprived. Include all steps of the Identify, Check, Calculate, Conclude framework.[44]

[44] CDC, Perceived Insufficient Rest or Sleep Among Adults — United States, 2008.

3.7 Hypothesis testing for $p_1 - p_2$

Does the approval rate of the 2010 healthcare law differ when asked in two different ways? Does the use of fish oils reduce heart attacks better than a placebo? How much evidence is there for each of these claims? In this section, we apply the hypothesis testing framework to a difference in population proportions.

Learning objectives

1. Identify and set up an appropriate testing method for a difference in population proportions $p_1 - p_2$.
2. Identify the null and alternative hypotheses for a difference between population proportions.
3. Justify the appropriateness of a hypothesis test for a difference between two population proportions by verifying conditions.
4. Calculate an appropriate test statistic and p-value for a hypothesis test for a difference in population proportions.
5. Interpret the p-value of a hypothesis test for a difference in population proportions.
6. Justify a claim about the difference in population proportions based on the results of the test.
7. Explain why the large counts condition and the standard error calculation are different for a hypothesis test and a confidence interval for a difference in population proportions.

3.7.1 Introducing hypothesis testing for a difference of proportions

Here we use a new example to examine a special estimate of the standard error when the null hypothesis is that two population proportions equal each other, i.e. H_0: $p_1 = p_2$. We investigate whether a survey question's phrasing can affect responses. Pew Research Center conducted a survey of adults in the US with the following question:

> As you may know, by 2014 nearly all Americans will be required to have health insurance. [People who do not buy insurance will pay a penalty] while [People who cannot afford it will receive financial help from the government]. Do you approve or disapprove of this policy?

For each randomly sampled respondent, the statements in brackets were randomized: either they were kept in the original order given above, or they were reversed. Results are presented in Figure 3.17.

G

GUIDED PRACTICE 3.67

Is this study an experiment or an observational study?[45]

[45]There is a random sample involved, but there are also two treatments. Half of the respondents are given the original statement order and the other half, randomly, are given the reversed statement order. This is an experiment because there are randomly assigned treatments.

	sample size	Approve law (%)	Disapprove law (%)	Other
"People who do not buy insurance will pay a penalty" is given first (original order)	771	47	49	3
"People who cannot afford it will receive financial help from the government" is given first (reversed order)	732	34	63	3

Figure 3.17: Results for a Pew Research Center poll where the ordering of two statements in a question regarding healthcare were randomized.

The approval percents of 47% and 34% seem far apart. However, could this difference be due to random chance? We will answer this question using a hypothesis test.

EXAMPLE 3.68

Set up hypotheses to test whether the two statement orders produce different responses.

Define the parameters of interest as follows:
p_1: proportion of adults in the US that would approve of policy with original statement ordering.
p_2: proportion of adults in the US that would approve of policy with reversed statement ordering.

The null claim is that the question order does not matter and the two proportions should be equal. The alternative claim, the one that bears the burden of proof, is that the question ordering does matter.

H_0: $p_1 = p_2$
H_A: $p_1 \neq p_2$

It is important to notice that:

$$p_1 = p_2 \text{ is equivalent to } p_1 - p_2 = 0, \text{ and}$$
$$p_1 \neq p_2 \text{ is equivalent to } p_1 - p_2 \neq 0.$$

We can now see that the hypotheses are really about a difference of proportions: $p_1 - p_2$. In the last section, we used a two-sample Z-interval to estimate the parameter $p_1 - p_2$; here, we will use a **two-sample Z-test for $p_1 - p_2$**, with null hypothesis: $p_1 - p_2 = 0$, i.e. $p_1 = p_2$. We will elaborate on the conditions needed for this test after introducing the concept of the pooled proportion.

3.7.2 Calculations and conditions for a test for a difference of proportions

For a Z-test for the survey question wording example, we will compute a Z-test statistic, which takes the familiar form:

$$Z = \frac{\text{point estimate} - \text{null value}}{SE \text{ of estimate}}$$

The parameter of interest is $p_1 - p_2$, so the point estimate will be the observed difference in sample proportions: $\hat{p}_1 - \hat{p}_2 = 0.47 - 0.34 = 0.13$. The null value depends on the null hypothesis. The null hypothesis is that the approval rate would be the same for both statement orderings, i.e. that the difference is 0, therefore, the null value is 0.

The SD of a difference in sample proportions has the form:

$$\sigma_{\hat{p}_1-\hat{p}_2} = \sqrt{\frac{p_1(1-p_1)}{n_1} + \frac{p_2(1-p_2)}{n_2}}$$

However, in a hypothesis test, the distribution of the point estimate is always examined assuming the null hypothesis is true, i.e. in this case, $p_1 = p_2$. Both the large counts check and the standard error formula should reflect this equality in the null hypothesis. We will use p_c to represent the common or combined proportion that support the healthcare law regardless of statement order:

$$\begin{aligned}\sigma_{\hat{p}_1-\hat{p}_2} &= \sqrt{\frac{p_c(1-p_c)}{n_1} + \frac{p_c(1-p_c)}{n_2}} \\ &= \sqrt{p_c(1-p_c)\left(\frac{1}{n_1} + \frac{1}{n_2}\right)}\end{aligned}$$

We don't know the true proportion p_c, but we can obtain a good estimate of it, $\hat{p}_c$, by *pooling* the results of both samples. We find the total number of "yeses" or "successes" and divide that by the total number of cases. This is equivalent to taking a weighted average of $\hat{p}_1$ and $\hat{p}_2$. We call $\hat{p}_c$ the **pooled sample proportion**, and we use it to check the large counts condition and to compute the standard error when the null hypothesis is $p_1 = p_2$. Here:

$$\hat{p}_c = \frac{771(0.47) + 732(0.34)}{771 + 732} = 0.407$$

POOLED SAMPLE PROPORTION

When the null hypothesis is $p_1 = p_2$, it is useful to find the pooled sample proportion:

$$\hat{p}_c = \frac{\text{number of "successes"}}{\text{number of cases}} = \frac{\text{x}_1 + \text{x}_2}{n_1 + n_2} = \frac{n_1\hat{p}_1 + n_2\hat{p}_2}{n_1 + n_2}$$

Here x_1 represents the number of successes in sample 1. If x_1 is not given, it can be computed as $n_1 \times \hat{p}_1$. Similarly, x_2 represents the number of successes in sample 2 and can be computed as $n_2 \times \hat{p}_2$.

EXAMPLE 3.69

Verify that conditions for using the normal distribution are met and find the SE of estimate for this hypothesis test. Recall that the pooled proportion $\hat{p}_c = 0.407$, $n_1 = 771$, and $n_2 = 732$.

The data do come from an experiment with two randomly assigned treatments. Here the treatments are the two different orderings of the question regarding healthcare. Because this is an experiment, the 10% condition does not need to be checked. Also, the large counts condition (minimums of 10) easily holds for each group.

$$771 \times 0.407 \geq 10 \qquad 771 \times (1 - 0.407) \geq 10 \qquad 732 \times 0.407 \geq 10 \qquad 732 \times (1 - 0.407) \geq 10$$

Because $H_0 : p_1 = p_2$, we compute the SE for the difference in sample proportions as:

$$SE_{\hat{p}_1 - \hat{p}_2} = \sqrt{\hat{p}_c(1 - \hat{p}_c)\left(\frac{1}{n_1} + \frac{1}{n_2}\right)} = \sqrt{0.407(1 - 0.407)\left(\frac{1}{771} + \frac{1}{732}\right)} = 0.025$$

To verify conditions are met for the two-sample Z-test for a difference in population proportions, check the following.

Independence. The data is collected from 2 independent random samples, where each sample size is less than 10% of the population size if done without replacement, or the data is collected from an experiment with two randomly assigned treatments (in this case the 10% condition is not relevant and does not need to be checked).

Large counts. The expected number of successes and failures should be at least 10 for both groups using the sample pooled proportion $\hat{p}_c$: $n_1\hat{p}_c \geq 10$, $n_1(1-\hat{p}_c) \geq 10$, $n_2\hat{p}_c \geq 10$, and $n_2(1-\hat{p}_c) \geq 10$

EXAMPLE 3.70

Complete the hypothesis test using a significance level of 0.01.

We have already set up the hypotheses and verified that the difference of proportions can be modeled using a normal distribution. We can now calculate the test statistic and p-value.

$$Z = \frac{\text{point estimate} - \text{null value}}{SE \text{ of estimate}} = \frac{(0.47 - 0.34) - 0}{0.025} = 5.2$$

This is a two-tailed test as H_A is that $p_1 \neq p_2$. We can find the area in one tail and double it. Here, the p-value ≈ 0. Because the p-value is smaller than $\alpha = 0.01$, we reject the null hypothesis and we have evidence that the order of the statements affects how likely a respondent is to support the 2010 healthcare law.

It might be unclear about when to use the pooled sample proportion versus the individual sample proportions for checking conditions for inference for a difference in proportions and computing the standard error. For confidence intervals, we use the individual sample proportions ($\hat{p}_1$ and $\hat{p}_2$) for these calculations, while for hypothesis tests use the pooled sample proportion, $\hat{p}_c$.

3.7.3 Summary and worked example

HYPOTHESIS TESTING FOR A DIFFERENCE IN PROPORTIONS

To carry out a complete hypothesis test to compare two population proportions,

Identify: Identify the test procedure, parameter, significance level, and hypotheses.

Use a **two-sample Z-test for a difference in population proportions $p_1 - p_2$**. Define the population proportions p_1 and p_2 in the context of the problem. Choose a significance level (α) and test the following hypotheses.

H_0: $p_1 = p_2$ $\quad$ ($p_1 - p_2 = 0$)

H_A: $p_1 \neq p_2$; $\; p_1 > p_2$; $\;$ or $\; p_1 < p_2$ $\quad$ ($p_1 - p_2 \neq 0$; $\; p_1 - p_2 > 0$; $\;$ or $\; p_1 - p_2 < 0$)

Check: Check conditions for the test statistic to be nearly normal, assuming H_0 is true.

1. Independence: Data come from 2 independent random samples or from a randomized experiment with two treatments. When sampling without replacement, check that the sample size is less than 10% of the population size for both samples.
2. Large counts: $n_1\hat{p}_c \geq 10$, $n_1(1 - \hat{p}_c) \geq 10$, $n_2\hat{p}_c \geq 10$, and $n_2(1 - \hat{p}_c) \geq 10$

Calculate: Calculate the Z-statistic and p-value.

$$Z = \frac{\text{point estimate} - \text{null value}}{SE \text{ of estimate}}$$

point estimate: $\hat{p}_1 - \hat{p}_2$, the difference in sample proportions

null value: 0

SE of estimate: $\sqrt{\hat{p}_c(1 - \hat{p}_c)\left(\frac{1}{n_1} + \frac{1}{n_2}\right)}$, where $\hat{p}_c$ is the pooled proportion

p-value = (based on the Z-statistic and the direction of H_A)

Conclude: Compare the p-value to α, and draw a conclusion in context.

If the p-value is $\leq \alpha$, reject H_0; there is sufficient evidence that [H_A in context].

If the p-value is $> \alpha$, do not reject H_0; there is not sufficient evidence that [H_A in context].

EXAMPLE 3.71

A 5-year experiment was conducted to evaluate the effectiveness of fish oils on reducing heart attacks, where each subject was randomized into one of two treatment groups. We'll consider heart attack outcomes in these patients:

	heart_attack	no_event	Total
fish_oil	145	12788	12933
placebo	200	12738	12938

Carry out a complete hypothesis test at the 10% significance level to test whether the use of fish oils is effective in reducing heart attacks.

Identify: Because we are testing whether the difference in proportions is 0, we choose the two-sample Z-test for a difference in population proportions $p_1 - p_2$, where

p_1 is the true proportion who would suffer a heart attack if given fish oil.

p_2 is the true proportion who would suffer a heart attack if given placebo.

We will test the following hypotheses at the $\alpha = 0.10$ significance level.

H_0: $p_1 = p_2$ Fish oil and placebo are equally effective.

H_A: $p_1 < p_2$ Fish oil is more effective in reducing heart attacks.

Check: We must verify that the difference in sample proportions can be modeled using a normal distribution. First we note that there is a randomized experiment with two treatments: fish oil and placebo. Since we have randomly assigned treatments, we do not need to check the 10% condition. Next, we calculate the pooled proportion as follows:

E

$$\hat{p}_c = \frac{x_1 + x_2}{n_1 + n_2} = \frac{145 + 200}{12933 + 12938} = 0.0133$$

We can now verify: $12933(0.0133) \geq 10$, $12933(1 - 0.0133) \geq 10$, $12938(0.0133) \geq 10$, and $12938(1 - 0.0133) \geq 10$, so both conditions are met.

Calculate: We will calculate the Z-statistic and the p-value.

$$Z = \frac{\text{point estimate} - \text{null value}}{SE \text{ of estimate}}$$

The point estimate is the difference in sample proportions: $\hat{p}_1 - \hat{p}_2 = 0.0112 - 0.0155 = -0.0043$.

SE of $\hat{p}_1 - \hat{p}_2$, assuming H_0 is true, is:
$\sqrt{\hat{p}_c(1 - \hat{p}_c)\left(\frac{1}{n_1} + \frac{1}{n_2}\right)} = \sqrt{0.0133(1 - 0.0133)\left(\frac{1}{12933} + \frac{1}{12938}\right)} = 0.00142.$

The null value is the hypothesized difference in population proportions, which is 0.

$$Z = \frac{(0.0112 - 0.0155) - 0}{\sqrt{0.0133(1 - 0.0133)\left(\frac{1}{12933} + \frac{1}{12938}\right)}} = \frac{-0.0043 - 0}{0.00142} = -3.0$$

Because H_A uses a less than, meaning that it is a lower-tail test, the p-value is the area to the *left* of $Z = -3.0$ under the standard normal distribution. This area can be found using technology or a normal table. The p-value = 0.0013.

Conclude: The p-value of 0.0013 is < 0.10, so we reject H_0; there is sufficient evidence that fish oil is effective in reducing heart attacks. Note that we are only allowed to use causal language here because there was an experiment.

3.7.4 Technology: the two-sample Z-interval and Z-test for $p_1 - p_2$

Figure 3.16, summarizes an experiment involving patients admitted to a hospital for a heart attack. 14 out of the 40 patients assigned to the treatment (blood thinner) group survived versus 11 out of 50 assigned to the control group. Calculate a 95% confidence interval for a difference in the proportion (treatment – control) that would survive. Also calculate the test statistic and p-value to test whether there is evidence that the true proportions *differ*. Conditions have been verified.

Desmos: Use the `zproptest(x1, n1, x2, n2)` function as explained below.

1. Click + in the upper left, then choose `inference`.
2. Choose `z-test for proportions` in the pop-up window.

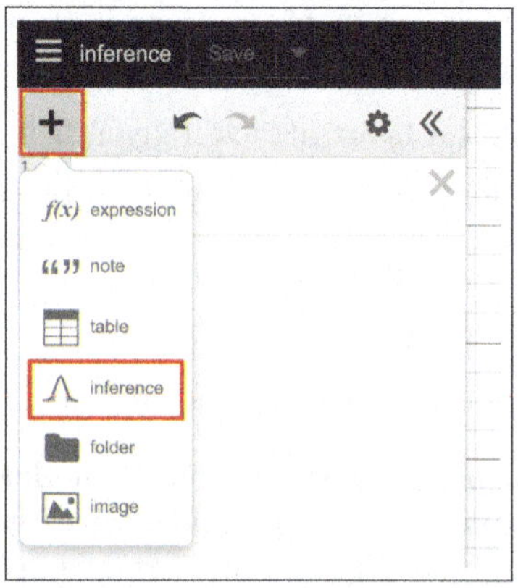

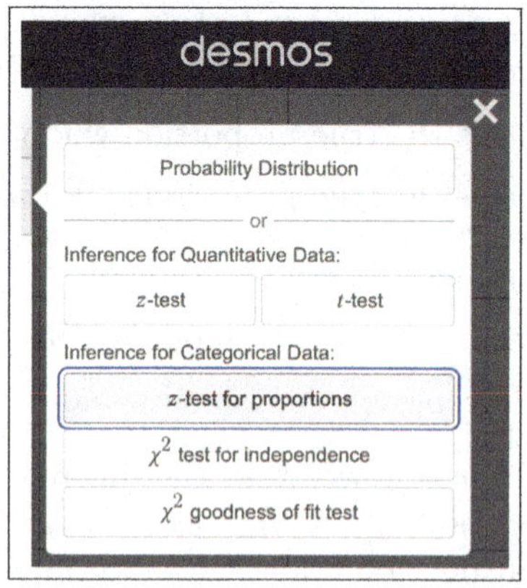

3. Under `SAMPLE 1`, enter `successes` (x_1) and `sample size` (n_1) for group 1. Click `SAMPLE 2`, enter `successes` (x_2) and `sample size` (n_2) for group 2. Click `Create Test`.

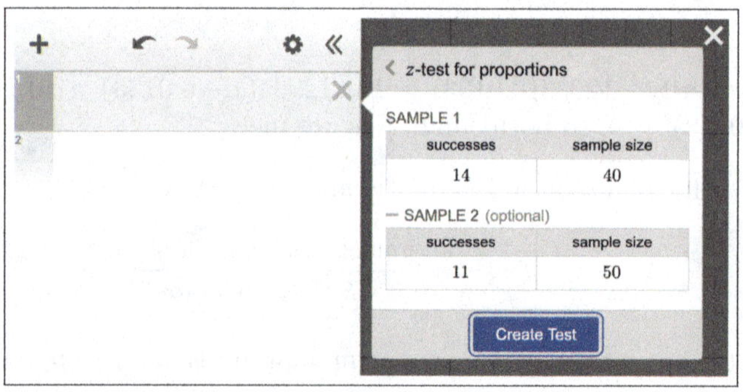

* You can type `zproptest(14, 40, 11, 50)` in place of steps 1-3 above.

4. Click the triangle next to `Confidence Interval` and input the desired `Confidence level`. Click the ⋮ to the right of the confidence interval to see additional information. Hover over the dot in the middle of the confidence interval to see the point estimate.
5. Click the triangle next to `Significance Test`. Enter the hypothesized value for the difference. Usually this will be 0. Select a value for `Tails` based on the direction of H_A.

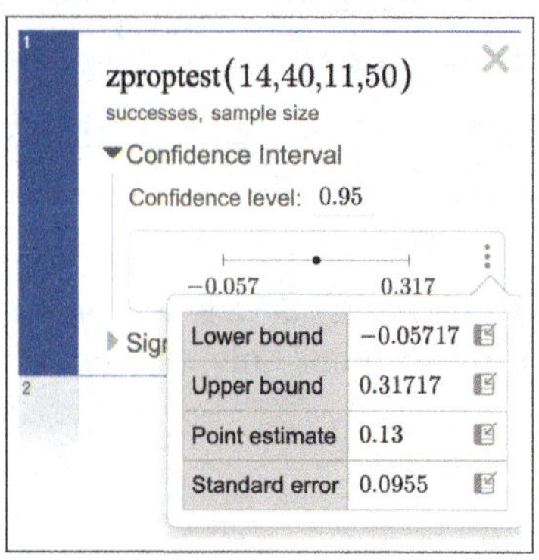

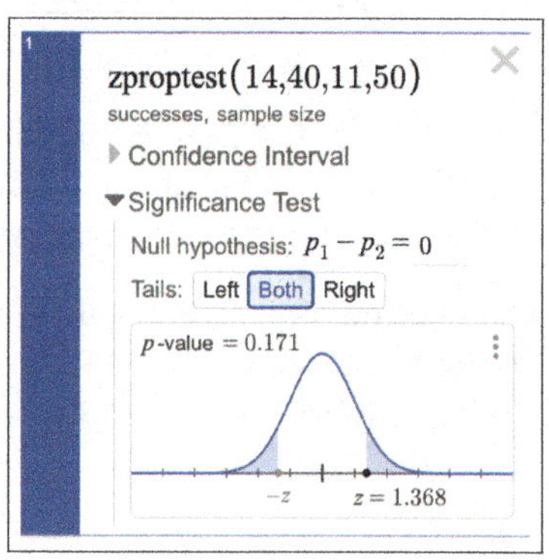

R: 2-proportion Z-interval/test for $p_1 - p_2$

Here we use `prop.test()` with `x = c(x1, x2)` and `n = c(n1, n2)`.

CONFIDENCE INTERVAL.

```
> prop.test(x=c(14, 11), n=c(40, 50), correct = FALSE, conf.level = 0.95)
2-sample test for equality of proportions without continuity correction
data:  c(14, 11) out of c(40, 50)
X-squared = 1.872, df = 1, p-value = 0.1712
alternative hypothesis:  two.sided
95 percent confidence interval:
-0.05716886 0.31716886
sample estimates:
prop 1 prop 2
0.35 0.22
```

HYPOTHESIS TEST.

```
> prop.test(x = c(14, 11), n=c(40, 50), correct = FALSE, alternative = "two.sided")
2-sample test for equality of proportions without continuity correction
data:  c(14, 11) out of c(40, 50)
X-squared = 1.872, df = 1, p-value = 0.1712
alternative hypothesis:  two.sided
95 percent confidence interval:
-0.05716886 0.31716886
sample estimates:
prop 1 prop 2
0.35 0.22
```

This test returns X-squared instead of Z.
Z = $+\sqrt{\texttt{X-squared}}$ or $-\sqrt{\texttt{X-squared}}$, depending on if sample (`prop 1 - prop 2`) is + or −.
Here (0.35 − 0.22) is positive, so Z = $\sqrt{1.872}$.

```
> Z = sqrt(1.872)
> Z
[1] 1.368211
```

Calculator: NumWorks calculator instructions and example output are included. For the TI-83/84 and Casio calculators, general instructions are provided, and worked example videos can be accessed via the 🎥 icon or at **openintro.org/ti** and **openintro.org/casio**.

NUMWORKS: 2-PROPORTION Z-INTERVAL

Use `OK` or `EXE` to make a selection.

1. From the home screen, select `Inference`, then `Intervals`, then `Two proportions`. If these options do not appear, click the ← button in the upper right as many times as needed.
2. Enter x_1, n_1, x_2, n_2, and `Confidence level`, then choose `Next`.
3. Click the down arrow to see all quantities returned and then choose `Next`.
4. In addition to seeing the confidence interval displayed in two ways, you can press the up and down arrows to quickly change confidence level and see the resulting interval and margin of error.

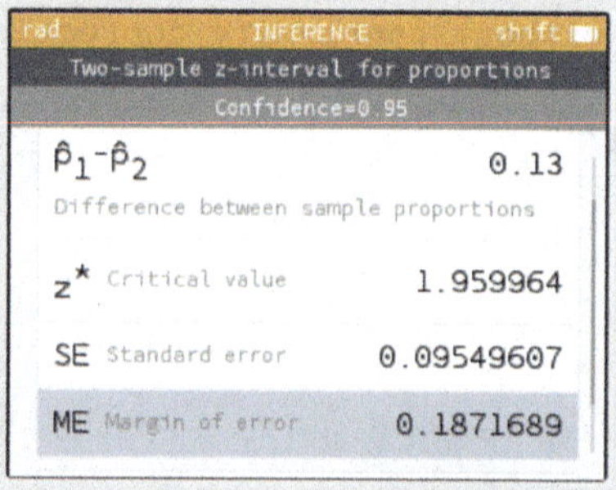

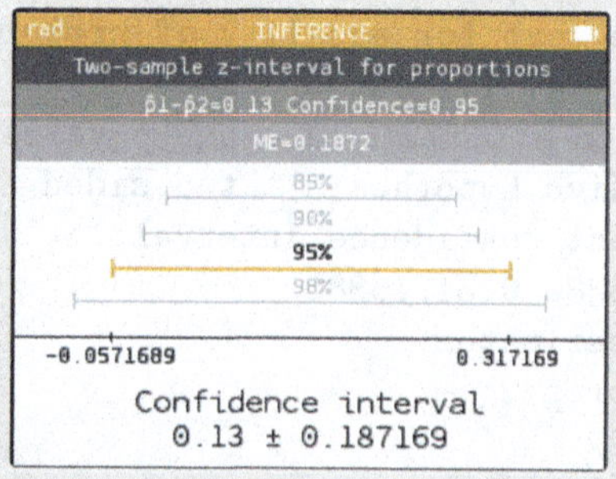

NUMWORKS: 2-PROPORTION Z-TEST

Use `OK` or `EXE` to make a selection.

1. From the home screen, select `Inference`, then `Tests`, then `Two proportions`. If these options do not appear, click the ← button in the upper right as many times as needed.
2. Enter the hypothesized difference. Usually this will be 0. Press the down arrow. Press `OK` and choose <, ≠ , or > for the Alternative hypothesis. Press the down arrow and choose `Next`.
3. Enter the values of x_1, n_1, x_2, n_2, and α. Hit the down arrow and choose `Next`.
4. Note the quantities returned. Click the down arrow to see all of the values, including the p-value, then choose `Next`.
5. On this screen, the p-value and alpha are shaded on the normal distribution and can be visually compared.

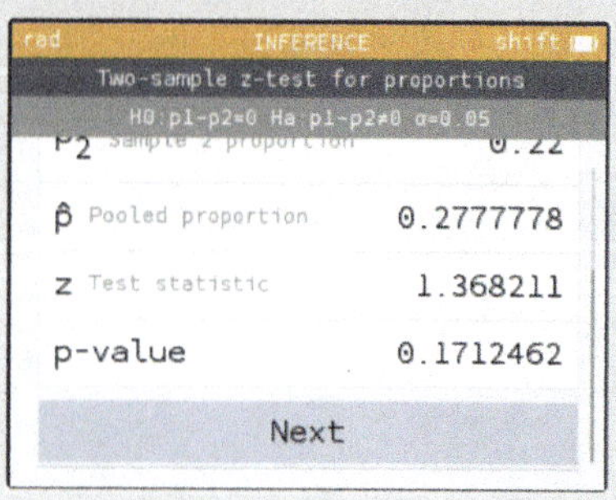

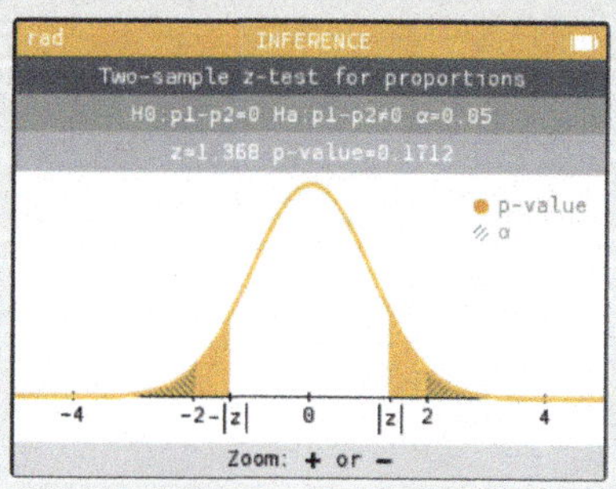

TI-83/84: 2-PROPORTION Z-INTERVAL

Use STAT, TESTS, 2-PropZInt.

1. Choose STAT.
2. Right arrow to TESTS.
3. Down arrow and choose B:2-PropZInt.
4. Let x1 be the *number* of yeses (must be an integer) in sample 1 and let n1 be the size of sample 1.
5. Let x2 be the *number* of yeses (must be an integer) in sample 2 and let n2 be the size of sample 2.
6. Let C-Level be the desired confidence level.
7. Choose Calculate and hit ENTER, which returns:

(__,__)	the confidence interval		
$\hat{p}_1$	sample 1 proportion	n_1	size of sample 1
$\hat{p}_2$	sample 2 proportion	n_2	size of sample 2

TI-83/84: 2-PROPORTION Z-TEST

Use STAT, TESTS, 2-PropZTest.

1. Choose STAT.
2. Right arrow to TESTS.
3. Down arrow and choose 6:2-PropZTest.
4. Let x1 be the *number* of yeses (must be an integer) in sample 1 and let n1 be the size of sample 1.
5. Let x2 be the *number* of yeses (must be an integer) in sample 2 and let n2 be the size of sample 2.
6. Choose ≠, <, or > to correspond to H_A.
7. Choose Calculate or Draw and hit ENTER. Draw shows the Z-statistic and p-value as well as a graph of the normal distribution with p-value shaded. Calculate returns:

z	Z-statistic	p	p-value
$\hat{p}_1$	sample 1 proportion	$\hat{p}$	pooled sample proportion
$\hat{p}_2$	sample 2 proportion		

CASIO FX-9750GII: 2-PROPORTION Z-INTERVAL

1. Navigate to `STAT` (`MENU` button, then hit the `2` button or select `STAT`).
2. Choose the `INTR` option (`F4` button).
3. Choose the `Z` option (`F1` button).
4. Choose the `2-P` option (`F4` button).
5. Specify the interval details:
   - Confidence level of interest for `C-Level`.
   - Enter the number of successes for each group, `x1` and `x2`.
   - Enter the sample size for each group, `n1` and `n2`.
6. Hit the `EXE` button, which returns

`Left`, `Right`	the ends of the confidence interval
`p̂1`, `p̂2`	the sample proportions
`n1`, `n2`	sample sizes

CASIO FX-9750GII: 2-PROPORTION Z-TEST

1. Navigate to `STAT` (`MENU` button, then hit the `2` button or select `STAT`).
2. Choose the `TEST` option (`F3` button).
3. Choose the `Z` option (`F1` button).
4. Choose the `2-P` option (`F4` button).
5. Specify the test details:
   - Specify the sidedness of the test using the `F1`, `F2`, and `F3` keys.
   - Enter the number of successes for each group, `x1` and `x2`.
   - Enter the sample size for each group, `n1` and `n2`.
6. Hit the `EXE` button, which returns

`z`	Z-statistic	`p̂1`, `p̂2`	sample proportions
`p`	p-value	`p̂`	pooled proportion
		`n1`, `n2`	sample sizes

Section summary

- The appropriate hypothesis testing procedure for a difference between two population proportions is a **two-sample Z-test for a difference in population proportions $p_1 - p_2$**. The parameters p_1 and p_2 should be identified in context.
- The parameters for a hypothesis test for a difference between two population proportions should reference the population parameters, the response variables, and the populations in context.
- The null hypotheses for a two-sample Z-test for a difference in population proportion indicates no difference and is written as:

  H_0: $p_1 = p_2$ (or equivalently H_0: $p_1 - p_2 = 0$).
- A one-sided alternative hypothesis is written as:

  H_A: $p_1 < p_2$ (or equivalently H_A: $p_1 - p_2 < 0$), or

  H_A: $p_1 > p_2$ (or equivalently H_A: $p_1 - p_2 > 0$).

  A two-sided alternative hypothesis is written as:

  H_A: $p_1 \neq p_2$ (or equivalently H_A: $p_1 - p_2 \neq 0$).
- The combined or pooled proportion is denoted $\hat{p}_c$ and is calculated as $\frac{x_1+x_2}{n_1+n_2}$ or $\frac{n_1\hat{p}_1+n_2\hat{p}_2}{n_1+n_2}$.
- The two-sample Z-test for $p_1 - p_2$ requires the following conditions be met:
  1. Independence: The data come from two independent random samples, each with sample size < 10% of its corresponding population size if sampling without replacement OR the data come from an experiment with two randomly assigned treatments.
  2. Large counts: the expected number of successes and failures for both samples, assuming H_0 is true, is at least 10, that is, $n_1\hat{p}_c \geq 10$, $n_1(1-\hat{p}_c) \geq 10$, $n_2\hat{p}_c \geq 10$ and $n_2(1-\hat{p}_c) \geq 10$.
- A **test statistic** has the form: test statistic $= \frac{\text{point estimate} - \text{null value}}{SE \text{ of estimate}}$.

  The test statistic for a two-sample Z-test for $p_1 - p_2$ is: $Z = \frac{(\hat{p}_1-\hat{p}_2)-0}{\sqrt{\hat{p}_c(1-\hat{p}_c)\left(\frac{1}{n_1}+\frac{1}{n_2}\right)}}$.

  The SE is calculated under the assumption that the H_0 is true and uses $\hat{p}_c$.
- The p-value for a Z-test corresponds to a lower tail, upper tail, or both tails of the standard normal distribution, depending on whether the direction of the alternate hypothesis is <, >, or ≠.
- The p-value for a two-sample Z-test for $p_1 - p_2$ is the probability of obtaining a Z-statistic as small or smaller, as large or larger, or as extreme or more extreme than the Z-statistic that was observed, depending on whether the direction of the alternate hypothesis is <, >, or ≠., assuming the null hypothesis is true (i.e. that the population proportions are equal).
- A formal decision explicitly compares the p-value to the significance level. If the p-value $\leq \alpha$, then reject the null hypothesis; if the p-value $> \alpha$, then fail to reject the null hypothesis. The conclusion should be stated in terms of the alternative hypothesis and should include context, referencing the parameters and the populations. Use non-causal language unless a well-designed experiment was conducted.
- The results of a hypothesis test for a difference between two population proportions can serve as the statistical reasoning to support the answer to an investigative question about the two populations that were sampled.

Exercises

3.49 Sleep deprived transportation workers. The National Sleep Foundation conducted a survey on the sleep habits of randomly sampled transportation workers and randomly sampled non-transportation workers that serve as a "control" for comparison. The results of the survey are shown below.[46]

		Transportation Professionals			
	Control	Pilots	Truck Drivers	Train Operators	Bus/Taxi/Limo Drivers
Less than 6 hours of sleep	35	19	35	29	21
6 to 8 hours of sleep	193	132	117	119	131
More than 8 hours	64	51	51	32	58
Total	292	202	203	180	210

Conduct a hypothesis test to evaluate if these data provide evidence of a difference between the proportion of truck drivers and non-transportation workers (the "control" group) who get less than 6 hours of sleep per day (i.e. are considered sleep deprived). Remember to Identify, Check, Calculate, and Conclude.

3.50 Sleep deprivation, CA vs. OR, Part II. Exercise 3.48 provides data on sleep deprivation rates of Californians and Oregonians. The proportion of California residents who reported insufficient rest or sleep during each of the preceding 30 days is 8.0%, while this proportion is 8.8% for Oregon residents. These data are based on simple random samples of 11,545 California and 4,691 Oregon residents.

(a) Conduct a full hypothesis test to determine if these data provide strong evidence that the rate of sleep deprivation is different for the two states. Remember to Identify, Check, Calculate and Conclude.

(b) It is possible the conclusion of the test in part (a) is incorrect. If this is the case, what type of error was made?

3.51 Malaria vaccine. With no currently licensed vaccines to inhibit malaria, good news was welcomed with a recent study reporting long-awaited vaccine success for children in Burkina Faso. With 450 children randomized to either one of two different doses of the malaria vaccine or a control vaccine, 89 of 292 malaria vaccine and 106 out of 147 control vaccine children contracted malaria within 12 months after the treatment.

(a) Conduct a full hypothesis test to determine if these data provide evidence that a lower proportion of children like those in the study would contract malaria if given the malaria vaccine than if given the control vaccine. Remember to Identify, Check, Calculate and Conclude.

(b) Interpret the p-value that you calculated in the context of the problem.

3.52 Prenatal vitamins and Autism. Researchers studying the link between prenatal vitamin use and autism surveyed the mothers of a random sample of children aged 24 - 60 months with autism and conducted another separate random sample for children with typical development. The table below shows the number of mothers in each group who did and did not use prenatal vitamins during the three months before pregnancy (periconceptional period).[47]

		Autism		
		Autism	Typical development	Total
Periconceptional	No vitamin	111	70	181
prenatal vitamin	Vitamin	143	159	302
	Total	254	229	483

(a) State appropriate hypotheses to test for a difference in autism rates between those whose mothers used prenatal vitamins during the three months before pregnancy and those whose mothers did not.

(b) Complete the hypothesis test and state an appropriate conclusion. (Reminder: Verify any necessary conditions for the test.)

(c) A New York Times article reporting on this study was titled "Prenatal Vitamins May Ward Off Autism". Do you find the title of this article to be appropriate? Explain your answer. Additionally, propose an alternative title.[48]

[46] National Sleep Foundation, 2012 Sleep in America Poll: Transportation Workers' Sleep, 2012.

[47] R.J. Schmidt et al. "Prenatal vitamins, one-carbon metabolism gene variants, and risk for autism". In: *Epidemiology* 22.4 (2011), p. 476.

[48] R.C. Rabin. "Patterns: Prenatal Vitamins May Ward Off Autism". In: *New York Times* (2011).

3.8 Goodness of fit using chi-square (special topic)

Are juries representative of the population in terms of race/ethnicity, or is there a bias in jury selection? Is the color distribution of actual M&M's consistent with what was reported on the Mars website? Do people choose rock, paper, scissors with the same likelihood, or is one choice favored over another? To answer such questions, we develop a method for assessing a null model when the data take on more than two categories.

Learning objectives

1. Describe chi-square distributions.
2. Identify and set up an appropriate testing method for assessing goodness of fit using a one-way table with a single categorical variable.
3. Calculate the expected counts and degrees of freedom for a one-way table, assuming the null hypothesis is true.
4. Verify whether the conditions for the chi-square goodness of fit are met.
5. Calculate the χ^2-test statistic, degrees of freedom and p-value for a goodness of fit test.

3.8.1 Creating a test statistic for one-way tables

Data is collected from a random sample of 275 jurors in a small county. Jurors identified their racial/ethnic group, as shown in Figure 3.18, and we would like to determine if these jurors are representative of the population with respect to race/ethnicity. If the jury is representative of the population, then the proportions in the sample should roughly reflect the population of eligible jurors, i.e. registered voters.

Race/Ethnicity	White	Black	Hispanic	Other	Total
Representation in juries	205	26	25	19	275
Registered voters	0.72	0.07	0.12	0.09	1.00

Figure 3.18: Representation by race in a city's juries and population.

While the proportions in the juries do not precisely represent the population proportions, it is unclear whether these data provide convincing evidence that the sample is not representative. If the jurors really were randomly sampled from the registered voters, we might expect small differences due to chance. However, unusually large differences may provide convincing evidence that the juries were not representative.

EXAMPLE 3.72

Of the people in the city, 275 served on a jury. If the individuals are randomly selected to serve on a jury, about how many of the 275 people would we expect to be White? How many would we expect to be Black?

About 72% of the population is White, so we would expect about 72% of the jurors to be White: $0.72 \times 275 = 198$.

Similarly, we would expect about 7% of the jurors to be Black, which would correspond to about $0.07 \times 275 = 19.25$ Black jurors.

GUIDED PRACTICE 3.73

Twelve percent of the population is Hispanic and 9% represent other racial/ethnic groups. How many of the 275 jurors would we expect to be Hispanic or from another racial/ethnic group? Answers can be found in Figure 3.19.

Race/Ethnicity	White	Black	Hispanic	Other	Total
Observed data	205	26	25	19	275
Expected counts	198	19.25	33	24.75	275

Figure 3.19: Actual and expected make-up of the jurors.

The sample proportion represented from each race/ethnicity among the 275 jurors was not a precise match for any ethnic group. While some sampling variation is expected, we would expect the sample proportions to be fairly similar to the population proportions if there is no bias on juries. We need to test whether the differences are strong enough to provide convincing evidence that the jurors are not a random sample. These ideas can be organized into hypotheses:

H_0: The jurors are a random sample, i.e. there is no racial/ethnic bias in who serves on a jury, and the observed counts reflect natural sampling fluctuation.

H_A: The jurors are not randomly sampled, i.e. there is racial/ethnic bias in juror selection.

To evaluate these hypotheses, we quantify how different the observed counts are from the expected counts. Strong evidence for the alternative hypothesis would come in the form of unusually large deviations in the groups from what would be expected based on sampling variation alone.

3.8.2 The chi-square test statistic

In previous hypothesis tests, we constructed a test statistic of the following form:

$$Z = \frac{\text{point estimate} - \text{null value}}{SE \text{ of point estimate}}$$

This construction was based on (1) identifying the difference between a point estimate and an expected value if the null hypothesis was true, and (2) standardizing that difference using the standard error of the point estimate. These two ideas will help in the construction of an appropriate test statistic for count data.

In this example we have four categories: White, Black, Hispanic, and other. Because we have four values rather than just one or two, we need a new tool to analyze the data. Our strategy will be to find a test statistic that measures the overall deviation between the observed and the expected counts.

We first find the difference between the observed and expected counts for the four groups:

	White	Black	Hispanic	Other
observed - expected	$205 - 198$	$26 - 19.25$	$25 - 33$	$19 - 24.75$

Next, we square the differences:

	White	Black	Hispanic	Other
$(\text{observed - expected})^2$	$(205 - 198)^2$	$(26 - 19.25)^2$	$(25 - 33)^2$	$(19 - 24.75)^2$

We must standardize each term. To know whether the squared difference is large, we compare it to what was expected. If the expected count was 5, a squared difference of 25 is very large. However, if the expected count was 1,000, a squared difference of 25 is very small. We will divide each of the squared differences by the corresponding expected count.

	White	Black	Hispanic	Other
$\frac{(\text{observed - expected})^2}{\text{expected}}$	$\frac{(205 - 198)^2}{198}$	$\frac{(26 - 19.25)^2}{19.25}$	$\frac{(25 - 33)^2}{33}$	$\frac{(19 - 24.75)^2}{24.75}$

Finally, to arrive at the overall measure of deviation between the observed counts and the expected counts, we add up the terms. This gives us a new test statistic called χ^2 (read as chi-square).

$$\begin{aligned}\chi^2 &= \sum \frac{(\text{observed - expected})^2}{\text{expected}} \\ &= \frac{(205 - 198)^2}{198} + \frac{(26 - 19.25)^2}{19.25} + \frac{(25 - 33)^2}{33} + \frac{(19 - 24.75)^2}{24.75}\end{aligned}$$

We can write an equation for χ^2 using the observed counts and expected counts:

$$\chi^2 = \frac{(\text{observed count}_1 - \text{expected count}_1)^2}{\text{expected count}_1} + \cdots + \frac{(\text{observed count}_4 - \text{expected count}_4)^2}{\text{expected count}_4}$$

The final number χ^2 summarizes how strongly the observed counts tend to deviate from the null counts.

Next, we will see that if the null hypothesis is true, then χ^2 follows a distribution called a *chi-square distribution*. Using this distribution, we will be able to obtain a p-value to evaluate whether there appears to be racial/ethnic bias in the juries for the city we are considering.

3.8.3 The chi-square distribution and finding areas

The **chi-square distribution** is sometimes used to characterize data sets and statistics that are always positive and typically right skewed. Recall a normal distribution had two parameters – mean and standard deviation – that could be used to describe its exact characteristics. The chi-square distribution has just one parameter called **degrees of freedom (df)**, which influences the shape, center, and spread of the distribution.

Figure 3.20 illustrates three general properties of chi-square distributions as the degrees of freedom increases: the distribution becomes more symmetric, the center moves to the right, and the variability inflates.

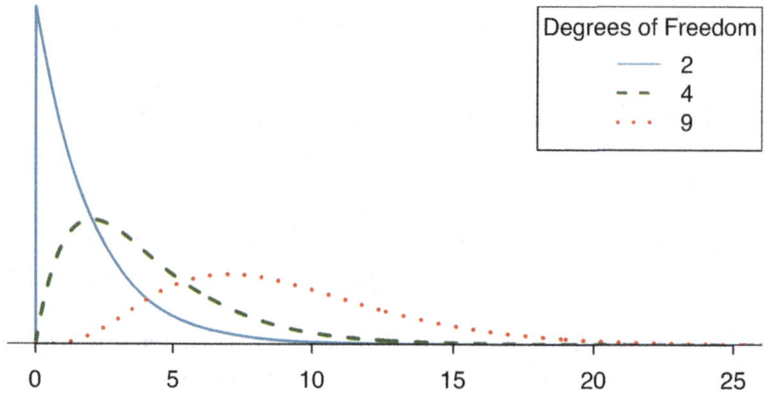

Figure 3.20: Three chi-square distributions with varying degrees of freedom. The distributions are non-negative and right skewed. As the degrees of freedom increase, chi-square distributions become less skewed and have a larger center and spread.

Our principal interest in the chi-square distribution is the calculation of p-values, which (as we have seen before) is related to finding the relevant area in the tail of a distribution.

In the previous section, Section 3.8.2, we identified a new test statistic (χ^2) within the context of assessing whether there was evidence of racial/ethnic bias in how jurors were sampled. We determined that a large χ^2 value would suggest strong evidence favoring the alternative hypothesis: that there was racial/ethnic bias. However, we could not quantify what the chance was of observing such a large test statistic ($\chi^2 = 5.89$) if the null hypothesis actually was true. This is where the chi-square distribution becomes useful. If the null hypothesis was true and there was no racial/ethnic bias, then χ^2 would follow a chi-square distribution, with three degrees of freedom in this case. Under certain conditions, the statistic χ^2 follows a chi-square distribution with $k - 1$ degrees of freedom, where k is the number of bins or categories of the variable.

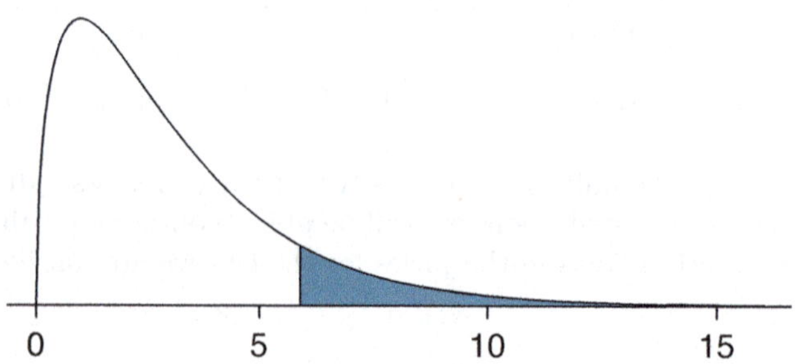

Figure 3.21: The p-value for the juror hypothesis test is shaded in the chi-square distribution with $df = 3$.

EXAMPLE 3.74

In our jury example, we have four categories of race/ethnicity. If the null hypothesis is true, the test statistic $\chi^2 = 5.89$ would be closely associated with a chi-square distribution with three degrees of freedom. Using this distribution and test statistic, identify the p-value and state whether or not there is evidence of racial/ethnic bias in the juror selection. See Section 3.9.7 for ways to find areas under a chi-square distribution using technology.

(E)

The chi-square distribution and p-value are shown in Figure 3.21. Because larger chi-square values correspond to stronger evidence against the null hypothesis, we shade the upper tail to represent the p-value. Using technology, we look at the chi-square distribution with 3 degrees of freedom and find the area to the right of $\chi^2 = 5.89$. This area, which corresponds to the p-value, is equal to 0.117. This p-value is larger than the default significance level of 0.05, so we do not reject the null hypothesis. In other words, the data do not provide convincing evidence of racial/ethnic bias in the juror selection.

The test that we just carried out regarding jury selection is known as the **χ^2 goodness of fit test**. It is called "goodness of fit" because we test whether or not the proposed or expected distribution is a good fit for the observed data.

Just like we checked conditions to use the normal model in earlier sections, we must also check conditions to safely model χ^2 with a chi-square distribution. Here we have a random sample, which is less than 10% of all potential jurors. For chi-square, we have a different sample size condition. Each expected count must be at least 5. In the juror example, the expected counts were 198, 19.25, 33, and 24.75, all easily above 5, so we can model the χ^2 test statistic using a chi-square distribution.

CONDITIONS FOR THE CHI-SQUARE GOODNESS OF FIT TEST

The chi-square goodness of fit test requires the test statistic to be well modeled by a chi-square distribution. This will be valid when the observations are independent and the expected counts are large. If these conditions are not met, the chi-square goodness of fit test should not be used.

Independence. The observations can be considered independent if the data come from a random process. If randomly sampling without replacement from a finite population, the observations can be considered independent when sampling less than 10% of the population.

Large expected counts. In order for the χ^2-statistic to follow the chi-square distribution, each particular bin or category must have at least 5 expected cases under the assumption that the null hypothesis is true.

3.8.4 Summary and worked example

GOODNESS OF FIT TEST FOR A ONE-WAY TABLE

When there is one sample and we are comparing the distribution of a categorical variable to a specified or population distribution,

Identify: Use a **χ^2 goodness of fit test** and the desired α significance level.

H_0: The distribution of [...] matches the specified or population distribution.

H_A: The distribution of [...] doesn't match the specified or population distribution.

Check: Check that the test statistic follows a chi-square distribution, assuming H_0 is true.

1. Independence: Data come from a random sample or random process. If sampling without replacement, check that sample size is less than 10% of the population size.
2. Expected counts: All expected counts are ≥ 5. (calculate and record expected counts).

Calculate: Calculate the χ^2-statistic, df, and p-value.

test statistic: $\chi^2 = \sum \frac{(\text{observed} - \text{expected})^2}{\text{expected}}$

df = # of categories − 1

p-value = (area to the *right* of χ^2-statistic under the chi-square distribution with the appropriate df)

Conclude: Compare the p-value to α, and draw a conclusion in context.

If the p-value is $\leq \alpha$, reject H_0; there is sufficient evidence that [H_A in context].

If the p-value is $> \alpha$, do not reject H_0; there is not sufficient evidence that [H_A in context].

EXAMPLE 3.75

Starting in 2016, a statistician named Rick Wicklin collected a sample of 712 M&M's to see if there was evidence that the actual color distribution of M&M's differed from the stated color distribution of M&M's on the Mars website, which had not been updated since 2008. Wickin's sample color distribution and the Mars website's stated color distribution are shown in the table below. (You can read about Wicklin's adventure in the Quartz article linked in the Data Appendix, which starts on page 460.)

	Blue	Orange	Green	Yellow	Red	Brown
website percentages (2008):	24%	20%	16%	14%	13%	13%
observed percentages:	18.7%	18.7%	19.5%	14.5%	15.1%	13.5%

Is there evidence at the 5% significance level that the distribution of M&M's in 2016 were different from the stated distribution on the website in 2008? Use the four-step framework to organize your work.

Identify: Because we have one variable (color), broken up into multiple categories, we choose the chi-square goodness of fit test. We will test the following hypotheses at the $\alpha = 0.05$ significance level.

H_0: The distribution of M&M colors is the same as the stated distribution in 2008.

H_A: The distribution of M&M colors is different than the stated distribution in 2008.

Check: We must verify that the test statistic follows a chi-square distribution. Note that there is only one sample here. The website percentages are considered fixed – they are not the result of a sample and do not have sampling variability associated with them. To carry out the chi-square goodness of fit test, we will have to assume that Wicklin's sample can be considered a random sample of M&M's. We note that the total population size of M&M's is much larger than 10 times the sample size of 712. Next, we need to find the expected counts. Here, $n = 712$. If H_0 is true, then we would expect 24% of the M&M's to be Blue, 20% to be Orange, etc. So the expected counts can be found as:

	Blue	Orange	Green	Yellow	Red	Brown
expected counts:	0.24(712) = 170.9	0.20(712) = 142.4	0.16(712) = 113.9	0.14(712) = 99.6	0.13(712) = 92.6	0.13(712) = 92.6

Calculate: We will calculate the chi-square statistic, degrees of freedom, and the p-value. To calculate the chi-square statistic, we need the observed counts as well as the expected counts. To find the observed counts, we use the observed percentages. For example, 18.7% of 712 = 0.187(712) = 133.

	Blue	Orange	Green	Yellow	Red	Brown
observed counts:	133	133	139	103	108	96
expected counts:	170.9	142.4	113.9	99.6	92.6	92.6

$$\begin{aligned}\chi^2 &= \sum \frac{(\text{observed} - \text{expected})^2}{\text{expected}} \\ &= \frac{(133-170.9)^2}{170.9} + \frac{(133-142.4)^2}{142.4} + \dots + \frac{(108-92.6)^2}{92.6} + \frac{(96-92.6)^2}{92.6} \\ &= 8.41 + 0.62 + 5.53 + 0.12 + 2.56 + 0.12 \\ &= 17.36\end{aligned}$$

Because there are six colors, the degrees of freedom is $6 - 1 = 5$. In a chi-square test, the p-value is always the area to the *right* of the chi-square statistic. Here, the area to the right of 17.36 under the chi-square distribution with 5 degrees of freedom is 0.004.

Conclude: The p-value of 0.004 is < 0.05, so we reject H_0; there is sufficient evidence that the distribution of M&M's does not match the stated distribution on the website in 2008.

3.8.5 Technology: the chi-square goodness of fit test

A spinner has four colors that are supposed to be equally likely: red, green, blue, yellow. Someone records the number of each color after many spins in a game. After 200 spins, there are 59 red, 44 green, 54 blue and 41 yellow. Use technology to find the test statistic, df, and p-value for a chi-square goodness of fit test, testing whether there is evidence that the colors do not have the same likelihood of coming up. Also find the expected values and chi-square contributions for the test. Assume conditions for the test are met.

Desmos: Use the `chisqtest()` function as explained below.

1. Click + in the upper left, then choose `inference`.
2. Choose χ^2 `goodness of fit test` in the pop-up window.

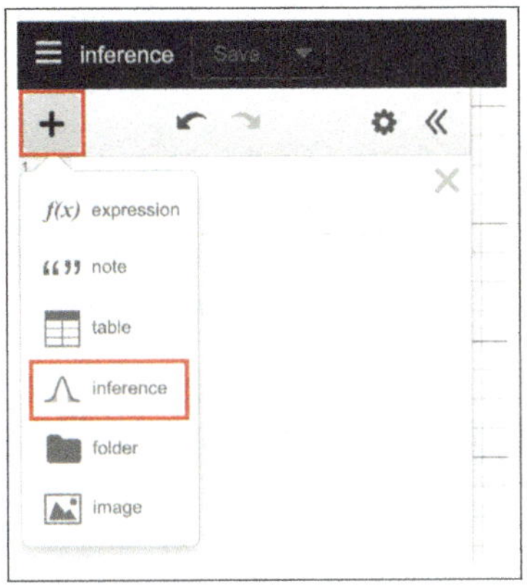

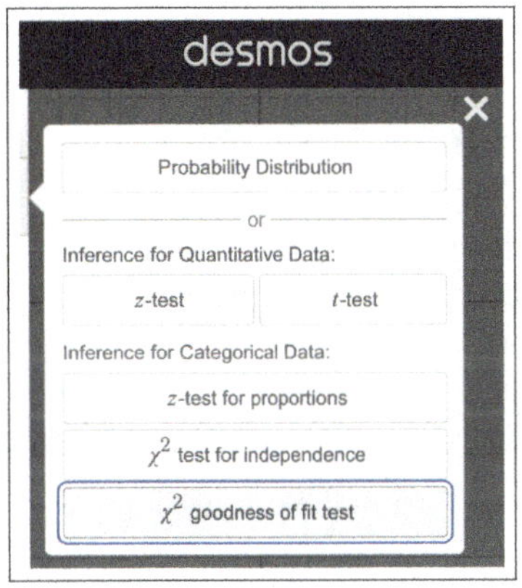

3. Enter the observed counts in the `Observed` column. If the null hypothesis is that the categories are equally likely, you can leave the `Expected` column blank. Otherwise, enter the expected values based on H_0. Click `Create Test`.

* You can type `chisqgof([59, 44, 54, 41])` in place of steps 1-3 above.

4. Click the triangle next to `Observed (Expected)`, then click the checkbox next to `Expected` to see the expected counts. Click the checkbox next to `Contributions` to see each value's contribution to the chi-square statistic. Click the checkbox next to `Totals` to see the column totals.
5. Click the triangle next to `Significance Test` to see the test statistic, df, and p-value.

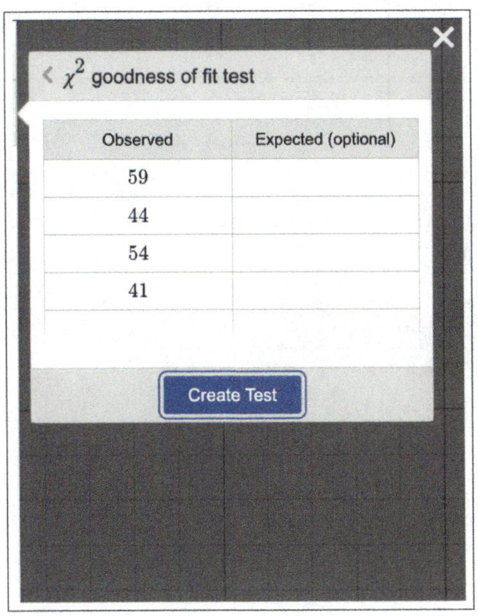

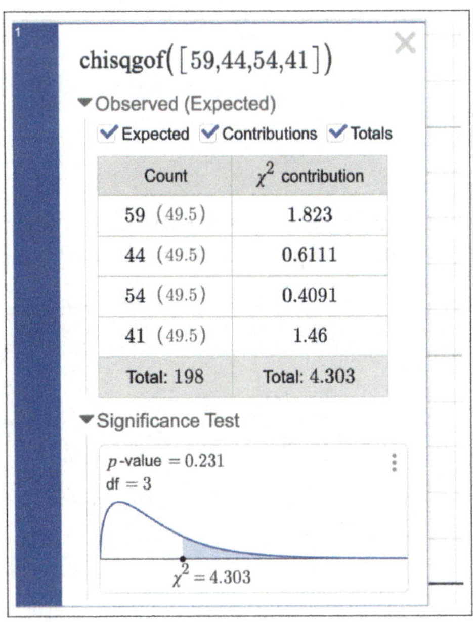

R: Chi-square goodness of fit test
Use `chisq.test(x = c(observed counts), p = c(expected proportions))`.

```
> X = chisq.test(c(59, 44, 54, 41), c(0.25, 0.25, 0.25, 0.25)) or
> X = chisq.test(x = c(59, 44, 54, 41), p = c(0.25, 0.25, 0.25, 0.25))

> X
Chi-squared test for given probabilities
data:  c(59, 44, 54, 41)
X-squared = 4.303, df = 3, p-value = 0.2305

> X$expected
[1] 49.5 49.5 49.5 49.5

> (X$residuals)^2
[1] 1.8232323 0.6111111 0.4090909 1.4595960
```

Calculator: NumWorks calculator instructions and example output are included. For the TI-83/84 and Casio calculators, general instructions are provided, and worked example videos can be accessed via the icon or at **openintro.org/ti** and **openintro.org/casio**.

NUMWORKS: CHI-SQUARE GOODNESS OF FIT TEST

Use `OK` or `EXE` to make a selection.

1. From the home screen, select `Inference`, then `Tests`, then `Chi-square`, then `Goodness of fit`. If these options do not appear, click the ← button in the upper right as many times as needed.
2. Enter the Observed counts, then enter the Expected counts. Use arrows as needed. Press the down arrow and record the df. Enter α, then choose `Next`.
3. On this screen, you will see the chi-square statistic and p-value. Use the up and right arrow and choose `Contributions` to see the chi-square contributions.

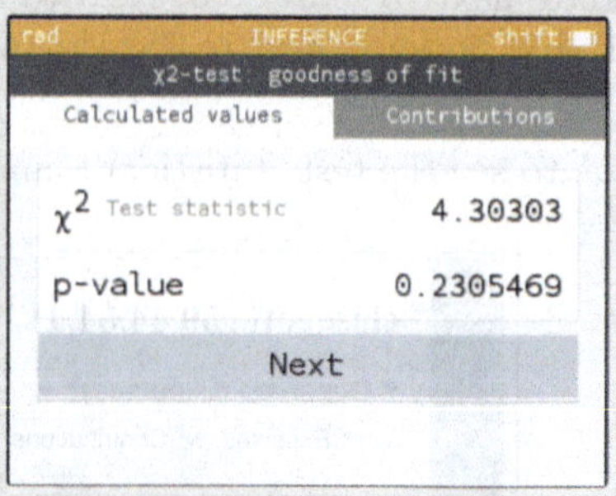

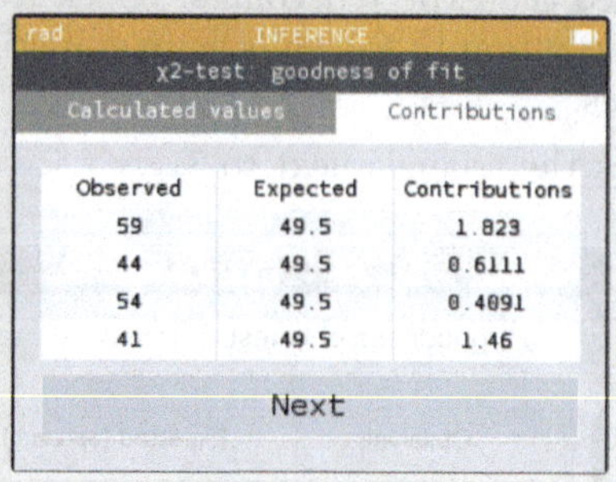

4. When you have recorded the information, press the down arrow and choose `Next` to see the chi-square distribution with the p-value shaded.

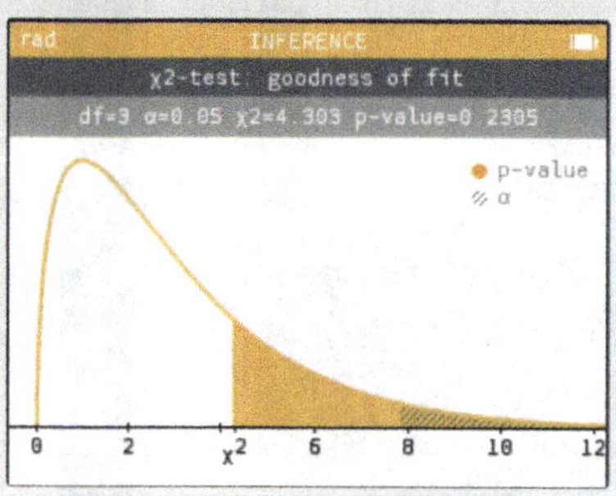

TI-84: CHI-SQUARE GOODNESS OF FIT TEST

Use STAT, TESTS, χ^2GOF-Test.

1. Enter the observed counts into list L1 and the expected counts into list L2.
2. Choose STAT.
3. Right arrow to TESTS.
4. Down arrow and choose D:χ^2GOF-Test.
5. Leave Observed: L1 and Expected: L2.
6. Enter the degrees of freedom after df:
7. Choose Calculate or Draw and hit ENTER. Draw shows the chi-square statistic and p-value as well as a graph of the chi-square distribution with p-value shaded. Calculate returns:

χ^2	chi-square test statistic
p	p-value
df	degrees of freedom

   CNTRB showing the chi-square contributions.
   Hit the right arrow to see all of the contributions.

TI-83: Unfortunately the TI-83 does not have this test built in. To carry out the test manually, make list L3 = (L1 - L2)^2 / L2 and do 1-Var-Stats on L3. The sum of L3 will correspond to the value of χ^2 for this test.

CASIO FX-9750GII: CHI-SQUARE GOODNESS OF FIT TEST

1. Navigate to STAT (MENU button, then hit the 2 button or select STAT).
2. Enter the observed counts into a list (e.g. List 1) and the expected counts into list (e.g. List 2).
3. Choose the TEST option (F3 button).
4. Choose the CHI option (F3 button).
5. Choose the GOF option (F1 button).
6. Adjust the Observed and Expected lists to the corresponding list numbers from Step 2.
7. Enter the degrees of freedom, df.
8. Specify a list where the contributions to the test statistic will be reported using CNTRB. This list number should be different from the others.
9. Hit the EXE button, which returns

χ^2	chi-square test statistic
p	p-value
df	degrees of freedom
CNTRB	list showing the test statistic contributions

Section summary

- While a normal distribution is defined by its mean and standard deviation, the chi-square distribution is defined by just one parameter called **degrees of freedom**. For a chi-square distribution, as the degrees of freedom increases: the center increases, the spread increases, and the shape becomes more symmetric and more normal.
- When we want to see if a hypothesized model is a good fit for observed data or if data is representative of a particular population, we can use a χ^2 **goodness of fit test**. This test requires one variable with multiple categories (bins) that can be arranged in a one-way table.
- The hypotheses for a χ^2 goodness of fit test can be written as:

  H_0: The distribution of [...] matches the specified or population distribution.

  H_A: The distribution of [...] doesn't match the specified or population distribution.
- For the χ^2 goodness of fit test, we check the following conditions to verify that the test statistic follows a chi-square distribution.
  1. Independence: Data come from a random sample or random process. When sampling without replacement, check that sample size is less than 10% of the population size.
  2. Expected counts: All expected counts are ≥ 5.
- We calculate the test statistic as follows:

  test statistic: $\chi^2 = \sum \frac{(\text{observed} - \text{expected})^2}{\text{expected}}$; $df = \#$ of categories $- 1$

  Always use whole numbers (counts) for the observed values, not proportions or percents. For each category, the expected counts can be found by multiplying the sample size by the expected proportion under the null hypothesis. Expected counts do *not* need to be integers.
- The p-value is the area to the *right* of the χ^2-statistic under the chi-square distribution with the appropriate df.
- A larger χ^2 represents greater deviation between the observed values and the expected values, relative to the expected values. For a fixed degrees of freedom, a larger χ^2 value leads to a smaller p-value, providing greater evidence against H_0.
- For a χ^2 test, the p-value corresponds to the probability of getting a test statistic as large as we got or larger, assuming the null hypothesis is true, i.e. that the population distribution is as hypothesized.

Exercises

3.53 Open source textbook. A professor using an open source introductory statistics book predicts that 60% of the students will purchase a hard copy of the book, 25% will print it out from the web, and 15% will read it online. At the end of the semester he asks his students to complete a survey where they indicate what format of the book they used. Of the 126 students, 71 said they bought a hard copy of the book, 30 said they printed it out from the web, and 25 said they read it online.

(a) State the hypotheses for testing if the professor's predictions were inaccurate.

(b) How many students did the professor expect to buy the book, print the book, and read the book exclusively online?

(c) List the conditions required for the chi-square goodness of fit test and discuss whether they are satisfied.

(d) Assume conditions are sufficiently met. Calculate the chi-square statistic, the degrees of freedom associated with it, and the p-value.

(e) Based on the p-value calculated in part (d), what is the conclusion of the hypothesis test? Interpret your conclusion in this context.

3.54 Barking deer. Microhabitat factors associated with forage and bed sites of barking deer in Hainan Island, China were examined. In this region woods make up 4.8% of the land, cultivated grass plot makes up 14.7%, and deciduous forests make up 39.6%. Of the 426 sites where the deer forage, 4 were categorized as woods, 16 as cultivated grassplot, and 61 as deciduous forests. The table below summarizes these data.[49]

Woods	Cultivated grassplot	Deciduous forests	Other	Total
4	16	61	345	426

Photo by Shrikant Rao (http://flic.kr/p/4Xjdkk) CC BY 2.0 license

(a) Do these data provide convincing evidence that barking deer prefer to forage in certain habitats over others? Conduct an appropriate hypothesis test to answer this research question, and acknowledge any assumptions you had to make to carry out this test. Remember to Identify, Check, Calculate, and Conclude.

(b) Interpret the calculated p-value in the context of the problem.

[49]Liwei Teng et al. "Forage and bed sites characteristics of Indian muntjac (Muntiacus muntjak) in Hainan Island, China". In: *Ecological Research* 19.6 (2004), pp. 675–681.

3.9 Chi-square tests for two-way tables

Does the distribution of successful versus unsuccessful web searches differ among different search algorithms? Is there an association between people's generation and whether or not they take action to help address climate change? In order to answer questions such as these, we revisit two-way tables, and we learn about two new and closely related chi-square tests.

Learning objectives

1. Describe chi-square distributions.
2. Identify and set up an appropriate testing method for comparing distributions in two-way tables of categorical data.
3. Identify the null and alternative hypotheses for a chi-square test for homogeneity or independence.
4. Justify the appropriateness of a hypothesis test for independence or homogeneity using a chi-square distribution by verifying conditions.
5. Calculate expected counts for two-way tables of categorical data, assuming a null hypothesis is true.
6. Calculate the appropriate test statistic, degrees of freedom and p-value for a chi-square test for homogeneity or independence.
7. Interpret the p-value for a chi-square test for homogeneity or independence.
8. Justify a claim about the population based on the results of a chi-square test for homogeneity or independence.
9. Explain the differences and similarities between the chi-square test for homogeneity and chi-square test for independence.

3.9.1 Introducing the chi-square test for homogeneity

Google is constantly running experiments to test new search algorithms. For example, Google might test three algorithms using a sample of 10,000 google.com search queries. Figure 3.22 shows an example of 10,000 queries randomly split into three algorithm groups.[50] The group sizes were specified before the start of the experiment to be 5000 for the current algorithm and 2500 for each test algorithm.

	Search algorithm			
	current	test 1	test 2	Total
Counts	5000	2500	2500	10000

Figure 3.22: Experiment breakdown of test subjects into three search groups.

In this experiment, the explanatory variable is the search algorithm. However, an outcome variable is also needed. This outcome variable should somehow reflect whether the search results satisfied the user. We will define the values of the `outcome` variable as "success": no new related search and "failure": follow-up related search.

[50] Google regularly runs experiments in this manner to help improve their search engine. It is entirely possible that if you perform a search and so does your friend, that you will have different search results. While the data presented in this section resemble what might be encountered in a real experiment, these data are simulated.

Figure 3.23 provides the results from the experiment. These data are very similar to the count data in Section 3.8. However, now the different combinations of two variables are binned in a *two-way* table. In examining these data, we want to evaluate whether there is strong evidence that at least one algorithm is more successful than the others.

		Search algorithm			
		current	test 1	test 2	Total
`outcome`	success	3511	1749	1818	7078
	failure	1489	751	682	2922
	Total	5000	2500	2500	10000

Figure 3.23: Results of the Google search algorithm experiment.

EXAMPLE 3.76

What is the ultimate goal of the Google experiment? What are the null and alternative hypotheses, in regular words?

The ultimate goal is to see whether there is a difference in the performance of the algorithms. The hypotheses can be described as the following:

H_0: There is no difference in the distribution of the `outcome` variable across the three search algorithms: current, test 1, and test 2.

H_A: There is a difference in the distribution of the `outcome` variable across the three search algorithms: current, test 1, and test 2.

Because the `outcome` variable has only two values: "success" and "failure", the hypotheses could also be worded as:

H_0: The proportion of successes would be the same regardless of which search algorithm is used.

H_A: The proportion of successes would not be the same for the different search algorithms.

The hypothesis test for this Google experiment is really about assessing whether there is statistically significant evidence that the likelihood of a successful search would be different based on the algorithm used. In other words, the goal is to check whether the three search algorithms perform differently. If there were only two algorithms, we could perform a two-sample Z-test for a difference of population proportions, as we did in Section 3.7. However, since we have three algorithms, we will need a new test that allows us to compare the distribution of a variable across multiple treatments or populations. We call this test a **chi-square test for homogeneity**.

Homogeneity means "same". The null hypothesis of this test says that the distribution of a variable is the *same* (or has no difference) across multiple treatments or populations. In our Google experiment example, the null claim says that the distribution of the `outcome` variable (or the likelihood of success) is the same for each of the three algorithms: current, test 1, and test 2.

3.9.2 Expected counts in two-way tables

When there are more than two treatments or populations to compare, instead of calculating a Z test statistic, we calculate a new test statistic called a chi-square test statistic. To do this, we must first calculate the expected count for each cell of the two-way table, under the assumption that the null hypothesis is true.

EXAMPLE 3.77

(E) From the experiment, we estimate the overall proportion of successful searches as $\frac{7078}{10000} = 0.7078$. If there really is no difference among the algorithms and 70.78% of searches, regardless of search algorithm, would result in "success", how many of the 5000 searches in the "current algorithm" group would be expected to result in "success"?

About 70.78% of the 5000 would result in "success":

$$0.7078 \times 5000 = 3539 \text{ searches}$$

That is, if there was no difference between the three algorithms, then we would expect 3539 of the current algorithm searches to result in "success".

GUIDED PRACTICE 3.78

(G) Using the same rationale described in Example 3.77, about how many searches in each algorithm group would result in "success" if the algorithms were equally helpful?[51]

We can compute the expected number of "success" for each group using the same strategy employed in Example 3.77 and Guided Practice 3.78. These expected counts are shown in Figure 3.24, alongside the observed counts from Figure 3.23.

		Search algorithm						
		current		test 1		test 2		Total
outcome	success	3511	(3539.0)	1749	(1769.5)	1818	(1769.5)	7078
	failure	1489	(1461.0)	751	(730.5)	682	(730.5)	2922
	Total	5000		2500		2500		10000

Figure 3.24: The observed counts and the (expected counts).

The examples and exercises above provided some help in computing expected counts. In general, expected counts for a two-way table may be computed using the row totals, column totals, and the table total. For instance, if there was no difference between the groups, then about 70.78% of each column should be in the first row:

$$0.7078 \times (\text{column 1 total}) = 3539$$
$$0.7078 \times (\text{column 2 total}) = 1769.5$$
$$0.7078 \times (\text{column 3 total}) = 1769.5$$

Looking back, we see that 0.7078 was computed as $\frac{7078}{10000}$, the fraction of successful outcomes. Therefore, these three expected counts could have been computed as:

$$\left(\frac{\text{row 1 total}}{\text{table total}}\right)(\text{column 1 total}) = 3539$$
$$\left(\frac{\text{row 1 total}}{\text{table total}}\right)(\text{column 2 total}) = 1769.5$$
$$\left(\frac{\text{row 1 total}}{\text{table total}}\right)(\text{column 3 total}) = 1769.5$$

[51] We would expect $0.7078 * 2500 = 1769.5$. It is okay that this is not an integer.

This leads us to a general formula for computing expected counts in a two-way table when we would like to test whether there is strong evidence of an association between the column variable and row variable.

COMPUTING EXPECTED COUNTS IN A TWO-WAY TABLE

To identify the expected count in a particular row and column, compute

$$\text{Expected Count} = \frac{(\text{row total}) \times (\text{column total})}{\text{table total}}$$

We can use mosaic plots, introduced in Section 2.1.2, to visually compare the expected counts to the observed counts.

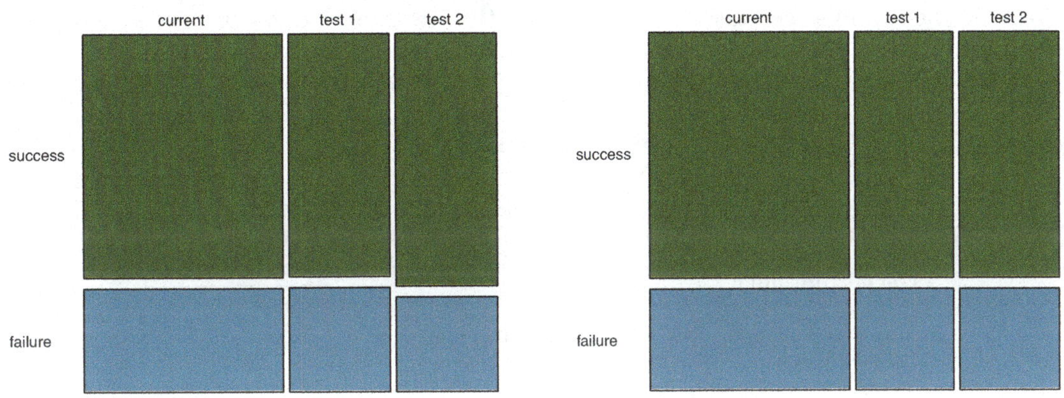

Figure 3.25: Mosaic plots using values from Figure 3.24. Left: observed counts; Right: expected counts assuming there is no difference in the distribution of the outcome variable across the three search algorithms: current, test 1, and test 2 in the population.

The difference between the mosaic plot of observed counts and of expected counts appears relatively small. However, even a small difference in this context could translate into a better experience for millions of users. To determine whether the difference is significant we will need to carry out a hypothesis test and calculate the p-value.

3.9.3 Verifying conditions and calculating the test statistic

When we have a two-way table of counts and the data is collected from multiple independent random samples or multiple randomly assigned treatments we would like to carry out what is called a **chi-square test for homogeneity**.

CONDITIONS FOR THE CHI-SQUARE TEST FOR HOMOGENEITY

There are two conditions that must be checked before performing a chi-square test for homogeneity. If these conditions are not met, this test should not be used.

Independence. The data must be arrived at by taking two or more independent random samples or two or more randomly assigned treatments. When sampling without replacement from a finite population, the sample sizes should be less than 10% of the corresponding population sizes.

Large expected counts. All of the cells in the two-way table must have at least 5 expected cases assuming the null hypothesis is true.

In our Google example, we have an experiment and the three algorithms are randomly assigned to the queries. Also, all the expected counts, assuming the null hypothesis is true, are well over 5. Therefore, the conditions for applying the chi-square test for homogeneity are met.

In previous hypothesis tests, we constructed a test statistic of the following form:

$$Z = \frac{\text{point estimate} - \text{null value}}{SE \text{ of point estimate}}$$

This construction was based on (1) identifying the difference between a point estimate and an expected value if the null hypothesis was true, and (2) standardizing that difference using the standard error of the point estimate. These two ideas will help in the construction of an appropriate test statistic for count data.

In this example we have six combinations of Search Algorithm and Result: (current & Success), (current & Failure), (test 1 & Success), (test 1 & Failure), (test 2 & Success), and (test 2 & Failure). Because we have six values rather than just one or two, we need a new tool to analyze the data. Our strategy will be to find a test statistic that measures the overall deviation between the observed and the expected counts. For each cell in the table, we first find the difference between the observed and expected counts. Then we square those differences. Next, we must standardize each term. To know whether the squared difference is large, we divide it by what was expected. Finally we add up all these terms and this sum gives us a new test statistic called χ^2 (read as chi-square).

The chi-square test statistic for a two-way table is found as follows:

General formula	$\dfrac{(\text{observed count} - \text{expected count})^2}{\text{expected count}}$
Row 1, Col 1	$\dfrac{(3511 - 3539)^2}{3539} = 0.222$
Row 1, Col 2	$\dfrac{(1749 - 1769.5)^2}{1769.5} = 0.237$
$\vdots$	$\vdots$
Row 2, Col 3	$\dfrac{(682 - 730.5)^2}{730.5} = 3.220$

Adding the computed value for each cell gives the chi-square test statistic:

$$\chi^2 = 0.222 + 0.237 + \cdots + 3.220 = 6.120$$

The final number χ^2 measures how strongly the observed counts deviate from the expected counts, relative to the expected counts. Moreover, each term in the sum tells us the **chi-square contribution** for that cell. The larger the chi-square contribution the farther the observed value is from the expected value, relative to expected. Here, the largest chi-square contribution is 3.220, which tells us that the observed count for (test 2 & Failure) has the largest relative difference from what would be expected under the null hypothesis of no difference. Also, the (observed count - expected count) for (test2 & Failure) is negative, telling us the observed is *less* than what would be expected under the null hypothesis.

3.9.4 Calculating and interpreting the p-value for a chi-square test

If the null hypothesis is true, then the chi-square test statistic follows a distribution called a *chi-square distribution*. This is the same distribution we saw in the "Goodness of fit using chi-square (special topic)" section. The **chi-square distribution** is sometimes used to characterize data sets and statistics that are always positive and typically right skewed. Recall a normal distribution had two parameters – mean and standard deviation – that could be used to describe its exact characteristics. The chi-square distribution has just one parameter called **degrees of freedom (df)**, which influences the shape, center, and spread of the distribution. For two way tables, the degrees of freedom is equal to (number of rows – 1)× (number of columns – 1).

COMPUTING DEGREES OF FREEDOM FOR A TWO-WAY TABLE

When using the chi-square test to a two-way table, we use

$$df = (\text{\# of rows } - 1) \times (\text{\# of columns } - 1)$$

Figure 3.26 illustrates three general properties of chi-square distributions as the degrees of freedom increases: the distribution becomes more symmetric, the center moves to the right, and the variability inflates.

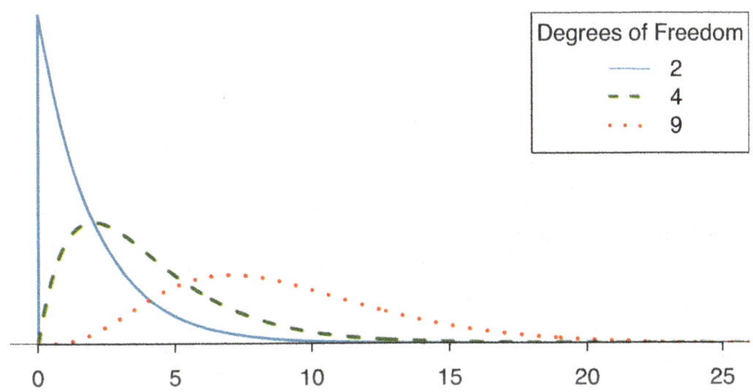

Figure 3.26: Three chi-square distributions with varying degrees of freedom. The distributions are non-negative and right skewed. As the degrees of freedom increase, chi-square distributions become less skewed and have a larger center and spread.

To compute the p-value in our example, we need to know the chi-square statistic and the degrees of freedom. We calculated the test statistic as $\chi^2 = 6.12$. The degrees of freedom for our 3×2 table is calculated as: $(3 - 1) \times (2 - 1) = 2$. If the null hypothesis is true (i.e. the algorithms are equally successful), then the test statistic $\chi^2 = 6.12$ closely follows a chi-square distribution with 2 degrees of freedom. To know how unlikely it is to get a test statistic at least as large as we got, assuming the null hypothesis is true, we find the area to the right of $\chi^2 = 6.12$ under the chi-square distribution with 2 degrees of freedom, as shown in Figure 3.27. While it is possible to find areas under a chi-square distribution using a chi-square table, such as the one found in Appendix C.4 on page 470, it is more common to use technology. See Section 3.9.7 for multiple ways to find areas under a chi-square distribution using technology.

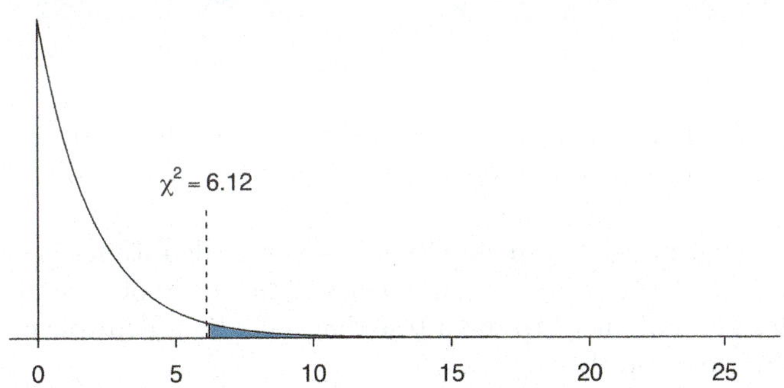

Figure 3.27: Computing the p-value for the Google hypothesis test.

EXAMPLE 3.79

Compute the p-value and draw a conclusion about whether the search algorithms have different success rates.

Using technology, we find that the p-value, which corresponds to the area under the chi-square distribution with 2 degrees of freedom to the *right* of $\chi^2 = 6.12$, equals 0.047. Using an $\alpha = 0.05$ significance level, the p-value $< \alpha$, so we reject H_0. The data provide convincing evidence that there is a difference in the distribution of the outcome variable (success rates), among the three algorithms.

EXAMPLE 3.80

Interpret the p-value for this test.

The p-value of 0.047 tells us that there is a 4.7% chance of getting a test statistic as large or larger than we got if H_0 is true, that is, if the algorithms do in fact perform equally well.

Notice that the conclusion of the test is that there is some difference in performance among the algorithms. This chi-square test does not tell us *which* algorithm performed better than the others. To compare the performance of the algorithms, we calculate the proportion of successes for each algorithm as follows:

$$\text{current: } \frac{3511}{5000} = 0.7022 \quad \text{test 1: } \frac{1749}{2500} = 0.700 \quad \text{test 2: } \frac{1818}{2500} = 0.7272$$

In Figure 3.28 we compare the proportion of successes and failures for each algorithm using a mosaic plot (left) and a simpler stacked bar chart (right).

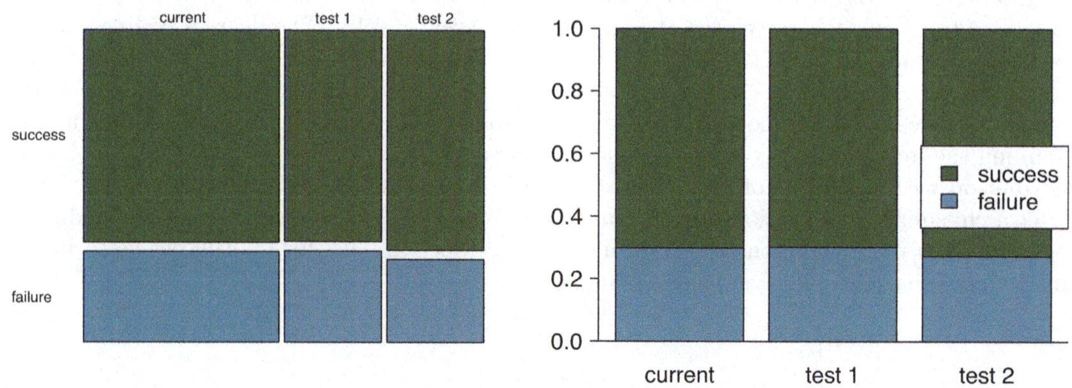

Figure 3.28: A mosaic plot (left) and a stacked bar chart (right) showing the proportion of successes and failures for each algorithm in the Google experiment.

Our calculations and our graphs suggest that the test 2 algorithm performed better than the current algorithm and test 1 algorithm, because it led higher rates of success; however, to formally test this specific claim we would need to use a test that includes a multiple comparisons correction, which is beyond the scope of this book.

A careful reader may have noticed that when there are exactly 2 random samples or treatments and the counts can be arranged in a 2×2 table, both a chi-square test for homogeneity *and* a two-sample Z-test for $p_1 - p_2$ could apply. In this case, the chi-square test for homogeneity and the two-sided two-sample Z-test for $p_1 - p_2$ are equivalent, meaning that they produce the same p-value.[52]

---

[52]Sometimes the large counts condition for the Z-test is weakened to require the expected number of successes and failures to be at least 5, making it consistent with the chi-square condition that expected counts must at least 5.

3.9.5 The chi-square test for independence in two-way tables

In Chapter 2 we determined whether two events A and B are independent by checking if $P(B|A) = P(B)$. Using the `family_college` data set, we saw that the probability a teenager attended college given that one of the teen's parents has a college degree is higher than the unconditional probability that a teenager attended college, making "teenager attended college" and "one of the teen's parents has a college degree" dependent. While $P(\texttt{teen college} \mid \texttt{parent degree})$ is not exactly equal to $P(\texttt{teen college})$, we may wonder – is this difference within the realm of expected variation or is this difference evidence of a real association between these two variables in the *population* from which the sample was taken? To answer this, we use a chi-square test for independence.

The chi-square test for independence tests for an association between two categorical variables within a population using data from a single random sample from that population. The null claim is always that the two variables are independent in the population, while the alternative claim is that the variables are dependent in the population. While the chi-square test for independence and the chi-square test for homogeneity are distinct, we will see that the calculations performed for these two tests are identical.

We begin by looking at a new example where the categorical variables each have three levels. Figure 3.29 summarizes the results of a Pew Research poll conducted in April-May of 2025. A random sample of adults in the US was taken, and each was asked how seriously they will consider buying an electric vehicle (EV) the next time they purchase a vehicle. Respondents were also classified as living in an Urban, Suburban, or Rural residential type. We would like to determine if residential type and consideration of buying an EV are associated.

	`Consideration of buying an EV`			
	Not too or not at all likely	Very or somewhat likely	Don't expect to buy a vehicle	Total
Urban	748	621	207	1576
Suburban	1466	978	353	2797
Rural	460	130	122	712
Total	2674	1729	682	5085

Figure 3.29: Results of a Pew Research poll from May 2025.

EXAMPLE 3.81

What are appropriate hypotheses for such a test?

The null claim always states that there is no association between the variables or that the variables are independent. We seek to find evidence for the alternative claim, that there is an association between the variables.

H_0: Residential type and consideration of buying an EV in the population of US adults not associated.

H_A: Residential type and consideration of buying an EV in the population of US adults are associated (dependent).

The null claim implies that there is no difference in consideration of buying an EV among Urban, Suburban, and Rural residing adults in the US. The alternative claim implies that there is some difference in consideration of buying an EV among Urban, Suburban, and Rural residing adults in the US, e.g. perhaps Urban residing adults in the US are more likely to consider buying an EV.

EXAMPLE 3.82

Using Figure 3.29, find the proportion of people in the sample who said they are are "Not too or not at all likely" to buy an EV within each residential type.

This is equivalent to finding conditional probabilities as follows:

$$P(\text{Not too or Not at all likely to buy an EV} \mid \text{Urban}) = \frac{748}{1576} = 0.475$$
$$P(\text{Not too or Not at all likely to buy an EV} \mid \text{Suburban}) = \frac{1466}{2797} = 0.524$$
$$P(\text{Not too or Not at all likely to buy an EV} \mid \text{Rural}) = \frac{460}{712} = 0.646.$$

Based on these calculations, we can see that within the data set, these variables are dependent, as those in Rural areas are more likely to say "Not too or not at all likely" to buy an EV than those in Suburban or Urban areas. However, a hypothesis test is always concerned with what is true in the *population* from which the sample was taken. We want to ask whether the association we see in the sample data could be explained by random variation or whether there is a real association in the greater population from which we sampled. To do a chi-square test of independence to answer this question, we need to check that the following conditions are met.

CONDITIONS FOR THE CHI-SQUARE TEST FOR INDEPENDENCE

A chi-square test for independence requires two categorical variables and the following conditions must be met.

Independence. The data must come from one random sample. When sampling without replacement, the sample size should be less than 10% of the population size.

Large expected counts. All of the cells in the two-way table must have expected counts of at least 5 under the assumption that the null hypothesis is true.

EXAMPLE 3.83

If the null hypothesis is true and residential type and consideration of buying an EV are independent, what proportion of each residential type would we expect to say that they are "Not too or not at all likely" to buy an EV? Use the data from Figure 3.29.

If residential type and consideration of buying an EV are independent, we would expect the proportion who would say that they are "Not too or not at all likely" to buy an EV to be the *same* for each residential type. To find this expected proportion, we use the overall proportion, which corresponds to the unconditional probability of saying"Not too or not at all likely" to buy an EV. To find this we take the column total that said "Not too or not at all likely" to buy an EV and divide by the table total. This gives: $\frac{2674}{5085} = 0.5259$.

EXAMPLE 3.84

Find the expected counts for the first column of Figure 3.29.

We multiply the row totals by the overall proportion who said "Not too or not at all likely" to buy an EV.

$$(\text{Urban row total}) \times 0.5259 = 1576 \times 0.5259 = 828.8$$
$$(\text{Suburban row total}) \times 0.5259 = 2797 \times 0.5259 = 1471.0$$
$$(\text{Rural row total}) \times 0.5259 = 712 \times 0.5259 = 374.4$$

We can use the same process for finding the expected counts for the second and third columns of the table. In each case, we will be multiplying: row total $\times \frac{\text{column total}}{\text{table total}}$. In other words, just as with the chi-square test for homogeneity, we calculate the expected counts as follows:

$$\text{Expected Count} = \frac{(\text{row total}) \times (\text{column total})}{\text{table total}}$$

Figure 3.30 show the table of observed counts, with the corresponding expected counts assuming the null hypothesis is true and the variables are independent. There are nine combinations of residential type and consideration of buying an EV.

	Consideration of buying an EV			
	Not too or not at all likely	Very or somewhat likely	Don't expect to buy a vehicle	Total
Urban	748 (828.8)	621 (535.9)	207 (211.4)	1576
Suburban	1466 (1471.0)	978 (951.0)	353 (375.1)	2797
Rural	460 (374.4)	130 (242.1)	122 (95.5)	712
Total	2674	1729	682	5085

Figure 3.30: The observed counts and the (expected counts).

Figure 3.31 offers a visual comparison of the observed and expected counts using mosaic plots. We see that the Urban, Suburban, and Rural residing group sizes are different, which explains variations in expected counts within the table. However, the expected *proportion* for each poll response is the same within each residential type, and the expected proportion of each residential type within each poll response is the same, under the assumption that that the variables are independent for adults in the US.

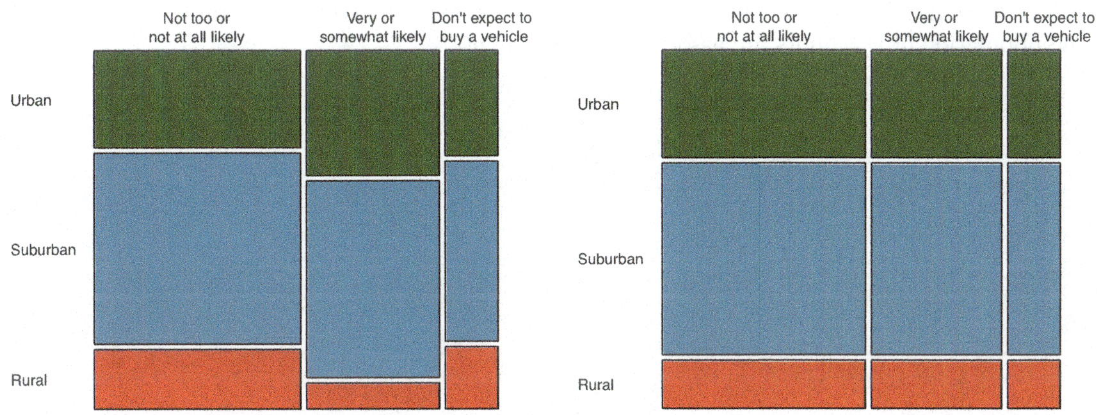

Figure 3.31: Mosaic plots for Figure 3.30. Left: observed counts; Right: expected counts assuming indpendence in the population.

All of the expected counts are at least 5. Also, in this Pew research poll, the data came from a random sample of adults in the US and the population size for adults in the US is much larger than 10 times the sample size of 5085. Therefore, conditions for the chi-square test for independence are met.

The chi-square test for independence and the chi-square test for homogeneity both involve counts in a two-way table. The expected counts, chi-square statistic, degrees of freedom, and p-value are calculated in the same way.

EXAMPLE 3.85

Calculate the chi-square statistic.

We calculate $\frac{(\text{obs}-\text{exp})^2}{\text{exp}}$ for each of the nine cells in the table. Adding the results of each cell gives the chi-square test statistic.

$$\begin{aligned}\chi^2 &= \sum \frac{(\text{obs} - \text{exp})^2}{\text{exp}}\\ &= \frac{(748 - 828.8)^2}{828.8} + \frac{(621 - 535.9)^2}{535.9} + \cdots + \frac{(122 - 95.5)^2}{95.5}\\ &= 7.9 + 13.5 + \cdots + 7.4\\ &= 102.4\end{aligned}$$

EXAMPLE 3.86

Find the p-value for the test and state the appropriate conclusion.

We must first find the degrees of freedom for this chi-square test. Because there are 3 rows and 3 columns, the degrees of freedom is $df = (3-1)\times(3-1) = 4$. We find the area to the right of $\chi^2 = 102.4$ under the chi-square distribution with $df = 4$. The p-value is extremely small, much less than 0.01, so we reject H_0. We have evidence that the residential type and consideration of buying an EV in the population of US adults are associated / dependent.

3.9.6 Comparing and applying the chi-square tests for two-way tables

We have seen that the chi-square test for homogeneity and for independence are both used for two-way tables of counts. The calculation of the expected counts, test statistic, and p-value is identical for the two tests. The main differences between them lie in *how the data is collected* and how we then present the hypotheses and check the random condition. We summarize the two tests here.

χ^2 TEST FOR HOMOGENEITY AND FOR INDEPENDENCE

When there are multiple samples or treatments and we are comparing the distribution of a categorical variable across several groups, e.g. comparing the distribution of rural/urban/suburban dwellers among 4 states,

Identify: Use a χ^2 **test for homogeneity** at the desired α significance level.

H_0: There is no difference in the distribution of [...] across populations or treatments.

H_A: There is a difference in the distribution of [...] across populations or treatments.

Check: Check that the test statistic follows a chi-square distribution, assuming H_0 is true.

1. Independence: Data come from multiple random samples or from a randomized experiment with multiple treatments. If sampling without replacement, check that the sample size is less than 10% of the population size for each sample.
2. Expected counts: All expected counts are ≥ 5 (calculate and record expected counts).

When there is one sample and we are looking for association or dependence between two categorical variables, e.g. testing for an association between gender and political party,

Identify: Use a χ^2 **test for independence** at the desired α significance level.

H_0: [variable 1] and [variable 2] in the population are independent.

H_A: [variable 1] and [variable 2] in the population are dependent.

Check: Check that the test statistic follows a chi-square distribution, assuming H_0 is true.

1. Independence: Data come from one random sample. If sampling without replacement, check that the sample size is less than 10% of the population size.
2. Expected counts: All expected counts are ≥ 5 (calculate and record expected counts).

The calculate and conclude steps for the chi-square test for homogeneity and test for independence are the same.

Calculate: Calculate the χ^2-statistic, df, and p-value.

test statistic: $\chi^2 = \sum \frac{(\text{observed} - \text{expected})^2}{\text{expected}}$

$df = (\#\text{ of rows} - 1) \times (\#\text{ of columns} - 1)$

p-value = (area to the *right* of χ^2-statistic under the chi-square distribution with the appropriate df)

Conclude: Compare the p-value to α, and draw a conclusion in context.

If the p-value is $\leq \alpha$, reject H_0; there is sufficient evidence that [H_A in context].

If the p-value is $> \alpha$, do not reject H_0; there is not sufficient evidence that [H_A in context].

EXAMPLE 3.87

In an experiment, each individual was asked to be a seller of an iPod (a product commonly used to store music on before smart phones). The participant received \$10 + 5% of the sale price for participating. The iPod they were selling had frozen twice in the past inexplicitly but otherwise worked fine. Unbeknownst to the participants who were the sellers in the study, the buyers were collaborating with the researchers to evaluate the influence of different questions on the likelihood of getting the sellers to disclose the past issues with the iPod. The scripted buyers ended with one of three questions:

- General: What can you tell me about it?
- Positive Assumption: It doesn't have any problems, does it?
- Negative Assumption: What problems does it have?

The outcome variable is whether the participant discloses or hides the problem with the iPod.

		Question Type		
		General	Positive Assump.	Negative Assump.
Response	Disclose	2	23	36
	Hide	71	50	37
	Total	73	73	73

E Is there evidence that the phrasing of the question affects how likely individuals are to disclose the problems with the iPod? Carry out an appropriate test at the 0.05 significance level.

Identify: We have an experiment with 3 treatments (question types) and we want to know if the distribution of disclose/hide is the same for each of the three question types, so we want a chi-square test for homogeneity. We test the following hypotheses at the $\alpha = 0.05$ significance level.

H_0: There is no difference in likelihood to disclose across the three question types.

H_A: There is a difference in likelihood to disclose across the three question types.

Check: This is an experiment in which there were three randomly assigned treatments (question types). All values in the following table of expected counts are ≥ 5.

		Question Type		
		General	Positive Assump.	Negative Assump.
Response	Disclose	20.3	20.3	20.3
	Hide	52.7	52.7	52.7

Calculate: Using technology, we get $\chi^2 = 40.1$, $df = (\# \text{ of rows} - 1) \times (\# \text{ of cols} - 1) = 2 \times 1 = 2$. The p-value is the area under the chi-square distribution with 2 degrees of freedom to the right of $\chi^2 = 40.1$. The p-value is almost 0.

Conclude: Because the p-value $\approx 0 < \alpha$, we reject H_0. We have strong evidence that there is a difference in likelihood to disclose across the three question types.

EXAMPLE 3.88

Which combination of Response and Question Type is the most underrepresent from what would be expected under the null hypothesis of no difference?

Using a technology option from Section 3.9.8, we find the chi-square contributions for each cell:

E

16.53 0.3497 12.07
6.382 0.135 4.66

General/Disclose has the largest chi-square contribution, so it deviated the most from what was expected under the null hypothesis of no difference. However, was it more or less than expected? Here we see that (observed − expected) = (2 − 20.3), which is negative, so we know that there were *fewer* in the General group that chose Disclose than we would have expected.

EXAMPLE 3.89

A 2021 Pew Research poll asked a random sample of US residents their generation and whether they have personally taken action to help address climate change within the last year. The data are shown below.

		Response		
		Took action	Didn't take action	Total
Generation	Gen Z	292	620	912
	Millenial	885	2,275	3,160
	Gen X	809	2,709	3,518
	Boomer & older	1,276	4,798	6,074
	Total	3,262	10,402	13,664

We can see that the percent in the sample from each generation that took action vary: 32% for Gen Z, 28% for Millenial, 23% for Gen X, and 21% for Boomer & older. However, could this be due to random variation based on who happened to end up in the sample? Carry out an appropriate test at the 0.05 significance level to see if there is an association between generation and taking action to help address climate change.

(E)

Identify: Two variables were recorded on the respondents: generation and whether or not they have taken action to help address climate change within the last year. We have one random sample and we want to know if these variables are associated / dependent, so we will carry out a chi-square test for independence. We will test the following hypotheses at the $\alpha = 0.05$ significance level.

H_0: Generation and taking action to help address climate change among US residents are independent.

H_A: Generation and taking action to help address climate change among US residents are dependent.

Check: According to the problem, there was one random sample taken. We note that the population of US residents is much larger than 10 times the sample size of 13,664. Also, all values in the table of expected counts are ≥ 5. Table of expected counts:

		Response	
		Took action	Didn't take action
Generation	Gen Z	217.72	694.28
	Millenial	754.39	2405.60
	Gen X	839.85	2678.10
	Boomer & older	1450.00	4624.00

Calculate: Using technology, we get $\chi^2 = 91.9$, $df = (\# \text{ of rows} - 1) \times (\# \text{ of columns} - 1) = 3 \times 1 = 3$. The p-value is the area under the chi-square distribution with 3 degrees of freedom to the right of $\chi^2 = 91.9$. Thus, the p-value $= 8.46\text{x}10^{-20} \approx 0$.

Conclude: Because the p-value $\approx 0 < \alpha$, we reject H_0. We have sufficient evidence that generation and taking action to help address climate change among US residents are dependent.

(G)

GUIDED PRACTICE 3.90

In context, interpret the p-value of the test from the above example.[53]

---

[53]Recall that the p-value in this test is calculated as the area to the right of $\chi^2 = 91.9$ under the chi-square distribution with 3 degrees of freedom. We interpret the p-value as follows: Assuming the null hypothesis is true, i.e. that generation and response really are independent, there is close to a 0% probability of getting a χ^2-statistic as large or larger than 91.9. Equivalently, it is the probability of our observed counts being this different from the expected counts, relative to the expected counts, if the null is true, i.e. that generation and response really are independent.

3.9.7 Technology: chi-square distribution probabilities

Use technology to find the upper tail area for a chi-square distribution with 5 degrees of freedom and a cutoff of 5.1.

Desmos: Use the `chsqdist(df)` function, replacing `df` with the degrees of freedom.

1. Type `chisqdist(5)`.
2. Click the triangle next to `Cumulative Probability`.
3. Choose `Inner`, `Outer`, `Left` or `Right` as illustrated below. Usually you will want Right.
4. Choose `Area` and enter the desired boundary value(s) as illustrated below.
5. Click the magnifying glass to Zoom Fit the graphing window.

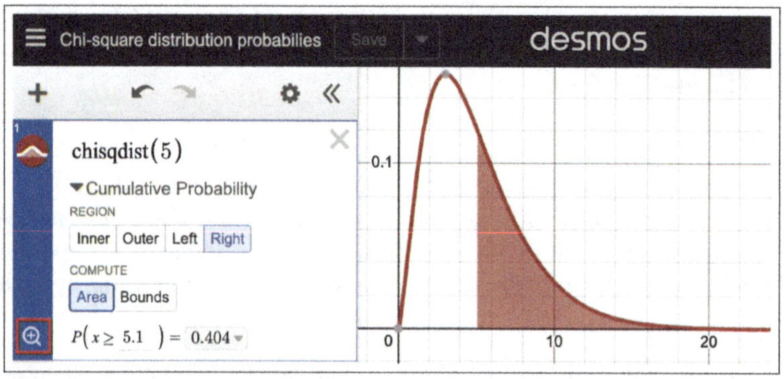

R:
`pchisq(q, df)` will give the area to the left of `q`, so we add `lower.tail = FALSE` to get the area to the right.

```
> pchisq(5.1, df = 5, lower.tail = FALSE)
[1] 0.4037985
```

Calculator: NumWorks calculator instructions and example output are included. For the TI-83/84 and Casio calculators, general instructions are provided, and worked example videos can be accessed via the icon or at **openintro.org/ti** and **openintro.org/casio**.

NUMWORKS: FINDING AREA UNDER THE CHI-SQUARE DISTRIBUTION

Use `OK` or `EXE` to make a selection.

1. From the home screen, select `Distributions`, arrow down and choose `Chi-square`. If a list of distributions does not appear, click the ↩ button in the upper right as many times as needed.
2. Enter the `Degrees of freedom`. Hit the down arrow and choose `Next`.
3. Hit the left arrow to highlight the graph. Hit the down arrow to choose whether you want left, inner, or right, then hit `OK`. Hit the right arrow and enter the boundary value(s), then hit `EXE`.

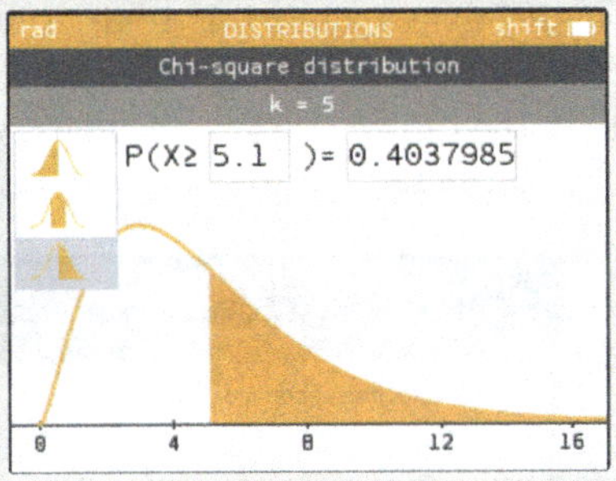

TI-84: FINDING AN UPPER TAIL AREA UNDER THE CHI-SQUARE DISTRIBUTION

Use the χ^2`cdf` command to find areas under the chi-square distribution.

1. Hit `2ND VARS` (i.e. `DISTR`).
2. Choose `8:`χ^2`cdf`.
3. Enter the lower bound, which is generally the chi-square value.
4. Enter the upper bound. Use a large number, such as 1000.
5. Enter the degrees of freedom.
6. Choose `Paste` and hit `ENTER`.

TI-83: Do steps 1-2, then type the lower bound, upper bound, and degrees of freedom separated by commas. e.g. χ^2`cdf(5, 1000, 3)`, and hit `ENTER`.

CASIO FX-9750GII: FINDING AN UPPER TAIL AREA UNDER THE CHI-SQ. CURVE

1. Navigate to `STAT` (`MENU` button, then hit the `2` button or select `STAT`).
2. Choose the `DIST` option (`F5` button).
3. Choose the `CHI` option (`F3` button).
4. Choose the `Ccd` option (`F2` button).
5. If necessary, select the `Var` option (`F2` button).
6. Enter the `Lower` bound (generally the chi-square value).
7. Enter the `Upper` bound (use a large number, such as 1000).
8. Enter the degrees of freedom, `df`.
9. Hit the `EXE` button.

3.9.8 Technology: the chi-square test for two-way tables

Consider the data from the search algorithm experiment introduced in Example 3.9.1. Use technology to find the test statistic, *df*, and p-value for a chi-square test for homogeneity, testing whether there is evidence that the algorithms do not perform equally well. Also find the expected values and chi-square contributions for this test. Conditions were verified to be met.

		Search algorithm			
		current	test 1	test 2	Total
outcome	success	3511	1749	1818	7078
	failure	1489	751	682	2922
	Total	5000	2500	2500	10000

Desmos: Use the `chisqtest()` function as explained below.

1. Click + in the upper left, then choose `inference`.
2. Choose χ^2 `test for independence` in the pop-up window. The calculations for the chi-square test for homogeneity and independence are the same and both use this test.
3. Enter the observed counts in the table. Use the down arrow and right arrow to enter additional row and column values. Do not enter the row and column totals. Click `Create Test`.

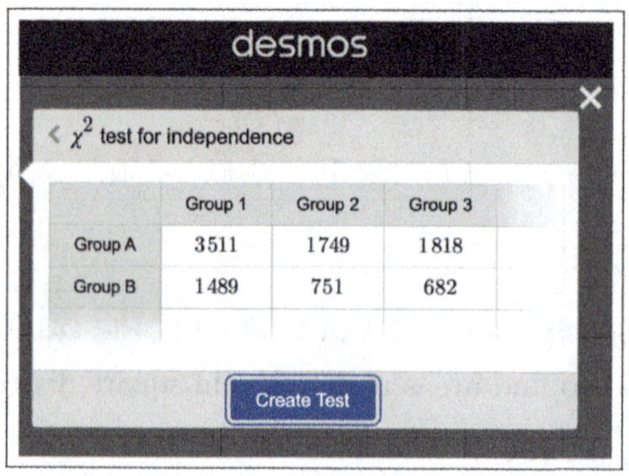

* You can type `chisqtest([3511,1489], [1749,751], [1818,682])` in place of steps 1-3.

4. Click the triangle next to `Observed (Expected)`. Click the checkbox next to `Expected` to see the expected counts in parentheses in each cell. Click the checkbox next to `Contributions` to see a graphical representation of the size of each value's contribution to the chi-square statistic. Click the checkbox next to `Totals` to see the row and column totals. Also, click any cell in the table to produce a pop-up box with the corresponding Observed value, Expected value, and the Contribution to the chi-square statistic.
5. Click the triangle next to `Significance Test` to see the test statistic, df, and p-value.

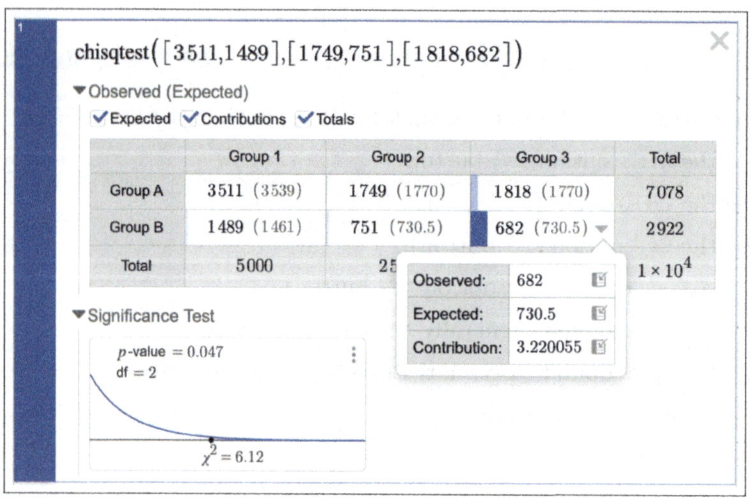

R: Chi-square test for two-way tables
Use `chisq.test()` with `cbind()` as illustrated below to bind the *columns* into a table.

```
> X = chisq.test(cbind(c(3511,1489), c(1749,751), c(1818,682)))
> X
Pearson's Chi-squared test
data:  cbind(c(3511, 1489), c(1749, 751), c(1818, 682))
X-squared = 6.1203, df = 2, p-value = 0.04688

> X$expected
[,1] [,2] [,3]
[,1] 3539 1769.5 1769.5
[,2] 1461 730.5 730.5
> (X$residuals)^2
[,1] [,2] [,3]
[1,] 0.2215315 0.2374965 1.329330
[2,] 0.5366188 0.5752909 3.220055
```

Calculator: NumWorks calculator instructions and example output are included. For the TI-83/84 and Casio calculators, general instructions are provided, and worked example videos can be accessed via the icon or at **openintro.org/ti** and **openintro.org/casio**.

NUMWORKS: CHI-SQUARE TEST FOR HOMOGENEITY AND INDEPENDENCE

Use `OK` or `EXE` to make a selection.

1. From the home screen, select `Inference`, then `Tests`, then `Chi-square`, then `Homogeneity/Indpendence`. If these options do not appear, click the ← button in the upper right as many times as needed.
2. Enter the observed counts, then α, then choose `Next`. Use arrows as needed.
3. On this screen, you will see the table of expected counts. Use the arrows to see any rows or columns that do not fit on the screen. To see the chi-square contributions, click the up and right arrow to choose `Contributions`. When you are finished, choose `Next`.

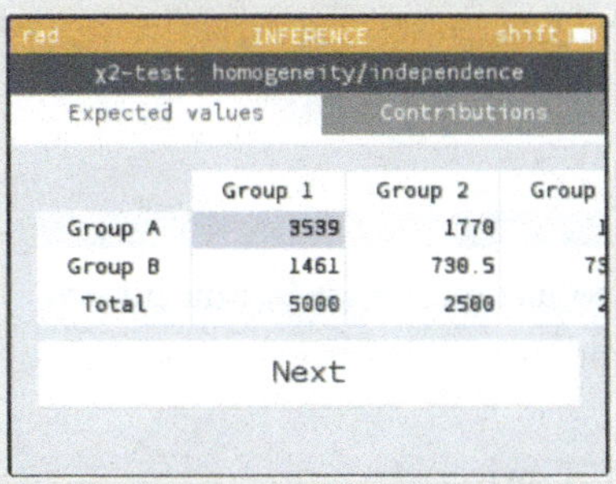

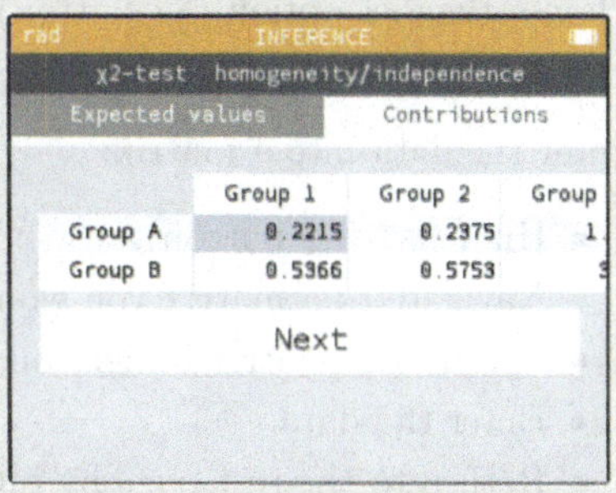

4. On this screen, you will see the test statistic, p-value and df. Choose `Next` to see the chi-square distribution with the p-value shaded.

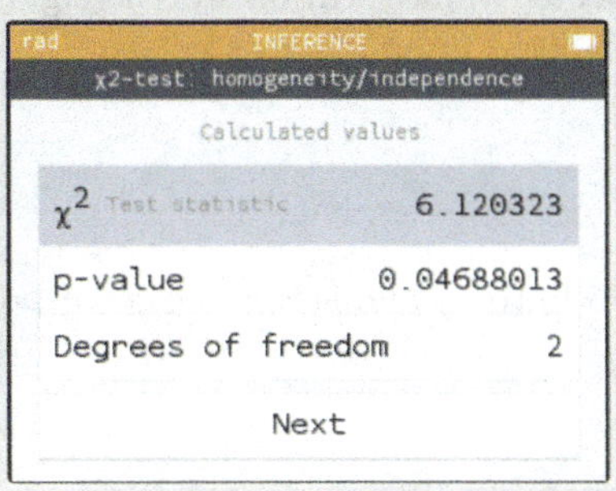

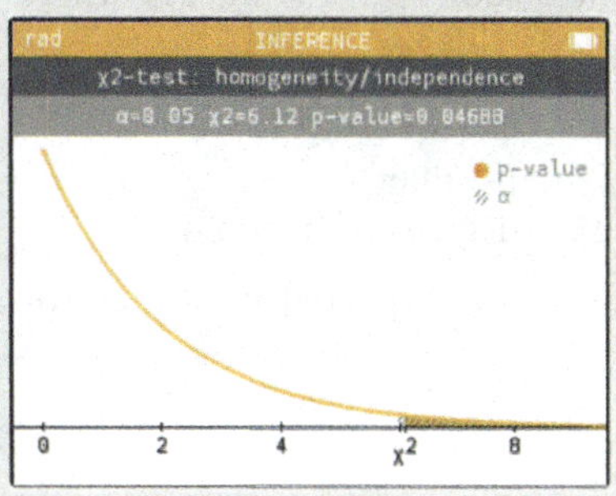

TI-83/84: CHI-SQUARE TEST FOR HOMOGENEITY AND INDEPENDENCE

First enter the counts in a two-way table.

1. Hit `2ND` x^{-1} (i.e. `MATRIX`).
2. Right arrow to `EDIT`.
3. Hit `1` or `ENTER` to select matrix `A`.
4. Enter the dimensions by typing #rows, `ENTER`, #columns, `ENTER`.
5. Enter the data from the two-way table.

Then use `STAT`, `TESTS`, χ^2-`Test`. Make sure you have entered the counts as described above.

1. Choose `STAT`.
2. Right arrow to `TESTS`.
3. Down arrow and choose `C:`χ^2`-Test`.
4. Down arrow, choose `Calculate` or `Draw` and hit `ENTER`. `Draw` shows the chi-square statistic and p-value as well as a graph of the chi-square distribution with p-value shaded. `Calculate` returns:

   χ^2 chi-square test statistic
   `p` p-value
   `df` degrees of freedom

Edit `Matrix B` to see the expected counts. Make sure you have already done the steps above.

1. Hit `2ND` x^{-1} (i.e. `MATRIX`).
2. Right arrow to `EDIT`.
3. Hit `2` to see matrix `B`. This matrix contains the expected counts.

CASIO FX-9750GII: CHI-SQUARE TEST FOR HOMOGENEITY AND INDEPENDENCE

1. Navigate to `STAT` (`MENU` button, then hit the `2` button or select `STAT`).
2. Choose the `TEST` option (`F3` button).
3. Choose the `CHI` option (`F3` button).
4. Choose the `2WAY` option (`F2` button).
5. Enter the data into a matrix:
   - Hit `▷MAT` (`F2` button).
   - Navigate to a matrix you would like to use (e.g. `Mat C`) and hit `EXE`.
   - Specify the matrix dimensions: `m` is for rows, `n` is for columns.
   - Enter the data.
   - Return to the test page by hitting `EXIT` twice.
6. Enter the `Observed` matrix that was used by hitting `MAT` (`F1` button) and the matrix letter (e.g. `C`).
7. Enter the `Expected` matrix where the expected values will be stored (e.g. `D`).
8. Hit the `EXE` button, which returns

   χ^2 chi-square test statistic
   `p` p-value
   `df` degrees of freedom
9. To see the expected values, go to `▷MAT` (`F6` button) and select the corresponding matrix.

Section summary

- The chi-square statistic measures the distance between observed and expected counts relative to expected counts.
- Chi-square distributions have positive values and are skewed right. Within this family of density curves, the skew becomes less pronounced with increasing degrees of freedom.
- To determine whether the distributions of a categorical variable across two or more populations are different, the appropriate test is the **chi-square test for homogeneity**. An example is testing for a difference in the distribution of rural/urban/suburban dwellers across 4 states.

  The hypotheses for a chi-square test for homogeneity are:

  H_0: There is no difference in the distribution of [...] across population or treatments.
  H_A: There is a difference in the distribution of [...] across populations or treatments.

  The conditions for a chi-square test for homogeneity are:

  1. Independence: The data come from two or more independent random samples, each with sample size < 10% of its corresponding population size if sampling without replacement OR the data come from an experiment with two or more randomly assigned treatments.
  2. Expected counts: all expected counts, assuming the null hypothesis is true, should be ≥ 5.
- To determine whether row and column variables in a two-way table of categorical data might be associated in the single population from which the data were sampled, the appropriate test is the **chi-square test for independence**. An example is testing for an association between gender and political party within a particular town.

  The hypotheses for a chi-square test for independence are:

  H_0: there is no association between two categorical variables in a given population or the two categorical variables in a given population are independent.
  H_A: there is an association between two categorical variables in a given population or the two categorical variables in a given population are not independent.

  The conditions for a chi-square test for independence are:

  1. Independence: The data come from *one* random sample, with sample size < 10% of the population size if sampling without replacement.
  2. Expected counts: all expected counts, assuming the null hypothesis is true, should be ≥ 5.
- The expected values (under the null hypothesis) in a particular cell of a two-way table of categorical data can be calculated using the formula: expected value $= \frac{\text{(row total)}\times\text{(column total)}}{\text{table total}}$.
- A chi-square test for homogeneity and for independence use a chi-square statistic: $\chi^2 = \sum \frac{(\text{observed count} - \text{expected count})^2}{\text{expected count}}$, where the sum is taken over all cells of the two-way table.
- The chi-square statistic has a chi-square distribution with $df = (\#\text{ of rows} - 1)(\#\text{ of cols} - 1)$.
- The p-value for a chi-square test is the area to the *right* of the χ^2-statistic under the chi-square distribution with the appropriate df. This can be found using technology.
- The p-value for a chi-square test is the probability of obtaining a χ^2 value as large or larger than the χ^2-statistic that was observed, assuming the null hypothesis is true.
- A formal decision explicitly compares the p-value to the significance level. If the p-value $\leq \alpha$, then reject the null hypothesis; if the p-value $> \alpha$, then fail to reject the null hypothesis. The conclusion should be stated in terms of the alternative hypothesis and should include context, using non-causal language.
- The results of a chi-square test for homogeneity or independence can serve as the statistical reasoning to support the answer to an investigative question about the population that was sampled (independence) or the populations that were sampled (homogeneity).

Exercises

3.55 True or false, Part I. Determine if the statements below are true or false. For each false statement, suggest an alternative wording to make it a true statement.

(a) The chi-square distribution, just like the normal distribution, has two parameters, mean and standard deviation.

(b) The chi-square distribution is always right skewed, regardless of the value of the degrees of freedom parameter.

(c) The chi-square statistic is always greater than or equal to 0.

(d) As the degrees of freedom increases, the shape of the chi-square distribution becomes more skewed.

3.56 True or false, Part II. Determine if the statements below are true or false. For each false statement, suggest an alternative wording to make it a true statement.

(a) As the degrees of freedom increases, the mean of the chi-square distribution increases.

(b) If you found $\chi^2 = 10$ with $df = 5$ you would fail to reject H_0 at the 5% significance level.

(c) When finding the p-value of a chi-square test, we always shade the tail areas in both tails.

(d) As the degrees of freedom increases, the variability of the chi-square distribution decreases.

3.57 Quitters. Does being part of a support group affect the ability of people to quit smoking? A county health department enrolled 300 smokers in a randomized experiment. 150 participants were randomly assigned to a group that used a nicotine patch and met weekly with a support group; the other 150 received the patch and did not meet with a support group. At the end of the study, 40 of the participants in the patch plus support group had quit smoking while only 30 smokers had quit in the other group.

(a) Create a two-way table presenting the results of this study.

(b) Answer each of the following questions under the null hypothesis that being part of a support group does not affect the ability of people to quit smoking, and indicate whether the expected values are higher or lower than the observed values.

i. How many subjects in the "patch + support" group would you expect to quit?

ii. How many subjects in the "patch only" group would you expect to not quit?

3.58 Full body scan, Part II. A news article reports that "Americans have differing views on two potentially inconvenient and invasive practices that airports could implement to uncover potential terrorist attacks." This news piece was based on a survey conducted among a random sample of 1,137 adults nationwide, where one of the questions on the survey was "Some airports are now using 'full-body' digital x-ray machines to electronically screen passengers in airport security lines. Do you think these new x-ray machines should or should not be used at airports?" Below is a summary of responses based on party affiliation.[54] The differences in each political group may be due to chance. Complete the following computations under the null hypothesis of independence between an individual's party affiliation and his support of full-body scans. It may be useful to first add on an extra column for row totals before proceeding with the computations.

		Party Affiliation		
		Republican	Democrat	Independent
Answer	Should	264	299	351
	Should not	38	55	77
	Don't know/No answer	16	15	22
	Total	318	369	450

(a) How many Republicans would you expect to not support the use of full-body scans?

(b) How many Democrats would you expect to support the use of full- body scans?

(c) How many Independents would you expect to not know or not answer?

[54] S. Condon. "Poll: 4 in 5 Support Full-Body Airport Scanners". In: *CBS News* (2010).

3.59 Offshore drilling. A survey asked 827 randomly sampled registered voters in California "Do you support? Or do you oppose? Drilling for oil and natural gas off the Coast of California? Or do you not know enough to say?" Below is the distribution of responses, separated based on whether or not the respondent has a college degree.[55] Complete a chi-square test for these data to test whether there is an association between opinions regarding offshore drilling for oil and having a college degree. Remember to Identify, Check, Calculate, and Conclude.

	College Degree	
	Yes	No
Support	154	132
Oppose	180	126
Do not know	104	131
Total	438	389

3.60 Parasitic worm. Lymphatic filariasis is a disease caused by a parasitic worm. Complications of the disease can lead to extreme swelling and other complications. Here we consider results from a randomized experiment that compared three different drug treatment options to clear people of the this parasite, which people are working to eliminate entirely. The results for the second year of the study are given below:[56]

	Clear at Year 2	Not Clear at Year 2
Three drugs	52	2
Two drugs	31	24
Two drugs annually	42	14

(a) Set up hypotheses for evaluating whether there is any difference in the performance of the treatments, and also check conditions.

(b) Statistical software was used to run a chi-square test, which output:

$$X^2 = 23.7 \qquad df = 2 \qquad \text{p-value} = 7.2\text{e-}6$$

Use these results to evaluate the hypotheses from part (a), and provide a conclusion in the context of the problem.

[55] Survey USA, Election Poll #16804, data collected July 8-11, 2010.

[56] Christopher King et al. "A Trial of a Triple-Drug Treatment for Lymphatic Filariasis". In: *New England Journal of Medicine* 379 (2018), pp. 1801–1810.

Chapter highlights

Calculating a confidence interval or a test statistic and p-value are generally done with statistical software. It is important, then, to focus not just on the calculations, but on the following components:

1. Choose the correct procedure.
2. Understand when the procedure does or does not apply by checking relevant conditions.
3. Interpret the results in context.

Choosing the correct procedure requires understanding the *method* of data collection and the *type* of data collected. All of the inference procedures in Chapter 3 are for categorical variables. Here we list the two confidence intervals and five hypothesis tests encountered in this chapter and when to use them.

- **one-sample Z-interval/test for a population proportion p**
  - *1 random sample*, a yes/no variable
  - Ask about the proportion of a yes/no variable in a single population; e.g. determine if there is evidence that the true approval rating for a governor is greater than 50%.
- **two-sample Z-interval/test for a difference in population proportions $p_1 - p_2$**
  - *2 independent random samples or randomly assigned treatments*, a yes/no variable
  - Compare two populations or treatments to each other with respect to one yes/no variable; e.g. comparing the proportion under age 18 in two different counties.
- **χ^2 goodness of fit test** (special topic)
  - *1 random sample*, a categorical variable (generally at least three categories)
  - Compare the distribution of a categorical variable to a fixed or known population distribution; e.g. comparing the distribution of color among M&M's to the published color distribution.
- **χ^2 test for homogeneity**:
  - *2+ independent random samples or randomly assigned treatments*, a categorical variable
  - Compare the distribution of a categorical variable across several populations or treatments; e.g. comparing party affiliation over various years or patient improvement across 3 treatments.
- **χ^2 test for independence**
  - *1 random sample*, 2 categorical variables
  - Determine if, in a single population, there is an association between two categorical variables; e.g. looking for association at a particular school between grade level and whether or not one plays a sport.

Even when the data type and data collection method correspond to a particular test, we must verify that conditions are met to see if the assumptions of the test are reasonable. All of the inferential procedures of this chapter require some type of random sample or process. In addition, the one-sample Z-test/interval for p and the two-sample Z-test/interval for $p_1 - p_2$ require that the large counts condition is met and the three χ^2 tests require that all expected counts are at least 5.

Finally, being able to accurately interpret a confidence interval or p-value and use them to justify claims about a population are essential.

Chapter exercises

3.61 Active learning. A teacher wanting to increase the active learning component of her course is concerned about student reactions to changes she is planning to make. She conducts a survey in her class, asking students whether they believe more active learning in the classroom (hands on exercises) instead of traditional lecture will helps improve their learning. She does this at the beginning and end of the semester and wants to evaluate whether students' opinions have changed over the semester. Can she used the methods we learned in this chapter for this analysis? Explain your reasoning.

3.62 Website experiment. The OpenIntro website occasionally experiments with design and link placement. We conducted one experiment testing three different placements of a download link for this textbook on the book's main page to see which location, if any, led to the most downloads. The number of site visitors included in the experiment was 701 and is captured in one of the response combinations in the following table:

	Download	No Download
Position 1	13.8%	18.3%
Position 2	14.6%	18.5%
Position 3	12.1%	22.7%

(a) Calculate the actual number of site visitors in each of the six response categories.

(b) Complete an appropriate hypothesis test to check whether there is evidence that there is a higher rate of site visitors clicking on the textbook link in any of the three groups. Remember to Identify, Check, Calculate and Conclude.

3.63 Shipping holiday gifts. A local news survey asked 500 randomly sampled Los Angeles residents which shipping carrier they prefer to use for shipping holiday gifts. The table below shows the distribution of responses by age group as well as the expected counts for each cell (shown in parentheses).

		Age			
		18-34	35-54	55+	Total
Shipping Method	USPS	72 (81)	97 (102)	76 (62)	245
	UPS	52 (53)	76 (68)	34 (41)	162
	FedEx	31 (21)	24 (27)	9 (16)	64
	Something else	7 (5)	6 (7)	3 (4)	16
	Not sure	3 (5)	6 (5)	4 (3)	13
	Total	165	209	126	500

(a) State the null and alternative hypotheses for testing for independence of age and preferred shipping method for holiday gifts among Los Angeles residents.

(b) Are the conditions for inference using a chi-square test satisfied?

3.64 The Civil War. A national survey conducted among a simple random sample of 1,507 adults shows that 56% of Americans think the Civil War is still relevant to American politics and political life.[57]

(a) Conduct a hypothesis test to determine if these data provide strong evidence that the majority of the Americans think the Civil War is still relevant.

(b) Interpret the p-value in this context.

(c) Calculate a 90% confidence interval for the proportion of Americans who think the Civil War is still relevant. Interpret the interval in this context, and comment on whether or not the confidence interval agrees with the conclusion of the hypothesis test.

[57] Pew Research Center Publications, Civil War at 150: Still Relevant, Still Divisive, data collected between March 30 - April 3, 2011.

3.65 College smokers. We are interested in estimating the proportion of students at a university who smoke. Out of a random sample of 200 students from this university, 40 students smoke.

(a) Calculate a 95% confidence interval for the proportion of students at this university who smoke, and interpret this interval in context. (Reminder: Check conditions.)

(b) If we wanted the margin of error to be no larger than 2% at a 95% confidence level for the proportion of students who smoke, how big of a sample would we need?

3.66 Acetaminophen and liver damage. It is believed that large doses of acetaminophen (the active ingredient in over the counter pain relievers like Tylenol) may cause damage to the liver. A researcher wants to conduct a study to estimate the proportion of acetaminophen users who have liver damage. For participating in this study, he will pay each subject $20 and provide a free medical consultation if the patient has liver damage.

(a) If he wants to limit the margin of error of his 98% confidence interval to 2%, what is the minimum amount of money he needs to set aside to pay his subjects?

(b) The amount you calculated in part (a) is substantially over his budget so he decides to use fewer subjects. How will this affect the width of his confidence interval?

3.67 Life after college. We are interested in estimating the proportion of graduates at a mid-sized university who found a job within one year of completing their undergraduate degree. Suppose we conduct a survey and find out that 348 of the 400 randomly sampled graduates found jobs. The graduating class under consideration included over 4500 students.

(a) Describe the population parameter of interest. What is the value of the point estimate of this parameter?

(b) Check if the conditions for constructing a confidence interval based on these data are met.

(c) Calculate a 95% confidence interval for the proportion of graduates who found a job within one year of completing their undergraduate degree at this university, and interpret it in the context of the data.

(d) What does "95% confidence" mean?

(e) Now calculate a 99% confidence interval for the same parameter and interpret it in the context of the data.

(f) Compare the widths of the 95% and 99% confidence intervals. Which one is wider? Explain.

3.68 Diabetes and unemployment. A Gallup poll surveyed Americans about their employment status and whether or not they have diabetes. The survey results indicate that 1.5% of the 47,774 employed (full or part time) and 2.5% of the 5,855 unemployed 18-29 year olds have diabetes.[58]

(a) Create a two-way table presenting the results of this study.

(b) State appropriate hypotheses to test.

(c) The sample difference is about 1%. If we completed the hypothesis test, we would find that the p-value is very small (about 0), meaning the difference is statistically significant. Use this result to explain the difference between statistically significant and practically significant findings.

3.69 Browsing on the mobile device. A survey of 2,254 randomly selected American adults indicates that 17% of cell phone owners browse the internet exclusively on their phone rather than a computer or other device.[59]

(a) According to an online article, a report from a mobile research company indicates that 38 percent of Chinese mobile web users only access the internet through their cell phones.[60] Conduct a hypothesis test to determine if these data provide strong evidence that the proportion of Americans who only use their cell phones to access the internet is different than the Chinese proportion of 38%.

(b) Interpret the p-value in this context.

(c) Calculate a 95% confidence interval for the proportion of Americans who access the internet on their cell phones, and interpret the interval in this context.

[58] Gallup Wellbeing, Employed Americans in Better Health Than the Unemployed, data collected Jan. 2, 2011 - May 21, 2012.

[59] Pew Internet, Cell Internet Use 2012, data collected between March 15 - April 13, 2012.

[60] S. Chang. "The Chinese Love to Use Feature Phone to Access the Internet". In: *M.I.C Gadget* (2012).

3.70 2010 Healthcare Law. On June 28, 2012 the U.S. Supreme Court upheld the much debated 2010 healthcare law, declaring it constitutional. A Gallup poll released the day after this decision indicates that 46% of 1,012 randomly sampled Americans agree with this decision. At a 95% confidence level, this sample has a 3% margin of error. Based on this information, determine if the following statements are true or false, and explain your reasoning.[61]

(a) We are 95% confident that between 43% and 49% of Americans in this sample support the decision of the U.S. Supreme Court on the 2010 healthcare law.

(b) We are 95% confident that between 43% and 49% of Americans support the decision of the U.S. Supreme Court on the 2010 healthcare law.

(c) If we considered many random samples of 1,012 Americans, and we calculated the sample proportions of those who support the decision of the U.S. Supreme Court, 95% of those sample proportions will be between 43% and 49%.

(d) The margin of error at a 90% confidence level would be higher than 3%.

3.71 Coffee and Depression. Researchers conducted a study investigating the relationship between caffeinated coffee consumption and risk of depression in women. They collected data on 50,739 women free of depression symptoms at the start of the study in the year 1996, and these women were followed through 2006. The researchers used questionnaires to collect data on caffeinated coffee consumption, asked each individual about physician- diagnosed depression, and also asked about the use of antidepressants. The table below shows the distribution of incidences of depression by amount of caffeinated coffee consumption.[62]

		Caffeinated coffee consumption					
		≤ 1 cup/week	2-6 cups/week	1 cup/day	2-3 cups/day	≥ 4 cups/day	Total
Clinical depression	Yes	670	373	905	564	95	2,607
	No	11,545	6,244	16,329	11,726	2,288	48,132
	Total	12,215	6,617	17,234	12,290	2,383	50,739

(a) What type of test is appropriate for evaluating if there is an association between coffee intake and depression?

(b) Write the hypotheses for the test you identified in part (a).

(c) Calculate the overall proportion of women who do and do not suffer from depression.

(d) Identify the expected count for the highlighted cell, and calculate the contribution of this cell to the test statistic, i.e. $(Observed - Expected)^2/Expected$.

(e) The test statistic is $\chi^2 = 20.93$. What is the p-value?

(f) What is the conclusion of the hypothesis test?

(g) One of the authors of this study was quoted on the NYTimes as saying it was "too early to recommend that women load up on extra coffee" based on just this study.[63] Do you agree with this statement? Explain your reasoning.

3.72 Government shutdown. The United States federal government shutdown of 2018–2019 occurred from December 22, 2018 until January 25, 2019, a span of 35 days. A Survey USA poll of 614 randomly sampled Americans during this time period reported that 48% of those who make less than $40,000 per year and 55% of those who make $40,000 or more per year said the government shutdown has not at all affected them personally. A 95% confidence interval for $(p_{<40K} - p_{\geq 40K})$, where p is the proportion of those who said the government shutdown has not at all affected them personally, is (-0.16, 0.02). Based on this information, determine if the following statements are true or false, and explain your reasoning if you identify the statement as false.[64]

(a) At the 5% significance level, the data provide convincing evidence of a real difference in the proportion who are not affected personally between Americans who make less than $40,000 annually and Americans who make $40,000 annually.

(b) We are 95% confident that 16% more to 2% fewer Americans who make less than $40,000 per year are not at all personally affected by the government shutdown compared to those who make $40,000 or more per year.

(c) A 90% confidence interval for $(p_{<40K} - p_{\geq 40K})$ would be wider than the $(-0.16, 0.02)$ interval.

(d) A 95% confidence interval for $(p_{\geq 40K} - p_{<40K})$ is (-0.02, 0.16).

[61] Gallup, Americans Issue Split Decision on Healthcare Ruling, data collected June 28, 2012.

[62] M. Lucas et al. "Coffee, caffeine, and risk of depression among women". In: *Archives of internal medicine* 171.17 (2011), p. 1571.

[63] A. O'Connor. "Coffee Drinking Linked to Less Depression in Women". In: *New York Times* (2011).

[64] Survey USA, News Poll #24568, data collected on April 21, 2019.

Chapter 4

Inference for numerical data: means

Chapter 3 summarized inference procedures for categorical data (counts and proportions), using the normal distribution and the chi-square distribution. In this chapter, we focus on inference procedures for numerical data and we encounter a new distribution called the t-distribution. In each case, the inference ideas remain the same: determine which point estimate or test statistic is useful, identify an appropriate distribution for the point estimate or test statistic, and apply the ideas of inference.

For videos, slides, and other resources, please visit
www.openintro.org/ahss

4.1 Sampling distribution of a sample mean

If the mean race time among all runners of a certain race is 94 minutes with a standard deviation of 16 minutes, what is the probability that the average race time of 30 randomly selected runners will be within 16 minutes of the mean? The answer is not 68%! To answer this question we must visualize and understand what is called the *sampling distribution* of a sample mean.

Learning objectives

1. Interpret and apply the concept of a sampling distribution in the context of a sample mean.
2. Distinguish between the standard deviation of a population and the standard deviation of a sampling distribution.
3. Calculate the mean and standard deviation of the sampling distribution of a sample mean.
4. Justify whether the independence condition is satisfied when considering properties of the sampling distribution of a sample mean.
5. Determine whether or not the shape of the sampling distribution of a sample mean is approximately normal.
6. Interpret the mean, standard deviation, and probabilities for the sampling distribution of a sample mean

4.1.1 Building a sampling distribution for a sample mean

In this section we consider a data set called `run17`, which represents all 19,961 runners who finished the 2017 Cherry Blossom 10 mile run in Washington, DC. Part of this data set is shown in Figure 4.1, and the variables are described in Figure 4.2.

ID	time	age	gender	state
1	92.25	38.00	M	MD
2	106.35	33.00	M	DC
⋮	⋮	⋮	⋮	⋮
16923	122.87	37.00	F	VA
16924	93.30	27.00	F	DC

Figure 4.1: Four observations from the `run17` data set.

variable	**description**
`time`	Ten mile run time, in minutes
`age`	Age, in years
`gender`	Gender (`M` for male, `F` for female)
`state`	Home state (or country if not from the US)

Figure 4.2: Variables and their descriptions for the `run17` data set.

These data are special because they include the results for the entire population of runners who finished the 2017 Cherry Blossom Run. We took a simple random sample of this population, which is represented in Figure 4.3. A histogram summarizing the time variable in the `run17samp` data set is shown in Figure 4.4.

ID	time	age	gender	state
1983	88.31	59	M	MD
8192	100.67	32	M	VA
⋮	⋮	⋮	⋮	⋮
1287	89.49	26	M	DC

Figure 4.3: Three observations for the `run17samp` data set, which represents a simple random sample of 100 runners from the 2017 Cherry Blossom Run.

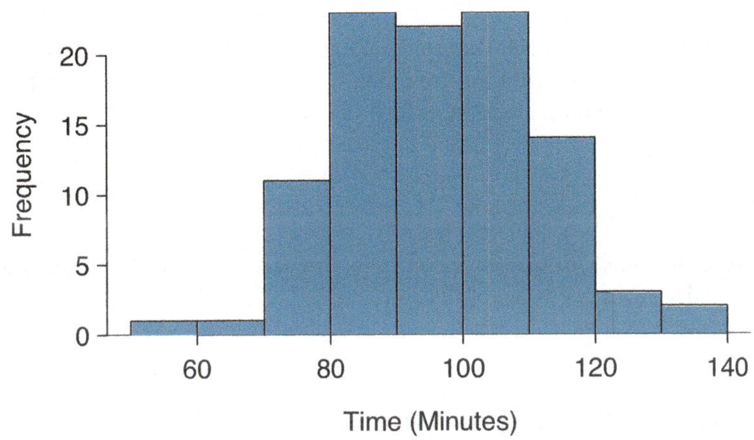

Figure 4.4: Histogram of `time` for a single sample of size 100. The average of the sample is in the mid-90s and the standard deviation of the sample $s \approx 17$ minutes.

From the random sample represented in `run17samp`, we use our sample mean to estimate that the average time it takes to run 10 miles is 95.61 minutes. Suppose we take another random sample of 100 individuals and take its mean: 95.30 minutes. Suppose we took another (93.43 minutes) and another (94.16 minutes), and so on. If we do this many many times – which we can do only because we have the entire population data set – we can build up a **sampling distribution** for the sample mean when the sample size is 100, shown in Figure 4.5.

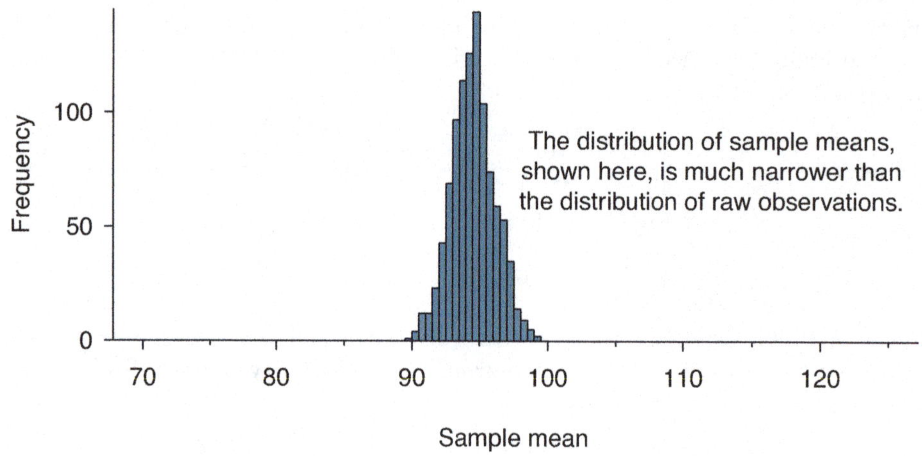

Figure 4.5: A histogram of 1000 sample means for run time, where the samples are of size $n = 100$. This histogram approximates the true sampling distribution of the sample mean, with mean $\mu_{\bar{x}}$ and standard deviation $\sigma_{\bar{x}}$.

SAMPLING DISTRIBUTION OF A SAMPLE MEAN

The sampling distribution of a sample mean $\bar{x}$ is the distribution of $\bar{x}$ values for all random samples of a given size from a given population.

4.1.2 The mean and standard deviation of $\bar{x}$

The sampling distribution shown in Figure 4.5 is unimodal and approximately symmetric. It is also centered exactly at the true population mean: $\mu = 94.52$. Intuitively, this makes sense. The sample mean should be an unbiased estimator of the population mean. We use $\mu_{\bar{x}}$ to denote the mean of the sampling distribution of $\bar{x}$.

We can see that the sample mean has some variability around the population mean, which can be quantified using the standard deviation of this distribution of sample means. The standard deviation of the sample mean tells us how far the typical estimate is away from the actual population mean, 94.52 minutes. It also describes the typical **error** of an estimate, and is denoted by the symbol $\sigma_{\bar{x}}$.

EXAMPLE 4.1

Looking at Figures 4.4 and 4.5, we see that the standard deviation of the sample mean with $n = 100$ is much smaller than the standard deviation of a single sample. Interpret this statement and explain why it is true.

(E)

The variation from one-sample mean to another sample mean is much smaller than the variation from one individual to another individual. This makes sense because when we average over 100 values, the large and small values tend to balance each other out. For instance, while many individuals have a time under 90 minutes, we can see in Figure 4.5 that it is unlikely for the *average* of 100 randomly sampled runners to be less than 90 minutes.

When considering how to calculate the standard deviation of a sample mean, there is one problem: there is no obvious way to estimate this from a single sample. However, statistical theory provides a helpful formula for this situation.

In the sample of 100 runners, the standard deviation of the sample mean is equal to one-tenth of the population standard deviation: 15.93/10 = 1.59. In other words, the standard deviation of the sample mean based on 100 observations is equal to

$$\sigma_{\bar{x}} = \frac{\sigma_x}{\sqrt{n}} = \frac{15.93}{\sqrt{100}} = 1.59$$

where σ_x is the standard deviation of the individuals in the population. This formula is not coincidental, and the standard deviation of the sample mean is generally equal equal to $\frac{\sigma_x}{\sqrt{n}}$.

MEAN AND STANDARD DEVIATION OF A SAMPLE MEAN

The mean and standard deviation of the sampling distribution of a sample mean describe the center and spread of the distribution of sample means $\bar{x}$ for all random samples of size n from a population of size N. Let μ represent the population mean. We find the mean and standard deviation of the sampling distribution of $\bar{x}$ as follows:

$$\mu_{\bar{x}} = \mu$$

$$\sigma_{\bar{x}} = \frac{\sigma}{\sqrt{n}} \quad \text{when } n < 0.10(N) \text{ if sampling without replacement}$$

EXAMPLE 4.2

We calculated the mean and standard deviation of the samping distribution of mean run time with $n = 100$ runners to be 94.52 minutes and 1.59 minutes, respectively. Interpret these quantities.

The mean run time among all random samples of size $n = 100$ runners is 94.52 minutes. Among all random samples of $n = 100$ runners, the sample mean would typically vary from the population mean of 94.52 minutes by about 1.59 minutes.

GUIDED PRACTICE 4.3

The average of the runners' ages is 35.05 years with a standard deviation of $\sigma = 8.97$. A simple random sample of 100 runners is taken. (a) What is the standard deviation of the sample mean? (b) Would you be surprised to get a sample of size 100 with an average of 36 years?[1]

GUIDED PRACTICE 4.4

(a) Would you be more trusting of a sample that has 100 observations or 400 observations? (b) We want to show mathematically that our estimate tends to be better when the sample size is larger. If the standard deviation of the individual observations is 10, what is our estimate of the standard deviation of the mean when the sample size is 100? What about when it is 400? (c) Explain how your answer to (b) mathematically justifies your intuition in part (a).[2]

[1](a) Assuming the sample size of 100 is less than 10% of the population size, that is that the number of total runners is greater than 1000, we apply the formula for the standard deviation of the sample mean as follows: $\sigma_{\bar{y}} = 8.97/\sqrt{100} = 0.90$ years. (b) It would not be surprising. 36 years is about 1 standard deviation from the true mean of 35.05. Based on the 68, 95 rule, we would get a sample mean at least this far away from the true mean approximately $100\% - 68\% = 32\%$ of the time.

[2](a) Extra observations are usually helpful in understanding the population, so a point estimate with 400 observations seems more trustworthy. (b) The standard deviation of the mean when the sample size is 100 is given by $10/\sqrt{100} = 1$. For 400: $10/\sqrt{400} = 0.5$. The larger sample has a smaller standard deviation of the mean. (c) The standard deviation of the mean of the sample with 400 observations is lower than that of the sample with 100 observations. The standard deviation of $\bar{x}$ describes the typical error, and because it is lower for the larger sample, this mathematically shows the estimate from the larger sample tends to be better – though it does not guarantee that every large sample will provide a better estimate than a particular small sample.

4.1.3 The Central Limit Theorem revisited

In Figure 4.5, the sampling distribution of the sample mean looks approximately normally distributed. Will the sampling distribution of a mean always be nearly normal? To address this question, we will investigate three cases to see roughly when the approximation is reasonable.

We consider three data sets: one from a *uniform* distribution, one from an *right skewed* distribution, and the other from a *normal* distribution. These distributions are shown in the top panels of Figure 4.6.

The left panel in the $n = 2$ row represents the sampling distribution of $\bar{x}$ if it is the sample mean of two observations from the uniform distribution shown. The dashed line represents the closest approximation of the normal distribution. Similarly, the center and right panels of the $n = 2$ row represent the respective distributions of $\bar{x}$ for data from right skewed and normal distributions.

EXAMPLE 4.5

(E) Examine the distributions in each row of Figure 4.6. What do you notice about the sampling distribution of the mean as the sample size, n, becomes larger?

The normal model becomes better as larger samples are used. However, in the case when the population distribution is normally distributed, the distribution of the sample mean is normal for all sample sizes.

GUIDED PRACTICE 4.6

(G) For the distributions in Figure 4.6, would a normal model for a sample mean be appropriate when the sample size is at least 30?[3]

DETERMINING IF THE SAMPLE MEAN IS NORMALLY DISTRIBUTED

If the population distribution is normal, the sampling distribution of $\bar{x}$ will be normal for any sample size.

The less normal the population, the larger n needs to be for the sampling distribution of $\bar{x}$ to be nearly normal. A good rule of thumb is that for most populations, the sampling distribution of $\bar{x}$ will be approximately normal if $n \geq 30$.

This brings us back to the **Central Limit Theorem**, introduced in Section 2.7. The Central Limit Theorem (CLT) is a fundamental theorem of Statistics because it allows us to understand the shape of certain sampling distributions, no matter the shape of the population from which the sample is drawn. This in turn allows us to easily find probabilities and, as we saw in Chapter 3, calculate p-values that we otherwise would not be able to.

CENTRAL LIMIT THEOREM (CLT)

For any population with a fixed mean and standard deviation, the shape of the sampling distribution of the mean of a random sample becomes more normal as the sample size n gets larger.

[3] Yes, the sampling distributions when $n = 30$ all look very much like a normal distribution.

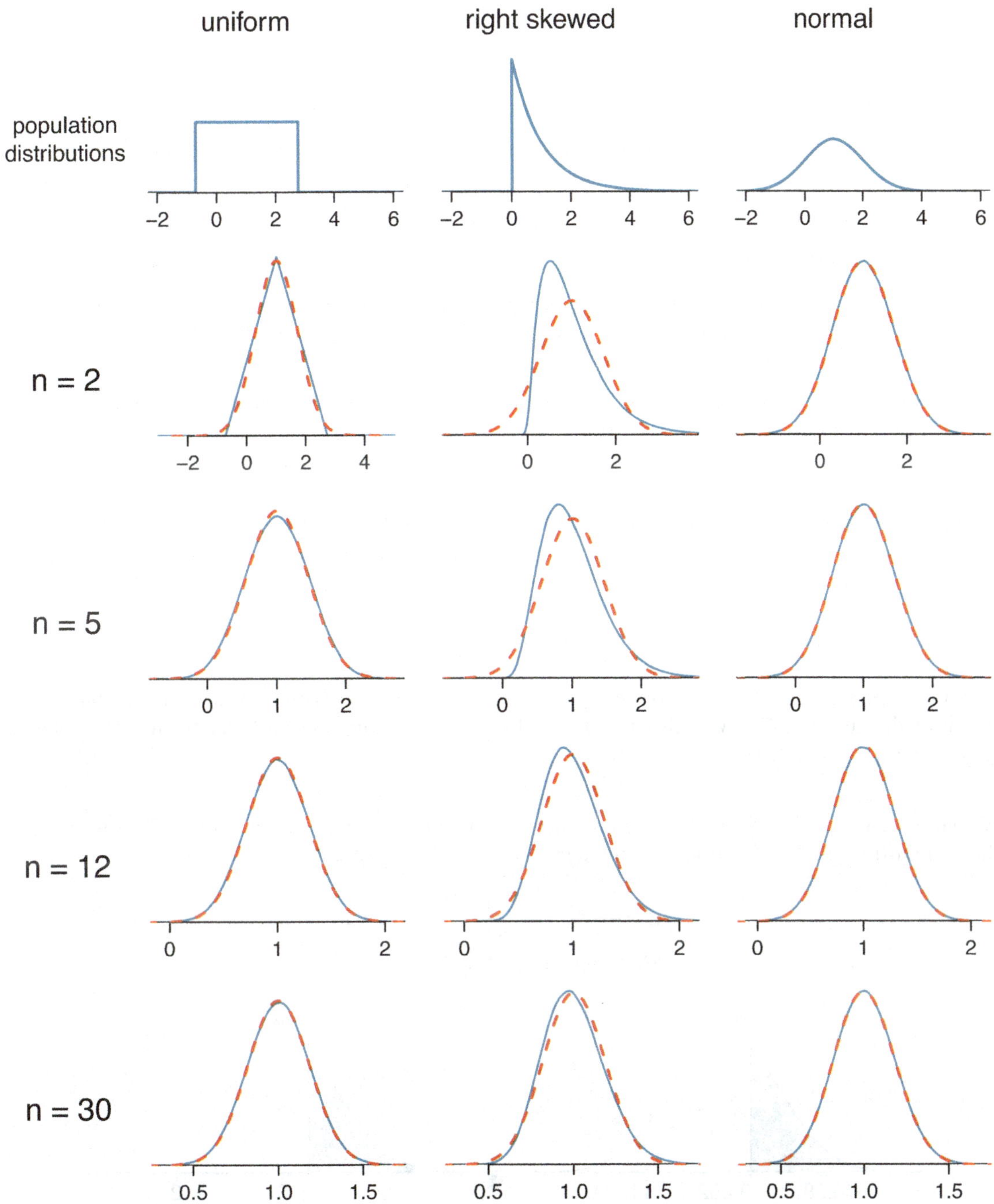

Figure 4.6: Sampling distributions for the mean at different sample sizes and for three different distributions. The dashed red lines show normal distributions. Note the change in scale of the horizontal axis for larger values of n.

4.1.4 Using a normal model for the sampling distribution of $\bar{x}$

We have already encountered a normal model for data and for sample proportions. When appropriate conditions are met, we can also use a normal approximation to estimate probabilities involving the distribution of a sample average. We must remember to verify that the conditions are met and use the mean $\mu_{\bar{x}}$ and standard deviation $\sigma_{\bar{x}}$ for the sampling distribution of the sample average.

THREE IMPORTANT FACTS ABOUT THE DISTRIBUTION OF A SAMPLE MEAN $\bar{x}$

When the observations can be considered independent, such as from a random sample of size n from a population of size N, the distribution of the sample mean can be described as follows.

1. The mean of a sample mean is denoted by $\mu_{\bar{x}}$, and it is equal to μ.
2. The SD of a sample mean is denoted by $\sigma_{\bar{x}}$, and it is equal to $\frac{\sigma}{\sqrt{n}}$, when $n < 0.10(N)$.
3. When the population distribution is nearly normal or when $n \geq 30$, the sample mean closely follows a normal distribution.

Before we apply a normal model to a sample mean, we review the use of normal approximation in the context of a population distribution.

EXAMPLE 4.7

In the 2017 Cherry Blossom 10 mile run, the average time for all of the runners is 94.52 minutes with a standard deviation of 8.97 minutes. The distribution of run times is approximately normal. Find the probability that a randomly selected runner completed the run in less than 90 minutes.

X: time of randomly selected runner (in minutes)

The problem tells us that X has an approximately normal distribution with mean 94.52 and standard deviation 8.97. We will use a technology option from Section 2.6.4 to find the relevant area under a normal distribution.

We can either use X is Normal($\mu = 94.52$, $\sigma = 8.97$) to find that $P(X < 90) = 0.307$, or we can use the standard normal distribution, Normal($\mu = 0$, $\sigma = 1$), and find:

$$Z = \frac{x - \mu}{\sigma} = \frac{90 - 94.52}{8.97} = -0.504$$
$$P(Z < -0.504) = 0.307$$

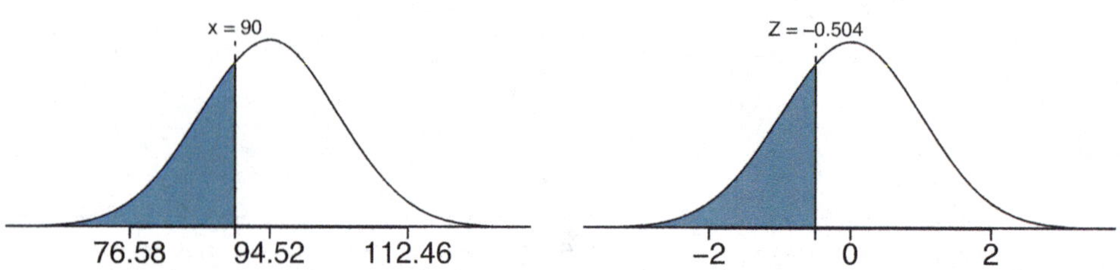

There is about a 30.7% probability that a randomly selected runner will completed the run in less than 90 minutes.

EXAMPLE 4.8

Using the information from Example 4.7, find the probabiliy that the average of 20 runners' times is less than 90 minutes.

Here we are interested in an *average*, so we use the sampling distribution of the sample average.

$\bar{x}$: sample mean of 20 randomly selected runners (in minutes)

$\mu_{\bar{x}} = \mu = 94.52$

The sample size 20 is less than 10% of all runners so we can calculate the standard deviation as follows: $\sigma_{\bar{x}} = \frac{\sigma}{\sqrt{n}} = \frac{8.97}{\sqrt{20}} = 2.01$.

The population distribution of run time is approximately normal, so we can use a normal model for $\bar{x}$. Given $\bar{x}$ is approximately Normal($\mu = 94.52$, $\sigma = 2.01$), we find that $P(\bar{x} < 90) = 0.012$.

Alternately, we could use the standard normal distribution, Normal($\mu = 0$, $\sigma = 1$), and find:

$$Z = \frac{\bar{x} - \mu_{\bar{x}}}{\sigma_{\bar{x}}} = \frac{90 - 94.52}{\frac{8.97}{\sqrt{20}}} = -2.25$$

$$P(Z < -2.25) = 0.0123$$

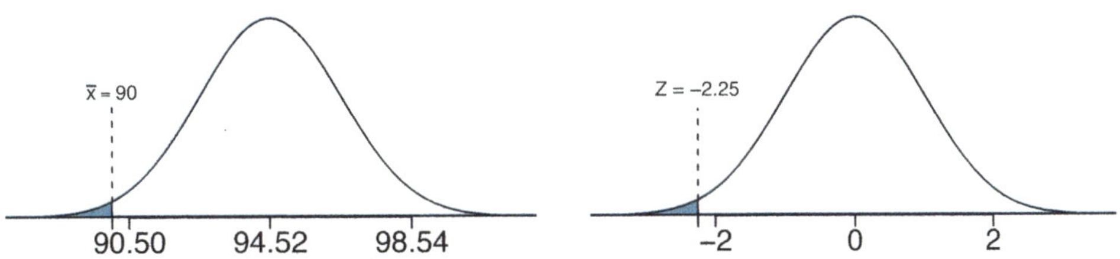

There is a 1.2% probability that the average run time of 20 randomly selected runners will be less than 90 minutes.

In Section 4.3, Hypothesis testing for a population mean, we will see parallels between the calculation of the Z-score shown above and the calculation of the test statistic.

GUIDED PRACTICE 4.9

Intuitively, why does it make sense that the probability found in Example 4.8 is smaller than the probability found in Example 4.7?[4]

REMEMBER TO DIVIDE BY $\sqrt{n}$

When finding the probability that an *average* or mean is greater or less than a particular value, remember to divide the standard deviation of the population by $\sqrt{n}$ to calculate the correct SD.

---

[4]When we average over 20 values, we expect less variability than when looking at individual values. Thus, if we are trying to estimate a population mean, our estimate will be more likely to be close to the true mean using a sample size of 20 than using a sample size of 1. Equivalently, our estimate will be *less* likely to be far from the true value using a sample size of 20 than using a sample size of 1.

E

EXAMPLE 4.10

The average of all the runners' ages is 35.05 years with a standard deviation of $\sigma = 8.97$. The distribution of age is somewhat skewed. What is the probability that a randomly selected runner is older than 37 years?

Because the distribution of age is skewed and is not normal, we cannot use normal approximation for this problem. In order to answer this question, we would need to look at all of the data.

G

GUIDED PRACTICE 4.11

What is the probability that the average of 50 randomly selected runners is greater than 37 years?[5]

The previous examples highlight the power of the Central Limit Theorem for sample means. Even if the population is not normally distributed, for large enough n, the sampling distribution of the *sample mean* will be approximately normally distributed, allowing us to calculate relevant probabilities using a normal model. When a normal model cannot be used, there may be no straightforward way to calculate these desired probabilities.

[5]Because $n = 50 \geq 30$, the sampling distribution of the sample mean is approximately normal, so we can use the normal approximation for $\bar{x}$. The population mean μ is given as 35.05 years.

$$\sigma_{\bar{x}} = \frac{\sigma}{\sqrt{n}} = \frac{8.97}{\sqrt{50}} = 1.27 \qquad Z = \frac{\bar{x} - \mu_{\bar{x}}}{\sigma_{\bar{x}}} = \frac{37 - 35.05}{1.27} = 1.535 \qquad P(Z > 1.535) = 0.062$$

There is a 6.2% chance that the average age of 50 runners will be greater than 37.

Section summary

- The symbol $\bar{x}$ denotes a sample average. $\bar{x}$ for any particular sample is a number. However, $\bar{x}$ can vary from sample to sample. The **sampling distribution** of a sample mean $\bar{x}$ is the distribution of values of $\bar{x}$ for all random samples of size n from a given population.
- When the observations can be considered independent, such as from a *random* sample:
  - The **mean** of the sampling distribution of a sample mean $\bar{x}$ is given by: $\mu_{\bar{x}} = \mu$, where μ is the population mean.
  - The **standard deviation** of the sampling distribution of a sample mean $\bar{x}$ is given by: $\sigma_{\bar{x}} = \frac{\sigma}{\sqrt{n}}$, where σ is the population standard deviation. When sampling without replacement, the sample size n should be less than 10% of the population size N, i.e. $n < 0.10(N)$, in order for this standard deviation formula to be used.
  - The **shape** of the sampling distribution of a sample mean $\bar{x}$ is approximately normal if the population distribution can be modeled by a normal distribution (regardless of the sample size) or the sample size $n \geq 30$.
- Consider taking a random sample from a population with a fixed mean and standard deviation. The **Central Limit Theorem** ensures that regardless of the shape of the original population, as the sample size increases, the distribution of the sample average $\bar{x}$ becomes more normal.
- $\mu_{\bar{x}}$, the mean of $\bar{x}$, describes the average of the sample means among all random samples of size n from a given population.
- $\sigma_{\bar{x}}$, the standard deviation of $\bar{x}$, measures how far the sample means typically vary from the population mean μ for all random samples of size n from a given population.
- The standard deviation of $\bar{x}$ will be *smaller* than the standard deviation of the population by a factor of $\sqrt{n}$. The larger the sample size, the better the estimate $\bar{x}$ tends to be for μ.
- To use a normal model to find probabilities involving a sample mean, first verify that the conditions for independence are met and that the sample size is at least 30 or the population distribution is approximately normal. Identify the distribution and its parameters, write the relevant probability statement, and answer the question in context.
- The mean, standard deviation, and probabilities for a sampling distribution of a sample mean should be interpreted within the context of a specific population

Exercises

4.1 Ages of pennies, Part I. The histogram below shows the distribution of ages of pennies at a bank.

(a) Describe the distribution.

(b) Sampling distributions for means from simple random samples of 5, 30, and 100 pennies is shown in the histograms below. Describe the shapes of these distributions and comment on whether they look like what you would expect to see based on the Central Limit Theorem.

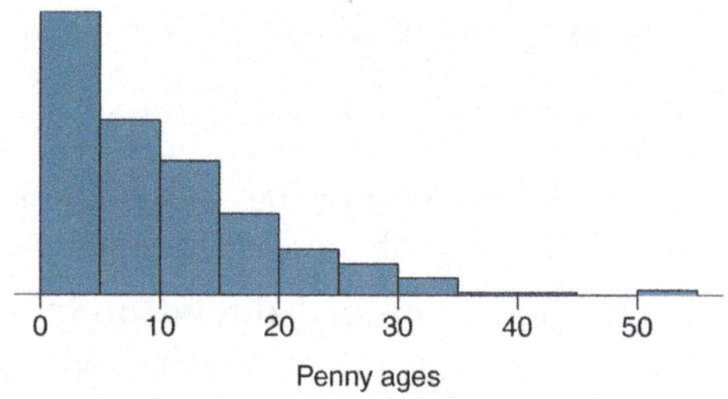

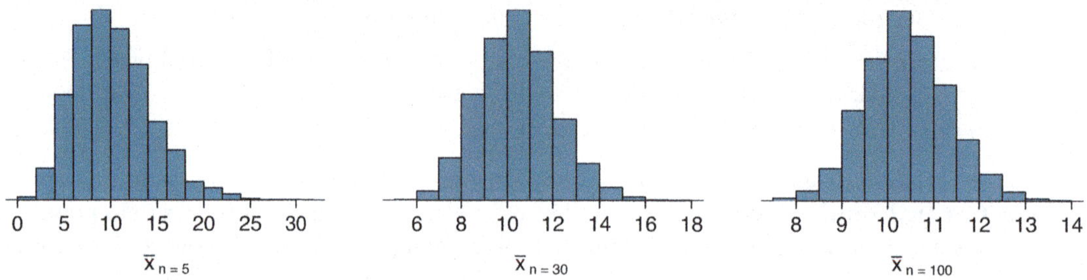

4.2 Ages of pennies, Part II. The mean age of the pennies from Exercise 4.1 is 10.44 years with a standard deviation of 9.2 years. Using the Central Limit Theorem, calculate the means and standard deviations of the distribution of the mean from random samples of size 5, 30, and 100. Comment on whether the sampling distributions shown in Exercise 4.1 agree with the values you compute.

4.3 Housing prices. A housing survey was conducted to determine the price of a typical home in Topanga, CA. The mean price of a house was roughly \$1.3 million with a standard deviation of \$300,000. There were no houses listed below \$600,000 but a few houses above \$3 million.

(a) Is the distribution of housing prices in Topanga symmetric, right skewed, or left skewed? *Hint:* Sketch the distribution.

(b) Would you expect most houses in Topanga to cost more or less than \$1.3 million?

(c) Can we estimate the probability that a randomly chosen house in Topanga costs more than \$1.4 million using the normal distribution?

(d) What is the probability that the mean of 60 randomly chosen houses in Topanga is more than \$1.4 million?

(e) How would doubling the sample size affect the standard deviation of the mean?

4.4 Stats final scores. Each year about 1500 students take the introductory statistics course at a large university. This year scores on the final exam are distributed with a median of 74 points, a mean of 70 points, and a standard deviation of 10 points. There are no students who scored above 100 (the maximum score attainable on the final) but a few students scored below 20 points.

(a) Is the distribution of scores on this final exam symmetric, right skewed, or left skewed?

(b) Would you expect most students to have scored above or below 70 points?

(c) Can we calculate the probability that a randomly chosen student scored above 75 using the normal distribution?

(d) What is the probability that the average score for a random sample of 40 students is above 75?

(e) How would cutting the sample size in half affect the standard deviation of the mean?

4.5 Identify distributions, Part I. Four plots are presented below. The plot at the top is a distribution for a population. The mean is 10 and the standard deviation is 3. Also shown below is a distribution of (1) a single random sample of 100 values from this population, (2) a distribution of 100 sample means from random samples with size 5, and (3) a distribution of 100 sample means from random samples with size 25. Determine which plot (A, B, or C) is which and explain your reasoning.

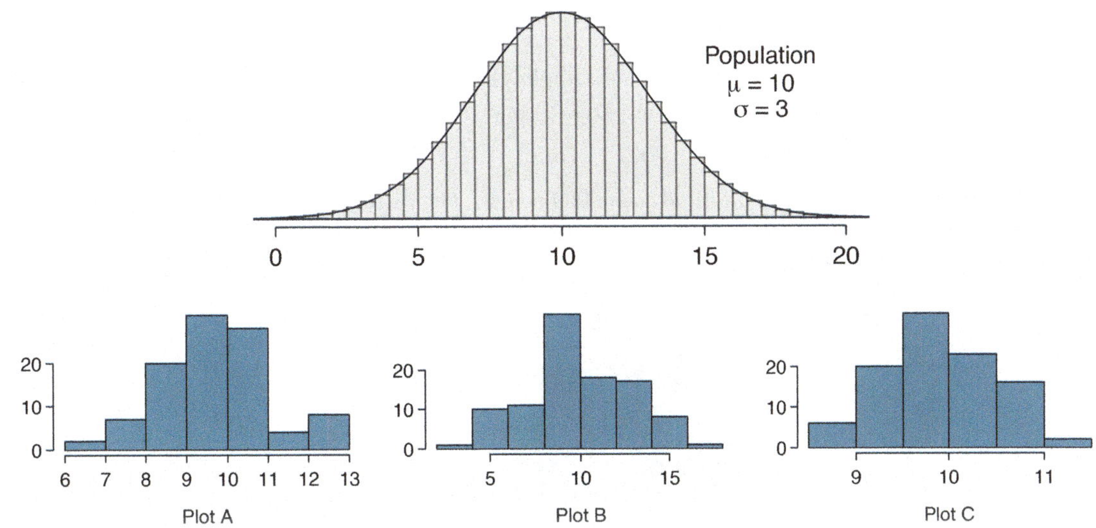

4.6 Identify distributions, Part II. Four plots are presented below. The plot at the top is a distribution for a population. The mean is 60 and the standard deviation is 18. Also shown below is a distribution of (1) a single random sample of 500 values from this population, (2) a distribution of 500 sample means from random samples of each size 18, and (3) a distribution of 500 sample means from random samples of each size 81. Determine which plot (A, B, or C) is which and explain your reasoning.

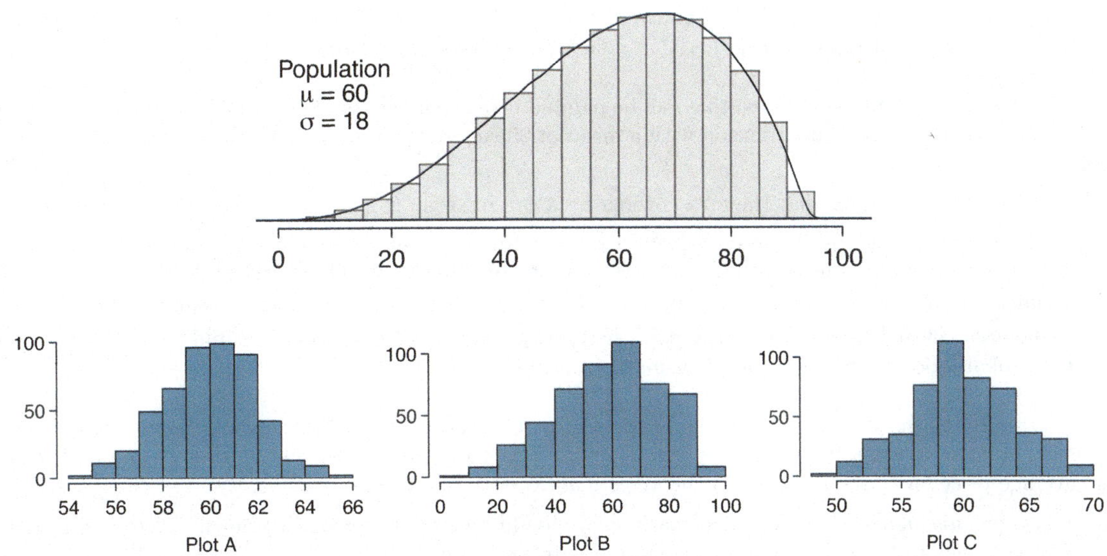

4.7 Weights of pennies. The distribution of weights of United States pennies is approximately normal with a mean of 2.5 grams and a standard deviation of 0.03 grams.

(a) What is the probability that a randomly chosen penny weighs less than 2.4 grams?

(b) Describe the sampling distribution of the mean weight of 10 randomly chosen pennies.

(c) What is the probability that the mean weight of 10 pennies is less than 2.4 grams?

(d) Sketch the two distributions (population and sampling) on the same scale.

(e) Could you estimate the probabilities from (a) and (c) if the weights of pennies had a skewed distribution?

4.8 CFLBs. A manufacturer of compact fluorescent light bulbs advertises that the distribution of the lifespans of these light bulbs is nearly normal with a mean of 9,000 hours and a standard deviation of 1,000 hours.

(a) What is the probability that a randomly chosen light bulb lasts more than 10,500 hours?

(b) Describe the distribution of the mean lifespan of 15 light bulbs.

(c) What is the probability that the mean lifespan of 15 randomly chosen light bulbs is more than 10,500 hours?

(d) Sketch the two distributions (population and sampling) on the same scale.

(e) Could you estimate the probabilities from parts (a) and (c) if the lifespans of light bulbs had a skewed distribution?

4.9 Songs on an iPod. Suppose an iPod has 3,000 songs. The histogram below shows the distribution of the lengths of these songs. We also know that, for this iPod, the mean length is 3.45 minutes and the standard deviation is 1.63 minutes.

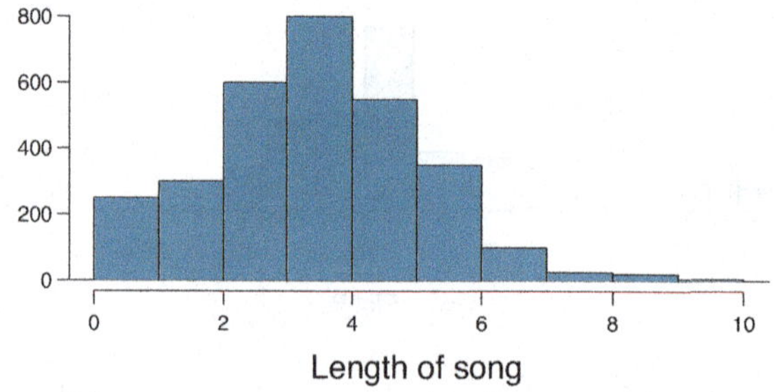

(a) Calculate the probability that a randomly selected song lasts more than 5 minutes.

(b) You are about to go for an hour run and you make a random playlist of 15 songs. What is the probability that your playlist lasts for the entire duration of your run? *Hint:* If you want the playlist to last 60 minutes, what should be the minimum average length of a song?

(c) You are about to take a trip to visit your parents and the drive is 6 hours. You make a random playlist of 100 songs. What is the probability that your playlist lasts the entire drive?

4.10 Spray paint. Suppose the area that can be painted using a single can of spray paint is slightly variable and follows a nearly normal distribution with a mean of 25 square feet and a standard deviation of 3 square feet.

(a) What is the probability that the area covered by a can of spray paint is more than 27 square feet?

(b) Suppose you want to spray paint an area of 540 square feet using 20 cans of spray paint. On average, how many square feet must each can be able to cover to spray paint all 540 square feet?

(c) What is the probability that you can cover a 540 square feet area using 20 cans of spray paint?

(d) If the area covered by a can of spray paint had a slightly skewed distribution, could you still calculate the probabilities in parts (a) and (c) using the normal distribution?

4.11 Wireless routers. John is shopping for wireless routers and is overwhelmed by the number of available options. In order to get a feel for the average price, he takes a random sample of 75 routers and finds that the average price for this sample is \$75 and the standard deviation is \$25.

(a) Based on this information, how much variability should he expect to see in the mean prices of repeated samples, each containing 75 randomly selected wireless routers?

(b) A consumer website claims that the average price of routers is \$80. Is a true average of \$80 consistent with John's sample?

4.12 Betting on dinner. A restaurant is having a promotion where prices of menu items are determined randomly following some underlying distribution. We are told that the price of basket of fries is drawn from a normal distribution with mean 6 and standard deviation of 2. You want to get 5 baskets of fries but you only have \$28 in your pocket. What is the probability that you would have enough money to pay for all five baskets of fries?

4.2 Confidence intervals for a population mean

How can we use sample data to estimate an unknown population mean, such as the true average mercury content in a particular type of dolphin? How can we calculate the margin of error for a given confidence level? While the concepts and the framework are the same as for estimating unknown population proportions, to estimate means we must investigate and use a new family of distributions, called t-distributions.

Learning objectives

1. Describe t-distributions.
2. Identify and set up an appropriate confidence interval procedure for a population mean μ or mean difference μ_d.
3. Justify the appropriateness of constructing a confidence interval for a population mean or population mean difference by verifying conditions.
4. Calculate the degrees of freedom and an appropriate confidence interval for a population mean or mean differences.
5. Calculate the standard error and margin of error for a sample size for a one-sample t-interval.
6. Interpret a confidence interval for a population mean or mean differences in context.
7. Justify a claim about a population mean or mean difference based on an appropriate confidence interval.
8. Identify the relationships among sample size, confidence interval width, confidence level, and margin of error for a population mean or population mean difference.

4.2.1 Using a normal distribution for inference when σ is known

In Section 4.1 we saw that the distribution of a sample mean is normal if the population distribution is normal or if the sample size is at least 30. In these problems, we used the population mean and population standard deviation to find a Z-score. However, in the case of inference, these values will be unknown. In rare circumstances we may know the standard deviation of a population, even though we do not know its mean. For example, in some industrial processes, the mean may be known to shift over time, while the standard deviation of the process remains the same. In these cases, we can use the normal model as the basis for our inference procedures. We use $\bar{x}$ as our point estimate for μ and the SD formula for a sample mean calculated in Section 4.1: $\sigma_{\bar{x}} = \frac{\sigma}{\sqrt{n}}$. That leads to a confidence interval and a test statistic as follows:

$$\text{CI: } \bar{x} \ \pm \ z^{\star}\frac{\sigma}{\sqrt{n}} \qquad\qquad Z = \frac{\bar{x} - \text{null value}}{\frac{\sigma}{\sqrt{n}}}$$

How do we evaluate $\sigma_{\bar{x}}$ if we do not know the population standard deviation σ, as is usually the case? The best we can do is use the sample standard deviation, denoted by s, to estimate σ. This gives us an estimate of $\sigma_{\bar{x}}$ which we call the Standard Error (SE) of $\bar{x}$.

$$SE_{\bar{x}} = \frac{s}{\sqrt{n}}$$

However, when we do this we run into a problem: when carrying out our inference procedures, we will be trying to estimate *two* quantities: both the mean and the standard deviation. Looking at the SD and SE formulas, we can make some important observations that will give us a hint as to what will happen when we use s instead of σ.

- For a given population, σ is a fixed number and does not vary.
- s, the standard deviation of a sample, will vary from one sample to the next and will not be exactly equal to σ.
- The larger the sample size n, the better the estimate s will tend to be for σ.

For this reason, the normal model may work reasonably well when the sample size is large. For smaller sample sizes, we run into a problem: our use of s, which is used when computing the standard error, tends to add more variability to our test statistic. It is this extra variability that leads us to a new distribution: the t-distribution.

4.2.2 Introducing the t-distribution

When we use the sample standard deviation s in place of the population standard deviation σ to standardize the sample mean, we get a new distribution - one that is similar to the normal distribution, but has greater spread. This distribution is known as the t-distribution. A t-distribution, shown as a solid line in Figure 4.7, has a bell shape. However, its tails are thicker than the normal model's. We can see that a greater proportion of the area under the t-distribution is beyond 2 standard units from 0 than under the normal distribution. These extra thick tails account for the extra variation introduced when we estimate σ with the sample standard deviation s.

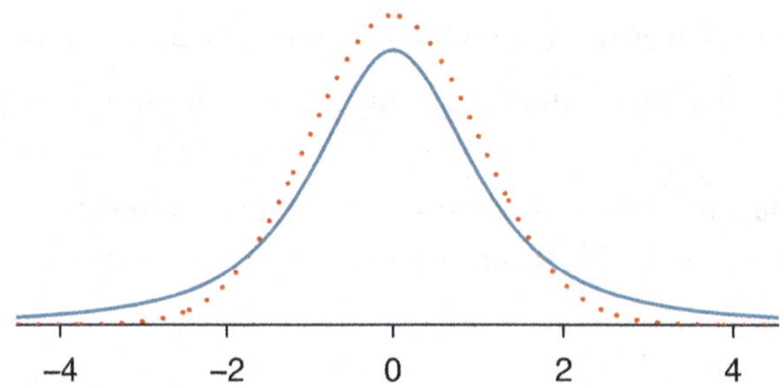

Figure 4.7: Comparison of a t-distribution (solid line) and a normal distribution (dotted line).

The t-distribution, always centered at zero, has a single parameter: degrees of freedom. The **degrees of freedom (df)** describes the precise form of the bell-shaped t-distribution. Several t-distributions are shown in Figure 4.8. When there are more degrees of freedom, the t-distribution looks more like the standard normal distribution.

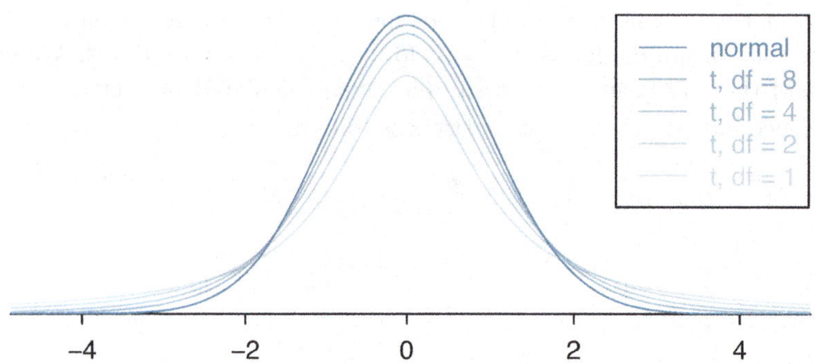

Figure 4.8: The larger the degrees of freedom, the more closely the t-distribution resembles the standard normal distribution.

DEGREES OF FREEDOM

t-distributions are identified using a parameter known as the degrees of freedom (df), which is based on the sample size(s). When the degrees of freedom are small, the t-distribution has a much narrower peak and fatter tails than a normal distribution. The larger the degrees of freedom, the more closely the t-distribution resembles the standard normal distribution.

When the degrees of freedom is large, about 30 or more, the t-distribution is nearly indistinguishable from the normal distribution. In Section 4.2.4, we will see how degrees of freedom relates to sample size.

We will find it useful to become familiar with the t-distribution, because it plays a very similar role to the normal distribution during inference. We use a **t-table**, partially shown in Figure 4.9, in place of the normal probability table when the population standard deviation is unknown, especially when the sample size is small. A larger table is presented in Appendix C.3.

	one tail	0.100	0.050	0.025	0.010	0.005
df	1	3.078	6.314	12.71	31.82	63.66
	2	1.886	2.920	4.303	6.965	9.925
	3	1.638	2.353	3.182	4.541	5.841
	⋮	⋮	⋮	⋮	⋮	
	17	1.333	1.740	2.110	2.567	2.898
	18	1.330	1.734	2.101	2.552	2.878
	19	1.328	1.729	2.093	2.539	2.861
	20	1.325	1.725	2.086	2.528	2.845
	⋮	⋮	⋮	⋮	⋮	
	1000	1.282	1.646	1.962	2.330	2.581
	∞	1.282	1.645	1.960	2.326	2.576
Confidence level C		80%	90%	95%	98%	99%

Figure 4.9: An abbreviated look at the t-table. Each row represents a different t-distribution. The columns describe the cutoffs for specific tail areas. The row with $df = 18$ has been highlighted.

Each row in the t-table represents a t-distribution with different degrees of freedom. The columns correspond to tail probabilities. For instance, if we know we are working with the t-distribution with $df = 18$, we can examine row 18, which is highlighted in Figure 4.9. If we want the value in this row that identifies the cutoff for an upper tail of 10%, we can look in the column where *one tail* is 0.100. This cutoff is 1.330. If we had wanted the cutoff for the lower 10%, we would use -1.330. Just like the normal distribution, all t-distributions are symmetric.

E

EXAMPLE 4.12

For the t-distribution with 18 degrees of freedom as shown in Figure 4.10, what percent of the distribution is contained between -1.734 and +1.734? Use the partial t-table shown in Figure 4.9.

Using row $df = 18$, we find 1.734 in the table. The area in each tail is 0.050 for a total of 0.100, which leaves 0.900 in the middle between -1.734 and +1.734. This corresponds to 90%, which can be found at the very bottom of that column.

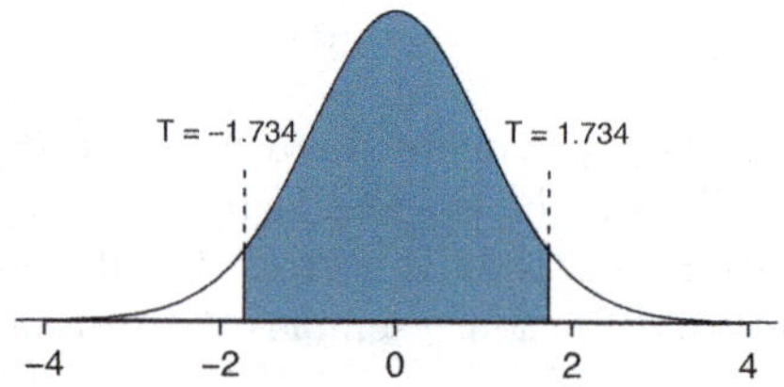

Figure 4.10: The t-distribution with 18 degrees of freedom. The area between -1.734 and 1.734 is shaded.

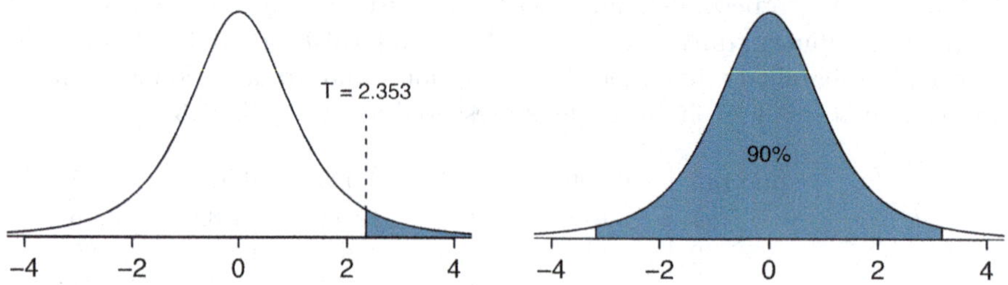

Figure 4.11: The t-distribution with 3 degrees of freedom is shown twice. Left: the area above 2.353 is shaded. Right: the middle 95% of the area is shaded.

EXAMPLE 4.13

For the t-distribution with 3 degrees of freedom, as shown in the left panel of Figure 4.11, what proportion of the distribution falls above 2.353?

To find this area, we identify the appropriate row: $df = 3$. Then we identify the column containing the absolute value of 2.353; it is the third column. Because we are looking for just one tail, we examine the top line of the table, which shows that a one tail area for a value in the third row corresponds to 0.05. That is, 5% of the distribution falls above 2.353.

EXAMPLE 4.14

For the t-distribution with 3 degrees of freedom, as shown in the right panel of Figure 4.11, what should the value of $t^\star$ be so that 95% of the area of the distribution falls between $-t^\star$ and $+t^\star$?

We can look at the column in the t-table that says 95% along the bottom row and trace it up to row $df = 3$ to find that $t^\star = 3.182$.

GUIDED PRACTICE 4.15

Without doing any calculations, will the area to the right of $Z = 3$ under the standard normal distribution be greater than, less than, or equal to the area to the right of $t = 3$ under the t-distribution with 35 degrees of freedom?[6]

When the desired degrees of freedom is not listed on the table, choose a conservative value: round the degrees of freedom down, i.e. move *up* to the previous row listed. Another option is to use technology to get a more precise answer. See Section 4.2.3 for finding areas and boundary values for t-distributions using technology.

[6]Because the t-distribution has greater spread and thicker tails than the normal distribution, we would expect the upper tail area to the right of $Z = 3$ under the standard normal distribution to be less than the upper tail area to the right of $t = 3$ under the t distribution with 35 degrees of freedom. One can confirm that the area to the right of $Z = 3$ is 0.0013, which is less than 0.0025. With a smaller degrees of freedom, this difference would be even more pronounced. Try it!

4.2.3 Technology: t-distribution probabilities and boundary values

Given a t-distribution with 3 degrees of freedom:

(i) Find the area to the right of $t = 1$.

(ii) Find the boundary value $t^\star$ such that 95% of the area is between $-t^\star$ and $t^\star$.

Desmos: Use the `tdist(df)` function, replacing `df` with the degrees of freedom.

1. Type `tdist(3)`.
2. Click the triangle next to `Cumulative Probability`.
3. Choose `Inner`, `Outer`, `Left` or `Right` as illustrated below.
   (i) To find a probability/area, choose `Area` and enter the boundary value(s).
   (ii) To find a boundary value, choose `Bounds` and enter the desired proportion to the right of the = sign.
4. Click the magnifying glass to Zoom Fit the graphing window.

(i) Finding probabilities/areas.

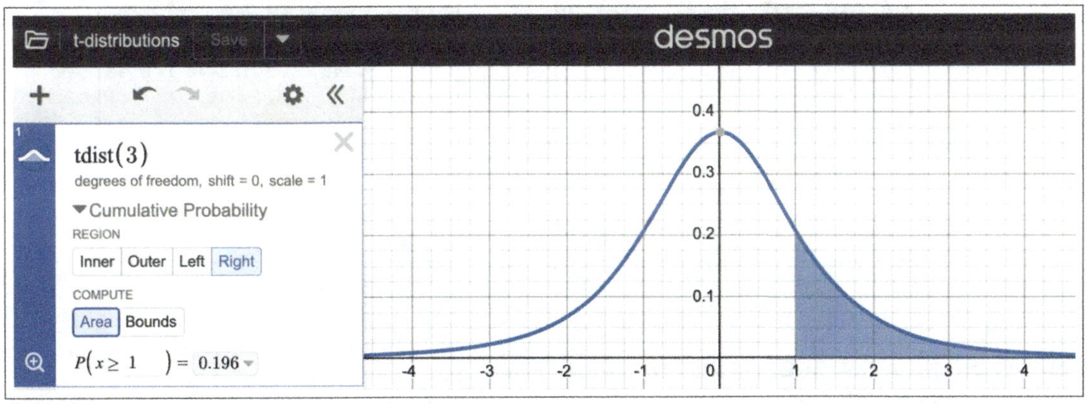

(ii) Finding boundary value(s).

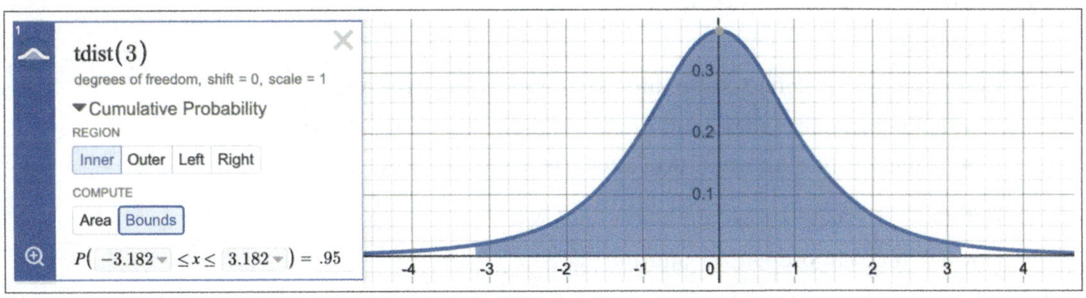

R: Probabilities and boundary values for a t-distribution with a given df.

(i) `pt(q, df)` gives the area to the left of `q`, so we add `lower.tail = FALSE` to get the area to the right.

```
> pt(1, df = 3, lower.tail = FALSE)
[1] 0.1955011
```

(ii) `qt(p, df)` gives the $t^\star$ value that has the probability `p` to the *left* of it, unless specified otherwise. To have 95% in the middle implies that there is 2.5% in the upper (and lower) tail.

```
> qt(0.025, df = 3, lower.tail = FALSE)
[1] 3.182446
```

Calculator: NumWorks calculator instructions and example output are included. For the TI-83/84 and Casio calculators, general instructions are provided, and worked example videos can be accessed via the icon or at **openintro.org/ti** and **openintro.org/casio**.

NUMWORKS: AREA AND BOUNDARY VALUES FOR THE T-DISTRIBUTION

Use `OK` or `EXE` to make a selection.

1. From the home screen, select `Distributions`, arrow down and choose `Student's t`. If a list of distributions does not appear, click the ↩ button in the upper right as many times as needed.
2. Enter the `Degrees of freedom`. Hit the down arrow and choose `Next`.
3. Hit the left arrow to highlight the graph. Hit the down arrow to choose whether you want left, inner, or right, then hit `OK`. Hit the right arrow to enter desired values.
   (i) For a probability, enter the boundary value(s), then hit `EXE`.
   (ii) For a boundary value, enter the desired area as a decimal to the right of the = sign, then hit `EXE`.

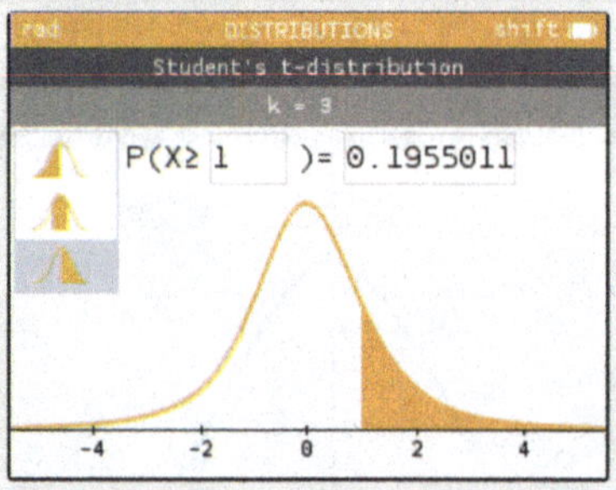

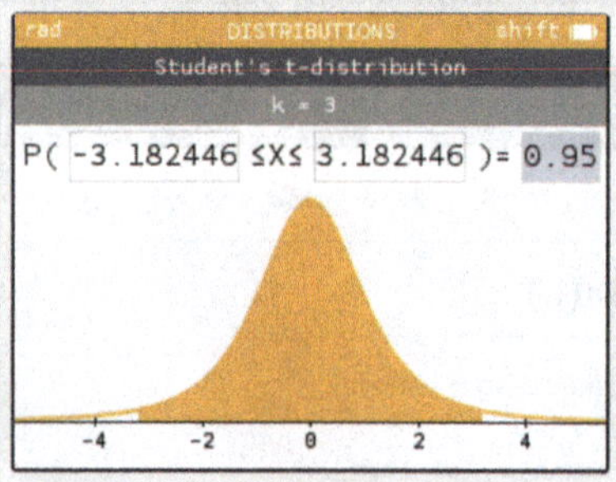

TI-84: FINDING AREA UNDER THE T-DISTRIBUTION

Use `2ND VARS`, `tcdf` to find an area/proportion/probability between two t-scores or to the left or right of a t-score.

1. Choose `2ND VARS` (i.e. `DISTR`).
2. Choose `6:tcdf`.
3. Enter the `lower` (left) t-score and the `upper` (right) t-score.
   - If finding just a lower tail area, set `lower` to `-∞` (−1E99).
   - If finding just an upper tail area, set `upper` to ∞ (−1E99).
4. Enter the degrees of freedom after `df:`.
5. Down arrow, choose `Paste`, and hit `ENTER`.

TI-83: Do steps 1-2, then enter lower bound, upper bound, and df separated by commas as follows: `tcdf(lower, upper, df)`. Then hit `ENTER`.

4.2.4 Checking conditions for inference on a mean using the t-distribution

Using the t-distribution for inference on a mean requires that the theoretical **sampling distribution of the sample mean $\bar{x}$ is nearly normal**. In practice, we check whether this assumption is reasonable by verifying that certain conditions are met.

Independence. Observations can be considered independent when the data are collected from a *random process*, such as rolling a die, or from a *random sample*. Without a random sample or process, the standard error formula would not apply, and it is unclear to what population the inference would apply. Recall that when sampling without replacement from a finite population, the observations can be considered independent when sampling less than 10% of the population.

Large sample / normal population. We saw in Section 4.1 that in order for the sampling distribution of a sample mean to be nearly normal, we also need the sample to be drawn from a nearly normal population or we need the sample size to be at least 30 ($n \geq 30$).

What should we do when the sample size is small and we are not sure whether the population distribution is nearly normal? In this case, a good practice is to check for excessive skew or clear outliers in the data, which would provide evidence that the population distribution from which we sampled is not nearly normal. If the data do not show obvious skew or outliers and we do not have reason to expect non-normality based on context-relevant knowledge (e.g. salaries are not normally distributed), then the idea of a nearly normal population is generally considered *reasonable*.

Note that by looking at a small data set, we cannot *prove* that the population distribution is nearly normal. However, the data can suggest to us whether the population distribution being nearly normal is an unreasonable assumption.

THE NORMALITY CONDITION WITH SMALL SAMPLES

If the sample is small and there is strong skew or extreme outliers in the data, the population from which the sample was drawn may not be nearly normal.

Ideally, we use a graph of the data to check for strong skew or outliers. When the full data set is not available, summary statistics can also be used.

For larger samples, it is less necessary to check for skew in the data. In general, when the sample size is 30 or more, it is no longer necessary that the population distribution be nearly normal. When the sample size is large, the Central Limit Theorem tells us that the sampling distribution of the sample mean will be nearly normal regardless of the distribution of the population.

EXAMPLE 4.16

Sometimes we do not know what the population distribution looks like. We have to infer it based on the distribution of a single sample. Figure 4.12 shows a histogram of 20 observations. These represent winnings and losses from 20 consecutive days of a professional poker player. Based on this sample data, can the normal approximation be applied to the distribution of the sample mean?

We should consider each of the required conditions.

(1) These are referred to as **time series data**, because the data arrived in a particular sequence. If the player wins on one day, it may influence how she plays the next. To make the assumption of independence we should perform careful checks on such data.

(2) The sample size is 20, which is smaller than 30.

(3) There are two outliers in the data, both quite extreme, which suggests the population may not be normal and instead may be very strongly skewed or have distant outliers. Outliers can play an important role and affect the distribution of the sample mean and the estimate of the standard deviation of the sample mean.

Since we should be skeptical of the independence of observations and the extreme upper outliers pose a challenge, we should not use the normal model for the sample mean of these 20 observations. If we can obtain a much larger sample, then the concerns about skew and outliers would no longer apply.

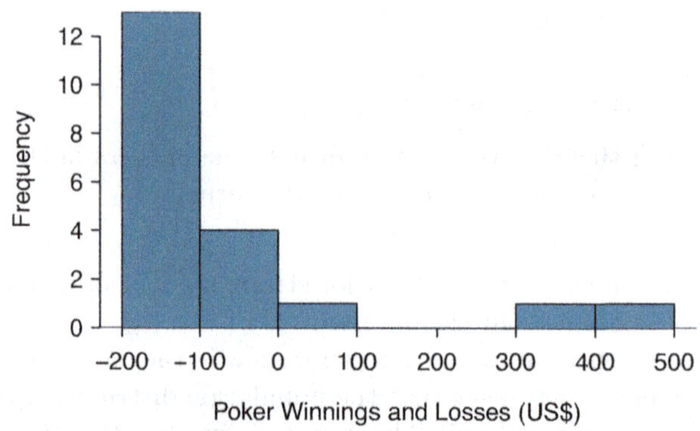

Figure 4.12: Sample distribution of poker winnings. These data include two very clear outliers. These are problematic when considering the normality of the sample mean. For example, outliers are often an indicator of very strong skew.

EXAMINE DATA STRUCTURE WHEN CONSIDERING INDEPENDENCE

Some data sets are collected in such a way that they have a natural underlying structure between observations, e.g. when observations occur consecutively. Be especially cautious about independence assumptions regarding such data sets.

WATCH OUT FOR STRONG SKEW AND OUTLIERS

Strong skew in the population is often identified by the presence of clear outliers in the data. If a data set has prominent outliers, then a larger sample size will be needed for the sampling distribution of $\bar{x}$ to be normal. There are no simple guidelines for what sample size is big enough for each situation. However, we can use the rule of thumb that, in general, an n of at least 30 is sufficient for most cases.

4.2.5 One-sample t-interval for a mean

Dolphins are at the top of the oceanic food chain, which causes dangerous substances such as mercury to concentrate in their organs and muscles. This is an important problem for both dolphins and other animals who eat them.

Figure 4.13: A Risso's dolphin.

We would like to create a confidence interval to estimate the average mercury content in dolphin muscles. We will use a sample of 19 Risso's dolphins from the Taiji area in Japan. The data are summarized in Figure 4.14.

n	$\bar{x}$	s	minimum	maximum
19	4.4	2.3	1.7	9.2

Figure 4.14: Summary of mercury content in the muscle of 19 Risso's dolphins from the Taiji area. Measurements are in μg/wet g (micrograms of mercury per wet gram of muscle).

Because we are estimating a mean, we would like to construct a t-interval, but first we must check whether the conditions for using a t-interval are met. We will start by assuming that the sample of 19 Risso's dolphins constitutes a random sample. Next, we note that the sample size is small (less than 30), and we do not know whether the distribution of mercury content for all dolphins is nearly normal. Therefore, we must look at the data. Since we do not have all of the data to graph, we look at the summary statistics provided in Figure 4.14. With such a small sample, these summary statistics do not suggest extreme skew or extreme outliers; all observations are within 2.5 standard deviations of the mean. Based on this evidence, we believe it is reasonable that the population distribution of mercury content in dolphins could be nearly normal.

With both conditions met, we will construct a 95% confidence interval. Recall that a confidence interval has the following form:

$$\text{point estimate} \pm \text{critical value} \times SE \text{ of estimate}$$

The point estimate is the sample mean and the SE of the sample mean is given by $s/\sqrt{n}$. What do we use for the critical value? Since we are using the t-distribution, we use a t-table or technology to find the critical value. We denote the critical value $t^\star$.

- For a 95% confidence interval, we want to find the cutoff $t^\star$ such that 95% of the t-distribution is between $-t^\star$ and $t^\star$.
- Using the t-table on page 329, we look at the row that corresponds to the degrees of freedom and the column that corresponds to the confidence level.

DEGREES OF FREEDOM FOR A SINGLE SAMPLE

If the sample has n observations and we are examining a single mean, then we use the t-distribution with $df = n - 1$.

EXAMPLE 4.17

Calculate a 95% confidence interval for the average mercury content in dolphin muscles based on this sample. Recall that $n = 19$, $\bar{x} = 4.4$ μg/wet g, and $s = 2.3$ μg/wet g.

To find the critical value $t^\star$ we use the t-distribution with $n - 1$ degrees of freedom. The sample size is 19, so $df = 19 - 1 = 18$ degrees of freedom. Using the t-table with row $df = 18$ and column corresponding to a 95% confidence level, we get $t^\star = 2.10$. The point estimate is the sample mean $\bar{x}$ and the standard error of a sample mean is given by $\frac{s}{\sqrt{n}}$. Now we have all the pieces we need to calculate a 95% confidence interval for the average mercury content in dolphin muscles.

$$\text{point estimate} \pm \text{critical value} \times SE \text{ of estimate}$$

$$\bar{x} \pm t^\star \times \frac{s}{\sqrt{n}} \qquad df = n - 1$$

$$4.4 \pm 2.10 \times \frac{2.3}{\sqrt{19}} \qquad df = 18$$

$$(3.29,\ 5.51)$$

EXAMPLE 4.18

How do we interpret this 95% confidence interval? To what population is it applicable?

A random sample of Risso's dolphins was taken from the Taiji area in Japan. The mercury content in the muscles of other types of dolphins and from dolphins from other regions may vary. Therefore, we can only make an inference to Risso's dolphins from this area. We are 95% confident that the interval (3.29, 5.51) contains the true average mercury content (μg/wet gram) in the muscles of Risso's dolphins in the Taiji area of Japan. We can also say that we are 95% confident the true average mercury content in the muscles of Risso's dolphins in the Taiji area of Japan is between 3.29 and 5.51 μg/wet gram.

EXAMPLE 4.19

Someone makes a claim that the mean mercury content in the muscles of Risso's dolphins in the Taiji area of Japan is 6.0 μg/wet gram. Based on the calculated confidence interval, do you have evidence against this claim?

Because 6.0 is not in the interval, we do have evidence, at the 95% confidence level, against this claim. Because the entire interval is below 6.0, we have evidence that the true mean mercury content is less than 6.0 μg/wet gram.

E

EXAMPLE 4.20

Intepret the confidence level of 95%

In repeated random sampling of this size from this population, approximately 95% of the confidence intervals created will capture the true mean mercury content in the muscles of Risso's dolphins in the Taiji area of Japan.

4.2.6 Estimating a mean of differences

Do course books tend to be cheaper on Amazon or at a college bookstore? How big is this difference, on average? We investigate a specific example involving books for UCLA courses and comparing their price on Amazon and at the UCLA Bookstore. In an earlier edition of this textbook, we found that Amazon prices were, on average, lower than those of the UCLA Bookstore for UCLA courses in 2010. It's been awhile, and many stores have adapted to the online market, so we wondered, how is the UCLA Bookstore doing today?

We sampled 201 UCLA courses. Of those, 68 required books that could be found on Amazon. A portion of the data set from these courses is shown in Figure 4.15, where prices are in U.S. dollars.

	subject	course_number	bookstore	amazon	price_difference
1	American Indian Studies	M10	47.97	47.45	0.52
2	Anthropology	2	14.26	13.55	0.71
3	Arts and Architecture	10	13.50	12.53	0.97
⋮	⋮	⋮	⋮	⋮	⋮
67	Korean	1	24.96	23.79	1.17
68	Jewish Studies	M10	35.96	32.40	3.56

Figure 4.15: Five cases of the `textbooks` data set.

Each textbook has two corresponding prices in the data set: one for the UCLA Bookstore and one for Amazon. Therefore, each textbook price from the UCLA bookstore has a natural correspondence with a textbook price from Amazon. When two sets of observations have this special correspondence, they are said to be **paired**.

PAIRED DATA

Two sets of observations are *paired* if each observation in one set has a special correspondence or connection with exactly one observation in the other data set.

To analyze paired data, it is often useful to look at the difference in outcomes of each pair of observations. In the `textbook` data set, we look at the differences in prices, which is represented as the `d` variable. Here, for each book, the differences are taken as

$$\text{UCLA Bookstore price} - \text{Amazon price}$$

It is important that we always subtract using a consistent order; here Amazon prices are always subtracted from UCLA prices. A histogram of these differences is shown in Figure 4.16. Using differences between paired observations is a common and useful way to analyze paired data.

G

GUIDED PRACTICE 4.21

The first difference shown in Figure 4.15 is computed as: 47.97 − 47.45 = 0.52. What does this difference tell us about the price for this textbook on Amazon versus the UCLA bookstore?[7]

[7]The difference is taken as UCLA Bookstore price − Amazon price. Because the difference is positive, it tells us that the UCLA Bookstore price was *greater* for this textbook. In fact, it was $0.52, or 52 cents, more expensive at the UCLA bookstore than on Amazon.

Figure 4.16: Histogram of the difference in price for each book sampled. These data are very strongly skewed.

$n_{_d}$	$\bar{x}_{_d}$	$s_{_d}$
68	3.58	13.42

Figure 4.17: Summary statistics for the price differences. There were 68 books, so there are 68 differences.

Paired data are often analyzed using the t-distribution, but before doing so it is important to verify that some conditions are met.

The observations are based on a random sample of books from a much larger population of books (more than 680 books), so independence is reasonable. While the distribution of the data is very strongly skewed, we do have $n = 68$ observations. This sample size is large enough that we do not have to worry about whether the population distribution for difference in price might be nearly normal or not. Because the conditions are satisfied, we can use the t-distribution in this setting.

We compute the standard error associated with $\bar{x}_d$ using the standard deviation of the differences ($s_{_d} = 13.42$) and the number of differences ($n_{_d} = 68$):

$$SE_{\bar{x}_d} = \frac{s_d}{\sqrt{n_d}} = \frac{13.42}{\sqrt{68}} = 1.63$$

We will construct a 95% confidence interval for the average price difference between books at the UCLA Bookstore and on Amazon. We must find the critical value, $t^\star$. Since $df = 67$ is not on the t-table, round the df down to 60 to get a $t^\star$ of 2.00 for 95% confidence, or use a technology option from Section 4.2.3 to get a $t^\star$ of 1.996 using 67 degrees of freedom. Plugging in the $t^\star$ value, point estimate, and standard error into the confidence interval formula, we get:

$$\text{point estimate} \ \pm \ t^\star \times SE \text{ of estimate}$$

$$3.58 \ \pm \ 1.996 \times \frac{13.42}{\sqrt{68}} \qquad df = 67$$

$$(0.33, 6.83)$$

We are 95% confident that the interval (0.33, 6.83) contains the true average price difference in UCLA course books (UCLA Bookstore – Amazon), that is, we are 95% confident that the UCLA bookstore is, on average, between \$0.33 and \$6.83 *more* expensive than Amazon for UCLA course books. Because our interval is entirely above 0, we have evidence that the true average difference is greater than zero, meaning we have evidence that, on average, books are more expensive at the UCLA Bookstore.

GUIDED PRACTICE 4.22

Based on the interval, should we recommend that UCLA students always buy their books on Amazon?[8]

4.2.7 Technology: the one-sample t-interval for μ

Section 4.3.4 demonstrates how to calculate the one-sample t-interval and the one-sample t-test (introduced in the next section) using Desmos, R, and the NumWorks, TI-83/84 and Casio calculator.

4.2.8 Summary and worked examples

CONSTRUCTING A CONFIDENCE INTERVAL FOR A MEAN

To carry out a complete confidence interval procedure to estimate a single population mean,

Identify: Identify the interval procedure, parameter, and confidence level.

Use a **one-sample t-interval for a population mean μ.** Define the population mean μ in words, referencing the population of interest. Choose a confidence level (C%).

When there is paired numerical data, use this same procedure to estimate μ_d, the mean of the population differences. In this case, use the mean and standard deviation of the sample differences, $\bar{x}_d$ and s_d, and the number of sample differences, n_d, when calculating the confidence interval.

Check: Check conditions for constructing a confidence interval using a t-distribution.

1. Independence: Data come from a random sample or random process. When sampling without replacement, check that sample size is less than 10% of the population size.
2. Large sample or normal population: $n \geq 30$ or the population distribution is nearly normal. If the sample size is less than 30 and the population distribution is unknown, check and confirm that there is no strong skew or outliers in the data in order to reasonably assume that the population distribution is nearly normal.

Calculate: Calculate the confidence interval and record it in interval form.

point estimate $\pm$ $t^{\star} \times SE$ of estimate, $df = n - 1$

point estimate: $\bar{x}$, the sample mean

SE of estimate: $\frac{s}{\sqrt{n}}$

$t^{\star}$: use technology or a t-table at row $df = n - 1$ and confidence level C%

(___, ___)

Conclude: Interpret the interval and, if applicable, draw a conclusion in context.

We are C% confident that the interval (___, ___) contains the true *mean* of [...]. A conclusion depends upon whether the interval is entirely above, is entirely below, or contains the value of interest.

[8]No, the fact that Amazon, is on average, less expensive does not imply that it is less expensive for *every* book. Examining the distribution in Figure 4.16, we see that there are many cases where the difference (UCLA Bookstore – Amazon) is negative, meaning that these books are *more* expensive on Amazon.

EXAMPLE 4.23

The FDA's webpage provides some data on mercury content of fish. Based on a sample of 15 croaker white fish (Pacific), a sample mean and standard deviation were computed as 0.287 and 0.069 ppm (parts per million), respectively. The 15 observations ranged from 0.18 to 0.41 ppm. Construct an appropriate 95% confidence interval for the true average mercury content of croaker white fish (Pacific). Is there evidence that the average mercury content is greater than 0.275 ppm? Use the four-step framework to organize your work.

Identify: Because the parameter to be estimated is a single mean, we will use a one-sample t-interval for a population mean μ. Here, μ is the true mean mercury content in croaker white fish (Pacific), and we will estimate this parameter at the 95% confidence level.

Check: We must check that the sampling distribution of the mean can be modeled using a normal distribution. We will assume that the sample constitutes a random sample of less than 10% of all croaker white fish (Pacific) and that independence is reasonable. The sample size n is small, but there are no obvious outliers; all observations are within 2 standard deviations of the mean. If there is skew, it is not too great. Therefore we think it is reasonable that the population distribution of mercury content in croaker white fish (Pacific) could be nearly normal.

(E) **Calculate:** We will calculate the interval:

$$\text{point estimate} \pm t^{\star} \times SE \text{ of estimate}$$

The point estimate is the sample mean: $\bar{x} = 0.287$.

SE of $\bar{x} = \frac{s}{\sqrt{n}} = \frac{0.069}{\sqrt{15}}$.

We find $t^{\star}$ for the one-sample t-interval using technology or using the t-table at row $df = n - 1$ and confidence level C%. For a 95% confidence level and $df = 15 - 1 = 14$, $t^{\star} = 2.145$. The 95% confidence interval is given by:

$$0.287 \pm 2.145 \times \frac{0.069}{\sqrt{15}} \qquad df = 14$$
$$0.287 \pm 2.145 \times 0.0178$$
$$(0.249,\ 0.325)$$

Conclude: We are 95% confident that the interval (0.249, 0.325) contains the true *average* mercury content of croaker white fish (Pacific). Because the interval contains 0.275 as well as values less than 0.275, we do not have evidence that the true average mercury content is greater than 0.275 ppm.

EXAMPLE 4.24

(E) Based on the interval calculated in Example 4.23 above, can we say that 95% of croaker white fish (Pacific) have mercury content between 0.249 and 0.325 ppm?

No. The interval estimates the *average* amount of mercury with 95% confidence. It is not trying to capture 95% of the values.

EXAMPLE 4.25

An SAT preparation company claims that its students' scores improve by over 100 points on average after their course. A consumer group would like to evaluate this claim, and they collect data on a random sample of 30 students who took the class. Each of these students took the SAT before and after taking the company's course, so we have a difference in scores for each student. We will examine these differences $x_1 = 57$, $x_2 = 133$, ..., $x_{30} = 140$ as a sample to evaluate the company's claim. The distribution of the differences has a mean of 135.9 and a standard deviation of 82.2. Construct a confidence interval to estimate the true average change in SAT after taking the company's course. Is there evidence at the 95% confidence level that students score an average of more than 100 points higher after the class? Use the four-step framework to organize your work.

Identify: Because we have paired data and the parameter to be estimated is a mean of differences, we will use a one-sample t-interval for a population mean μ_d. Here, μ_d represents the true mean of (SAT score after course – SAT score before course) for all students who would take the company's SAT prep course. We will estimate this parameter at the 95% confidence level.

Check: We have a random sample of students with paired observations on them. We will assume that these 30 students represent less than 10% of the total number of such students. Finally, the number of differences is $n_d = 30 \geq 30$, so we can proceed with the one-sample t-interval.

Calculate: We will calculate the confidence interval as follows.

$$\text{point estimate} \pm t^{\star} \times SE \text{ of estimate}$$

The point estimate is the sample mean of differences: $\bar{x}_d = 135.9$.

SE of $\bar{x}_d = \frac{s_d}{\sqrt{n_d}} = \frac{82.2}{\sqrt{30}} = 15.0$.

We find $t^{\star}$ for the one-sample case using the t-table at row $df = n - 1$ and confidence level C%. For a 95% confidence level and $df = 30 - 1 = 29$, $t^{\star} = 2.045$.

The 95% confidence interval is given by:

$$135.9 \pm 2.045 \times \frac{82.2}{\sqrt{30}} \qquad df = 29$$
$$135.9 \pm 2.045 \times 15.0$$
$$(105.2,\ 166.6)$$

Conclude: We are 95% confident that the interval (105.2, 166.6) contains the true *average* change in SAT score following the company's course. There is sufficient evidence that students score greater than 100 points higher, on average, after the company's course because the entire interval is above 100.

EXAMPLE 4.26

Based on the interval calculated in Example 4.25, can a random student be 95% confident that their SAT score will be 100 points higher if they take the company's course than if they do not take the company's course?

No, the interval estimates the *average* increase, not the increase of an individual student. Moreover, this is not an experiment - we did not randomize some students to take the SAT course and some to not take it and then compare the scores between the two groups. Instead, all students took the SAT course and each student's SAT score after the course was compared to their SAT score before the course. It is possible that scores just tend to go up when taking the SAT a second time.

Section summary

- t-distributions form a family of symmetric, bell-shaped, standardized distributions and are identified using a parameter known as the degrees of freedom (df), which is based on the sample size(s). When the degrees of freedom are small, the t-distribution has a much narrower peak and fatter tails than a normal distribution. As the degrees of freedom increase, the t-distributions more closely resemble the standard normal distribution.

- t-distributions are used for finding critical values and test statistics for inferences about a population mean μ when the population standard deviation σ is unknown and the sample standard deviation s must be used instead.

- The appropriate confidence interval for a population mean μ with unknown population standard deviation σ is a **one-sample t-interval for a population mean μ.** The parameter μ should be identified in context.

- Paired data can come from a random sample or a matched pairs experiment. For a matched pairs design, the appropriate analysis calculates *differences* between pairs of values to produce one sample of differences. The confidence interval procedure for the matched pairs design is a one-sample t-interval for a population mean difference μ_d. Use $\bar{x}_d$ for the sample mean difference, s_d for the standard deviation of sample differences, and n_d for the number of sample differences.

- The one-sample t-interval for μ or μ_d requires the following conditions are met.
  1. Independence: The data come from a random sample or random process. When sampling without replacement, check that the sample size is less than 10% of the population size.
  2. Large sample or normal population: $n \geq 30$ or population distribution is nearly normal. If the sample size is less than 30 and the population distribution is unknown, check and confirm that there is no strong skew or outliers in the data in order to reasonably assume that the population distribution is nearly normal.

- The general form for a confidence interval is: point estimate ± critical value × SE of estimate.

- A C% one-sample t-interval for μ can be written as: $\bar{x} \pm t^\star \frac{s}{\sqrt{n}}$, with $df = n - 1$. $t^\star$ is the critical value for the central C% of a t-distribution with $n - 1$ degrees of freedom.

- The SE of a sample mean is: $\frac{s}{\sqrt{n}}$.

- The margin of error of a sample mean is: $t^\star \frac{s}{\sqrt{n}}$.

- Because the confidence interval is based on a sample, the point estimate has associated error and the confidence interval may or may not contain the true value of the population mean.

- The interpretation of the confidence level C% is that in repeated random sampling with the same sample size from the same population, approximately C% of confidence intervals created will capture the population mean or population mean difference.

- We say we are C% confident that a particular interval (___, ___) contains the true population mean or population mean difference.

- A confidence interval provides a range of plausible values for a parameter and can be used as evidence to justify a claim about a population proportion. At a particular confidence level, we say that values are considered reasonable if they are inside the confidence interval and that values are considered unreasonable if they are outside the confidence interval.

- For a given sample, increasing the confidence level will result in a larger critical value, a larger margin of error, and a wider confidence interval.

- Increasing the sample size n decreases the standard error of $\bar{x}$ and, when all other things remain the same, decreases the width of a confidence interval for μ. The width of the interval is approximately proportional to $\frac{1}{\sqrt{n}}$.

Exercises

4.13 Identify the critical t. A random sample is selected from an approximately normal population with unknown standard deviation. Find the degrees of freedom and the critical t-value (t^*) for the given sample size and confidence level.

(a) $n = 6$, CL = 90%
(b) $n = 21$, CL = 98%
(c) $n = 29$, CL = 95%
(d) $n = 12$, CL = 99%

4.14 t-distribution. The figure on the right shows three unimodal and symmetric curves: the standard normal (z) distribution, the t-distribution with 5 degrees of freedom, and the t-distribution with 1 degree of freedom. Determine which is which, and explain your reasoning.

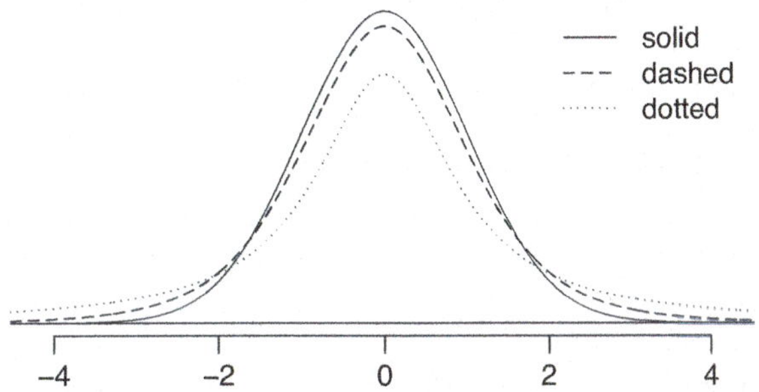

4.15 Working backwards, Part I. A 95% confidence interval for a population mean, μ, is given as (18.985, 21.015). This confidence interval is based on a simple random sample of 36 observations. Calculate the sample mean and standard deviation. Assume that all conditions necessary for inference are satisfied. Use the t-distribution in any calculations.

4.16 Working backwards, Part II. A 90% confidence interval for a population mean is (65, 77). The population distribution is approximately normal and the population standard deviation is unknown. This confidence interval is based on a simple random sample of 25 observations. Calculate the sample mean, the margin of error, and the sample standard deviation.

4.17 Exclusive relationships. A survey conducted on a reasonably random sample of 203 undergraduates asked, among many other questions, about the number of exclusive relationships these students have been in. The histogram below shows the distribution of the data from this sample. The sample average is 3.2 with a standard deviation of 1.97.

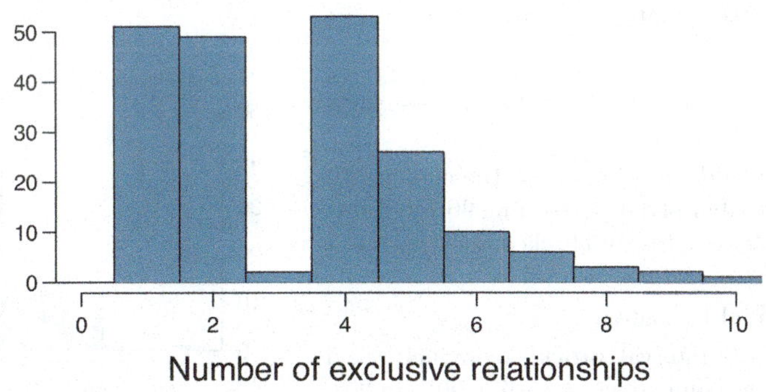

Estimate the average number of exclusive relationships Duke students have been in using a 90% confidence interval and interpret this interval in context. Check any conditions required for inference, and note any assumptions you must make as you proceed with your calculations and conclusions.

4.18 Forest management. Forest rangers wanted to better understand the rate of growth for younger trees in the park. They took measurements of a random sample of 50 young trees in 2009 and again measured those same trees in 2019. The data below summarize their measurements, where the heights are in feet:

	2009	2019	Differences
$\bar{x}$	12.0	24.5	12.5
s	3.5	9.5	7.2
n	50	50	50

Construct a 99% confidence interval for the average growth of (what had been) younger trees in the park over 2009-2019.

4.19 Paired or not? Part I. In each of the following scenarios, determine if the data are paired.

(a) Compare pre- (beginning of semester) and post-test (end of semester) scores of students.

(b) Assess gender-related salary gap by comparing salaries of randomly sampled men and women.

(c) Compare artery thicknesses at the beginning of a study and after 2 years of taking Vitamin E for the same group of patients.

(d) Assess effectiveness of a diet regimen by comparing the before and after weights of subjects.

4.20 Paired or not? Part II. In each of the following scenarios, determine if the data are paired.

(a) We would like to know if Intel's stock and Southwest Airlines' stock have similar rates of return. To find out, we take a random sample of 50 days, and record Intel's and Southwest's stock on those same days.

(b) We randomly sample 50 items from Target stores and note the price for each. Then we visit Walmart and collect the price for each of those same 50 items.

(c) A school board would like to determine whether there is a difference in average SAT scores for students at one high school versus another high school in the district. To check, they take a simple random sample of 100 students from each high school.

4.21 Sample size and pairing. Determine if the following statement is true or false, and if false, explain your reasoning: If comparing means of two groups with equal sample sizes, always use a paired test.

4.22 t^* vs. z^*. For a given confidence level, $t^\star_{df}$ is larger than $z^\star$. Explain how $t^\star_{df}$ being slightly larger than $z^\star$ affects the width of the confidence interval.

4.23 Global warming, Part I. Consider a limited set of climate data, examining temperature differences in 1948 vs 2018. We sampled 197 locations from the National Oceanic and Atmospheric Administration's (NOAA) historical data, where the data was available for both years of interest. We want to know: were there more days with temperatures exceeding 90°F in 2018 or in 1948?[9] The difference in number of days exceeding 90°F (number of days in 2018 - number of days in 1948) was calculated for each of the 197 locations. The average of these differences was 2.9 days with a standard deviation of 17.2 days. We are interested in determining whether these data provide strong evidence that there were more days in 2018 that exceeded 90°F from NOAA's weather stations.

(a) Discuss whether the conditions are met for a one-sample t-interval for a mean.

(b) Calculate a 90% confidence interval for the average difference between number of days exceeding 90°F between 1948 and 2018. We've already checked the conditions for you.

(c) Interpret the interval in context.

(d) Does the confidence interval provide convincing evidence that there were more days exceeding 90°F in 2018 than in 1948 at NOAA stations? Explain.

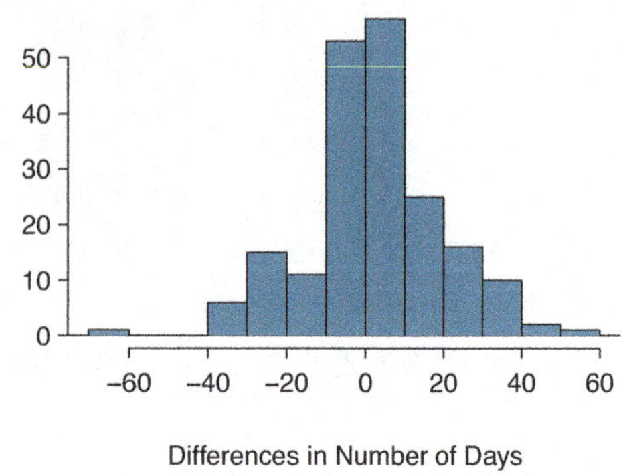

[9] NOAA, www.ncdc.noaa.gov/cdo-web/datasets, April 24, 2019.

4.24 High School and Beyond, Part I. The National Center of Education Statistics conducted a survey of high school seniors, collecting test data on reading, writing, and several other subjects. Here we examine a simple random sample of 200 students from this survey. Side-by-side box plots of reading and writing scores as well as a histogram of the differences in scores for those students are shown below. The mean and standard deviation of the differences are $\bar{x}_{read-write} = -0.545$ and $s_{read-write} = 8.887$ points.

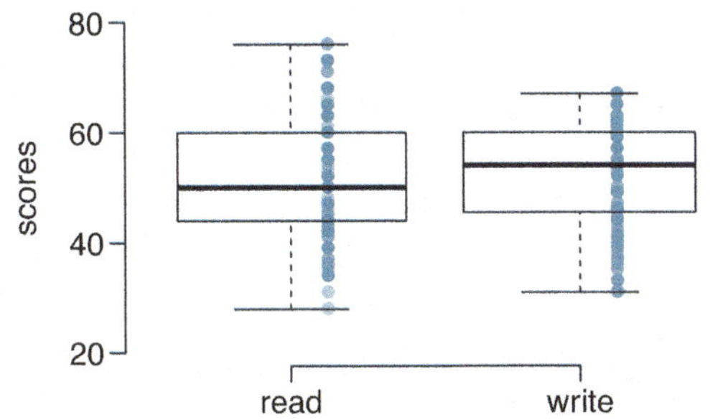

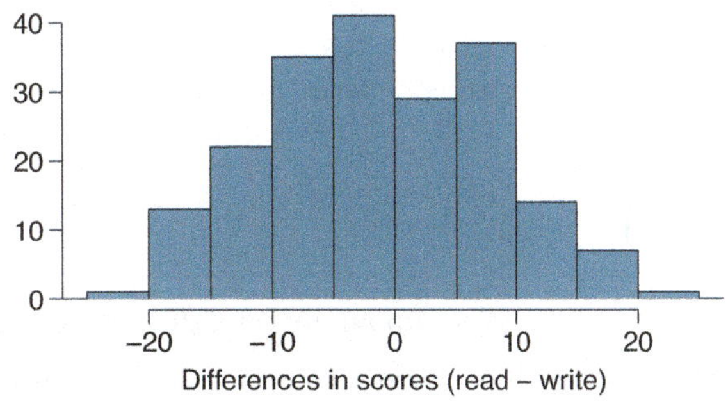

(a) Discuss whether conditions are met for a one-sample t-interval for a mean.

(b) Calculate a 95% confidence interval for the average difference between the reading and writing scores of all students.

(c) Interpret this interval in context.

(d) Does the confidence interval provide convincing evidence that there is a real difference in the average scores? Explain.

4.3 Hypothesis testing for a population mean

Is there evidence that the mean speed of U.S. runners has changed over time? Is there evidence that, on average, the Amazon price for course books is lower than the price at a college bookstore for those same books? In this section, we consider hypothesis testing for a mean and for a mean of differences. As with confidence intervals for a mean, we will use the t-distribution.

Learning objectives

1. Identify and set up an appropriate test for a population mean μ or mean difference μ_d.
2. Identify the null and alternative hypotheses for a population mean or population mean difference with unknown σ.
3. Justify the appropriateness of a hypothesis test for a population mean or mean difference by verifying conditions.
4. Calculate the test statistic, degrees of freedom and p-value for a test for a population mean or mean difference.
5. Interpret the p-value of a hypothesis test for a population mean or mean difference.
6. Justify a claim about a population mean or mean difference based on the results of a test.

4.3.1 Intro to hypothesis testing for a single mean

Is the typical U.S. runner getting faster or slower over time? Technological advances in shoes, training, and diet might suggest runners would be faster. An opposing viewpoint might say that with the average body mass index on the rise, people tend to run slower. In fact, all of these components might be influencing run time.

We consider this question in the context of the Cherry Blossom Race, which is a 10-mile race in Washington, DC each spring. The average time for all runners who finished the Cherry Blossom Race in 2006 was 93.3 minutes (93 minutes and about 18 seconds). We want to determine using data from 100 participants in the 2017 Cherry Blossom Race whether runners in this race are getting faster or slower, versus the other possibility that there has been no change. Figure 4.18 shows run times for 100 randomly selected participants.

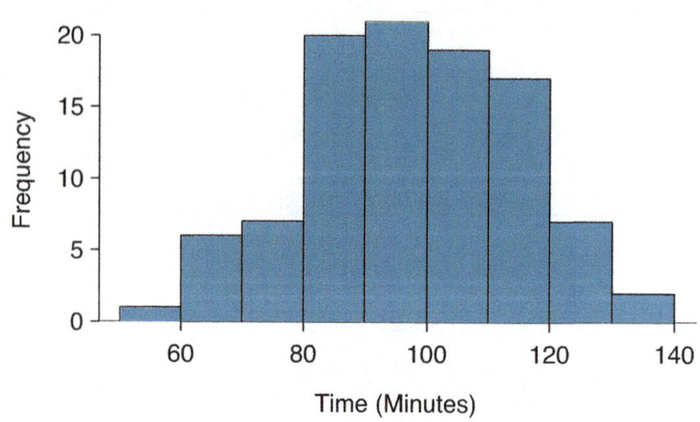

Figure 4.18: A histogram of `time` for the sample Cherry Blossom Race data.

EXAMPLE 4.27

What are appropriate hypotheses for this context?

We know that the average run time for all runners in 2006 was 93.3 minutes. We have a sample of times from the 2017 race. We are interested in whether the average run time has *changed*, so we will use a two-sided H_A.

Let μ represent the average 10-mile run time of all participants in 2017, which is unknown to us.

H_0: $\mu = 93.3$ minutes.
H_A: $\mu \neq 93.3$ minutes.

The data come from a random sample from a large population, so the observations are independent. Do we need to check for skew in the data? No – with a sample size of 100, well over 30, the Central Limit Theorem tells us that the sampling distribution of $\bar{x}$ will be nearly normal.

With independence satisfied and slight skew not a concern for this large of a sample, we can proceed with performing a hypothesis test using the t-distribution.

The sample mean and sample standard deviation of the 100 runners from the 2017 Cherry Blossom Race are 97.3 and 17.0 minutes, respectively. We want to know whether the observed sample mean of 97.3 is far enough away from 93.3 to provide convincing evidence of a real difference, or if it is within the realm of expected variation for a sample of size 100.

To answer this question we will find the test statistic and p-value for the hypothesis test. Since we will be using a sample standard deviation in our calculation of the test statistic, we will need to use a t-distribution, just as we did with confidence intervals for a mean. We call the test statistic a T-statistic. It has the same general form as a Z-statistic.

$$T = \frac{\text{point estimate} - \text{null value}}{SE \text{ of estimate}}$$

As we saw before, when carrying out inference on a single mean, the degrees of freedom is given by $n - 1$.

THE T-STATISTIC

The **T-statistic** (or T-score) is analogous to a Z-statistic (or Z-score). Both represent how many standard errors the observed value is from the null value.

EXAMPLE 4.28

Calculate the test statistic and degrees of freedom for this test.

Here, our point estimate is the sample mean, $\bar{x} = 97.3$ minutes.

The SE of $\bar{x} = \frac{s}{\sqrt{n}} = \frac{17.0}{\sqrt{100}} = 1.7$ minutes.

The null value is the value hypothesized in the null hypothesis: $\mu_0 = 93.3$ minutes.

$$T = \frac{97.3 - 93.3}{1.7} = 2.35 \qquad df = 100 - 1 = 99$$

EXAMPLE 4.29

Calculate the p-value and then interpret the p-value in context.

H_A is $\mu \neq 93.3$, so this is a two-tailed test. Using a technology option from Section 4.2.3, we find the p-value, which corresponds to the area below −2.35 plus the area above 2.35 under the *t*-distribution with 99 degrees of freedom. The p-value = 0.021.

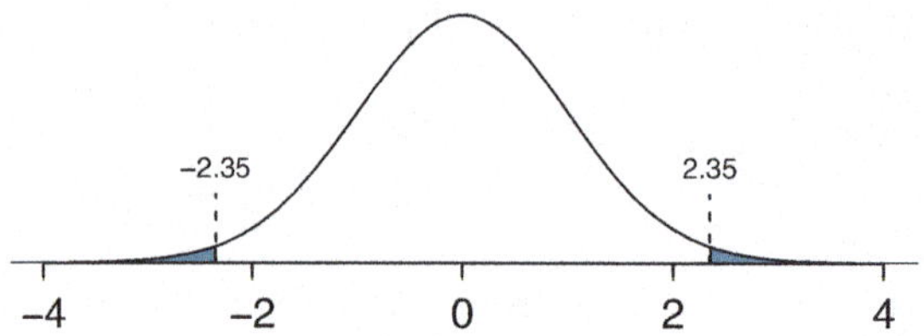

The p-value is the probability of getting data as extreme as we got assuming H_0 is true. In context, we can say that there is a 2.1% chance of getting a test statistic larger than 2.35 or less than −2.35 assuming the average 10-mile run time of all participants in 2017 really is 93.3 minutes.

EXAMPLE 4.30

Does the data provide sufficient evidence that the average Cherry Blossom Run time in 2017 is different than 93.3 min (the known value in 2006)?

This depends upon the desired significance level. Since the p-value = $0.02 < 0.05$, there is sufficient evidence at the 5% significance level. However, as the p-value of $0.02 > 0.01$, there is not sufficient evidence at the 1% significance level. This is why it is important to choose a significance level before seeing seeing the data and beginning the analysis.

EXAMPLE 4.31

Would you expect the hypothesized value of 93.3 to fall inside or outside of a 95% confidence interval? What about a 99% confidence interval?

Because the hypothesized value of 93.3 was rejected by the two-sided $\alpha = 0.05$ test, we would expect it to be outside the 95% confidence interval. However, because the hypothesized value of 93.3 was not rejected by the two-sided $\alpha = 0.01$ test, we would expect it to fall inside the (wider) 99% confidence interval.

4.3.2 Hypothesis testing for a mean of differences

Consider again the table summarizing data on: (UCLA Bookstore price − Amazon price), for each of the 68 books sampled.

n_d	$\bar{x}_d$	s_d
68	3.58	13.42

Figure 4.19: Summary statistics for the price differences. There were 68 books, so there are 68 differences.

We will set up and implement a hypothesis test to determine whether, on average, there is a difference in textbook prices between Amazon and the UCLA bookstore. We are considering two scenarios: there is no difference in prices or there is some difference in prices.

H_0: $\mu_d = 0$. On average, there is no difference in textbook prices.

H_A: $\mu_d \neq 0$. On average, there is some difference in textbook prices.

Conditions were checked in the previous section. We noted that the observations are based on a random sample of books from a large population of books (more than 680 books) and that the sample size is well above 30, thus satisfying the conditions for a one-sample t-test.

Next we compute the test statistic. The point estimate is the observed value of $\bar{x}_d$. The null value is the value hypothesized under the null hypothesis. Here, the null hypothesis is that the true mean of the differences is 0.

$$T = \frac{\text{point estimate} - \text{null value}}{SE \text{ of estimate}} = \frac{3.58 - 0}{\frac{13.42}{\sqrt{68}}} = \frac{3.58 - 0}{1.63} = 2.20$$

The degrees of freedom are $df = 68 - 1 = 67$. To visualize the p-value, the sampling distribution for $\bar{x}_d$ is drawn as though H_0 is true. This is shown in Figure 4.20. Equivalently, we can draw T distribution with 67 degrees of freedom, shading the area to the left of $T = -2.20$ and to the right of $T = 2.20$. Because this is a two-sided test, the p-value corresponds to the area in both tails. Using statistical software, we find the area in the tails to be 0.0312.

Because the p-value of 0.0312 is less than 0.05, we reject the null hypothesis. We have evidence that, on average, there is a difference in textbook prices. In particular, we can say that, on average, Amazon prices are lower than the UCLA Bookstore prices for UCLA course books.

EXAMPLE 4.32

The p-value for this two-sided test is 0.0312. Interpret this quantity in context.

(E) The p-value is the probability of getting data as extreme as we got assuming H_0 is true. In context, we can say that there is a 3.12% chance of getting a test statistic larger than 2.20 or less than −2.20 assuming there really is no difference, on average, between book prices at UCLA Bookstore and on Amazon. Equivalently, we can say that there is a 3.12% chance of getting a sample mean difference $\bar{x}_d$ greater than \$2.98 or less than −\$2.98 assuming there really is no difference, on average, between book prices at UCLA Bookstore and on Amazon.

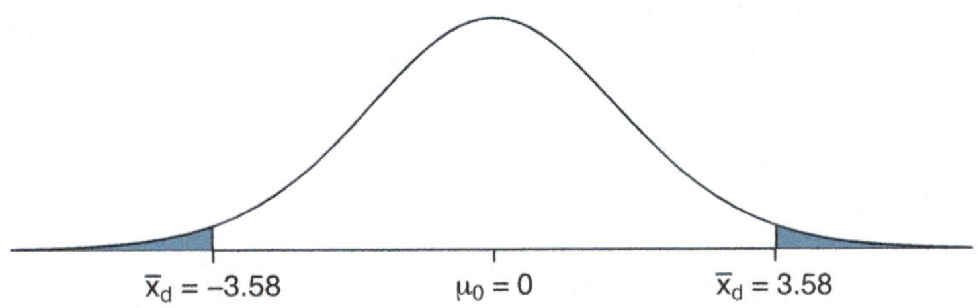

Figure 4.20: Sampling distribution of the mean difference in book prices, if the true average difference is zero. $\bar{x}$ values at least as extreme as our $\bar{x}$ of 3.58 are shaded.

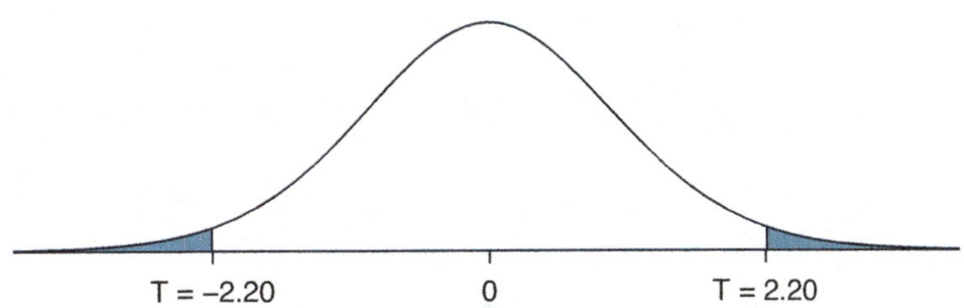

Figure 4.21: The T distribution with 67 degrees of freedom. T values at least as extreme as our T-statistic of 2.20 are shaded.

4.3.3 Summary and worked examples

HYPOTHESIS TEST FOR A MEAN

To carry out a complete hypothesis test to evaluate a claim about a population mean,

Identify: Identify the test procedure, parameter, significance level, and hypotheses.

Use a **one-sample t-test for a population mean μ.** Define the population mean μ in words, referencing the population of interest. Choose a significance level (α) and test the following hypotheses.

H_0: $\mu = \mu_0$

H_A: $\mu \neq \mu_0$; $\mu > \mu_0$; or $\mu < \mu_0$ (μ_0 is the null or hypothesized mean)

When there is paired numerical data, use this same procedure to test whether μ_d, the mean of the population differences, is zero. In this case, use the mean and standard deviation of the sample differences, $\bar{x}_d$ and s_d, and the number of sample differences, n_d, when calculating the test statistic.

Check: Check conditions for the test statistic to have a t-distribution, assuming H_0 is true.

1. Independence: Data come from a random sample or random process. When sampling without replacement, check that sample size is less than 10% of the population size.
2. Large sample or normal population: $n \geq 30$ or the population distribution is nearly normal. If the sample size is less than 30 and the population distribution is unknown, check and confirm that there is no strong skew or outliers in the data in order to reasonably assume that the population distribution is nearly normal.

Calculate: Calculate the t-statistic, df, and p-value.

$$T = \frac{\text{point estimate} - \text{null value}}{SE \text{ of estimate}}, \qquad df = n - 1$$

point estimate: $\bar{x}$, the sample mean

SE of estimate: $\frac{s}{\sqrt{n}}$

null value: μ_0

p-value = (based on the t-statistic, the df, and the direction of H_A)

Conclude: Compare the p-value to α, and draw a conclusion in context.

If the p-value is $\leq \alpha$, reject H_0; there is sufficient evidence that [H_A in context].

If the p-value is $> \alpha$, do not reject H_0; there is not sufficient evidence that [H_A in context].

EXAMPLE 4.33

In Section 4.2.6, we discussed an example involving the mercury content in croaker white fish (Pacific). Based on a sample of size 15, a sample mean and standard deviation were computed as 0.287 and 0.069 ppm (parts per million), respectively. Carry out an appropriate test to determine if 0.25 is a reasonable value for the average mercury content of croaker white fish (Pacific) using a 5% significance level. Use the four-step method to organize your work.

Identify: Because we are hypothesizing about a single mean we choose the one-sample t-test for a population mean μ. Here, μ is the true mean mercury content in croaker white fish (Pacific), and we test the following hypotheses at the $\alpha = 0.05$ significance level.

H_0: $\mu = 0.25$
H_A: $\mu \neq 0.25$

Check: The conditions were checked previously, namely – the data come from a random sample of less than 10% of the population of all croaker white fish (Pacific), and because n is less than 30, we checked and confirmed that there is no strong skew or outliers in the data, so the assumption that the population distribution of mercury is nearly normally distributed is reasonable.

Calculate: We will calculate the t-statistic, df, and p-value.

$$T = \frac{\text{point estimate} - \text{null value}}{SE \text{ of estimate}}$$

The point estimate is the sample mean: $\bar{x} = 0.287$, and the null value is: $\mu_0 = 0.25$.

SE of $\bar{x} = \frac{s}{\sqrt{n}} = \frac{0.069}{\sqrt{15}} = 0.0178$. We calculate the test statistic as follows:

$$T = \frac{0.287 - 0.25}{\frac{0.069}{\sqrt{15}}} = \frac{0.287 - 0.25}{0.0178} = 2.07 \qquad df = 15 - 1 = 14$$

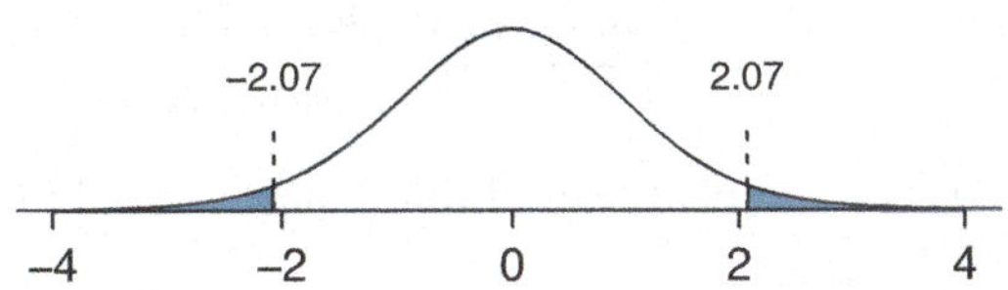

Because H_A is a two-tailed test ($\neq$), the p-value corresponds to the area to the right of 2.07 plus the area to the left of −2.07 under the t-distribution with 14 degrees of freedom. The p-value $= 2 \times 0.029 = 0.058$.

Conclude: The p-value of $0.058 > 0.05$, so we do not reject the null hypothesis. We do not have sufficient evidence that the average mercury content in croaker white fish (Pacific) is not 0.25.

GUIDED PRACTICE 4.34

Recall that the 95% confidence interval for the average mercury content in croaker white fish was (0.249, 0.325). Discuss whether the conclusion of the hypothesis test in the previous example is consistent or inconsistent with the conclusion of the confidence interval.[10]

---

[10]It is consistent because 0.25 is located (just barely) inside the confidence interval, so it is considered a reasonable value. Our hypothesis test did not reject the hypothesis that $\mu = 0.25$, also implying that it is a reasonable value. Note that the p-value was just over the cutoff of 0.05. This is consistent with the value of 0.25 being just inside the confidence interval. Also note that the hypothesis test did not *prove* that $\mu = 0.25$. The value 0.25 is just one of many reasonable values for the true mean.

EXAMPLE 4.35

An SAT preparation company claims that its students' scores improve by over 100 points on average after their course. A consumer group would like to evaluate this claim, and they collect data on a random sample of 30 students who took the class. Each of these students took the SAT before and after taking the company's course, so we have a difference in scores for each student. We will examine these differences $x_1 = 57$, $x_2 = 133$, ..., $x_{30} = 140$. The distribution of the differences has a mean of 135.9, a standard deviation of 82.2, and is shown below. Do the data provide convincing evidence to back up the company's claim? Use the four-step framework to organize your work.

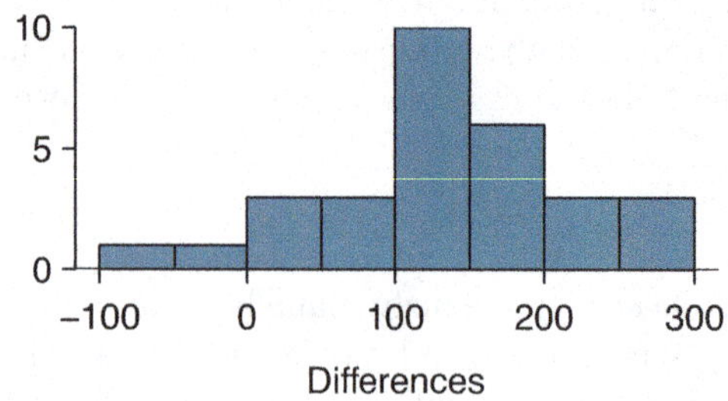

E

Identify: Because we have paired data and the parameter of interest is a mean of differences, we will use a one-sample t-test for a population mean μ_d. Here μ_d is the true mean of (SAT score after course − SAT score before course) for all students who would take the company's SAT prep course. We will test the following hypotheses at the $\alpha = 0.05$ level.

H_0: $\mu_d = 100$. On average, student scores improve by 100 points.

H_A: $\mu_d > 100$. On average, student scores improve by more than 100 points.

Check: We have a random sample of students and have paired data on them. We will assume that this sample of size 30 represents less than 10% of the total population of such students. Finally, the number of differences is $n_d = 30 \geq 30$, so we can proceed with the one-sample t-test.

Calculate: We will calculate the test statistic, df, and p-value.

$$T = \frac{\text{point estimate} - \text{null value}}{SE \text{ of estimate}}$$

The point estimate is the sample mean of differences: $\bar{x}_d = 135.9$

SE of $\bar{x}_d = \frac{s_d}{\sqrt{n_d}} = \frac{82.2}{\sqrt{30}} = 15.0$

$$T = \frac{135.9 - 100}{\frac{82.2}{\sqrt{30}}} = \frac{135.9 - 100}{15.0} = 2.4 \qquad df = 30 - 1 = 29$$

The p-value is the area to the right of 2.4 under the t-distribution with 29 degrees of freedom. The p-value $= 0.012$.

Conclude: p-value $= 0.012 < \alpha$ so we reject the null hypothesis. The data provide convincing evidence to support the company's claim that students' scores improve by more than 100 points, on average, following the class.

GUIDED PRACTICE 4.36

G

Because we found evidence to support the company's claim, does this mean that a student will score more than 100 points higher on the SAT if they take the class than if they do not take the class?[11]

[11]No. First, this is an observational study, so we cannot make a causal conclusion. Maybe SAT test takers tend to improve their score over time even if they don't take this SAT class. Second, the test considers the average. It does not imply that each student improved. With a sample standard deviation of 82.2 and a mean of 135.9, some students did worse after the SAT class, as shown in the histogram in Example 4.35.

4.3.4 Technology: the one-sample t-interval and t-test for μ

The data set `loan50`, introduced in Chapter 1, contains information on randomly sampled loans. Download the `loan50` CSV file from openintro.org/data. Open it and calculate a 95% confidence interval for the true mean of `loan_amount`. Also find the test statistic, df, and p-value for a test with the alternative hypothesis that the true mean is less than $20,000.

Desmos: Use the `ttest([data])` or `ttest(n, mean, stdev)` function as explained below.

1. Click + in the upper left, then choose `inference`.
2. Choose `t-test` in the pop-up window.

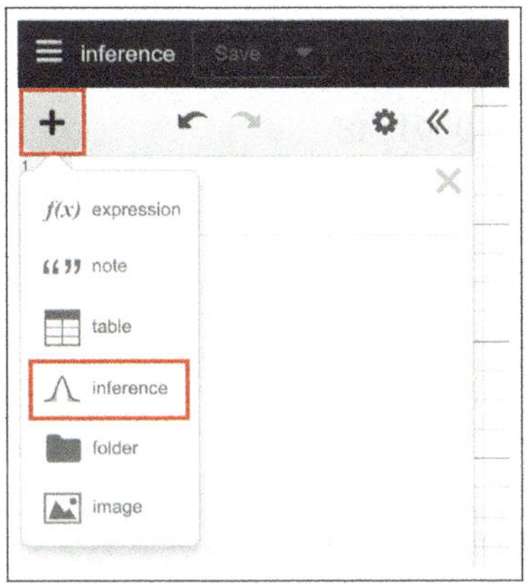

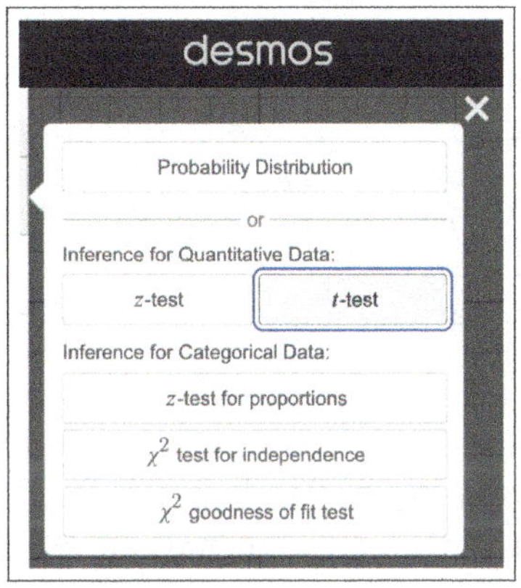

3. If you have all the data, enter the data separated by commas or copy and paste it in the box. Here we highlight the `loan_amount` column and paste it in the box. If you have the summary stats, click on `Stats` and then enter the `sample size`, `mean` and `stdev`. Click `Create Test`.

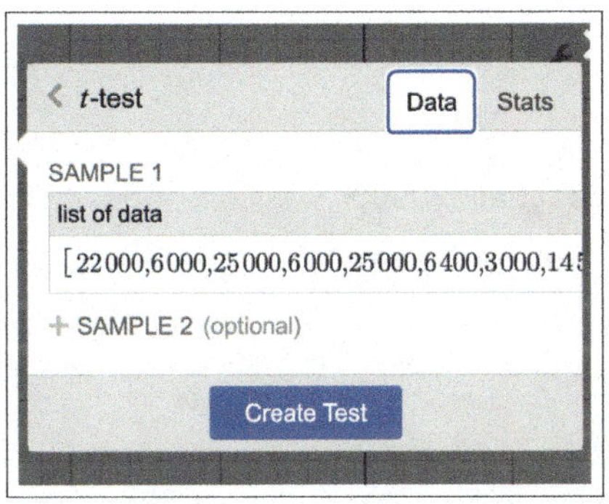

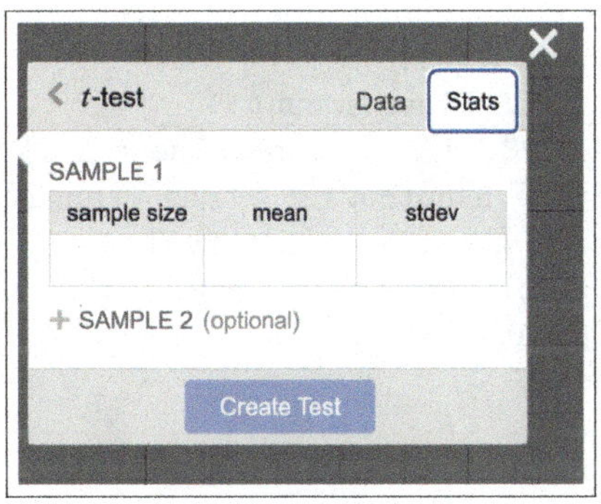

4. Click the triangle next to `Confidence Interval` and input the desired `Confidence level`. Here we use 0.95, which is entered by default. Click the ⋮ to the right of the confidence interval to more information.
5. Click the triangle next to `Significance Test`. Enter the hypothesized value for `mu` and select `Tails` to be `Left`, `Right` or `Both` depending on the direction of the alternative hypothesis. Here the hypothesized value of μ is 20,000 and H_A uses a <, so we select Tails to be Left.

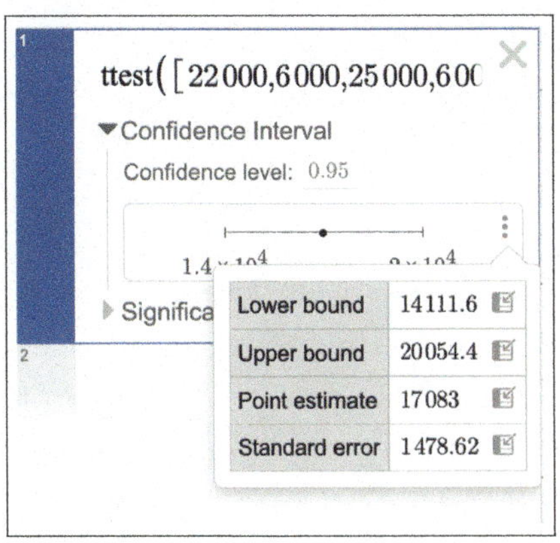

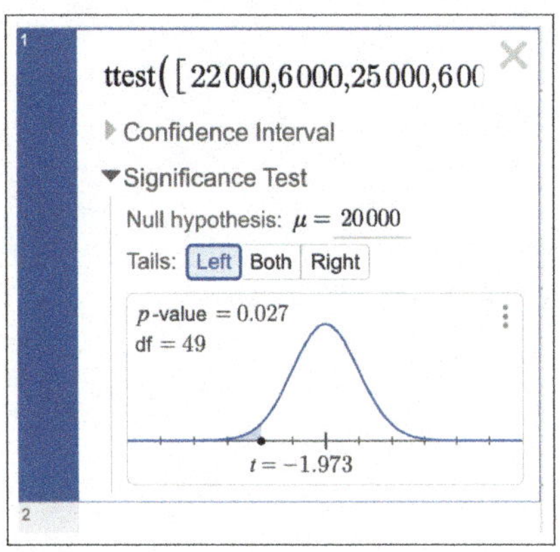

R: 1-sample t-interval/test for μ

First store the data into a variable as described on page 52.
`> loan_amount = scan()` Hit return, paste the numerical data, then hit return and return again.

You can also manually type in data as follows:
`> loan_amount = c(22000, 6000, 25000, 6000...)`

CONFIDENCE INTERVAL.
`t.test(data, conf.level = )`

```
> t.test(loan_amount, conf.level = 0.95)
One Sample t-test
data:  loan_amount
t = 11.553, df = 49, p-value = 0.000000000000001348
alternative hypothesis:  true mean is not equal to 0
95 percent confidence interval:
 14111.59 20054.41
sample estimates:
mean of x
17083
```

HYPOTHESIS TEST.
`t.test(data, mu = , alternative = "two.sided","greater","less")`

If a hypothesized value for μ is not entered, a default of 0 is used. If `alternative` is not specified, a default of `"two.sided"` is used. Here our alternative hypothesis is $\mu < 20,000$.

```
> t.test(loan_amount, mu = 20000, alternative = "less")
One Sample t-test
data:  loan50$loan_amount
t = -1.9728, df = 49, p-value = 0.02709
alternative hypothesis:  true mean is less than 20000
95 percent confidence interval:
-Inf 19561.99
sample estimates:
mean of x
17083
```

If you have installed the `openintro` package as described on page 52, you can skip the first step of storing the data into the variable `loan_amount`. Instead, simply reference `loan50$loan_amount`:
`> t.test(loan50$loan_amount, conf.level = 0.95)`

Calculator: NumWorks calculator instructions and example output are included. For the TI-83/84 and Casio calculators, general instructions are provided, and worked example videos can be accessed via the icon or at **openintro.org/ti** and **openintro.org/casio**.

NUMWORKS: 1-SAMPLE T-INTERVAL.

Use `OK` or `EXE` to make a selection.

1. From the home screen, select `Inference`, then `Intervals`, then `One mean`, then `t-interval`. If these options do not appear, click the ↩ button in the upper right as many times as needed.
2. Choose `Input statistics` or `Use a dataset` and enter the needed values. Then use the down arrow and choose `Next`.
3. Note the quantities returned. Press the down arrow and choose `Next`.
4. In addition to seeing the confidence interval displayed in two ways, you can press the up and down arrows to quickly change the confidence level and see the resulting interval and margin of error.

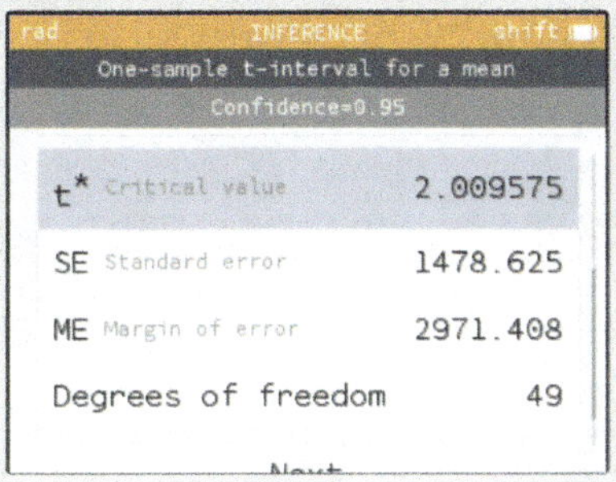

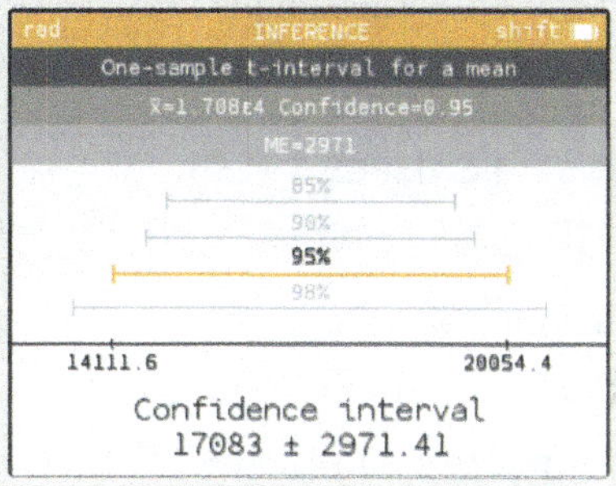

NUMWORKS: 1-SAMPLE T-TEST

Use `OK` or `EXE` to make a selection.

1. From the home screen, select `Inference`, then `Tests`, then `One mean`, then `t-test`. If these options do not appear, click the ↩ button in the upper right as many times as needed.
2. Enter the value of the hypothesized mean for the Null hypothesis. Press the down arrow. Press `OK` and choose <, ≠, or > for the Alternative hypothesis. Press the down arrow and choose `Next`.
3. Choose `Input statistics` or `Use a dataset` and enter the needed values. Then use the down arrow and choose `Next`.
4. Note the quantities returned. Click the down arrow and choose `Next`.
5. On this screen, the p-value and alpha are shaded on the t-distribution and can be visually compared.

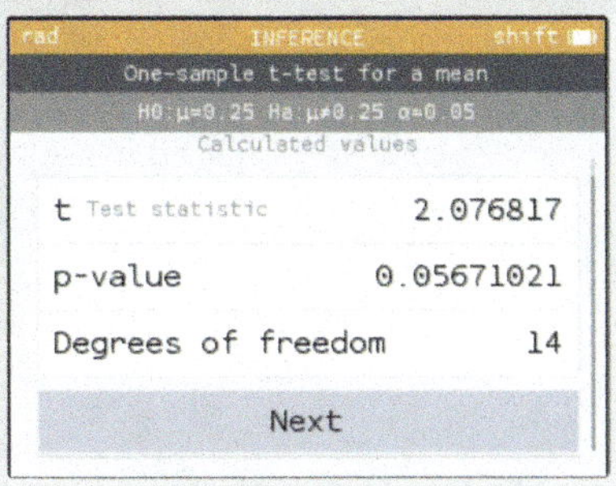

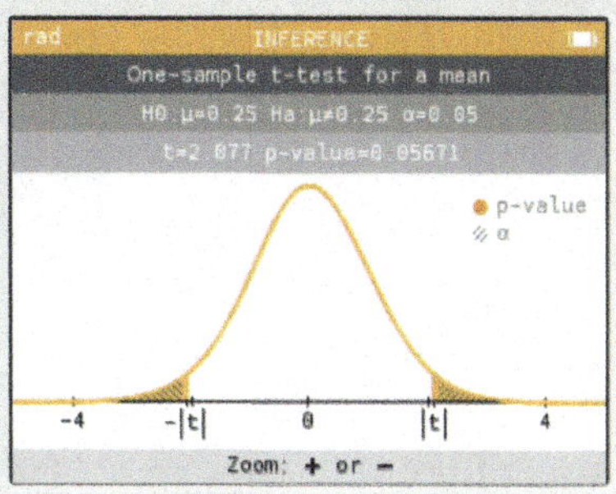

TI-83/84: 1-SAMPLE T-INTERVAL

Use STAT, TESTS, TInterval.

1. Choose STAT.
2. Right arrow to TESTS.
3. Down arrow and choose 8:TInterval.
4. Choose Data if you have all the data or Stats if you have the mean and standard deviation.
   - If you choose Data, let List be L1 or the list in which you entered your data (don't forget to enter the data!) and let Freq be 1.
   - If you choose Stats, enter the mean, *SD*, and sample size.
5. Let C-Level be the desired confidence level.
6. Choose Calculate and hit ENTER, which returns:

(__,__)	the confidence interval
x̄	the sample mean
Sx	the sample *SD*
n	the sample size

TI-83/84: 1-SAMPLE T-TEST

Use STAT, TESTS, T-Test.

1. Choose STAT.
2. Right arrow to TESTS.
3. Down arrow and choose 2:T-Test.
4. Choose Data if you have all the data or Stats if you have the mean and standard deviation. Note: When carrying out a test for a mean of differences, make sure to use the sample *differences* or the summary statistics for the *differences*.
5. Let μ_0 be the null or hypothesized value of μ.
   - If you choose Data, let List be L1 or the list in which you entered your data (don't forget to enter the data!) and let Freq be 1.
   - If you choose Stats, enter the mean, *SD*, and sample size.
6. Choose ≠, <, or > to correspond to H_A.
7. Choose Calculate or Draw and hit ENTER. Draw shows the t-statistic and p-value as well as a graph of the t-distribution with p-value shaded. Calculate returns:

t	T-statistic	Sx	the sample standard deviation
p	p-value	n	the sample size
x̄	the sample mean		

CASIO FX-9750GII: 1-SAMPLE T-INTERVAL

1. Navigate to STAT (MENU button, then hit the 2 button or select STAT).
2. If necessary, enter the data into a list.
3. Choose the INTR option (F3 button), t (F2 button), and 1-S (F1 button).
4. Choose either the Var option (F2) or enter the data in using the List option.
5. Specify the interval details:
   - Confidence level of interest for C-Level.
   - If using the Var option, enter the summary statistics. If using List, specify the list and leave Freq value at 1.
6. Hit the EXE button, which returns

Left, Right	ends of the confidence interval
x̄	sample mean
sx	sample standard deviation
n	sample size

CASIO FX-9750GII: 1-SAMPLE T-TEST

1. Navigate to STAT (MENU button, then hit the 2 button or select STAT).
2. If necessary, enter the data into a list. Note: if doing a test for a mean of differences, enter the computed differences.
3. Choose the TEST option (F3 button).
4. Choose the t option (F2 button).
5. Choose the 1-S option (F1 button).
6. Choose either the Var option (F2) or enter the data in using the List option.
7. Specify the test details:
   - Specify the sidedness of the test using the F1, F2, and F3 keys.
   - Enter the null value, μ0.
   - If using the Var option, enter the summary statistics. If using List, specify the list and leave Freq values at 1.
8. Hit the EXE button, which returns

	alternative hypothesis	x̄	sample mean
t	T-statistic	sx	sample standard deviation
p	p-value	n	sample size

Section summary

- The appropriate hypothesis testing procedure for a population mean μ with unknown population standard deviation σ is a **one-sample t-test for a population mean μ**. The parameter μ should be identified in context.

- For a matched pairs design, the appropriate analysis calculates *differences* between pairs of values to produce one sample of differences. The hypothesis testing procedure for a matched pairs design is a one-sample t-test for a population mean difference, where we use μ_d for the population mean difference, $\bar{x}_d$ for the mean of sample differences, s_d for the standard deviation of sample differences, and n_d for the number of sample differences.

- The null hypotheses for a one-sample t-test for a population mean μ is:

  H_0: $\mu = \mu_0$, where μ_0 is the null hypothesized value for the population mean.

- The alternative hypothesis may be one-sided (< or >) or two-sided (≠).

  H_A: $\mu < \mu_0$. The p-value will correspond to a lower tail.

  H_A: $\mu > \mu_0$. The p-value will correspond to an upper tail.

  H_A: $\mu \neq \mu_0$. The p-value will correspond to both tails.

- The one-sample t-test for a population mean or mean difference has the same conditions as the one-sample t-interval. We check that the following conditions are met.

  1. Independence: The data come from a random sample or random process. When sampling without replacement, check that the sample size is less than 10% of the population size.
  2. Large sample or normal population: $n \geq 30$ or population distribution is nearly normal. If the sample size is less than 30 and the population distribution is unknown, check and confirm that there is no strong skew or outliers in the data in order to reasonably assume that the population distribution is nearly normal.

- A test statistic has the form:

$$\text{test statistic} = \frac{\text{point estimate} - \text{null value}}{SE \text{ of estimate}}.$$

- The test statistic for a one-sample t-test for μ is:

$$T = \frac{\bar{x} - \mu_0}{\frac{s}{\sqrt{n}}}, \quad df = n - 1$$

  When the null hypothesis is true, the t-statistic follows a t-distribution with $df = n - 1$.

- The p-value for a one-sample t-test for μ corresponds to a lower tail, upper tail, or both tails of the t-distribution with $n - 1$ degrees of freedom, depending on whether the direction of the alternate hypothesis is <, >, or ≠.

- The p-value for a one-sample t-test for μ is the probability of obtaining a t-statistic as small or smaller, as large or larger, or as extreme or more extreme than the t-statistic that was observed, depending on whether the direction of the alternate hypothesis is <, >, or ≠, assuming the null hypothesis is true (i.e. that the population mean really equals μ_0).

- A formal decision explicitly compares the p-value to the significance level. If the p-value $\leq \alpha$, then reject the null hypothesis; if the p-value $> \alpha$, then fail to reject the null hypothesis. The conclusion should be stated in terms of the alternative hypothesis and should include context, referencing the parameters and the populations, using non-causal language.

Exercises

4.25 Find the p-value, Part I. A random sample is selected from an approximately normal population with an unknown standard deviation. Find the p-value for the given sample size and test statistic. Also determine if the null hypothesis would be rejected at $\alpha = 0.05$.

(a) $n = 11$, $T = 1.91$

(b) $n = 17$, $T = -3.45$

(c) $n = 7$, $T = 0.83$

(d) $n = 28$, $T = 2.13$

4.26 Find the p-value, Part II. A random sample is selected from an approximately normal population with an unknown standard deviation. Find the p-value for the given sample size and test statistic. Also determine if the null hypothesis would be rejected at $\alpha = 0.01$.

(a) $n = 26$, $T = 2.485$

(b) $n = 18$, $T = 0.5$

4.27 Online communication. A study suggests that the average college student spends 10 hours per week communicating with others online. You believe that this is an underestimate and decide to collect your own sample for a hypothesis test. You randomly sample 60 students from your dorm and find that on average they spent 13.5 hours a week communicating with others online. A friend of yours, who offers to help you with the hypothesis test, comes up with the following set of hypotheses. Indicate any errors you see.

$$H_0 : \bar{x} < 10 \ hours$$
$$H_A : \bar{x} > 13.5 \ hours$$

4.28 Mean travel time to work. Suppose it is known that the mean travel time to work for adults in the US is 25 minutes. A social scientist thinks this value is different for the adults in her county. She takes a random sample of 100 adults in her county and finds that their average travel time to work is 26.4 minutes. Below is how she set up her hypotheses to test if her county average is different than the US average. Indicate any errors you see.

$$H_0 : \bar{x} \neq 26.4 \ minutes$$
$$H_A : \bar{x} = 26.4 \ minutes$$

4.29 Sleep habits of New Yorkers. New York is known as "the city that never sleeps". A random sample of 25 New Yorkers were asked how much sleep they get per night. Statistical summaries of these data are shown below. The point estimate suggests New Yorkers sleep less than 8 hours a night on average. Evaluate the claim that New York is the city that never sleeps keeping in mind that, despite this claim, the true average number of hours New Yorkers sleep could be less than 8 hours or more than 8 hours.

n	$\bar{x}$	s	min	max
25	7.73	0.77	6.17	9.78

(a) Write the hypotheses in symbols and in words.

(b) Check conditions, then calculate the test statistic, T, and the associated degrees of freedom.

(c) Find and interpret the p-value in this context. Drawing a picture may be helpful.

(d) What is the conclusion of the hypothesis test?

(e) If you were to construct a 90% confidence interval that corresponded to this hypothesis test, would you expect 8 hours to be in the interval?

4.30 Heights of adults. Researchers studying anthropometry collected body girth measurements and skeletal diameter measurements, as well as age, weight, height and gender, for 507 physically active individuals. The mean height is 171.1 centimeters with a standard deviation of 9.4 centimeters. The minimum height is 147.2 centimeters and the maximum height is 198.1 centimeters.

(a) If we wanted to conduct a hypothesis test to determine whether there is evidence that the average height of physically active individuals is greater than 160 cm, what conditions would need to be met? Do these conditions seem to be met here?

(b) If we wanted to construct a confidence interval interval to estimate the mean height of physically active individuals, would the conditions be any different? If so, which ones?

4.31 Play the piano. Georgianna claims that in a small city renowned for its music school, the average child takes less than 5 years of piano lessons. We have a random sample of 20 children from the city, with a mean of 4.6 years of piano lessons and a standard deviation of 2.2 years.

(a) Evaluate Georgianna's claim using a hypothesis test. Remember to Identify, Check, Calculate, and Conclude.

(b) Use an appropriate procedure to estimate the average number of years students in this city take piano lessons with 95% confidence. Identify, Check, Calculate, and Conclude.

(c) Do your results from the hypothesis test and the confidence interval agree? Explain your reasoning.

4.32 Auto exhaust and lead exposure. Researchers interested in lead exposure due to car exhaust sampled the blood of 52 police officers subjected to constant inhalation of automobile exhaust fumes while working traffic enforcement in a primarily urban environment. The blood samples of these officers had an average lead concentration of 124.32 μg/l and a SD of 37.74 μg/l; a previous study of individuals from a nearby suburb, with no history of exposure, found an average blood level concentration of 35 μg/l.[12]

(a) Write down the hypotheses that would be appropriate for testing if the police officers appear to have been exposed to a different concentration of lead.

(b) Explicitly state and check all conditions necessary for inference on these data.

(c) Regardless of your answers in part (b), test the hypothesis that the downtown police officers have a higher lead exposure than the group in the previous study. Interpret your results in context.

4.33 Global warming, Part II. Consider a limited set of climate data, examining temperature differences in 1948 vs 2018. We sampled 197 locations from the National Oceanic and Atmospheric Administration's (NOAA) historical data, where the data was available for both years of interest. We want to know: were there more days with temperatures exceeding 90°F in 2018 or in 1948?[13] The difference in number of days exceeding 90°F (number of days in 2018 - number of days in 1948) was calculated for each of the 197 locations. The average of these differences was 2.9 days with a standard deviation of 17.2 days. We are interested in determining whether these data provide strong evidence that there were more days in 2018 that exceeded 90°F from NOAA's weather stations.

(a) Is there a relationship between the observations collected in 1948 and 2018? Or are the observations in the two groups independent? Explain.

(b) Write hypotheses for this research in symbols and in words.

(c) Check the conditions required to complete this test. A histogram of the differences is given to the right.

(d) Calculate the test statistic, degrees of freedom and p-value.

(e) Use $\alpha = 0.05$ to evaluate the test, and interpret your conclusion in context.

(f) What type of error might we have made? Explain in context what the error means.

(g) Based on the results of this hypothesis test, would you expect a confidence interval for the average difference between the number of days exceeding 90°F from 1948 and 2018 to include 0? Explain your reasoning.

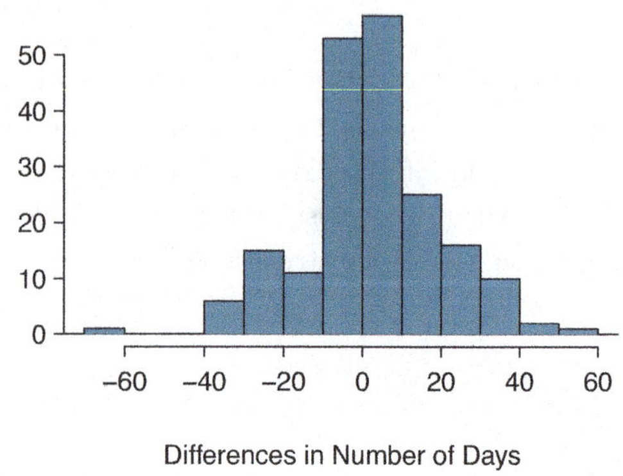

[12] WI Mortada et al. "Study of lead exposure from automobile exhaust as a risk for nephrotoxicity among traffic policemen." In: *American journal of nephrology* 21.4 (2000), pp. 274–279.

[13] NOAA, www.ncdc.noaa.gov/cdo-web/datasets, April 24, 2019.

4.34 High School and Beyond, Part II. The National Center of Education Statistics conducted a survey of high school seniors, collecting test data on reading, writing, and several other subjects. Here we examine a simple random sample of 200 students from this survey. Side-by-side box plots of reading and writing scores as well as a histogram of the differences in scores are shown below.

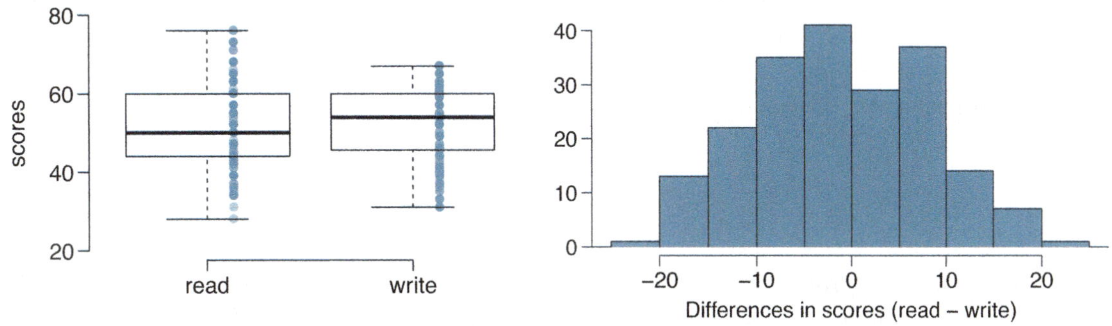

(a) Is there a clear difference in the average reading and writing scores?

(b) Are the reading and writing scores of each student independent of each other?

(c) Create hypotheses appropriate for the following research question: is there an evident difference in the average scores of students in the reading and writing exam?

(d) Check the conditions required to complete this test.

(e) The average observed difference in scores is $\bar{x}_{read-write} = -0.545$, and the standard deviation of the differences is 8.887 points. Do these data provide convincing evidence of a difference between the average scores on the two exams?

(f) What type of error might we have made? Explain what the error means in the context of the application.

(g) Based on the results of this hypothesis test, would you expect a confidence interval for the average difference between the reading and writing scores to include 0? Explain your reasoning.

4.4 Sampling distribution for $\bar{x}_1 - \bar{x}_2$

If two populations means are the same, how much variability can we expect in sample means based on random samples of a certain size? In this section, we describe the sampling distribution for a difference in sample means, and we find the probability that a sample difference would be greater than a certain value due to chance variation.

Learning objectives

1. Calculate and interpret the mean and standard deviation of a sampling distribution for a difference in sample means.
2. Determine if the sampling distibution for a difference in sample means is approximately normal.
3. If appropriate, use the normal distribution to estimate probabiliies involving a difference in sample means and interpret these quantities.

4.4.1 A sampling distribution for a difference in sample means

In hypothesis testing, we would like to ask: assuming two population means are the same, what is the likelihood that two sample means will be as different as we observed in our samples? To be able to calculate this p-value, we need to understand the properties of the sampling distribution for the difference of sample means.

In Section 4.1 we started with all of the data from the 2017 Cherry Blossom Run, and we considered what the sampling distribution for a mean would look like for random samples of size n. The population mean for all the runners is 94.52 minutes and the population standard deviation is 15.93 minutes. Now imagine taking two independent random samples of size 50 from this population. What is the likelihood that the sample means from these two independent random samples (from populations with the same mean) would differ by at least 3 minutes?

In Section 3.5, we conducted a simulation for the sampling distribution of $\hat{p}_1 - \hat{p}_2$. Here, we will conduct a simulation to approximate the sampling distribution of $\bar{x}_1 - \bar{x}_2$. We take two separate and independent random samples of size 50 from the population of run time values, and the we find the difference in the sample means, rounded to the nearest 0.5. We repeat this 300 times, giving us 300 values of $\bar{x}_1 - \bar{x}_2$. These 300 sample differences are graphed in Figure 4.22.

We see that the distribution in Figure 4.22 is centered on 0, which makes sense because the samples come from the same population so the the sampling distributions for the mean have the same center. Each dot in Figure 4.22 represents one value of $\bar{x}_1 - \bar{x}_2$.

Given that the samples are from the same population and that their expected means are the same, what is the likelihood that the sample means from these two independent random samples would differ by at least 3 minutes? We can count that there are 61 values at or below −3 and 58 values at or above 3 so, based on the simulation, we estimate that there is a $\frac{119}{300}$, or about a 39.7% chance that the sample means will differ by at least 3 minutes.

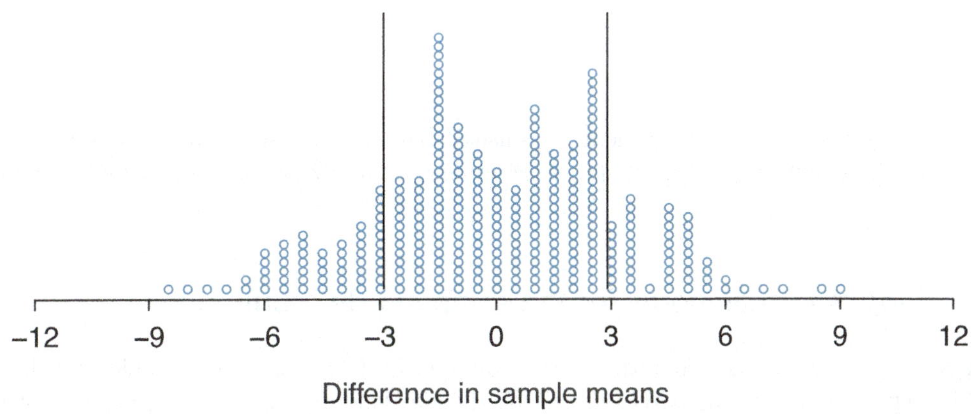

Figure 4.22: 300 simulated differences in sample means.

4.4.2 Mean and standard deviation for a difference in sample means

We would like to be able to find the mean (expected value) and the standard deviation for a difference in sample means, $\bar{x}_1 - \bar{x}_2$. We again use the formulas discussed in Section 3.5.2 for a difference in two independent random variables, $X - Y$, but this time we apply it to a difference in sample means:

$$\mu_{\bar{x}_1-\bar{x}_2} = \mu_{\bar{x}_1} - \mu_{\bar{x}_2} = \mu_1 - \mu_2$$

$$\sigma_{\bar{x}_1-\bar{x}_2} = \sqrt{(\sigma_{\bar{x}_1})^2 + (\sigma_{\bar{x}_2})^2} = \sqrt{\frac{\sigma_1^2}{n_1} + \frac{\sigma_2^2}{n_2}}.$$

That is, for two independent random samples, the distribution of values of $\bar{x}_1 - \bar{x}_2$ for all random samples of size n_1 and n_2 from given populations is centered on the true difference $\mu_1 - \mu_2$ and the typical distance or error of $\bar{x}_1 - \bar{x}_2$ from $\mu_1 - \mu_2$ is given by $\sqrt{\frac{\sigma_1^2}{n_1} + \frac{\sigma_2^2}{n_2}}$.

MEAN AND STANDARD DEVIATION OF A DIFFERENCE IN SAMPLE MEANS

The mean and standard deviation of the sampling distribution for a difference in sample means describe the center and spread of the distribution of $\bar{x}_1 - \bar{x}_2$ values for all random samples of size n_1 and n_2 from the given populations. Given population means μ_1 and μ_2, population sizes N_1 and N_2, individual population standard deviations σ_1 and σ_2, and independent random samples of size n_1 and n_2, we have the following:

$$\mu_{\bar{x}_1-\bar{x}_2} = \mu_1 - \mu_2$$

$$\sigma_{\bar{x}_1-\bar{x}_2} = \sqrt{\frac{\sigma_1^2}{n_1} + \frac{\sigma_2^2}{n_2}} \qquad \text{when } n_1 < 0.10(N_1) \text{ and } n_2 < 0.10(N_2)$$

When sampling without replacement, as is usually the case, the standard deviation formula will provide a good estimate when the sample sizes are less than 10% of the corresponding population sizes.

4.4.3 Using a normal model for the sampling distribution for $\bar{x}_1 - \bar{x}_2$

In Section 3.5.2, we saw that the sum or difference of two random variables will be nearly normal if each variable is itself nearly normal and the two random variables are independent of each other. We use this principle to identify that the difference of sample means can be modeled using a normal distribution when each sample mean can be modeled using a normal distribution.

Independence. The observations should be independent within and between groups. The independence condition is satisfied if the data is collected from 2 independent random samples, where each sample size is less than 10% of the population size if done without replacement. We also consider the independence condition satisfied if the data is collected from an experiment with two randomly assigned treatments (in this case the 10% condition is not relevant and does not need to be checked).

Large sample / normal population. Each population distribution should be nearly normal or each sample size should be at least 30. As before, if the sample sizes are small and the population distributions are not known to be nearly normal, we look at the data for strong skew or outliers. If we do not find strong skew or outliers in either group, the assumption that the populations are nearly normal is typically considered reasonable.

EXAMPLE 4.37

Let's return to the Cherry Blossom Run application. We have that the population mean for all the runners in the 2017 Cherry Blossom Run is 94.52 minutes and the population standard deviation is 15.93 minutes. If we take two independent random samples of 50 runners, what is the probability that the sample means from these two samples will differ by at least 3 minutes?

Since either sample mean could be at least 3 minutes greater than the other sample mean, we want to find: $P(\bar{x}_1 - \bar{x}_2 \le -3) + P(\bar{x}_1 - \bar{x}_2 \ge 3)$.

First, we find the mean of $\bar{x}_1 - \bar{x}_2$:

$$\mu_{\bar{x}_1-\bar{x}_2} = \mu_1 - \mu_2 = 94.52 - 94.52 = 0$$

Because the two random samples are independent and the sample sizes of $n_1 = 50$ and $n_2 = 50$ are both less than 10% of the total number of runners, we calculate the standard deviation of $\bar{x}_1 - \bar{x}_2$ as:

$$\sigma_{\bar{x}_1-\bar{x}_2} = \sqrt{\frac{\sigma_1^2}{n_1} + \frac{\sigma_2^2}{n_2}} = \sqrt{\frac{15.93^2}{50} + \frac{15.93^2}{50}} = 3.186$$

Because the sample sizes n_1 and n_2 are both 50, and $50 \ge 30$, the distribution of $\bar{x}_1 - \bar{x}_2$ is nearly normal whether or not the population distribution of run times is nearly normal. Therefore, $\bar{x}_1 - \bar{x}_2$ is approximately Normal($\mu = 0$, $\sigma = 3.186$).

Using technology, we find that $P(\bar{x}_1 - \bar{x}_2 \le -3) + P(\bar{x}_1 - \bar{x}_2 \ge 3) = 0.173 + 0.173 = 0.346$.

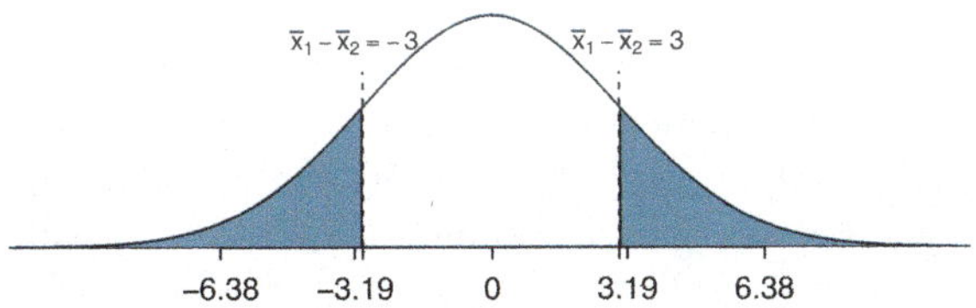

Even though the samples are from the same population of runners, there is still about a 34.6% probability that the sample means will differ by more than 3 minutes.

We can also find the probability above using the Z-score and the standard normal distribution, Normal($\mu = 0$, $\sigma = 1$), as follows:

$$Z = \frac{3 - 0}{\sqrt{\frac{15.93^2}{50} + \frac{15.93^2}{50}}} = -0.942$$

$$P(Z \le -0.942) + P(Z \ge 0.942) = 0.346.$$

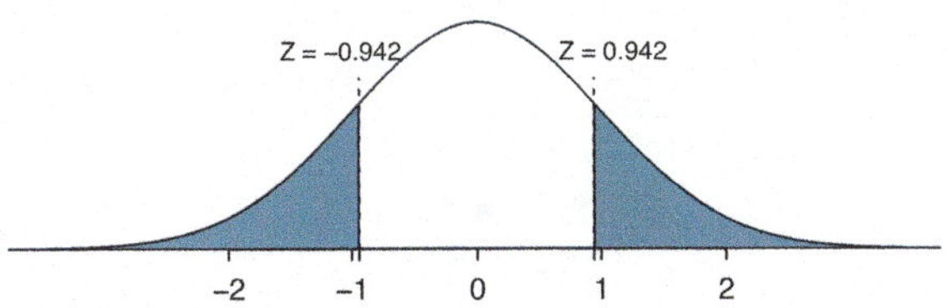

We arrive at the same answer that there is about a 34.6% chance that the sample means will differ by at least 3 minutes. In Section 4.6, Hypothesis testing for a difference in population means, we will see parallels between the calculation of the Z-score above and the calculation of the test statistic.

Section summary

- $\bar{x}_1 - \bar{x}_2$ represents a difference in sample means and can take on different values for different samples. For two independent populations, the sampling distribution of $\bar{x}_1 - \bar{x}_2$ is the distribution of values of $\bar{x}_1 - \bar{x}_2$ for all random samples of size n_1 and n_2 from given populations.

- When the observations can be treated as independent, such as from two independent random samples or two randomly assigned treatments:

  - The **mean** of the sampling distribution of $\bar{x}_1 - \bar{x}_2$ is given by: $\mu_{\bar{x}_1-\bar{x}_2} = \mu_1 - \mu_2$, where μ_1 and μ_2 are population means.
  - The **standard deviation** of the sampling distribution of $\bar{x}_1 - \bar{x}_2$ is given by: $\sigma_{\bar{x}_1-\bar{x}_2} = \sqrt{\frac{\sigma_1^2}{n_1} + \frac{\sigma_2^2}{n_2}}$, where σ_1 and σ_2 are population standard deviations. If randomly sampling without replacement, with each sample size should be less than the corresponding population size, i.e. $n_1 < 0.10(N_1$ and $n_2 < 0.10(N_2)$, for this standard deviation formula to be used. If data is collected from an experiment with two randomly assigned treatments, the 10% condition does not need to be checked.
  - The **shape** of the sampling distribution of a difference in sample means is approximately normal when the distributions for both populations are nearly normal or when $n_1 \geq 30$ and $n_2 \geq 30$.

- $\mu_{\bar{x}_1-\bar{x}_2}$, the mean of $\bar{x}_1 - \bar{x}_2$, describes the average of values of $\bar{x}_1 - \bar{x}_2$ for all random samples of size n_1 and n_2 from the given populations.

- $\sigma_{\bar{x}_1-\bar{x}_2}$, the standard deviation of $\bar{x}_1 - \bar{x}_2$, describes the typical variation in values of $\bar{x}_1 - \bar{x}_2$ from $\mu_1 - \mu_2$ for all random samples of size n_1 and n_2 from the given populations.

- To use a normal model to find probabilities involving a difference in sample means, first verify that the conditions for independence are met and that the distributions for both populations are nearly normal or $n_1 \geq 30$ and $n_2 \geq 30$. Identify the distribution and its parameters, write the relevant probability statement, and answer the question in context.

- The mean, standard deviation, and probabilities for the sampling distribution for a difference between two sample means should be interpreted within the context of two specific populations.

Exercises

4.35 Difference of means, Part 1. Suppose we will collect two random samples from the following distributions:

	Mean	Standard Deviation	Sample Size
Sample 1	15	20	50
Sample 2	20	10	30

In each of the parts below, consider the sample means $\bar{x}_1$ and $\bar{x}_2$ that we might observe from these two samples.

(a) What is the associated mean and standard deviation of $\bar{x}_1$?

(b) What is the associated mean and standard deviation of $\bar{x}_2$?

(c) Calculate and interpret the mean and standard deviation associated with the difference in sample means for the two groups, $\bar{x}_2 - \bar{x}_1$.

(d) How are the standard deviations from parts (a), (b), and (c) related?

4.36 Difference of means, Part 2. Suppose we will collect two random samples from the following distributions:

	Mean	Standard Deviation	Sample Size
Sample 1	15	20	50
Sample 2	20	10	30

In each of the parts below, consider the sample means $\bar{x}_1$ and $\bar{x}_2$ that we might observe from these two samples.

(a) What distribution is associated with the difference $\bar{x}_2 - \bar{x}_1$? Justify your answer.

(b) Determine the probability that $\bar{x}_2 - \bar{x}_1$ will be larger than 7.

(c) Determine the probability that $\bar{x}_2 - \bar{x}_1$ will be smaller than 3.

(d) Determine the probability that $\bar{x}_2 - \bar{x}_1$ will be smaller than 0.

4.5 Confidence intervals for $\mu_1 - \mu_2$

Can a name affect how much an employer is willing to pay an applicant? Are faculty willing to pay someone named "John" more than someone named "Jennifer"? If so, how much more? In this section we will learn a confidence interval procedure for estimating a difference between two means.

Learning objectives

1. Determine when it is appropriate to use a one-sample t-procedure versus a two-sample t-procedure.
2. Identify and set up an appropriate confidence interval procedure for estimating the difference in population means $\mu_1 - \mu_2$.
3. Verify whether conditions for a confidence interval for a difference in population means using a t-distribution are met.
4. Calculate an appropriate confidence interval for a difference in population means.
5. Calculate the standard error and margin of error for a confidence interval for a difference in population means.
6. Interpret a confidence interval for a difference in population means.
7. Justify a claim about the difference in population means based on an appropriate confidence interval

4.5.1 Estimating a difference of means

What's in a name? Are employers more likely to offer interviews or higher pay to prospective employees when the name on a resume suggests the candidate is a man versus a woman? This is a challenging question to tackle, because employers are influenced by many aspects of a resume. Thinking back to Chapter 1, we could imagine a host of confounding factors associated with name and gender. How could we possibly isolate just the factor of name? We would need an experiment in which name was the only variable and everything else was held constant.

Researchers at Yale carried out precisely this experiment. Their results were published in the Proceedings of the National Academy of Sciences (PNAS). The researchers sent out resumes to faculty at academic institutions for a lab manager position. The resumes were identical, except that on half of them the applicant's name was John and on the other half, the applicant's name was Jennifer. They wanted to see if faculty, specifically faculty trained in conducting scientifically objective research, held implicit gender biases.

Unlike in a matched pairs scenario, each faculty member received only one resume. We are interested in comparing the mean salary offered to John relative to the mean salary offered to Jennifer. Instead of taking the average of a set of differences, we find the average of each group separately and take their difference. Let

$$\bar{x}_1 : \text{mean salary offered to John}$$
$$\bar{x}_2 : \text{mean salary offered to Jennifer}$$

We will use $\bar{x}_1 - \bar{x}_2$ as our point estimate for $\mu_1 - \mu_2$. The data is given in the table below.

Name	n	$\bar{x}$	s
John	63	\$30,238	\$5567
Jennifer	64	\$26,508	\$7247

We can calculate the difference as

$$\bar{x}_1 - \bar{x}_2 = 30,238 - 26,508 = 3730.$$

EXAMPLE 4.38

Interpret the point estimate 3730. Why might we want to construct a confidence interval?

E The average salary offered to John was \$3,730 higher than the average salary offered to Jennifer. Because there is randomness in which faculty ended up in the John group and which faculty ended up in the Jennifer group, there is error in our estimate. To measure the typical error we calculate the SE for the difference in sample means.

We calculate the SE for a difference in sample means as follows:

$$SE_{\bar{x}_1 - \bar{x}_2} = \sqrt{\frac{s_1^2}{n_1} + \frac{s_2^2}{n_2}}$$

Note that the standard error for a difference in sample means follows the same structure as the standard deviation for a difference in sample means calculated in the previous section, except that we replace the unknown population standard deviations σ_1 and σ_2 with the sample standard deviations s_1 and s_2.

EXAMPLE 4.39

Calculate and interpret the SE for a difference in sample means.

E

$$SE_{\bar{x}_1 - \bar{x}_2} = \sqrt{\frac{s_1^2}{n_1} + \frac{s_2^2}{n_2}} = \sqrt{\frac{(5567)^2}{63} + \frac{(7247)^2}{64}} = 1151$$

Using samples of size $n_1 = 63$ and $n_2 = 64$, the typical error when using $\bar{x}_1 - \bar{x}_2$ to estimate $\mu_1 - \mu_2$, the real difference in mean salary that the faculty would offer John versus Jennifer, is \$1151.

We see that the difference in sample means of \$3,730 is more than 3 SE above 0, which makes us think that the difference being 0 is unreasonable. We would like to construct a 95% confidence interval for the theoretical difference in mean salary that would be offered to John versus Jennifer. For this, we need the degrees of freedom associated with a two-sample t-interval.

For the one-sample t-procedure, the degrees of freedom is given by the simple expression $n - 1$, where n is the sample size. For the two-sample t-procedures, however, there is a complex formula for calculating the degrees of freedom, which is based on the two sample sizes and the two sample standard deviations. In practice, we find the degrees of freedom using technology (see Section 4.5.3). If this is not possible, the alternative is to use the smaller of $n_1 - 1$ and $n_2 - 1$. The degrees of freedom will fall between $n_1 + n_2 - 2$ and the smaller of $n_1 - 1$ and $n_2 - 1$.

DEGREES OF FREEDOM FOR TWO-SAMPLE T-PROCEDURES

Use statistical software or a calculator to compute the degrees of freedom for two-sample t-procedures. The degrees of freedom will fall between $n_1 + n_2 - 2$ and the smaller of $n_1 - 1$ and $n_2 - 1$.

4.5.2 Conditions and calculations for a confidence interval for a difference of means

When performing inference on a difference of means, $\mu_1 - \mu_2$, we use the t-distribution just as we did for inference on a single mean. In order to use the t-distribution, we need to check the following conditions.

Independence. The independence condition is satisfied if the data is collected from 2 independent random samples, where each sample size is less than 10% of the population size if done without replacement. We also consider the independence condition satisfied if the data is collected from an experiment with two randomly assigned treatments (in this case the 10% condition is not relevant and does not need to be checked).

Large sample / normal population. Each population distribution should be nearly normal or each sample size should be at least 30. As before, if the sample sizes are small and the population distributions are not known to be nearly normal, we look at the data for strong skew or outliers. If we do not find strong skew or outliers in either group, the assumption that the populations are nearly normal is typically considered reasonable.

EXAMPLE 4.40

Verify that conditions are met for a two-sample t-test. Then, construct the 95% confidence interval for a difference of means.

E

We noted previously that this is an experiment and that the two treatments (name Jennifer and name John) were randomly assigned. Also, both sample sizes are well over 30, so conditions for using a t-interval are met. Using technology, we find that $df = 118.1$. Because 118.1 is not on the t-table, we round the degrees of freedom down to 100. Using a t-table at row $df = 100$ with 95% confidence, we get a $t^\star = 1.984$. We calculate the confidence interval as follows.

$$\begin{aligned} \text{point estimate} &\pm t^\star \times SE \text{ of estimate} \\ 3730 &\pm 1.984 \times 1151 \\ 3730 &\pm 2284 \\ &(1446, 6014) \end{aligned}$$

GUIDED PRACTICE 4.41

G

Instead of using a t-table, use technology to calculate the 95% confidence interval for the previous example.[14]

[14] Using technology, we get the 95% confidence interval: (1461, 5999).

We are 95% confident that the interval ($1446, $6014) contains the difference (John – Jennifer) in mean salaries that faculty like the ones in this study would offer for a lab manager position. That is, we are 95% confident that the mean salary faculty like the ones in this study would offer John for a lab manager position is between $1,446 and $6,014 *more* than the mean salary they would offer Jennifer for the position.[15]

EXAMPLE 4.42

Given that this was a well-designed experiment, can we say *which* faculty discriminated in their salary offer?

E

No - each faculty member received only one of the resumes. A faculty member that offered "Jennifer" a very low salary may have also offered "John" a very low salary. It is only possible to say that overall there is evidence that faculty are willing to offer John more money for the lab manager position than Jennifer. Finding proof of bias for individual cases is a persistent challenge in enforcing anti-discrimination laws.

GUIDED PRACTICE 4.43

G

We might imagine an experiment in which each faculty received both resumes, so that we could compare how much they would offer someone named John versus someone named Jennifer. This would be a matched pairs experiment. Is a matched pairs experiment feasible in this context? Why or why not?[16]

4.5.3 Technology: the two-sample t-interval for $\mu_1 - \mu_2$

Section 4.6.4 demonstrates how to calculate the two-sample t-interval and the two-sample t-test (introduced in the next section) using Desmos, R, and the NumWorks, TI-83/84 and Casio calculator.

[15] A similar study sent out identical resumes with different names to investigate the importance of perceived race. Resumes with a name commonly perceived to be for a White person (e.g. Emily) were 50% more likely to receive a callback than the same resume with a name commonly perceived to be for a Black person (e.g. Lakisha). More information is given in Appendix B – see the `resume` data set.

[16] No, because what makes the experiment work is that the resumes are *exactly the same* except for the name. An employer would notice something fishy if they received two identical resumes.

4.5.4 Summary and worked example

CONSTRUCTING A CONFIDENCE INTERVAL FOR A DIFFERENCE IN MEANS

To carry out a complete confidence interval procedure to estimate the difference in population means,

Identify: Identify the interval procedure, parameter, and confidence level.

Use a **two-sample t-interval for a difference in population means $\mu_1 - \mu_2$**. Define the difference in population means $\mu_1 - \mu_2$ in words, referencing the populations of interest. Choose a confidence level (C%).

Check: Check conditions for constructing a confidence interval based on a t-distribution.

1. Independence: Data come from 2 independent random samples or from a randomized experiment with 2 treatments. When sampling without replacement, check that the sample size is less than 10% of the population size for each sample.
2. Large samples or normal populations: $n_1 \geq 30$ and $n_2 \geq 30$ or both population distributions are nearly normal. If the sample sizes are less than 30 and the population distributions are unknown, there should be no strong skew or outliers in either data set (this makes us believe that it is reasonable that both population distributions could be nearly normal).

Calculate: Calculate the confidence interval and record it in interval form.

point estimate $\pm$ $t^\star \times SE$ of estimate, df: use technology to calculate

point estimate: $\bar{x}_1 - \bar{x}_2$, the difference in sample means

SE of estimate: $\sqrt{\frac{s_1^2}{n_1} + \frac{s_2^2}{n_2}}$

$t^\star$: use technology or use a t-table at row df and confidence level C%

(____, ____)

Conclude: Interpret the interval and, if applicable, draw a conclusion in context.

We are C% confident that the interval (____, ____) contains the difference (specify order) in the true mean [...]. If applicable, draw a conclusion based on whether the interval is entirely above, is entirely below, or contains the value 0.

EXAMPLE 4.44

An instructor decided to run two slight variations of the same exam. Prior to passing out the exams, she shuffled the exams together to ensure each student received a random version. Summary statistics for how students performed on these two exams are shown in Figure 4.44. Anticipating complaints from students who took Version B, she would like to evaluate whether the difference observed in the groups is so large that it provides convincing evidence that Version B was more difficult (on average) than Version A. Use a 95% confidence interval to estimate the difference (version A − version B) in average score.

Version	n	$\bar{x}$	s	min	max
A	30	79.4	14	45	100
B	30	74.1	20	32	100

Identify: Because we are estimating the difference between two means, we will use a two-sample t-interval for a difference in population means $\mu_1 - \mu_2$. We define $\mu_1 - \mu_2$ as the difference (Version A − Version B) in average score, and we will estimate this parameter at the 95% confidence level.

Check: The data was collected from a randomized experiment with two treatments: Version A and Version B of test. The 10% condition does not need to be checked here because we are not sampling from a population. There were 30 students in each group, so the condition that both group sizes are at least 30 is met.

E

Calculate: We will calculate the confidence interval as follows.

$$\text{point estimate} \pm t^{\star} \times SE \text{ of estimate}$$

The point estimate is the difference in sample means: $\bar{x}_1 - \bar{x}_2 = 79.4 - 74.1 = 5.3$.

$$SE \text{ of } \bar{x}_1 - \bar{x}_2 = \sqrt{\frac{s_1^2}{n_1} + \frac{s_2^2}{n_2}} = \sqrt{\frac{14^2}{30} + \frac{20^2}{30}} = 4.46.$$

Using technology, we find $df = 51.9$ and $t^{\star} = 2.007$. The 95% confidence interval is given by:

$$(79.4 - 74.1) \pm 2.007 \times \sqrt{\frac{14^2}{30} + \frac{20^2}{30}} \qquad df = 51.9$$
$$5.3 \pm 2.007 \times 4.46$$
$$(-3.66,\ 14.26)$$

Conclude: We are 95% confident that the interval (−3.66, 14.26) contains the difference (Version A −Version B) in average score that students like those in the study would receive if given Version A or Version B. Because the interval contains both positive and negative values, the data do not convincingly show that one exam version is more difficult than the other, and the teacher should not be convinced that she should add points to the Version B exam scores.

Section summary

- This section introduced inference for a difference of means, which is distinct from inference for a mean difference. To calculate a difference of means, $\bar{x}_1 - \bar{x}_2$, we first calculate the mean of each group, then we take the difference between those two statistics. To calculate a mean difference, $\bar{x}_d$, we first calculate all of the differences, then we find the mean of those differences.
- The appropriate confidence interval to estimate a difference between two population means μ_1 and μ_2 is a **two-sample t-interval for a difference in population means $\mu_1 - \mu_2$**. The parameters μ_1 and μ_2 should be identified in context.
- The two-sample t-interval for a difference in population means requires the follow conditions be met:
  1. Independence: The data come from two independent random samples, each with sample size <10% of its corresponding population size if sampling without replacement OR the data come from a randomized experiment with two randomly assigned treatments.
  2. Large sample or normal population: both sample sizes are at least 30 or both population distributions are nearly normal. If either sample size is less than 30 and the population distributions are unknown, check and confirm that there is no strong skew or outliers in either data set in order to reasonably assume that the population distributions are nearly normal.
- The general form for a C% confidence interval is:

$$\text{point estimate} \pm \text{margin of error, or}$$
$$\text{point estimate} \pm \text{critical value} \times SE \text{ of estimate.}$$

- A two-sample t-interval for a difference in population means $\mu_1 - \mu_2$ can be written as follows:

$$(\bar{x}_1 - \bar{x}_2) \pm t^\star\sqrt{\frac{s_1^2}{n_1} + \frac{s_2^2}{n_2}}, \quad df\text{: use technology.}$$

  df is calculated using technology and will fall between $n_1 + n_2 - 2$ and the smaller of $n_1 - 1$ and $n_2 - 1$. $t^\star$ is the critical value for the middle C% of a t-distribution with the appropriate degrees of freedom.
- The SE of $\bar{x}_1 - \bar{x}_2$ is: $\sqrt{\frac{s_1^2}{n_1} + \frac{s_2^2}{n_2}}$.
- The margin of error of $\bar{x}_1 - \bar{x}_2$ is: $t^\star\sqrt{\frac{s_1^2}{n_1} + \frac{s_2^2}{n_2}}$.
- The interpretation of the confidence level C% is: In repeated random sampling with the same sample size from the same populations, approximately C% of confidence intervals created will capture the true difference between the two population means.
- When interpreting a C% confidence interval for a difference between two population means, we say we are C% confident that the interval (___, ___) contains the value of the difference in the population means.

Exercises

4.37 Air quality. Air quality measurements were collected in a random sample of 25 country capitals in 2013, and then again in the same cities in 2014. We would like to use these data to compare average air quality between the two years. Should we use a paired or non-paired test? Explain your reasoning.

4.38 Fuel efficiency of manual and automatic cars, Part II. Each year the US Environmental Protection Agency (EPA) releases fuel economy data on cars manufactured in that year. Below are summary statistics on fuel efficiency (in miles/gallon) from random samples of cars with manual and automatic transmissions. Use these statistics to calculate a 98% confidence interval for the difference between average highway mileage of manual and automatic cars, and interpret this interval in the context of the data.[17]

	Hwy MPG	
	Automatic	Manual
Mean	22.92	27.88
SD	5.29	5.01
n	26	26

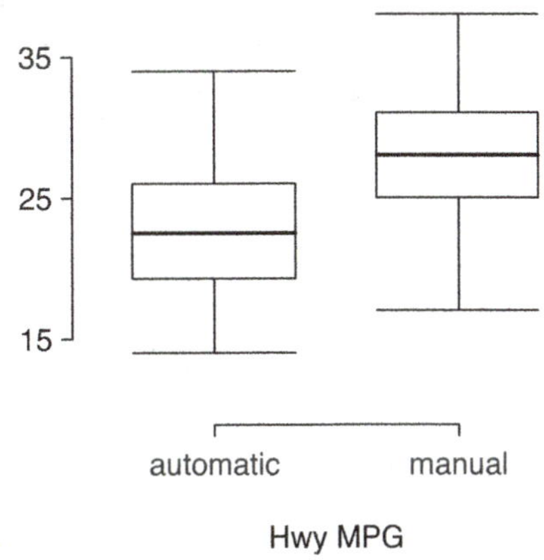

4.39 Chicken diet and weight, Part I. Chicken farming is a multi-billion dollar industry, and any methods that increase the growth rate of young chicks can reduce consumer costs while increasing company profits, possibly by millions of dollars. An experiment was conducted to measure and compare the effectiveness of various feed supplements on the growth rate of chickens. Newly hatched chicks were randomly allocated into six groups, and each group was given a different feed supplement. Below are some summary statistics from this data set along with box plots showing the distribution of weights by feed type.[18]

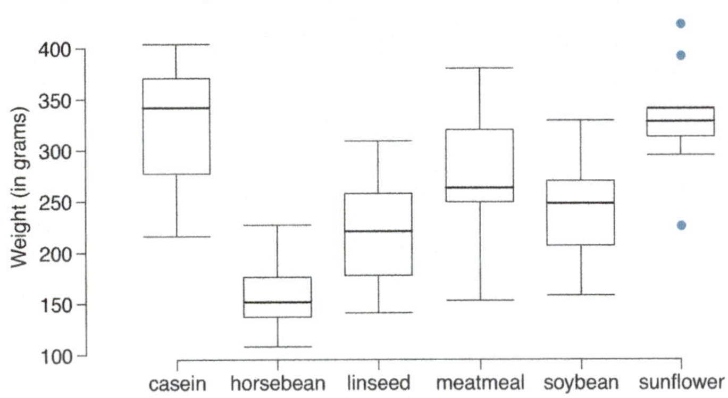

	Mean	SD	n
casein	323.58	64.43	12
horsebean	160.20	38.63	10
linseed	218.75	52.24	12
meatmeal	276.91	64.90	11
soybean	246.43	54.13	14
sunflower	328.92	48.84	12

Casein is a common weight gain supplement for humans. How does it compare to soybean as a feed supplement? Construct a 95% confidence interval to estimate the difference between average weight of chickens that would be fed casein and average weight of chickens that would be fed soybean. Use the Identify, Check, Calculate, Conclude framework. Is there evidence that average weight of chickens that would be fed casein is higher than average weight of chickens that would be fed soybean? Justify your answer based on the confidence interval.

[17] U.S. Department of Energy, Fuel Economy Data, 2012 Datafile.

[18] Chicken Weights by Feed Type, from the `datasets` package in R..

4.40 Work hours and education. The General Social Survey collects data on demographics, education, and work, among many other characteristics of US adults.[19] Below are the summary statistics for hours worked for those with less than a high school degree and for those with a Bachelor's degree.

	Less than HS	Bachelor's
Mean	38.67	42.55
SD	15.81	13.62
n	121	253

(a) We would like to estimate the difference in hours worked between US adults with less than a high school degree and US adults with a Bachelor's degree. Discuss whether conditions for a two-sample t-interval for a difference in means are met.

(b) Assuming conditions are met, construct a 95% confidence interval to estimate the difference in hours worked between US adults with less than a high school degree and US adults with a Bachelor's degree. Based on the interval, is there evidence that the average hours worked for these two groups differ?

[19] National Opinion Research Center, General Social Survey, 2018.

4.6 Hypothesis testing for $\mu_1 - \mu_2$

How do we measure how much evidence we have of a difference in means between two treatments or populations? For example, how much evidence is there that using embryonic stem cells helps improve heart function following a heart attack? Is there a significance difference in the average weight of newborns between mothers who smoke versus mothers who do not smoke? In this section, we apply the hypothesis testing framework to a difference of population means.

Learning objectives

1. Determine when it is appropriate to use a one-sample t-procedure versus a two-sample t-procedure.
2. Identify and set up an appropriate testing method for a difference in population means $\mu_1 - \mu_2$.
3. Verify whether conditions for the hypothesis test for a difference in population means using a t-distribution are met.
4. Calculate the t-statistic, degrees of freedom and p-value for a hypothesis test for a difference in population means.
5. Interpret the p-value of a hypothesis test for a difference in population means.
6. Justify a claim about the difference in population means based on the results of the test.

4.6.1 Introducing hypothesis testing for a difference of means

Four cases from a data set called `ncbirths`, which represents mothers and their newborns in North Carolina, are shown in Figure 4.23. We are particularly interested in two variables: `weight` and `smoke`. The `weight` variable represents the weights of the newborns and the `smoke` variable describes which mothers smoked during pregnancy. We would like to know, is there convincing evidence that newborns from mothers who smoke have a different average birth weight than newborns from mothers who don't smoke? The smoking group includes a random sample of 50 cases and the nonsmoking group contains a random sample of 100 cases, represented in Figure 4.24.

	fAge	mAge	weeks	weight	sex	smoke
1	NA	13	37	5.00	female	nonsmoker
2	NA	14	36	5.88	female	nonsmoker
3	19	15	41	8.13	male	smoker
⋮	⋮	⋮	⋮	⋮	⋮	
150	45	50	36	9.25	female	nonsmoker

Figure 4.23: Four cases from the `ncbirths` data set. The value "NA", shown for the first two entries of the first variable, indicates pieces of data that are missing.

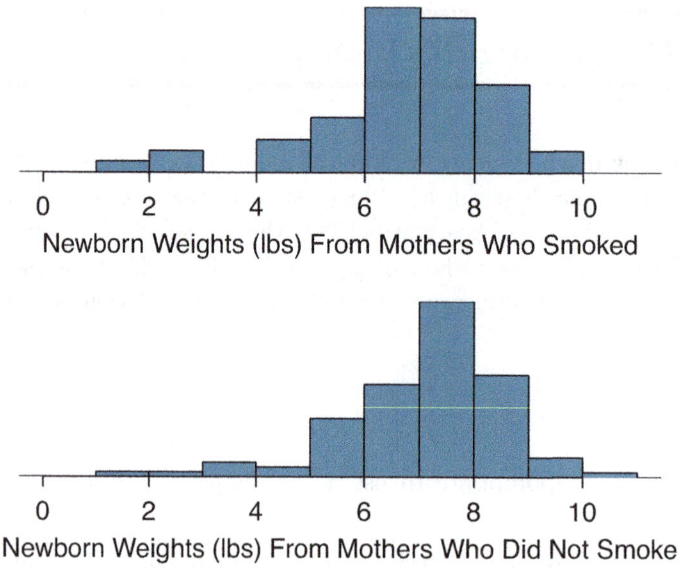

Figure 4.24: The top panel represents birth weights for infants whose mothers smoked. The bottom panel represents the birth weights for infants whose mothers who did not smoke. The distributions exhibit moderate-to-strong and strong skew, respectively.

E

EXAMPLE 4.45

Set up appropriate hypotheses to evaluate whether there is a relationship between a mother smoking and average birth weight.

We define our parameters as follows:
μ_1: mean birth weight of newborns from North Carolina mothers who did smoke during pregnancy
μ_2: mean birth weight of newborns from North Carolina mothers who did not smoke during pregnancy

H_0: $\mu_1 - \mu_2 = 0$. There is no difference in average birth weight for newborns from mothers who did and did not smoke.

H_A: $\mu_1 - \mu_2 \neq 0$. There is some difference in average newborn weights from mothers who did and did not smoke.

4.6.2 Checking conditions for a hypothesis test for a difference of means

The conditions for a two-sample t-test for $\mu_1 - \mu_2$ are exactly the same as the conditions for a two-sample t-interval for $\mu_1 - \mu_2$.

Independence. The independence condition is satisfied if the data is collected from 2 independent random samples, where each sample size is less than 10% of the population size if done without replacement. We also consider the independence condition satisfied if the data is collected from an experiment with two randomly assigned treatments (in this case the 10% condition is not relevant and does not need to be checked).

Large sample / normal population. Each population distribution should be nearly normal or each sample size should be at least 30. As before, if the sample sizes are small and the population distributions are not known to be nearly normal, we look at the data for strong skew or outliers. If we do not find strong skew or outliers in either group, the assumption that the populations are nearly normal is typically considered reasonable.

Let's check the two conditions necessary to use the t-distribution to the difference in sample means for the North Carolina births data set. (1) We will assume that we have two independent random samples and that the populations they were sampled from are much larger than 10 times the sample sizes of 50 and 100. (2) The sample sizes of 50 and 100 are well over 30, so we do not worry about the distributions of the original populations. Since both conditions are satisfied, we can use the two-sample t-test for $\mu_1 - \mu_2$.

	smoker	nonsmoker
mean	6.78	7.18
st. dev.	1.43	1.60
samp. size	50	100

Figure 4.25: Summary statistics for the `ncbirths` data set.

EXAMPLE 4.46

We will use the summary statistics in Figure 4.25 for this exercise.
(a) What is the point estimate of the population difference, $\mu_1 - \mu_2$?
(b) Compute the standard error of the point estimate from part (a).

(a) The point estimate is the difference in sample means: $\bar{x}_1 - \bar{x}_2 = 6.78 - 7.18 = -0.40$ pounds.
(b) The standard error formula for a difference in sample means looks like the standard deviation formula for a difference in sample means, but with the sample standard deviations used in place of the population standard deviations.

$$SE_{\bar{x}_1-\bar{x}_2} = \sqrt{\frac{s_1^2}{n_1} + \frac{s_2^2}{n_2}} = \sqrt{\frac{1.43^2}{50} + \frac{1.60^2}{100}} = 0.26 \text{ pounds}$$

EXAMPLE 4.47

Compute the test statistic.

We have already found the point estimate and the SE of estimate. The null hypothesis is that the two means are equal, or that their difference equals 0. The null value for the difference, therefore is 0. We now have everything we need to compute the test statistic.

$$T = \frac{\text{point estimate} - \text{null value}}{SE \text{ of estimate}} = \frac{(6.78 - 7.18) - 0}{\sqrt{\frac{1.43^2}{50} + \frac{1.60^2}{100}}} = \frac{-0.40 - 0}{0.26} = -1.54$$

EXAMPLE 4.48

Calculate the p-value for this hypothesis test.

We want to find tail areas of a t-distribution, but we need to know the degrees of freedom for the t-distribution. We saw previously that we can use the smaller of $n_1 - 1$ and $n_2 - 1$ for the degrees of freedom and then use a t-table to get the p-value if technology is not available. Alternately, we can use a technology option from Section 4.6.4 to find that the appropriate degrees of freedom for this test is 108.5. For the t-distribution with $df = 108.5$, shown in Figure 4.26, the area to the left of $T = -1.54$ is 0.062. Because this is a two-sided test, we care about both tails so we double this and find that the p-value is 0.124.

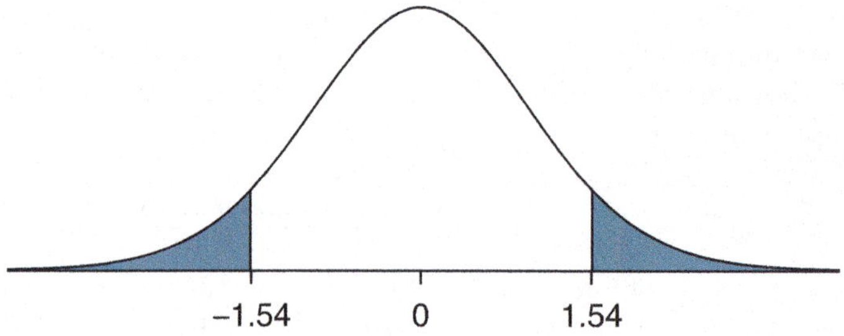

Figure 4.26: A t-distribution with 108.5 degrees of freedom. Values at least as extreme as our test statistic of -1.54 are shaded.

E

EXAMPLE 4.49

Interpret the p-value of 0.124 in the context of the problem.

In this context, a p-value of 0.124 means that there is a 12.4% probability of getting a T-statistic less than or equal to -1.54 or greater than or equal to 1.54 assuming there really is no difference in mean birth weight of newborns from North Carolina mothers who did smoke during pregnancy and newborns from North Carolina mothers who did not smoke during pregnancy. Equivalently, we can say that there is a 12.4% probability of getting a difference in sample means as small or smaller than -0.40 or as large or larger than 0.40 with random samples of these sizes assuming there really is no difference in mean birth weight of newborns from North Carolina mothers who did smoke during pregnancy and newborns from North Carolina mothers who did not smoke during pregnancy. In both cases, we are assessing the probability of getting a difference as extreme as we observed, under the assumption that H_0 is true.

E

EXAMPLE 4.50

What can we conclude from this p-value? Use a significance level of $\alpha = 0.05$.

This p-value of 0.124 is larger the significance level of 0.05, so we do not reject the null hypothesis. There is not sufficient evidence to say there is a difference in average birth weight of newborns from North Carolina mothers who did smoke during pregnancy and newborns from North Carolina mothers who did not smoke during pregnancy.

E

EXAMPLE 4.51

Does the conclusion to Example 4.47 mean that smoking and average birth weight are unrelated?

Not necessarily. It is possible that there is some difference but that we did not detect it. The result must be considered in light of other evidence and research. In fact, larger data sets do tend to show that women who smoke during pregnancy have smaller newborns.

G

GUIDED PRACTICE 4.52

If we made an error in our conclusion, which type of error could we have made: Type I or Type II?[20]

G

GUIDED PRACTICE 4.53

If we made a Type II Error and there is a difference, what could we have done differently in data collection to be more likely to detect the difference?[21]

---

[20]Since we did not reject H_0, it is possible that we made a Type II Error. It is possible that there is some difference but that we did not detect it.

[21]We could have collected more data. If the sample sizes are larger, we tend to have a better shot at finding a difference if one exists. In other words, increasing the sample size increases the power of the test.

4.6.3 Summary and worked example

HYPOTHESIS TEST FOR A DIFFERENCE IN MEANS

To carry out a complete hypothesis test to compare two population means,

Identify: Identify the test procedure, parameter, significance level, and hypotheses.

Use a **two-sample *t*-test for a difference in population means $\mu_1 - \mu_2$**. Define the population means μ_1 and μ_2 in words, referencing the populations of interest. Choose a significance level (α) and test the following hypotheses.

H_0: $\mu_1 = \mu_2$ $\quad(\mu_1 - \mu_2 = 0)$

H_A: $\mu_1 \neq \mu_2$; $\mu_1 > \mu_2$; or $\mu_1 < \mu_2$ $\quad(\mu_1 - \mu_2 \neq 0$; $\mu_1 - \mu_2 > 0$; or $\mu_1 - \mu_2 < 0)$

Check: Check conditions for the test statistic to have a t-distribution, assuming H_0 is true.

1. Independence: Data come from 2 independent random samples or from a randomized experiment with 2 treatments. When sampling without replacement, check that the sample size is less than 10% of the population size for each sample.
2. Large samples or normal populations: $n_1 \geq 30$ and $n_2 \geq 30$ or both population distributions are nearly normal. If the sample sizes are less than 30 and the population distributions are unknown, there should be no strong skew or outliers in either data set (this makes us believe that it is reasonable that both population distributions could be nearly normal).

Calculate: Calculate the t-statistic, df, and p-value.

$T = \frac{\text{point estimate} - \text{null value}}{SE \text{ of estimate}}$ $\quad df$: use technology to calculate

point estimate: $\bar{x}_1 - \bar{x}_2$, the difference in sample means

SE of estimate: $\sqrt{\frac{s_1^2}{n_1} + \frac{s_2^2}{n_2}}$

null value: 0

p-value = (based on the t-statistic, the df, and the direction of H_A)

Conclude: Compare the p-value to α, and draw a conclusion in context.

If the p-value is $\leq \alpha$, reject H_0; there is sufficient evidence that [H_A in context].

If the p-value is $> \alpha$, do not reject H_0; there is not sufficient evidence that [H_A in context].

EXAMPLE 4.54

Do embryonic stem cells (ESCs) help improve heart function following a heart attack? The following table and figure summarize results from an experiment to test ESCs in sheep that had a heart attack.

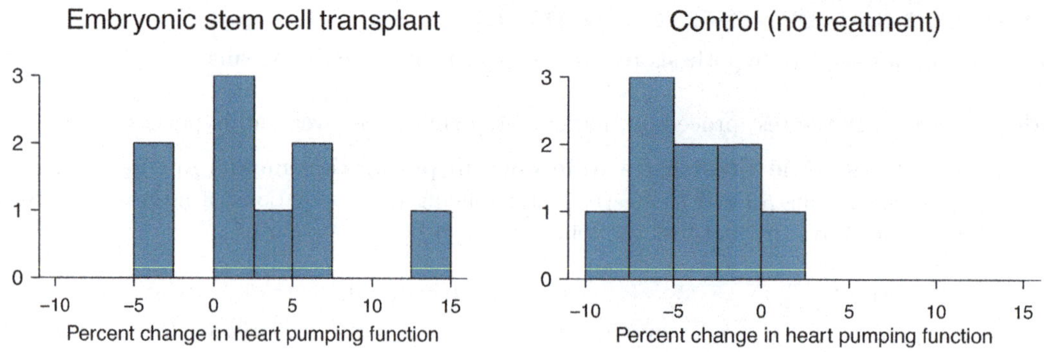

	n	$\bar{x}$	s
ESCs	9	3.50	5.17
control	9	-4.33	2.76

Each of these sheep was randomly assigned to the ESC or control group, and the change in their hearts' pumping capacity was measured. A positive value generally corresponds to increased pumping capacity, which suggests a stronger recovery. The sample data is also graphed. Use the given information and an appropriate statistical test to answer the research question.

E

Identify: Because we are hypothesizing about a difference of means we use a two-sample t-test for a difference in population means $\mu_1 - \mu_2$. Here, μ_1 is the mean percent change for sheep that would receive ESC and μ_2 is the mean percent change for sheep that would be in the control group. We will test the following hypotheses at the $\alpha = 0.05$ significance level.

$$H_0: \mu_1 = \mu_2$$
$$H_A: \mu_1 > \mu_2$$

Check: The data come from a randomized experiment with two treatment groups: ESC and control. Because this is an experiment, we do not need to check the 10% condition. The group sizes are small, but the data show no strong skew or outliers, so the assumption that the population distributions are nearly normal is reasonable.

Calculate: We will calculate the t-statistic and the p-value.

$$T = \frac{\text{point estimate} - \text{null value}}{SE \text{ of estimate}}$$

The point estimate is the difference in sample means: $\bar{x}_1 - \bar{x}_2 = 3.50 - (-4.33) = 7.83$.

$$SE \text{ of } \bar{x}_1 - \bar{x}_2 = \sqrt{\frac{s_1^2}{n_1} + \frac{s_2^2}{n_2}} = \sqrt{\frac{(5.17)^2}{9} + \frac{(2.76)^2}{9}} = 1.95.$$

The null value is the hypothesized difference in population means, which is 0.

$$T = \frac{(3.50 - (-4.33)) - 0}{\sqrt{\frac{(5.17)^2}{9} + \frac{(2.76)^2}{9}}} = \frac{7.83 - 0}{1.95} = 4.01$$

Because H_A is an upper tail test (>), the p-value corresponds to the area to the right of $t = 4.01$ with the appropriate degrees of freedom. Using technology, we find $df = 12.2$ and p-value $= 8.4 \times 10^{-4} = 0.00084$.

Conclude: The p-value is much less than 0.05, so we reject the null hypothesis. There is sufficient evidence that embryonic stem cells improve the heart's pumping function in sheep that have suffered a heart attack.

4.6.4 Technology: the two-sample t-interval and t-test for $\mu_1 - \mu_2$

Use technology to find a 95% confidence interval for a difference in true average scores between Version A and Version B of the exam, as described in Example 4.5.4. Also find the test statistic and p-value for a two-sided test to evaluate whether there is evidence that the true means differ. Conditions were verified to be met.

Version	n	$\bar{x}$	s	min	max
A	30	79.4	14	45	100
B	30	74.1	20	32	100

Desmos: Use the `ttest([data],[data])` or `ttest(n1, mean1, stdev1, n2, mean1, stdev2)` function as explained below.

1. Click + in the upper left, then choose `inference`.
2. Choose `t-test` in the pop-up window.

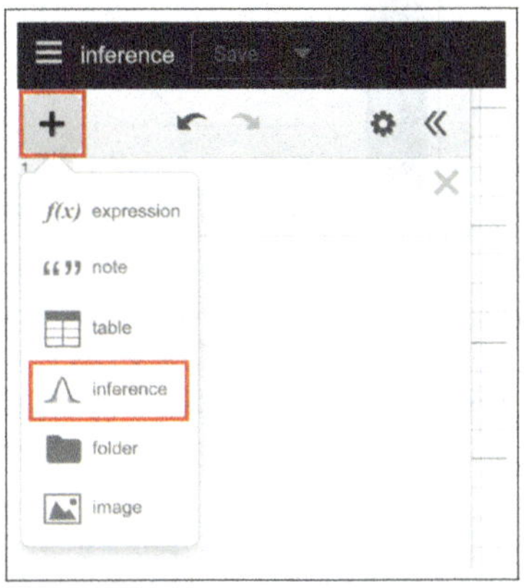

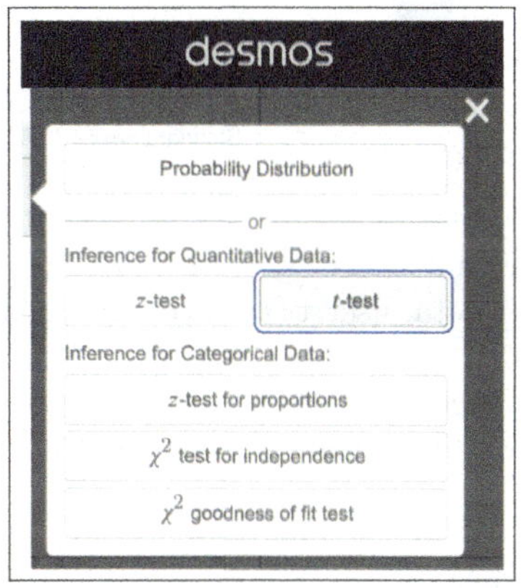

3. If you have all the data: under `SAMPLE 1` enter the first data set separated by commas or copy and paste it in the box, then click on `SAMPLE 2` and enter the second data set separated by commas or copy and paste it in the box.
   If you have the summary statistics: click on `Stats` and under `SAMPLE 1`, enter the first group's `sample size`, `mean` and `stdev`, then click on `SAMPLE 2` and enter the second group's `sample size`, `mean` and `stdev`. Then click `Create Test`. Here we have summary statistics, so we click `Stats`, enter the relevant values from the table above, then click Create Test.

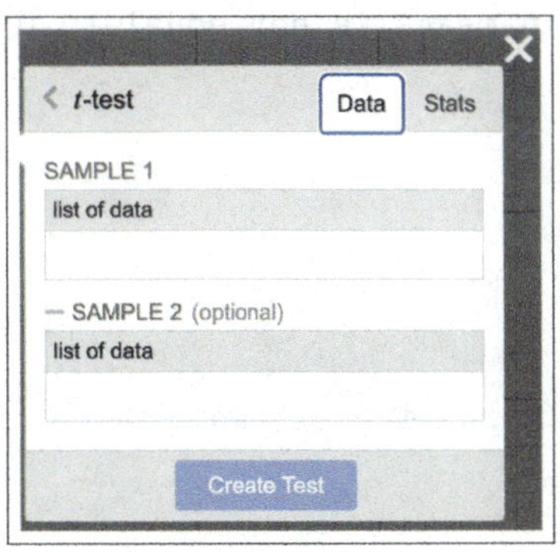

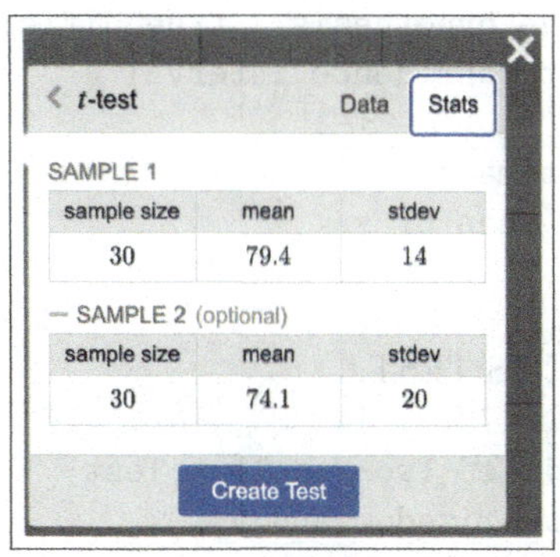

* You can type `ttest(30, 79.4, 14, 30, 74.1, 20)` in place of steps 1-3 above.

4. Click the triangle next to `Confidence Interval` and input the desired `Confidence level`. Here we use 0.95, which is entered by default. Click on ⋮ to the right of the confidence interval to see additional information. You can also hover over the dot in the middle of the confidence interval to see the point estimate. Find *df* by clicking Significance Test.
5. Click the triangle next to `Significance Test`. Enter the hypothesized difference for the null hypothesis. Select `Tails` to be `Left`, `Right` or `Both` depending on the direction of the alternative hypothesis. Here the hypothesized difference 0 and H_A uses a ≠, so we select Tails to be Both.

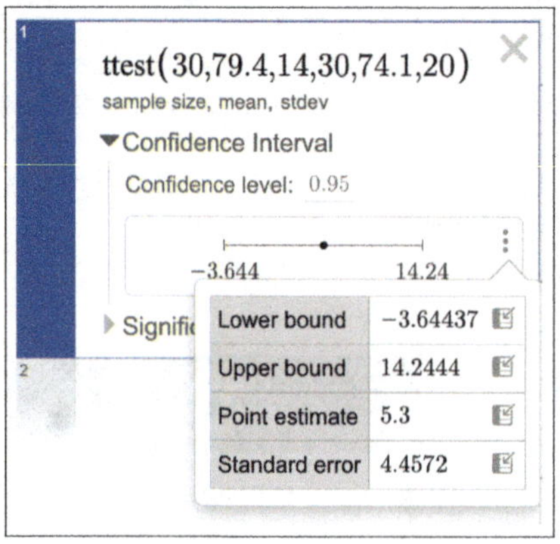

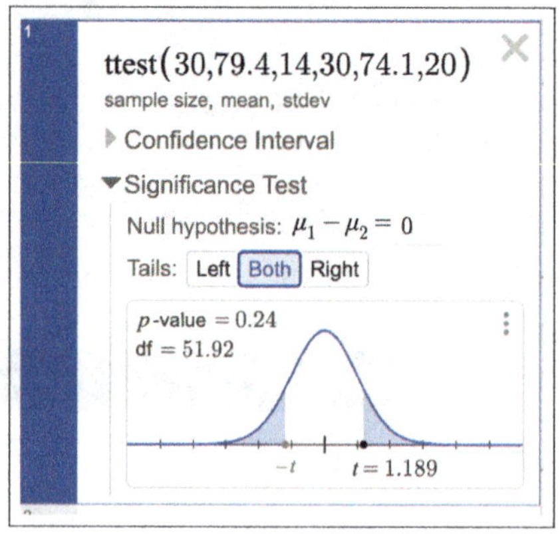

R: 2-sample t-interval/test for $\mu_1 - \mu_2$
With all the data, use: `t.test(data 1, data 2, conf.level = , alternative = )`

Here we do not have all the data, so we install the BSDA (Basic Statistics and Data Analysis) package. You only need to enter the first line once on a computer and the second line once per R session.

```
> install.packages("BSDA")
> library(BSDA)
```

This allow us to use:
`tsum.test(mean.x, sd.x, n.x, mean.y, sd.y, n.y, conf.level = , alternative = )`

CONFIDENCE INTERVAL.

```
> tsum.test(79.4, 14, 30, 74.1, 20, 30, conf.level = 0.95)
Welch Modified Two-Sample t-Test
data:  Summarized x and y
t = 1.1891, df = 51.918, p-value = 0.2398
alternative hypothesis:  true difference in means is not equal to 0
95 percent confidence interval:
 -3.644372 14.244372
sample estimates:
mean of x mean of y
79.4 74.1
```

HYPOTHESIS TEST.

```
> tsum.test(79.4, 14, 30, 74.1, 20, 30, alternative = "two.sided")
Welch Modified Two-Sample t-Test
data:  Summarized x and y
t = 1.1891, df = 51.918, p-value = 0.2398
alternative hypothesis:  true difference in means is greater than 0
95 percent confidence interval:
-2.164646 NA
sample estimates:
mean of x mean of y
79.4 74.1
```

Note that `conf.level = 0.95` and `alternative = "two.sided"` are the default values if those arguments are omitted. You can also use `alternative = "less"` or `alternative = "greater"` .

Calculator: NumWorks calculator instructions and example output are included. For the TI-83/84 and Casio calculators, general instructions are provided, and worked example videos can be accessed via the 🎥 icon or at **openintro.org/ti** and **openintro.org/casio**.

NUMWORKS: 2-SAMPLE T-INTERVAL.

Use `OK` or `EXE` to make a selection.

1. From the home screen, select `Inference`, then `Intervals`, then `Two means`, then `t-interval`. If these options do not appear, click the ← button in the upper right as many times as needed.
2. Choose `Input statistics` or `Use a dataset` and enter the needed values. Then use the down arrow and choose `Next`.
3. Note the quantities returned. Press the down arrow and choose `Next`.
4. In addition to seeing the confidence interval displayed in two ways, you can press the up and down arrows to quickly change confidence level and see the resulting interval and margin of error.

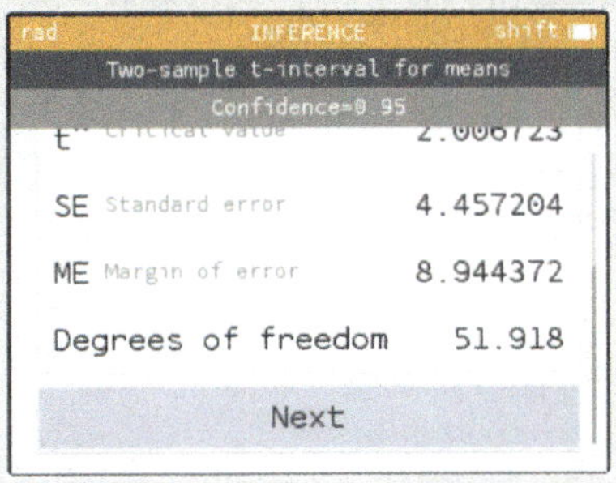

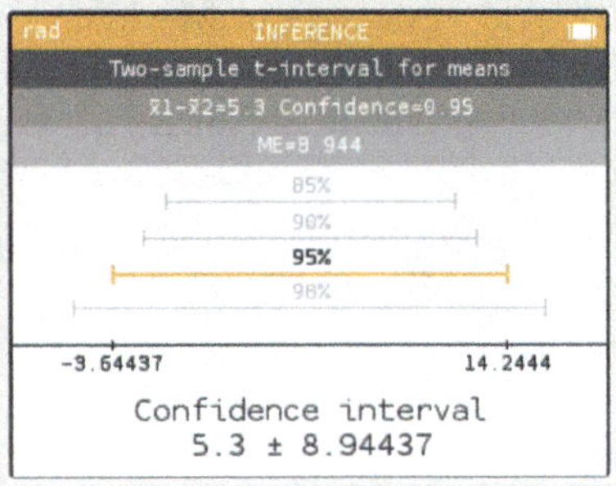

NUMWORKS: 2-SAMPLE T-TEST

Use `OK` or `EXE` to make a selection.

1. From the home screen, select `Inference`, then `Tests`, then `Two means`, then `t-test`. If these options do not appear, click the ← button in the upper right as many times as needed.
2. Enter the hypothesized difference for the Null hypothesis. Usually this will be 0. Press the down arrow. Press `OK` and choose <, ≠ , or > for the Alternative hypothesis. Press the down arrow and choose `Next`.
3. Choose `Input statistics` or `Use a dataset` and enter the needed values. Then use the down arrow and choose `Next`.
4. Note the quantities returned. Click the down arrow and choose `Next`.
5. On this screen, the p-value and alpha are shaded on the t-distribution and can be visually compared.

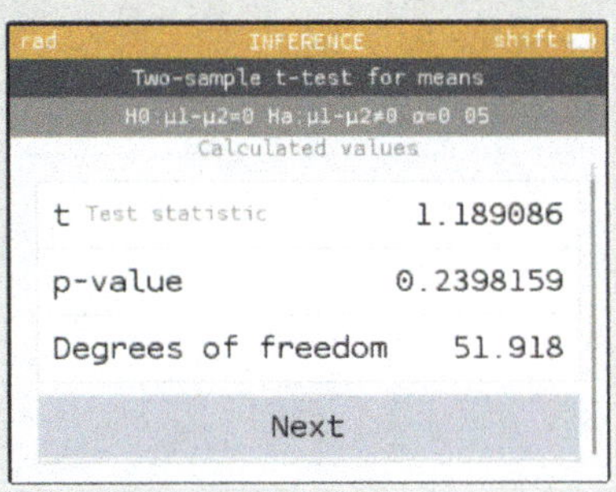

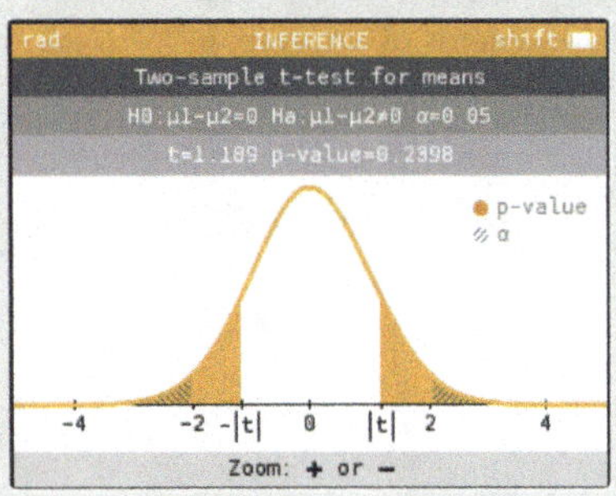

TI-83/84: 2-SAMPLE T-INTERVAL

Use STAT, TESTS, 2-SampTInt.

1. Choose STAT.
2. Right arrow to TESTS.
3. Down arrow and choose 0:2-SampTTInt.
4. Choose Data if you have all the data or Stats if you have the means and standard deviations.
    - If you choose Data, let List1 be L1 or the list that contains sample 1 and let List2 be L2 or the list that contains sample 2 (don't forget to enter the data!). Let Freq1 and Freq2 be 1.
    - If you choose Stats, enter the mean, SD, and sample size for sample 1 and for sample 2.
5. Let C-Level be the desired confidence level and let Pooled be No.
6. Choose Calculate and hit ENTER, which returns:

(__,__)	the confidence interval	Sx1	SD of sample 1
df	degrees of freedom	Sx2	SD of sample 2
$\bar{x}_1$	mean of sample 1	n1	size of sample 1
$\bar{x}_2$	mean of sample 2	n2	size of sample 2

TI-83/84: 2-SAMPLE T-TEST

Use STAT, TESTS, 2-SampTTest.

1. Choose STAT.
2. Right arrow to TESTS.
3. Choose 4:2-SampTTest.
4. Choose Data if you have all the data or Stats if you have the means and standard deviations.
    - If you choose Data, let List1 be L1 or the list that contains sample 1 and let List2 be L2 or the list that contains sample 2 (don't forget to enter the data!). Let Freq1 and Freq2 be 1.
    - If you choose Stats, enter the mean, SD, and sample size for sample 1 and for sample 2
5. Choose ≠, <, or > to correspond to H_A.
6. Let Pooled be NO.
7. Choose Calculate or Draw and hit ENTER. Draw shows the t-statistic and p-value as well as a graph of the t-distribution with p-value shaded. Calculate returns:

t	T-statistic	Sx1	SD of sample 1
p	p-value	Sx2	SD of sample 2
df	degrees of freedom	n1	size of sample 1
$\bar{x}_1$	mean of sample 1	n2	size of sample 2
$\bar{x}_2$	mean of sample 2		

CASIO FX-9750GII: 2-SAMPLE T-INTERVAL

1. Navigate to `STAT` (`MENU` button, then hit the `2` button or select `STAT`).
2. If necessary, enter the data into a list.
3. Choose the `INTR` option (`F4` button).
4. Choose the `t` option (`F2` button).
5. Choose the `2-S` option (`F2` button).
6. Choose either the `Var` option (`F2`) or enter the data in using the `List` option.
7. Specify the test details:
   - Confidence level of interest for `C-Level`.
   - If using the `Var` option, enter the summary statistics for each group. If using `List`, specify the lists and leave `Freq` values at `1`.
   - Choose whether to pool the data or not.
8. Hit the `EXE` button, which returns

`Left`, `Right`	ends of the confidence interval
`df`	degrees of freedom
`x̄1`, `x̄2`	sample means
`sx1`, `sx2`	sample standard deviations
`n1`, `n2`	sample sizes

CASIO FX-9750GII: 2-SAMPLE T-TEST

1. Navigate to `STAT` (`MENU` button, then hit the `2` button or select `STAT`).
2. If necessary, enter the data into a list.
3. Choose the `TEST` option (`F3` button).
4. Choose the `t` option (`F2` button).
5. Choose the `2-S` option (`F2` button).
6. Choose either the `Var` option (`F2`) or enter the data in using the `List` option.
7. Specify the test details:
   - Specify the sidedness of the test using the `F1`, `F2`, and `F3` keys.
   - If using the `Var` option, enter the summary statistics for each group. If using `List`, specify the lists and leave `Freq` values at `1`.
   - Choose whether to pool the data or not.
8. Hit the `EXE` button, which returns

μ1 __ μ2	alt. hypothesis	`x̄1`, `x̄2`	sample means
`t`	T-statistic	`sx1`, `sx2`	sample standard deviations
`p`	p-value	`n1`, `n2`	sample sizes
`df`	degrees of freedom		

Section summary

- The appropriate hypothesis testing procedure for a difference between two population means μ_1 and μ_2 is a **two-sample *t*-test for a difference in population means $\mu_1 - \mu_2$**. The parameters μ_1 and μ_2 should be identified in context.
- The null hypotheses for a two-sample t-test for a difference in population means indicates no difference and is written as:

  H_0: $\mu_1 = \mu_2$ (or equivalently H_0: $\mu_1 - \mu_2 = 0$).
- A one-sided alternative hypothesis is written as:

  H_A: $\mu_1 < \mu_2$ (or equivalently H_A: $\mu_1 - \mu_2 < 0$), or
  H_A: $\mu_1 > \mu_2$ (or equivalently H_A: $\mu_1 - \mu_2 > 0$).

  A two-sided alternative hypothesis is written as:

  H_A: $\mu_1 \neq \mu_2$ (or equivalently H_A: $\mu_1 - \mu_2 \neq 0$).
- The null hypotheses for a two-sample t-test for a difference in population means can be written as: H_0: $\mu_1 = \mu_2$ or H_0: $\mu_1 - \mu_2 = 0$. The one-sided alternative hypothesis is: H_A: $\mu_1 < \mu_2$ (or equivalently H_A: $\mu_1 - \mu_2 < 0$) or H_A: $\mu_1 > \mu_2$ (or equivalently H_A: $\mu_1 - \mu_2 > 0$). The two-sided alternative hypothesis is: H_A: $\mu \neq \mu_0$ (or equivalently H_A: $\mu_1 - \mu_2 \neq 0$).
- The two-sample t-test for a difference in population means has the same conditions as the two-sample t-interval for a difference in population means.

  1. Independence: The data come from two independent random samples, each with sample size < 10% of its corresponding population size if sampling without replacement OR the data come from a randomized experiment with two randomly assigned treatments.
  2. Large sample or normal population: both sample sizes are at least 30 or population distributions are nearly normal. If either sample size is less than 30 and the population distributions are unknown, check and confirm that there is no strong skew or outliers in either data set in order to reasonably assume that the population distributions are nearly normal.
- A test statistic has the form:

$$\text{test statistic} = \frac{\text{point estimate} - \text{null value}}{SE \text{ of estimate}}.$$

- The test statistic for a two-sample t-test for $\mu_1 - \mu_2$ is:

$$T = \frac{(\bar{x}_1 - \bar{x}_2) - 0}{\sqrt{\frac{s_1^2}{n_1} + \frac{s_2^2}{n_2}}}, \quad df\text{: use technology}$$

  When the null hypothesis is true, the test statistics has a t-distribution with degrees of freedom that can be calculated using technology. The degrees of freedom fall between $n_1 + n_2 - 2$ and the smaller of $n_1 - 1$ and $n_2 - 1$.
- The p-value of a t-test corresponds to a lower tail, upper tail, or both tails of the t-distribution with the appropriate degrees of freedom, depending on whether the direction of the alternate hypothesis is <, >, or ≠.
- The p-value for a two-sample t-test for $\mu_1 - \mu_2$ is the probability of obtaining t-statistic as small or smaller, as large or larger, or as extreme or more extreme than the t-statistic that was observed, depending on whether the direction of the alternative hypothesis is <, >, or ≠, assuming the null hypothesis is true (i.e. that the population means are equal to each other).
- A formal decision explicitly compares the p-value to the significance level. If the p-value $\leq \alpha$, then reject the null hypothesis; if the p-value $> \alpha$, then fail to reject the null hypothesis. The conclusion should be stated in terms of the alternative hypothesis and should include context, referencing the parameters and the populations. Use non-causal language unless a well-designed experiment was conducted.

Exercises

4.41 Chicken diet and weight, Part II. In Exercise 4.39, we learned about an experiment that was conducted to measure and compare the effectiveness of various feed supplements on the growth rate of chickens. Newly hatched chicks were randomly allocated into six groups, and each group was given a different feed supplement. Below are some summary statistics from this data set along with box plots showing the distribution of weights by feed type.[22]

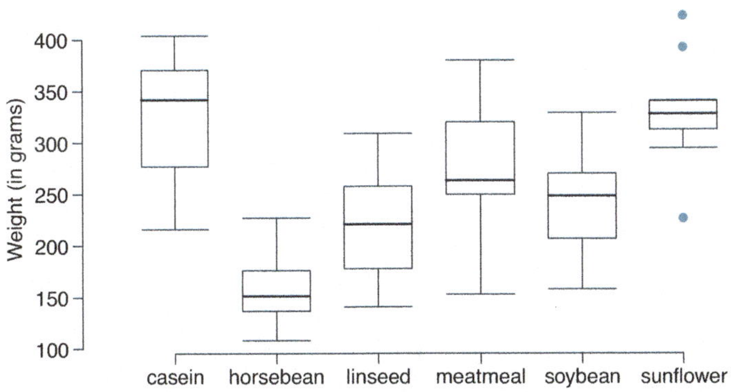

	Mean	SD	n
casein	323.58	64.43	12
horsebean	160.20	38.63	10
linseed	218.75	52.24	12
meatmeal	276.91	64.90	11
soybean	246.43	54.13	14
sunflower	328.92	48.84	12

(a) Describe the distributions of weights of chickens that were fed linseed and horsebean.

(b) Do these data provide strong evidence that the average weights of chickens that would be fed linseed and horsebean are different? Use a 5% significance level.

(c) What type of error might we have committed? Explain.

(d) Would your conclusion change if we used $\alpha = 0.01$?

4.42 Fuel efficiency of manual and automatic cars, City. The table provides summary statistics on highway fuel economy of the same 52 cars from Exercise 4.38. Do these data provide strong evidence of a difference between the average fuel efficiency of cars with manual and automatic transmissions in terms of their average city mileage?[23]

	City MPG	
	Automatic	Manual
Mean	16.12	19.85
SD	3.58	4.51
n	26	26

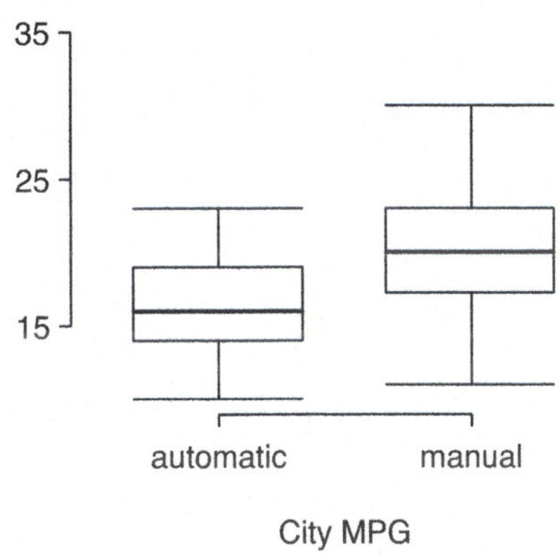

4.43 Gaming and distracted eating, Part I. A group of researchers are interested in the possible effects of distracting stimuli during eating, such as an increase or decrease in the amount of food consumption. To test this hypothesis, they monitored food intake for a group of 44 patients who were randomized into two equal groups. The treatment group ate lunch while playing solitaire, and the control group ate lunch without any added distractions. Patients in the treatment group ate 52.1 grams of biscuits, with a standard deviation of 45.1 grams, and patients in the control group ate 27.1 grams of biscuits, with a standard deviation of 26.4 grams. Do these data provide convincing evidence that the average food intake (measured in amount of biscuits consumed) is different for the patients in the treatment group? Assume that conditions for inference are satisfied.[24]

[22] Chicken Weights by Feed Type, from the `datasets` package in R..

[23] U.S. Department of Energy, Fuel Economy Data, 2012 Datafile.

[24] R.E. Oldham-Cooper et al. "Playing a computer game during lunch affects fullness, memory for lunch, and later snack intake". In: *The American Journal of Clinical Nutrition* 93.2 (2011), p. 308.

4.44 Gaming and distracted eating, Part II. The researchers from Exercise 4.43 also investigated the effects of being distracted by a game on how much people eat. The 22 patients in the treatment group who ate their lunch while playing solitaire were asked to do a serial-order recall of the food lunch items they ate. The average number of items recalled by the patients in this group was 4. 9, with a standard deviation of 1.8. The average number of items recalled by the patients in the control group (no distraction) was 6.1, with a standard deviation of 1.8. Do these data provide strong evidence that the average number of food items recalled by the patients in the treatment and control groups are different?

Chapter highlights

We've reviewed a wide set of inference procedures over the last 2 chapters. Let's revisit each and discuss the similarities and differences among them. The following confidence intervals and tests are structurally the same – they all involve inference for a population parameter, where that parameter is a proportion, a difference of proportions, a mean, a mean of differences, or a difference of means.

- one-sample z-test/interval for p
- two-sample z-test/interval for $p_1 - p_2$
- one-sample t-test/interval for μ or μ_d
- two-sample t-test/interval for $\mu_1 - \mu_2$

The above inferential procedures all involve a **point estimate**, a **standard error** of the estimate, and an assumption about the **shape of the sampling distribution** for the point estimate.

In Chapter 3, we were also introduced to two χ^2 tests for two-way tables:

- χ^2 test for homogeneity: compares a categorical variable across multiple groups.
- χ^2 test for independence: looks for association between two categorical variables.

χ^2 is a measure of *overall* deviation between observed values and expected values, relative to expected values. These tests stand apart from the other tests because when using χ^2 there is not a parameter of interest. For this reason there are no confidence intervals using χ^2. Also, for χ^2 tests, the hypotheses are usually written in words, because they are not about a single parameter.

While formulas and conditions vary, all of these procedures follow the same basic logic and process.

- Identify the appropriate procedure and the parameter of interest (if applicable). For a confidence interval, also identify the confidence level; for a hypothesis test, identify the significance level and the hypotheses to be tested.
- Check that all conditions for the procedure are met.
- Calculate the confidence interval or the test statistic and p-value, as well as the *df* if applicable.
- Conclude by interpreting the results in context and drawing a conclusion based on the data.

For a summary of these hypothesis test and confidence interval procedures, see the Inference Guide in Appendix D.2.

Chapter exercises

4.45 Hen eggs. The distribution of the number of eggs laid by a certain species of hen during their breeding period has a mean of 35 eggs with a standard deviation of 18.2. Suppose a group of researchers randomly samples 45 hens of this species, counts the number of eggs laid during their breeding period, and records the sample mean. They repeat this 1,000 times, and build a distribution of sample means.

(a) What is this distribution called?

(b) Would you expect the shape of this distribution to be symmetric, right skewed, or left skewed? Explain your reasoning.

(c) Calculate the variability of this distribution and state the appropriate term used to refer to this value.

(d) Suppose the researchers' budget is reduced and they are only able to collect random samples of 10 hens. The sample mean of the number of eggs is recorded, and we repeat this 1,000 times, and build a new distribution of sample means. How will the variability of this new distribution compare to the variability of the original distribution?

4.46 College credits. A college counselor is interested in estimating how many credits a student typically enrolls in each semester. The counselor decides to randomly sample 100 students by using the registrar's database of students. The histogram below shows the distribution of the number of credits taken by these students. Sample statistics for this distribution are also provided.

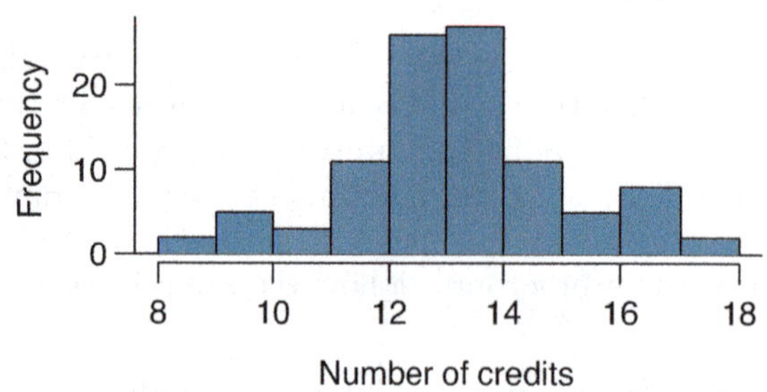

Min	8
Q1	13
Median	14
Mean	13.65
SD	1.91
Q3	15
Max	18

(a) What is the point estimate for the average number of credits taken per semester by students at this college? What about the median?

(b) What is the point estimate for the standard deviation of the number of credits taken per semester by students at this college? What about the IQR?

(c) Is a load of 16 credits unusually high for this college? What about 18 credits? Explain your reasoning.

(d) The college counselor takes another random sample of 100 students and this time finds a sample mean of 14.02 units. Should she be surprised that this sample statistic is slightly different than the one from the original sample? Explain your reasoning.

(e) The sample means given above are point estimates for the mean number of credits taken by all students at that college. What measure do we use to quantify the variability of this estimate? Compute this quantity using the data from the original sample.

4.47 Air quality. Air quality measurements were collected in a random sample of 25 country capitals in 2013, and then again in the same cities in 2014. We would like to use these data to compare average air quality between the two years. Should we use a paired or non-paired test? Explain your reasoning.

4.48 True / False: paired. Determine if the following statements are true or false. If false, explain.

(a) In a paired analysis we first take the difference of each pair of observations, and then we do inference on these differences.

(b) Two data sets of different sizes cannot be analyzed as paired data.

(c) Consider two sets of data that are paired with each other. Each observation in one data set has a natural correspondence with exactly one observation from the other data set.

(d) Consider two sets of data that are paired with each other. Each observation in one data set is subtracted from the average of the other data set's observations.

4.49 Find the mean. You are given the following hypotheses:

$$H_0 : \mu = 60$$
$$H_A : \mu < 60$$

We know that the sample standard deviation is 8 and the sample size is 20. For what sample mean would the p-value be equal to 0.05? Assume that all conditions necessary for inference are satisfied.

4.50 Diamonds, Part I. Prices of diamonds are determined by what is known as the 4 Cs: cut, clarity, color, and carat weight. The prices of diamonds go up as the carat weight increases, but the increase is not smooth. For example, the difference between the size of a 0.99 carat diamond and a 1 carat diamond is undetectable to the naked human eye, but the price of a 1 carat diamond tends to be much higher than the price of a 0.99 diamond. In this question we use two random samples of diamonds, 0.99 carats and 1 carat, each sample of size 23, and compare the average prices of the diamonds. In order to be able to compare equivalent units, we first divide the price for each diamond by 100 times its weight in carats. That is, for a 0.99 carat diamond, we divide the price by 99. For a 1 carat diamond, we divide the price by 100. The distributions and some sample statistics are shown below.[25]

Conduct a hypothesis test to evaluate if there is a difference between the average standardized prices of 0.99 and 1 carat diamonds. Remember to Identify, Check, Calculate, and Conclude.

	0.99 carats	1 carat
Mean	\$44.51	\$56.81
SD	\$13.32	\$16.13
n	23	23

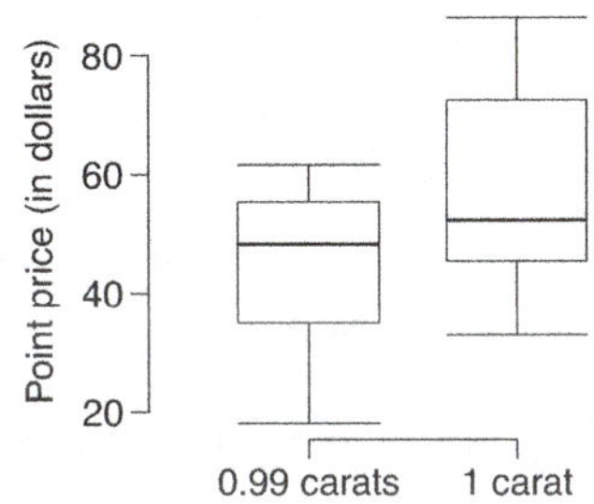

4.51 Friday the 13th, Part I. In the early 1990's, researchers in the UK collected data on traffic flow, number of shoppers, and traffic accident related emergency room admissions on Friday the 13th and the previous Friday, Friday the 6th. The histograms below show the distribution of number of cars passing by a specific intersection on Friday the 6th and Friday the 13th for many such date pairs. Also given are some sample statistics, where the difference is the number of cars on the 6th minus the number of cars on the 13th.[26]

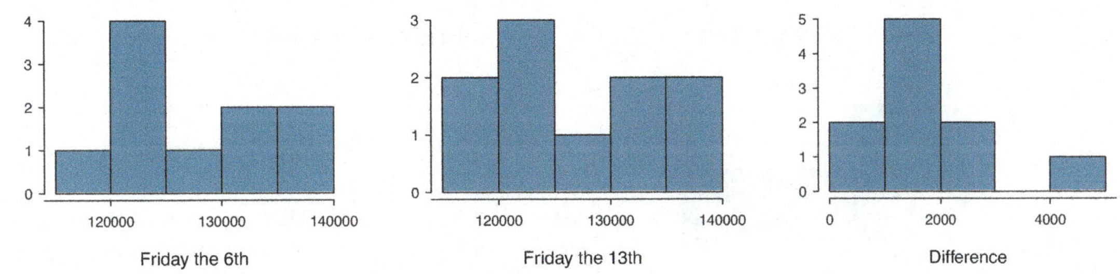

	6th	13th	Diff.
$\bar{x}$	128,385	126,550	1,835
s	7,259	7,664	1,176
n	10	10	10

(a) Are there any underlying structures in these data that should be considered in an analysis? Explain.

(b) What are the hypotheses for evaluating whether the number of people out on Friday the 6th is different than the number out on Friday the 13th?

(c) Check conditions to carry out the hypothesis test from part (b).

(d) Calculate the test statistic and the p-value.

(e) What is the conclusion of the hypothesis test?

(f) Interpret the p-value in this context.

(g) What type of error might have been made in the conclusion of your test? Explain.

[25] H. Wickham. *ggplot2: elegant graphics for data analysis.* Springer New York, 2009.

[26] T.J. Scanlon et al. "Is Friday the 13th Bad For Your Health?" In: *BMJ* 307 (1993), pp. 1584–1586.

4.52 Age at first marriage, Part I. The National Survey of Family Growth conducted by the Centers for Disease Control gathers information on family life, marriage and divorce, pregnancy, infertility, use of contraception, and men's and women's health. One of the variables collected on this survey is the age at first marriage. The histogram below shows the distribution of ages at first marriage of 5,534 randomly sampled women between 2006 and 2010. The average age at first marriage among these women is 23.44 with a standard deviation of 4.72.[27]

Estimate the average age at first marriage of women using a 95% confidence interval, and interpret this interval in context. Discuss any relevant assumptions.

4.53 Friday the 13th, Part II. The Friday the 13th study reported in Exercise 4.51 also provides data on traffic accident related emergency room admissions. The distributions of these counts from Friday the 6th and Friday the 13th are shown below for six such paired dates along with summary statistics. You may assume that conditions for inference are met.

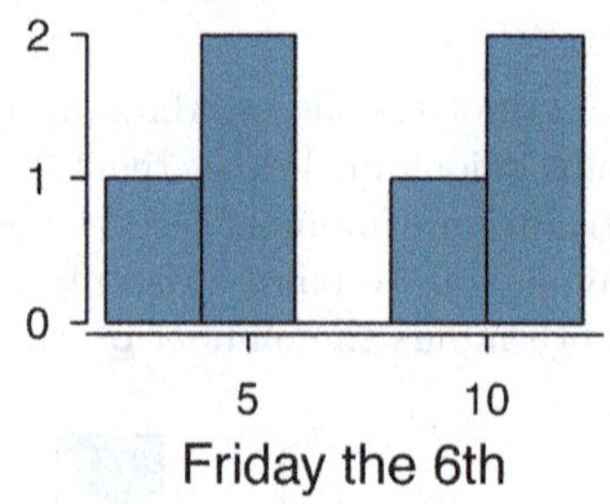

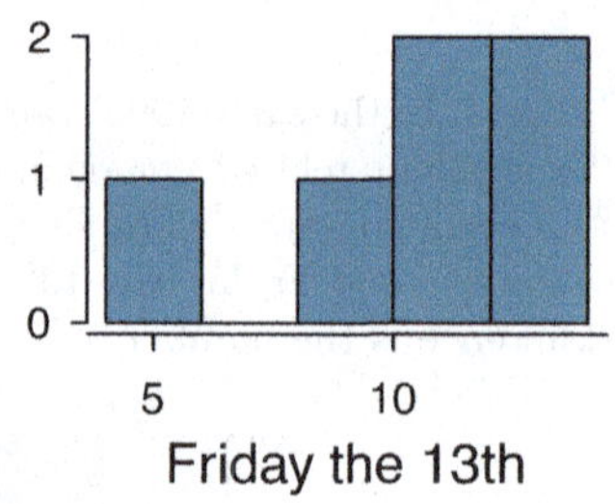

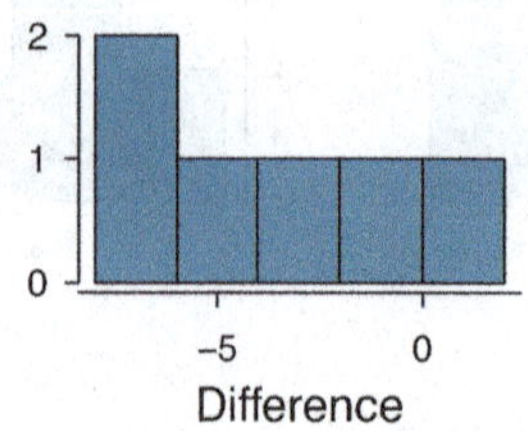

	6th	13th	diff
Mean	7.5	10.83	-3.33
SD	3.33	3.6	3.01
n	6	6	6

(a) Conduct a hypothesis test to evaluate if there is a difference between the average numbers of traffic accident related emergency room admissions between Friday the 6th and Friday the 13th.

(b) Calculate a 95% confidence interval for the difference between the average numbers of traffic accident related emergency room admissions between Friday the 6th and Friday the 13th.

(c) The conclusion of the original study states, "Friday 13th is unlucky for some. The risk of hospital admission as a result of a transport accident may be increased by as much as 52%. Staying at home is recommended." Do you agree with this statement? Explain your reasoning.

[27] Centers for Disease Control and Prevention, National Survey of Family Growth, 2010.

4.54 Diamonds, Part II. In Exercise 4.50, we discussed diamond prices (standardized by weight) for diamonds with weights 0. 99 carats and 1 carat. See the table for summary statistics, and then use a 95% confidence interval procedure to estimate the difference in means between the standardized prices of 0.99 and 1 carat diamonds. Remember to Identify, Check, Calculate, and Conclude.

	0.99 carats	1 carat
Mean	\$44.51	\$56.81
SD	\$13.32	\$16.13
n	23	23

Chapter 5

Regression Analysis

Linear regression is a very powerful statistical technique. Many people have some familiarity with regression just from reading the news, where graphs with straight lines are overlaid on scatterplots. Linear models can be used to see trends and to make predictions.

For videos, slides, and other resources, please visit
www.openintro.org/ahss

5.1 Summarizing bivariate numerical data

In this section, we explore scatterplots and describe the relationship between two numerical variables. We use the `loan50` data set and a new data set on US counties, called `county_2023`. We also introduce a new bivariate summary called the *correlation coefficient*.

Learning objectives

1. Use scatterplots to depict the relationship between two numerical variables.
2. Describe the characteristics of a scatterplot.
3. Justify a claim using scatterplots depicting the distribution of two numerical variables.
4. Interpret the correlation for a linear relationship.

5.1.1 Scatterplots for paired data

Scatterplots are one type of graph used to study the relationship between two numerical variables. A **scatterplot** is a 2-dimensional graph where each case is plotted as a point. In a practical sense, it provides a case-by-case view of data that illustrates the relationship between two numerical variables. Consider the `loan50` data set introduced earlier in the book. Figure 5.1 displays rows 1, 2, 3, and 50 of the data set for 50 randomly sampled loans offered through Lending Club.

	loan_amount	interest_rate	term	grade	state	total_income	homeownership
1	7500	7.34	36	A	MD	70000	rent
2	25000	9.43	60	B	OH	254000	mortgage
3	14500	6.08	36	A	MO	80000	mortgage
⋮	⋮	⋮	⋮	⋮	⋮	⋮	⋮
50	3000	7.96	36	A	CA	34000	rent

Figure 5.1: Four rows from the `loan50` data matrix.

A scatterplot is shown in Figure 5.2, showing the total income of a borrower (`total_income`) and the amount they borrowed (`loan_amount`) for the `loan50` data set. In any scatterplot, each point represents a single case. For this example, each point represents a single loan. Since there are 50 loans in `loan50`, there are 50 points in Figure 5.2.

Looking at Figure 5.2, we see that there are many borrowers with an income below $100,000 on the left side of the graph, while there are only a few borrowers with income above $250,000. We can also see the range of loan amounts by looking at the vertical axis. While it is more difficult to discern the exact distributions of each of these variables, what we can easily observe from the scatterplot is the *relationship* or association between the variables.

G

GUIDED PRACTICE 5.1

Examine the variables in the `loan50` data set summarized in Figure 5.1, Create two questions about possible relationships between variables in `loan50` that are of interest to you.[1]

[1] Two example questions: (1) What is the relationship between loan amount and interest rate? (2) If someone's income is above the average, will their interest rate tend to be above or below the average?

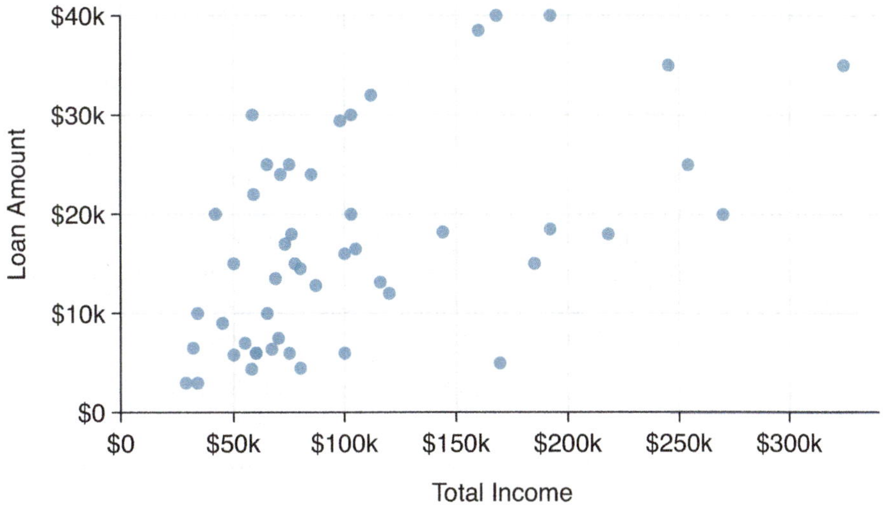

Figure 5.2: A scatterplot of `loan_amount` versus `total_income` for the `loan50` data set.

EXAMPLE 5.2

A scatterplot requires **bivariate**, or **paired data**. What does paired data mean?

We say observations are *paired* when the two observations correspond to the same case or individual. In unpaired data, there is no such correspondence. In our example the two observations correspond to a particular loan.

The variable that is suspected to be the response variable (or dependent variable) is plotted on the vertical (y) axis and the variable that is suspected to be the explanatory variable is plotted on the horizontal (x) axis. In this example, we suspect the loan amount is dependent on total income, not the reverse.

DRAWING SCATTERPLOTS

(1) Decide which variable should go on each axis, and draw and label the two axes.
(2) Note the range of each variable, and add tick marks and scales to each axis.
(3) Plot the dots as you would on an (x, y) coordinate plane.

GUIDED PRACTICE 5.3

Why are scatterplots useful? What do they show us that a list of (x, y) data points does not?[2]

[2]Answers may vary. Scatterplots are helpful in quickly spotting associations between variables.

5.1.2 Describing the relationship between two numerical variables

Many analyses are motivated by a researcher looking for a relationship between two or more variables. We will investigate relationships between variables using a new data set, the `county_2023` data set. These data come from the US Census Bureau's 2023 American Community Survey (ACS).

Unlike the Decennial Census, which takes place every 10 years and attempts to collect basic demographic data from every resident of the US, the ACS is an ongoing survey that is sent to approximately 3.5 million households per year. As stated on the ACS website, these data help communities "plan for hospitals and schools, support school lunch programs, improve emergency services, build bridges, and inform businesses looking to add jobs and expand to new markets, and more."[3] A social scientist may like to answer some of the following questions:

(1) How strongly associated are median household income and share of adults with a bachelor's degree?

(2) Is the relationship between median household income and poverty rate for the counties linear or nonlinear?

(3) How useful a predictor is share of adults with a bachelor's degree for mean travel time to work?

When describing a scatterplot or describing the association between two numerical variables, it is helpful to discuss *form*, *strength*, and *direction*. Form can be described as **linear**, where the points follow a trend that changes at an approximately constant rate, or **nonlinear**, where the points follow a trend that is curved rather than constant. We describe the strength of the association as weak, moderate, or strong, depending on how closely the points follow the general pattern.

When two variables show some connection with one another, they are said to be associated. The direction of association between two variables can be **positive** or **negative**, or there can be no association. Positive association means that larger values of the first variable are associated with larger values of the second variable.

EXAMPLE 5.4

What would it mean for two variables to have a *negative* association? What about *no* association?

Negative association implies that larger values of the first variable are associated with smaller values of the second variable. No association implies that the values of the second variable tend to be independent of differences in the first variable.

Figure 5.3 shows the association between the percent with a bachelor's degree among those 25 years and older and median household income for each of the 3,144 US counties. Each point on the plot represents a single county. For instance, the highlighted dot corresponds to County 1456 in the `county_2023` data set: Oktibbeha County, Mississippi, which has 45.8% of those ages 25 and older with a bachelor's degree and a median household income of $43,482.

The scatterplot in Figure 5.3 shows a roughly linear, constant trend, with a moderate association. It also shows a positive relationship between the two variables: counties with a higher median household income tend to have higher percent of those age 25 and older with a bachelor's degree. In this case, it is a choice which variable to put on the horizontal axis. It is true that households with higher income tend to be more able to afford college, but it is also true that those with a college degree tend to earn higher incomes.

Oktibbeha County, Mississippi, identified in Figure 5.3 does not quite follow the trend. This county has a lower than average household income, but a higher than average percent of bachelor's degrees among those 25 years and older. One possible factor is the presence of Mississippi State University (MSU) in Oktibbeha County and its adjoining county, which attracts and graduates many students in the area.

[3] `https://www.census.gov/programs-surveys/acs/about.html`

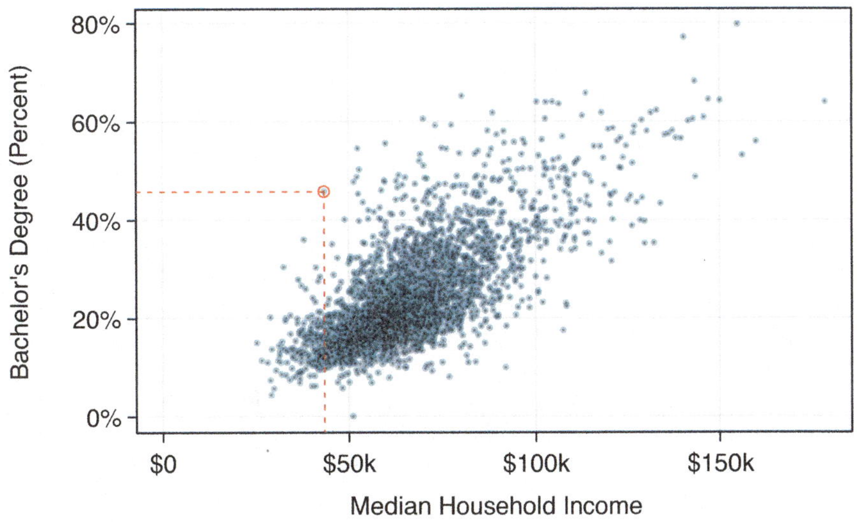

Figure 5.3: A scatterplot of percent with a bachelor's degree among those 25 years and older versus median household income for the `county_2023` data set. The highlighted dot represents Oktibbeha County, Mississippi, which has median household income of $43,482 and 45.8% of those age 25 and older with a bachelor's degree.

EXAMPLE 5.5

Figure 5.4 shows a scatterplot of median household income versus the poverty rate for the 3,144 counties in the US. Each point represents one county. Describe the relationship between these two variables.

The relationship between median household income and poverty rate is nonlinear, as highlighted by the dashed line. This is different from the scatterplot in Figure 5.3, which did not show much, if any, curvature in the trend. The association seems fairly strong as the points follow the curved trend pretty closely. There is also a negative association here, as higher rates of poverty tend to be associated with lower median household income.

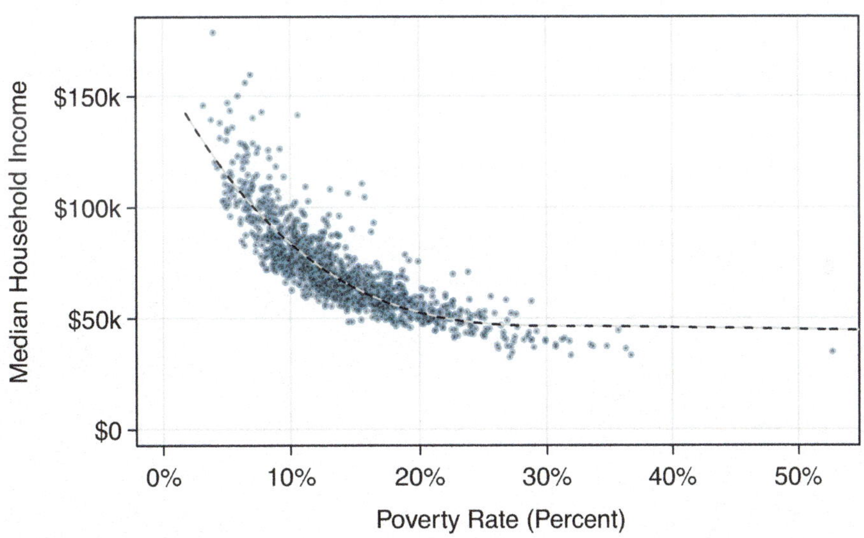

Figure 5.4: A scatterplot of the median household income against the poverty rate for the `county_2023` data set. A statistical model has also been fit to the data and is shown as a dashed line.

EXAMPLE 5.6

Figure 5.5 shows a scatterplot of mean travel time to work versus percent with a bachelor's degree among those 25 years and older for counties in the US. The mean travel time to work across all counties is shown as a dashed line. Based on the scatterplot, how much help does knowing the percent with a bachelor's degree in a county provide for predicting that county's mean travel time to work?

Ⓔ There seems to be almost no association between mean travel time to work and the percent with a bachelor's degree. Therefore, knowing the percent with a bachelor's degree in a county offers almost no help in trying to predict that county's mean travel time to work. For example, the mean travel time to work among those counties with a low percent with a bachelor's degree seems to be about the same as the mean travel time to work among those counties with a high percent with a bachelor's degree. On the other hand, in Figure 5.4, we see a fairly strong negative association between median household income and poverty rate. If a county has a low poverty rate, we would tend to predict a higher median household income, whereas if a county has a high poverty rate, we would tend to predict a lower median income.

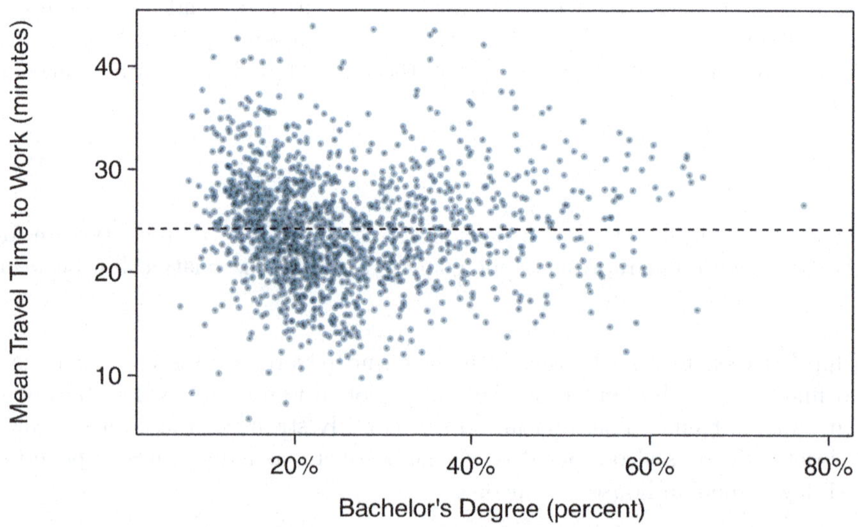

Figure 5.5: A scatterplot of mean travel time to work versus the percent with a bachelor's degree among those 25 years for the `county_2023` data set.

GUIDED PRACTICE 5.7

Ⓖ Describe two variables that would have a horseshoe-shaped association in a scatterplot (∩ or ∪).[4]

---

[4]Consider the case where your vertical axis represents something "good" and your horizontal axis represents something that is only good in moderation. Health and water consumption fit this description: we require some water to survive, but consume too much and it becomes toxic and can kill a person. If health was represented on the vertical axis and water consumption on the horizontal axis, then we would create a ∩ shape.

GUIDED PRACTICE 5.8

Consider the dot plots in Figure 5.6. Do these graphs tell us anything about the association between unemployment rates in the two states?[5]

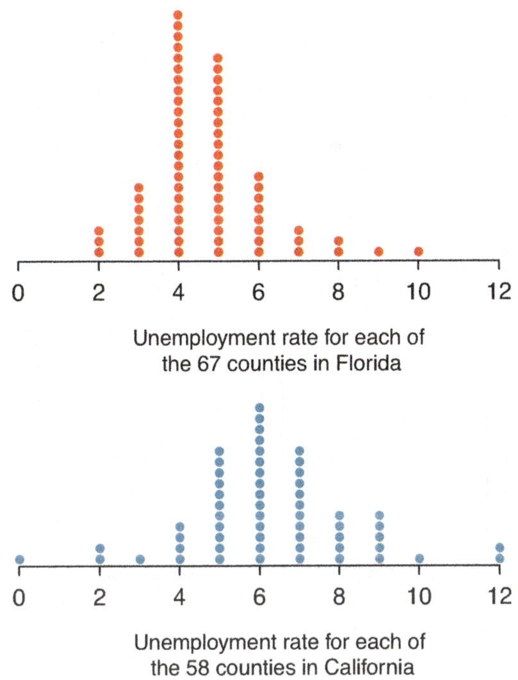

Figure 5.6: Dot plots for unemployment rate, rounded to the nearest percent, for counties in Florida and counties in California from the `county_2023` data set.

Describing association between two numerical variables is different than comparing distributions. When comparing distributions, we ask questions such as, "Which distribution has a greater average?" and "How do the shapes of the distribution differ?" The number of individuals in each data set need not be the same. When we look at association, we are interested in form, direction, and strength of the association between the variables. This requires data sets of equal length that are essentially paired (e.g. unemployment rate and poverty rate measured for each county or loan amount and corresponding interest rate for the loan).

COMPARING DISTRIBUTIONS VERSUS LOOKING AT ASSOCIATION

We compare two distributions with respect to center, spread, and shape. To compare the distributions visually, we use comparative graphs, such as two histograms, two dot plots, parallel box plots, or a back-to-back stem-and-leaf. When describing association between two variables, we comment on type, strength, and direction of the association. To see association visually, we require a scatterplot.

[5]No, to see association we require a scatterplot. Moreover, these data are not paired in any way, so the discussion of association does not make sense here.

5.1.3 Describing linear relationships with correlation

When a linear relationship exists between two variables, we can quantify the strength and direction of the linear relation with the **correlation coefficient**, or just **correlation** for short. Figure 5.7 shows eight plots and their corresponding correlations.

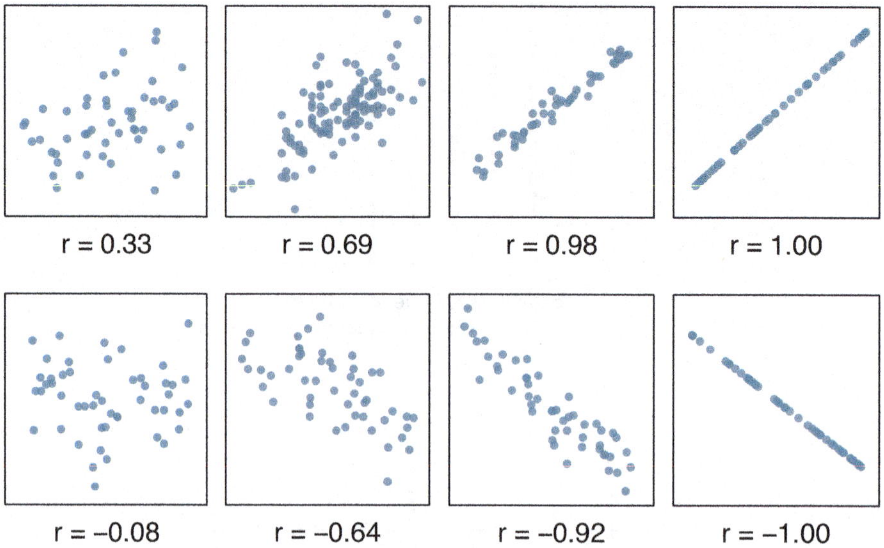

Figure 5.7: Sample scatterplots and their correlations. The first row shows variables with a positive relationship, represented by the trend up and to the right. The second row shows variables with a negative trend, where a large value in one variable is associated with a low value in the other.

Only when the relationship is perfectly linear is the correlation coefficient either −1 or 1. If the linear relationship is strong and positive, the correlation coefficient will be near +1. If it is strong and negative, it will be near −1. If there is no apparent linear relationship between the variables, then the correlation coefficient will be near zero.

THE CORRELATION COEFFICIENT MEASURES THE STRENGTH OF A LINEAR RELATIONSHIP

The **correlation coefficient**, which always takes values between -1 and 1, describes the direction and strength of the linear relationship between two numerical variables. The strength can be strong, moderate, or weak.

We compute the correlation using a formula, just as we did with the sample mean and standard deviation. Formally, we can compute the correlation for observations (x_1, y_1), (x_2, y_2), ..., (x_n, y_n) using the formula

$$r = \frac{1}{n-1}\sum\left(\frac{x_i - \bar{x}}{s_x}\right)\left(\frac{y_i - \bar{y}}{s_y}\right)$$

where $\bar{x}$, $\bar{y}$, s_x, and s_y are the sample means and standard deviations for each variable. This formula is rather complex, and we generally perform the calculations on a computer or calculator. We can note, though, that the computation involves taking, for each point, the product of the Z-scores that correspond to the x and y values.

EXAMPLE 5.9

Take a look at Figure 5.5 on page 402. How would the correlation coefficient between Mean Travel Time to Work and Percent with a Bachelor's degree change if travel time was recorded in hours rather than seconds?

Here, changing the units of y corresponds to dividing all the y values by a certain number. This would change the mean and the standard deviation of y, but it would not change the correlation coefficient. To see this, imagine dividing every number on the vertical axis by 60. The units of y are now in hours rather than minutes, but the graph has remained exactly the same. The units of y have changed, by the relative distance of the y values about the mean are the same; that is, the Z-scores corresponding to the y values have remained the same.

CHANGING UNITS OF X AND Y DOES NOT AFFECT THE CORRELATION COEFFICIENT

The correlation coefficient, r, between two variables is not dependent upon the units in which the variables are recorded. The correlation coefficient itself has no units.

The correlation coefficient is intended to quantify the strength of a linear trend. Nonlinear trends, even when strong, sometimes produce correlations coefficients that do not reflect the strength of the relationship; see three such examples in Figure 5.8.

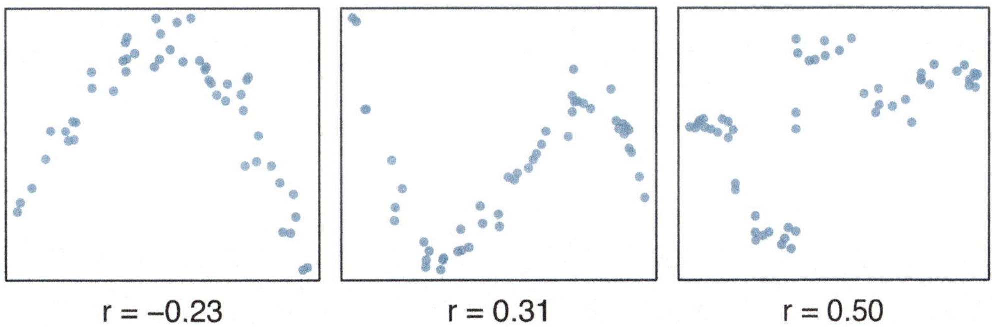

Figure 5.8: Sample scatterplots and their correlation coefficients. In each case, there is a strong relationship between the variables. However, the correlation is not very strong, and the relationship is not linear.

EXAMPLE 5.10

Consider the four scatterplots in Figure 5.9. In which scatterplot is the correlation between x and y the strongest?

E

All four data sets have the exact same correlation coefficient of $r = 0.816$ as well as the same equation for the best fit line! This group of four graphs, known as Anscombe's Quartet, remind us that knowing the value of the correlation coefficient does not tell us what the corresponding scatterplot looks like. It is always important to first graph the data.

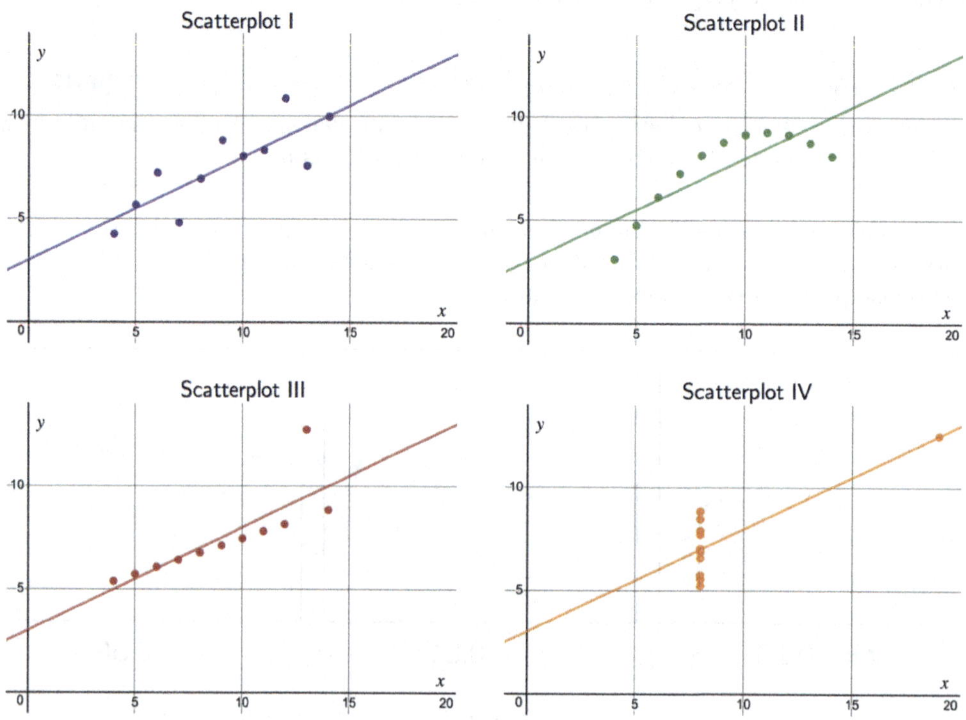

Figure 5.9: Four scatterplots with best fit line drawn in, made in Desmos. Investigate Anscombe's Quartet at: desmos.com/calculator/paknt6oneh

Section summary

- A bivariate quantitative data set consists of observations of ordered pairs from two quantitative variables, collected from the same individuals in a sample or population, and can be used to construct a scatterplot.
- A **scatterplot** shows the relationship between two quantitative variables for each observation, one corresponding to the value on the x-axis and one corresponding to the value on the y-axis. The explanatory variable is placed on the x-axis and is the variable whose values are used to explain or predict the corresponding values for the response variable, which is placed on the y-axis.
- A description of the *association* shown in a scatterplot includes form, direction, strength, and unusual features.
  - The form of the association shown in a scatterplot, if any, can be described as linear or nonlinear.
  - The direction of the association shown in a scatterplot, if any, can be described as positive or negative. A positive association means that as values of the explanatory variable increase, the values of the response variable tend to increase. A negative association means that as values of the explanatory variable increase, the values of the response variable tend to decrease. Note that the terms "increase" and "decrease" here are descriptive and do not imply a causal relationship.
  - The strength of the association shown in a scatterplot is how closely the points follow the general pattern. Strength can be described as strong, moderate, or weak.
  - Unusual features of a scatterplot include clusters of individual points or points that don't fit in the general pattern of association between the two variables.
- Scatterplots depicting the distribution of two numeric variables may reveal information that can be used to justify claims about the variable in context.
- The **correlation coefficient** r, summarizes the strength and direction of the linear association between two quantitative variables. The correlation coefficient r is unit-free and always between -1 and 1, inclusive. A negative correlation coefficient value indicates a negative association, and a positive correlation coefficient value indicates a positive association.
- The strength of the linear association is determined by how close the correlation coefficient is to -1 or 1. A value of $r = 0$ indicates that there is no *linear* association, though there may be a different type of association such as a quadratic association. A value of $r = -1$ or $r = 1$ indicates that there is a perfect linear association.
- A correlation coefficient close to -1 or 1 does not necessarily mean that a linear model is appropriate.
- A perceived or real relationship between two variables does not mean that changes in one variable cause changes in the other. That is, correlation does not necessarily imply causation.

Exercises

5.1 ACS, Part I. Each year, the US Census Bureau surveys about 3.5 million households with The American Community Survey (ACS). Data collected from the ACS have been crucial in government and policy decisions, helping to determine the allocation of federal and state funds each year. Some of the questions asked on the survey are about their income, age (in years), and gender. The table below contains this information for a random sample of 20 respondents to the 2012 ACS.[6]

	Income	Age	Gender
1	53,000	28	male
2	1600	18	female
3	70,000	54	male
4	12,800	22	male
5	1,200	18	female
6	30,000	34	male
7	4,500	21	male
8	20,000	28	female
9	25,000	29	female
10	42,000	33	male

	Income	Age	Gender
11	670	34	female
12	29,000	55	female
13	44,000	33	female
14	48,000	41	male
15	30,000	47	female
16	60,000	30	male
17	108,000	61	male
18	5,800	50	female
19	50,000	24	female
20	11,000	19	male

(a) Use technology to create a scatterplot of income vs. age, and describe the relationship between these two variables.

(b) Now create two scatterplots: one for income vs. age for males and another for females.

(c) How, if at all, do the relationships between income and age differ for males and females?

5.2 MLB stats. A baseball team's success in a season is usually measured by their number of wins. In order to win, the team has to have scored more points (runs) than their opponent in any given game. As such, number of runs is often a good proxy for the success of the team. The table below shows number of runs, home runs, and batting averages for a random sample of 10 teams in the 2014 Major League Baseball season.[7]

	Team	Runs	Home runs	Batting avg.
1	Baltimore	705	211	0.256
2	Boston	634	123	0.244
3	Cincinnati	595	131	0.238
4	Cleveland	669	142	0.253
5	Detroit	757	155	0.277
6	Houston	629	163	0.242
7	Minnesota	715	128	0.254
8	NY Yankees	633	147	0.245
9	Pittsburgh	682	156	0.259
10	San Francisco	665	132	0.255

(a) Use technology to create a scatterplot of runs vs. home runs.

(b) Now create a scatterplot of runs vs. batting averages.

(c) Are home runs or batting averages more strongly associated with number of runs? Explain your reasoning.

[6] United States Census Bureau. Summary File. 2012 American Community Survey. U.S. Census Bureau's American Community Survey Office, 2013. Web.

[7] ESPN: MLB Team Stats - 2014.

5.3 Mammal life spans. Data were collected on life spans (in years) and gestation lengths (in days) for 62 mammals. A scatterplot of life span versus length of gestation is shown below.[8]

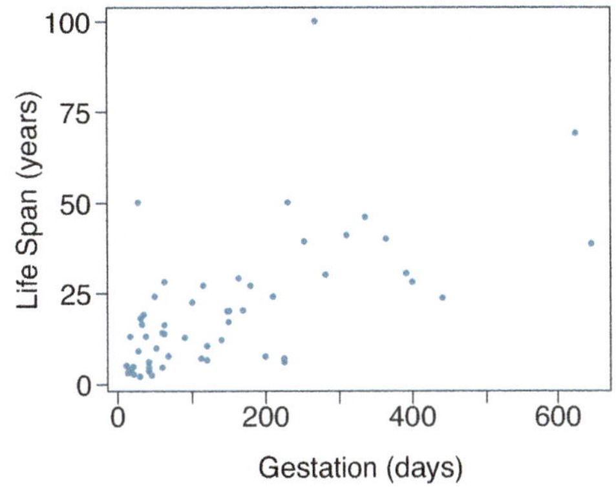

(a) What type of an association is apparent between life span and length of gestation?

(b) What type of an association would you expect to see if the axes of the plot were reversed, i.e. if we plotted length of gestation versus life span?

(c) Are life span and length of gestation independent? Explain your reasoning.

5.4 Associations. Indicate which of the plots show (a) a positive association, (b) a negative association, or (c) no association. Also determine if the positive and negative associations are linear or nonlinear. Each part may refer to more than one plot.

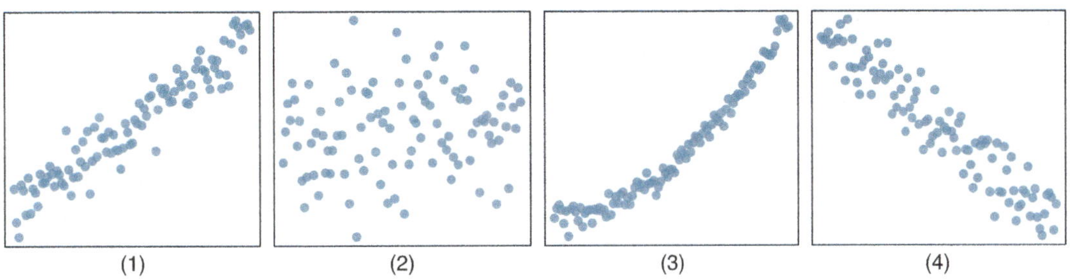

5.5 Exams and grades. The two scatterplots below show the relationship between final and mid-semester exam grades recorded during several years for a Statistics course at a university.

(a) Based on these graphs, which of the two exams has the strongest correlation with the final exam grade? Explain.

(b) Can you think of a reason why the correlation between the exam you chose in part (a) and the final exam is higher?

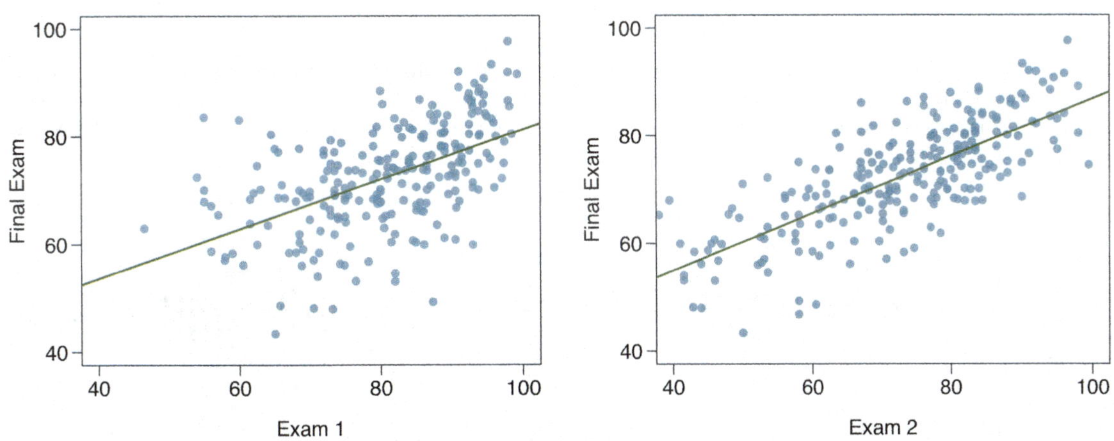

[8] T. Allison and D.V. Cicchetti. "Sleep in mammals: ecological and constitutional correlates". In: *Arch. Hydrobiol* 75 (1975), p. 442.

5.6 Spouses, Part I. The Great Britain Office of Population Census and Surveys once collected data on a random sample of 170 married women in Britain, recording the age (in years) and heights (converted here to inches) of the women and their spouses.[9] The scatterplot on the left shows the spouse's age plotted against the woman's age, and the plot on the right shows spouse's height plotted against the woman's height.

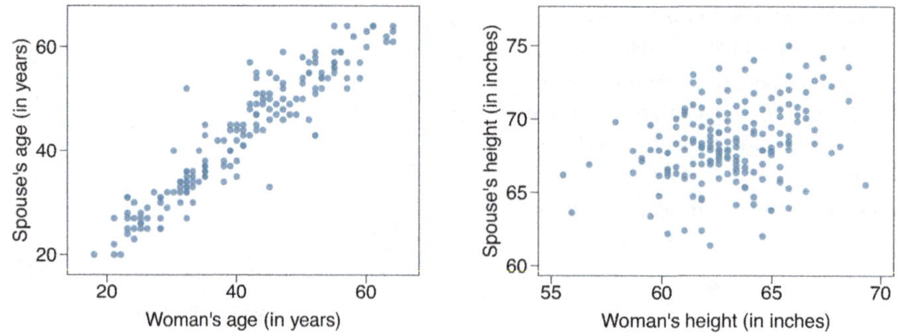

(a) Describe the relationship between the ages of women in the sample and their spouses' ages.

(b) Describe the relationship between the heights of women in the sample and their spouses' heights.

(c) Which plot shows a stronger correlation? Explain your reasoning.

(d) Data on heights were originally collected in centimeters, and then converted to inches. Does this conversion affect the correlation between heights of women in the sample and their spouses' heights?

5.7 Match the correlation, Part I. Match each correlation to the corresponding scatterplot.

(a) $r = -0.7$
(b) $r = 0.45$
(c) $r = 0.06$
(d) $r = 0.92$

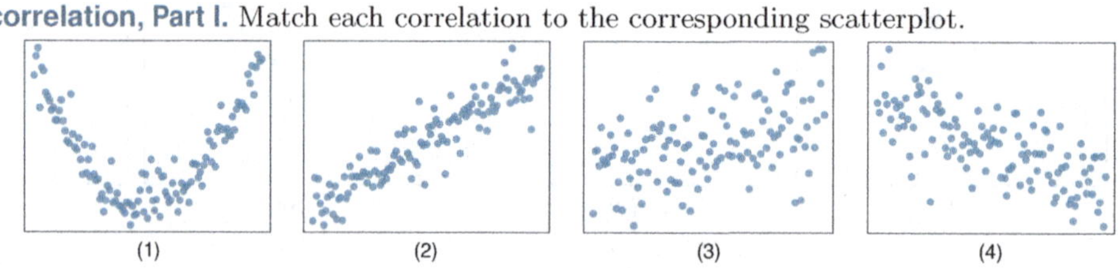

5.8 Match the correlation, Part II. Match each correlation to the corresponding scatterplot.

(a) $r = 0.49$
(b) $r = -0.48$
(c) $r = -0.03$
(d) $r = -0.85$

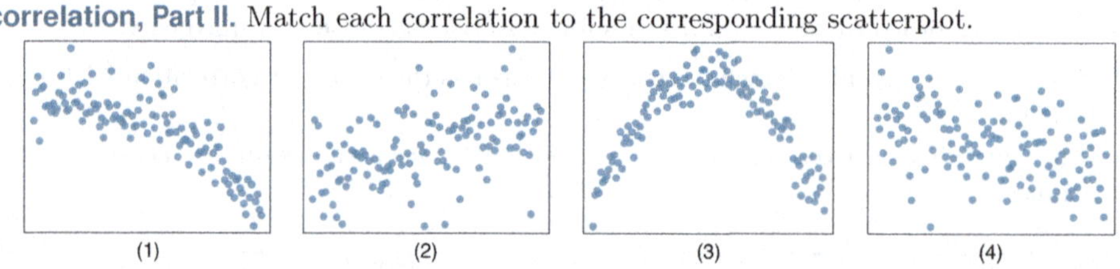

5.9 Speed and height. 1,302 UCLA students were asked to fill out a survey where they were asked about their height, fastest speed they have ever driven, and gender. The scatterplot on the left displays the relationship between height and fastest speed, and the scatterplot on the right displays the breakdown by gender in this relationship.

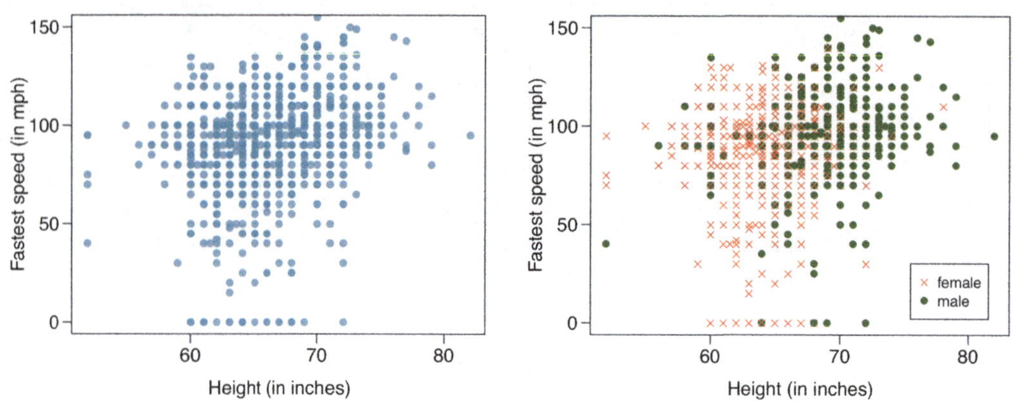

(a) Describe the relationship between height and fastest speed.

(b) Why do you think these variables are positively associated?

(c) What role does gender play in the relationship between height and fastest driving speed?

[9] D.J. Hand. *A handbook of small data sets*. Chapman & Hall/CRC, 1994.

5.10 Guess the correlation. Eduardo and Rosie are both collecting data on number of rainy days in a year and the total rainfall for the year. Eduardo records rainfall in inches and Rosie in centimeters. How will their correlation coefficients compare?

5.11 The Coast Starlight, Part I. The Coast Starlight Amtrak train runs from Seattle to Los Angeles. The scatterplot below displays the distance between each stop (in miles) and the amount of time it takes to travel from one stop to another (in minutes).

(a) Describe the relationship between distance and travel time.

(b) How would the relationship change if travel time was instead measured in hours, and distance was instead measured in kilometers?

(c) The correlation between travel time (in minutes) and distance (in miles) is $r = 0.636$. What are the units for the correlation in this context?

(d) Suppose we had instead measured travel time in hours and measured distance in kilometers (km). What would be the correlation in these different units?

(e) How would the correlation change if the x and y variables were swapped?

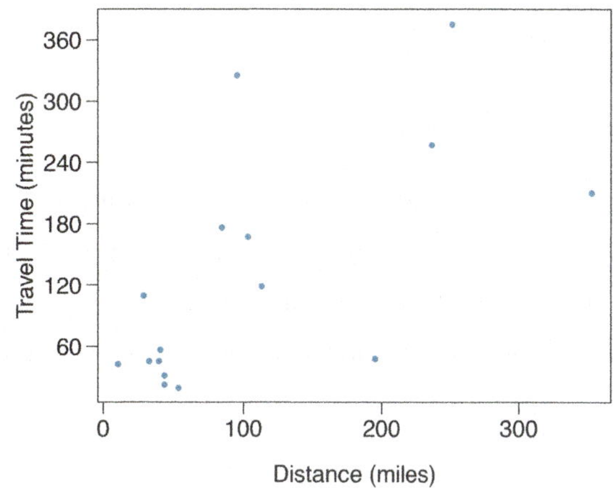

5.12 Crawling babies. A study conducted at the University of Denver investigated whether babies take longer to learn to crawl in cold months, when they are often bundled in clothes that restrict their movement, than in warmer months.[10] Infants born during the study year were split into twelve groups, one for each birth month. We consider the average crawling age of babies in each group against the average temperature when the babies are six months old (that's when babies often begin trying to crawl). Temperature is measured in degrees Fahrenheit (°F) and age is measured in weeks.

(a) Describe the relationship between temperature and crawling age.

(b) How would the relationship change if temperature was measured in degrees Celsius (°C) and age was measured in months?

(c) What are

(d) The correlation between temperature in °F and age in weeks was $r = -0.70$. What are the units for the correlation in this context?

(e) If we converted the temperature to °C and age to months, what would the correlation be?

(f) How would the correlation change if the x and y variables were swapped?

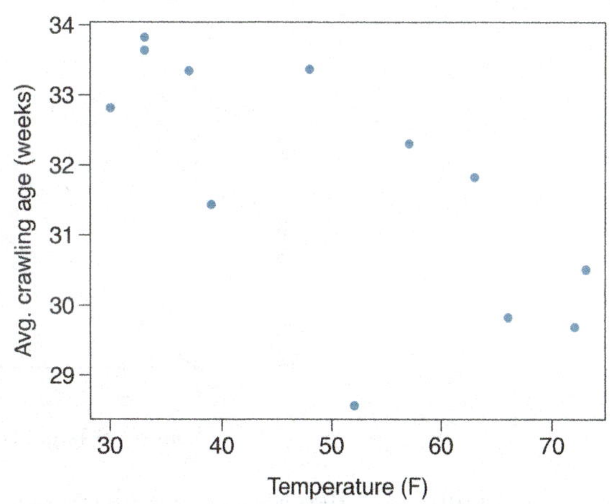

[10] J.B. Benson. "Season of birth and onset of locomotion: Theoretical and methodological implications". In: *Infant behavior and development* 16.1 (1993), pp. 69–81. ISSN: 0163-6383.

5.2 Line fitting and residuals

How can we use information about one variable to estimate or predict another variable? How do we determine when these predictions will be reasonable or unreasonable? In this section we continue our investigation of the relationship between bivariate, numerical data. We introduce the linear regression model and the concept of residuals as error between an actual y-value and a predicted y-value.

Learning objectives

1. Calculate a predicted response value using a linear regression model.
2. Calculate the differences between the observed and predicted values.
3. Interpret the differences between the observed and predicted values.
4. Describe the form of association of bivariate data using residual plots.

5.2.1 Fitting a line to data

Requests from twelve separate buyers were simultaneously placed with a trading company to purchase Target Corporation stock (ticker TGT, April 26th, 2012). We let x be the number of stocks to purchase and y be the total cost. Because the cost is computed using a linear formula, the linear fit is perfect, and the equation for the line is: $y = 5 + 57.49x$. If we know the number of stocks purchased, we can determine the cost based on this linear equation with no error. Additionally, we can say that each additional share of the stock cost \$57.49 and that there was a \$5 fee for the transaction.

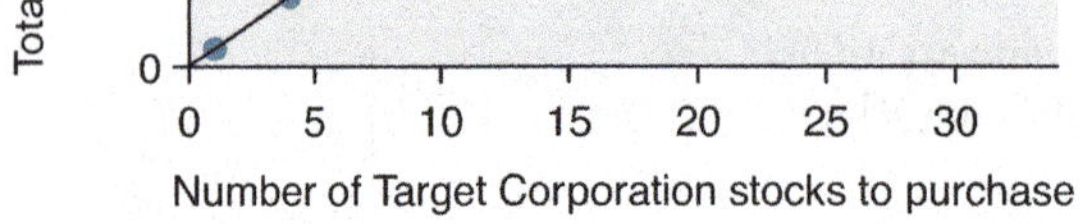

Figure 5.10: Requests from twelve separate buyers were simultaneously placed with a trading company to purchase Target Corporation stock (ticker TGT, December 28th, 2018), and the total cost of the shares were reported. Because the cost is computed using a linear formula, the linear fit is perfect.

Perfect linear relationships are unrealistic in almost any natural process. For example, if we took family income (x), this value would provide some useful information about how much financial support a college may offer a prospective student (y). However, the prediction would be far from perfect, because other factors play a role in financial support beyond a family's income.

It is rare for all of the data to fall perfectly on a straight line. Instead, it's more common for data to appear as a *cloud of points*, such as those shown in Figure 5.11. In each case, the data fall around a straight line, even if none of the observations fall exactly on the line. The first plot shows a relatively strong downward linear trend, where the remaining variability in the data around the line is minor relative to the strength of the relationship between x and y. The second plot shows an upward trend that, while evident, is not as strong as the first. The last plot shows a very weak downward trend in the data, so slight we can hardly notice it.

In each of these examples, we can consider how to draw a "best fit line". For instance, we might wonder, should we move the line up or down a little, or should we tilt it more or less? As we move forward in this chapter, we will learn different criteria for line-fitting, and we will also learn about the uncertainty associated with estimates of model parameters.

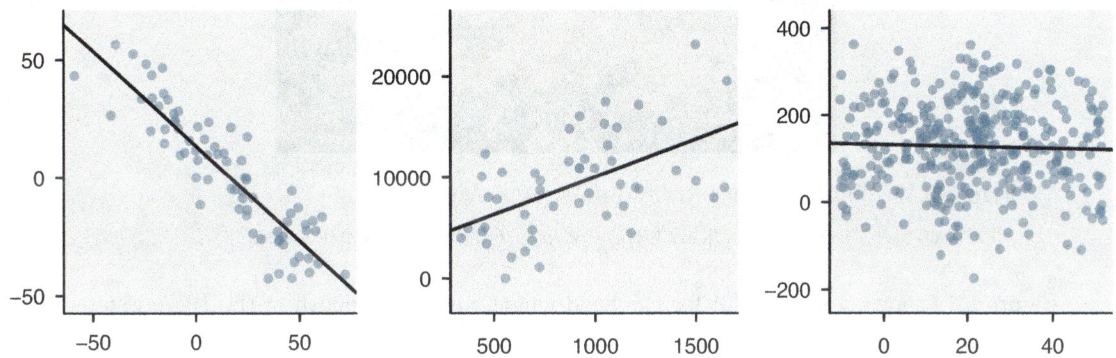

Figure 5.11: Three data sets where a linear model may be useful even though the data do not all fall exactly on the line.

There are also cases where fitting a straight line to the data, even if there is a clear relationship between the variables, is not helpful. One such case is shown in Figure 5.12 where there is a very clear relationship between the variables even though the trend is not linear.

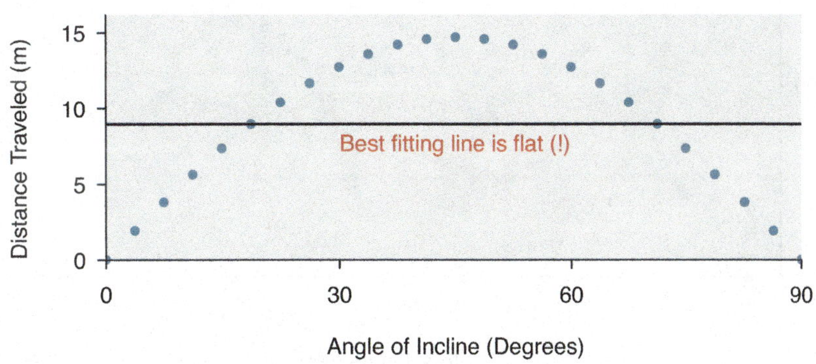

Figure 5.12: A linear model is not useful in this nonlinear case. These data are from an introductory physics experiment.

5.2.2 Using linear regression to make predictions

Brushtail possums are a marsupial that lives in Australia. A photo of one is shown in Figure 5.13. Researchers captured 104 of these animals and took body measurements before releasing the animals back into the wild. We consider two of these measurements: the total length of each possum, from head to tail, and the length of each possum's head.

Figure 5.13: The common brushtail possum of Australia.

Figure 5.14 shows a scatterplot for the head length and total length of the 104 possums. Each point represents a single case (possum) from the data.

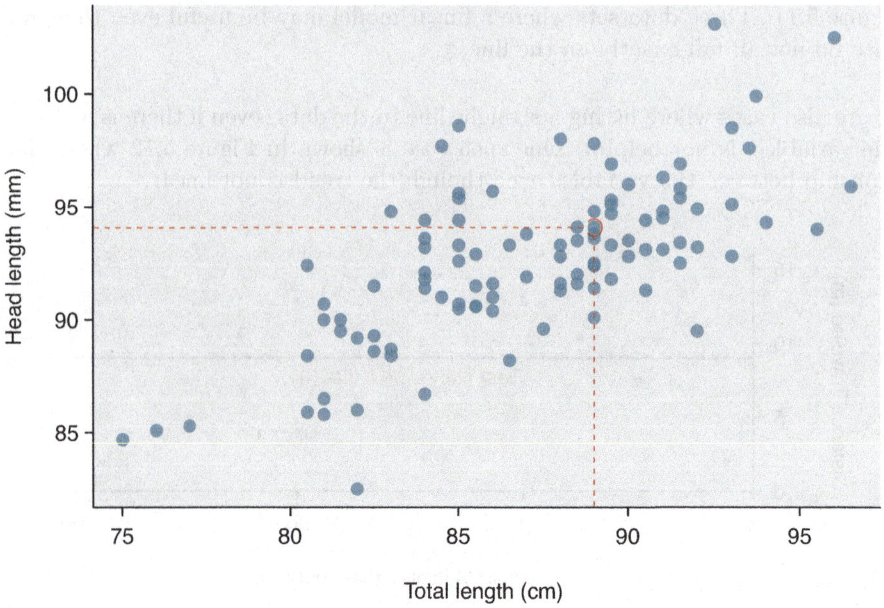

Figure 5.14: A scatterplot showing head length against total length for 104 brushtail possums. A point representing a possum with head length 94.1 mm and total length 89 cm is highlighted.

The head and total length variables are associated: possums with an above average total length also tend to have above average head lengths. While the relationship is not perfectly linear, it could be helpful to partially explain the connection between these variables with a straight line. We will use the total length, x, to explain or predict a possum's head length, y. When we use x to predict y, we usually call x the **explanatory variable** or predictor variable, and we call y the **response variable**. We could fit the linear relationship by eye, as in Figure 5.15. We call this the *regression line* and write it in the form $\hat{y} = a + bx$, where a is the y-intercept of the line and b is the slope of the line. The equation for this regression line that was fit by eye is

$$\hat{y} = 41 + 0.59x$$

A "hat" on y is used to signify that this is a predicted value, not an observed value. We can use this line to discuss properties of possums. For instance, the equation predicts a possum with a total length of 80 cm will have a head length (in mm) of

$$\begin{aligned}\hat{y} &= 41 + 0.59(80)\\ &= 88.2\end{aligned}$$

The value $\hat{y}$ may be viewed as an average: the equation predicts that possums with a total length of 80 cm will have an average head length of 88.2 mm. The value $\hat{y}$ is also a prediction: absent further information about an 80 cm possum, this is our best prediction for a the head length of a single 80 cm possum.

5.2.3 Extrapolation is treacherous

When those blizzards hit the East Coast this winter, it proved to my satisfaction that global warming was a fraud. That snow was freezing cold. But in an alarming trend, temperatures this spring have risen. Consider this: On February 6th *it was 10 degrees. Today it hit almost 80. At this rate, by August it will be 220 degrees. So clearly folks the climate debate rages on.*

Stephen Colbert
April 6th, 2010 [11]

Linear models can be used to approximate the relationship between two variables. However, these models have real limitations. Linear regression is simply a modeling framework. The truth is almost always much more complex than our simple line. For example, we do not know how the data outside of our limited window will behave. In general, **interpolation**, which is predicting a response value using an x-value that is within the range of the x-values in the data set, is safer than **extrapolation**, which is predicting a response value using an x-value that is outside the range of the x-values in the data set.

5.2.4 Residuals

Residuals are the leftover variation in the response variable after fitting a model. Each observation will have a residual, and three of the residuals for the linear model we fit for the `possum` data are shown in Figure 5.15. If an observation is above the regression line, then its residual, the vertical distance from the observation to the line, is positive. Observations below the line have negative residuals. One goal in picking the right linear model is for these residuals to be as small as possible.

Let's look closer at the three residuals featured in Figure 5.15. The observation marked by an "×" has a small, negative residual of about -1; the observation marked by "+" has a large residual of about +7; and the observation marked by "△" has a moderate residual of about -4. The size of a residual is usually discussed in terms of its absolute value. For example, the residual for "△" is larger than that of "×" because $|-4|$ is larger than $|-1|$.

[11] *The Colbert Report* on April 6th, 2010.

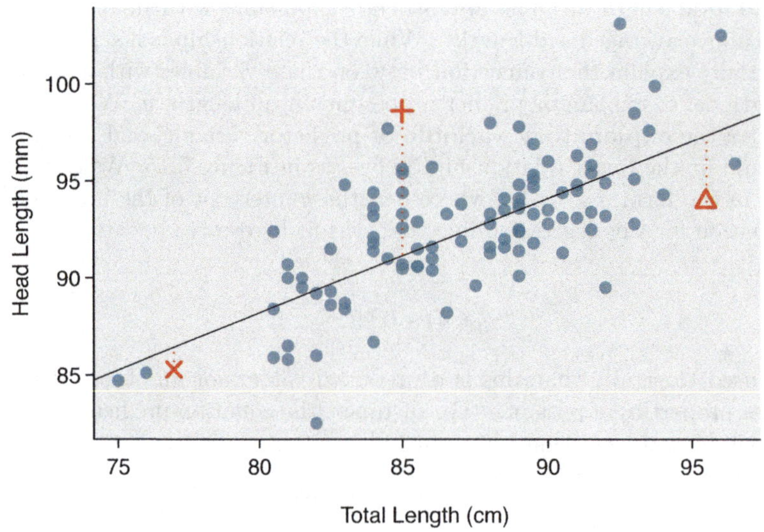

Figure 5.15: A reasonable linear model was fit to represent the relationship between head length and total length.

RESIDUAL: DIFFERENCE BETWEEN OBSERVED AND EXPECTED

The residual for a particular observation (x, y) is the difference between the observed response and the response we would predict based on the model:

$$\begin{aligned} \text{residual} &= \text{observed } y - \text{predicted } y \\ &= y - \hat{y} \end{aligned}$$

We typically identify $\hat{y}$ by plugging x into the model.

EXAMPLE 5.11

The linear fit shown in Figure 5.15 is given as $\hat{y} = 41 + 0.59x$. Based on this line, compute and interpret the residual of the observation (77.0, 85.3). This observation is denoted by "×" on the plot. Recall that x is the total length measured in cm and y is head length measured in mm.

We first compute the predicted value based on the model:

$$\begin{aligned} \hat{y} &= 41 + 0.59x \\ &= 41 + 0.59(77.0) \\ &= 86.4 \end{aligned}$$

E

Next we compute the difference of the actual head length and the predicted head length:

$$\begin{aligned} residual &= y - \hat{y} \\ &= 85.3 - 86.4 \\ &= -1.1 \end{aligned}$$

The residual for this point is -1.1 mm, which is very close to the visual estimate of -1 mm. For this particular possum with total length of 77 cm, the model's prediction for its head length was 1.1 mm *too high*.

GUIDED PRACTICE 5.12

G

If a model underestimates an observation, will the residual be positive or negative? What about if it overestimates the observation?[12]

(G)

GUIDED PRACTICE 5.13

Compute the residual for the observation (95.5, 94.0), denoted by "△" in the figure, using the linear model: $\hat{y} = 41 + 0.59x$.[13]

Residuals are helpful in evaluating how well a linear model fits a data set. We often display the residuals in a **residual plot** such as the one shown in Figure 5.16. Here, the residuals are calculated for each x value, and plotted versus x. For instance, the point (85.0, 98.6) had a residual of 7.45, so in the residual plot it is placed at (85.0, 7.45). Creating a residual plot is sort of like tipping the scatterplot over so the regression line is horizontal.

From the residual plot, we can better estimate the **standard deviation of the residuals**, often denoted by the letter s. The standard deviation of the residuals tells us typical size of the residuals. As such, it is a measure of the typical deviation between the y values and the model predictions. In other words, it tells us the typical prediction error using the model.[14]

(E)

EXAMPLE 5.14

Estimate the standard deviation of the residuals for predicting head length from total length using the line: $\hat{y} = 41 + 0.59x$ using Figure 5.16. Also, interpret the quantity in context.

To estimate this graphically, we use the residual plot. The approximate 68, 95 rule for standard deviations applies. Approximately 2/3 of the points are within ± 2.5 and approximately 95% of the points are within ± 5, so 2.5 is a good estimate for the standard deviation of the residuals. The typical error when predicting head length using this model is about 2.5 mm.

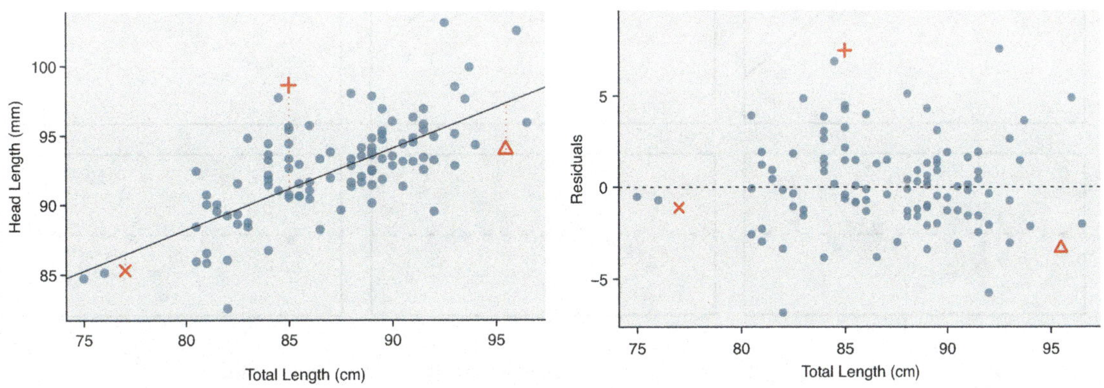

Figure 5.16: Left: Scatterplot of head length versus total length for 104 brushtail possums. Three particular points have been highlighted. Right: Residual plot for the model shown in left panel.

---

[12]If a model underestimates an observation, then the model estimate is below the actual. The residual, which is the actual observation value minus the model estimate, must then be positive. The opposite is true when the model overestimates the observation: the residual is negative.

[13]First compute the predicted value based on the model, then compute the residual.

$$\hat{y} = 41 + 0.59x = 41 + 0.59(95.50) = 97.3$$

$$residual = y - \hat{y} = 94.0 - 97.3 = -3.3$$

The residual is -3.3, so the model *overpredicted* the head length for this possum by 3.3 mm.

[14]The standard deviation of the residuals is calculated as: $s = \sqrt{\frac{\sum (y_i - \hat{y})^2}{n-2}}$.

EXAMPLE 5.15

One purpose of residual plots is to identify characteristics or patterns still apparent in data after fitting a model. Figure 5.17 shows three scatterplots with linear models in the first row and residual plots in the second row. Can you identify any patterns remaining in the residuals?

In the first data set (first column), the residuals show no obvious patterns. The residuals appear to be scattered randomly around the dashed line that represents 0.

The second data set shows a pattern in the residuals. There is some curvature in the scatterplot, which is more obvious in the residual plot. We should not use a straight line to model these data. Instead, a more advanced technique should be used.

The last plot shows very little upwards trend, and the residuals also show no obvious patterns. It is reasonable to try to fit a linear model to the data. However, it is unclear whether there is statistically significant evidence that the slope parameter is different from zero. The slope of the sample regression line is not zero, but we might wonder if this could be due to random variation.

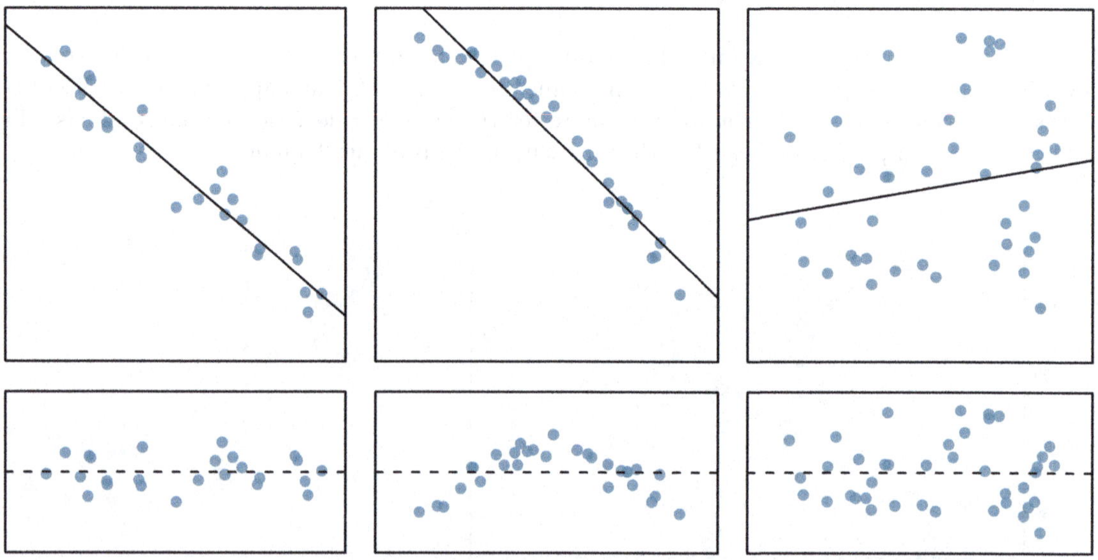

Figure 5.17: Sample data with their best fitting lines (top row) and their corresponding residual plots (bottom row).

Section summary

- If the form of the relationship between x and y appears linear, we can approximate the relationship between x and y using a **linear regression model**, which is a linear equation that uses an explanatory variable, x, to predict the response variable, y. Linear models should not be used if the trend between the variables is curved.
- In a linear regression model, the *predicted* response value, denoted by $\hat{y}$, is calculated as $\hat{y} = a + bx$, where a is the y-intercept, b is the slope of the regression line, and x is the explanatory variable.
- **Extrapolation** is predicting a response value using a value for the explanatory variable that is beyond the interval of x-values used to determine the regression line. The predicted value is less reliable the further the estimate is extrapolated.
- **Interpolation** is predicting a response value using a value for the explanatory variable that is within the interval of x-values used to determine the regression line.
- A **residual** is the difference between the observed response value and the predicted response value for the given value of the explanatory variable: residual $= y - \hat{y}$ or (residual = observed y − predicted y).
- If the residual is positive, the model underpredicts (underestimates) the value of the response variable. If the residual is negative, the model overpredicts (overestimates) the value of the response variable.
- A **residual plot** is a scatterplot of the residuals versus the predicted response values (or the explanatory variable values).
- Residual plots can be used to investigate the appropriateness of the linear regression model for the observed data.
- The linear regression model should only be fit to the data if the data exhibit a linear trend. Apparent randomness in a residual plot for a linear regression model is confirmation of a linear form in the association between the two variables and indicates that the simple linear regression model is an appropriate model for the data.
- Curvature in the residual plot for a linear regression model suggests that the linear model is not the most appropriate model for the data.

Exercises

5.13 Visualize the residuals. The scatterplots shown below each have a superimposed regression line. If we were to construct a residual plot (residuals versus x) for each, describe what those plots would look like.

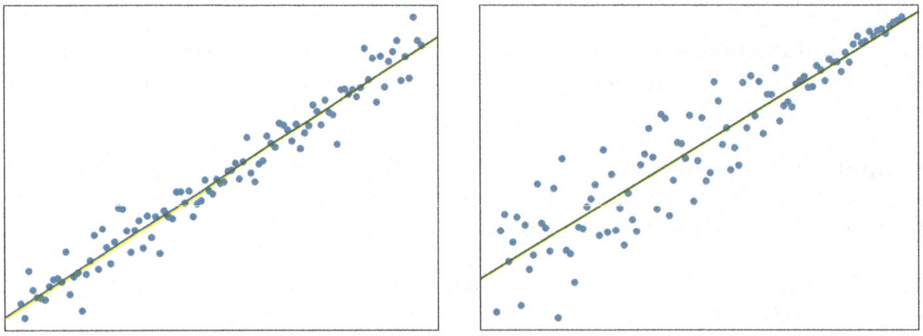

(a) (b)

5.14 Trends in the residuals. Shown below are two plots of residuals remaining after fitting a linear model to two different sets of data. Describe important features and determine if a linear model would be appropriate for these data. Explain your reasoning.

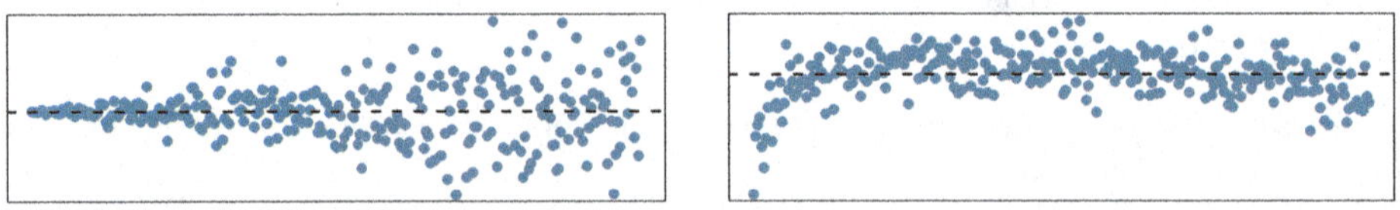

(a) (b)

5.15 Units of regression. Consider a regression predicting weight (kg) from height (cm) for a sample of adult males. What are the units of the correlation coefficient, the intercept, and the slope?

5.16 Which is higher? Determine if I or II is higher or if they are equal. Explain your reasoning. For a regression line, the uncertainty associated with the slope estimate, b, is higher when

I. there is a lot of scatter around the regression line or

II. there is very little scatter around the regression line

5.17 Over-under, Part I. Suppose we fit a regression line to predict the shelf life of an apple based on its weight. For a particular apple, we predict the shelf life to be 4.6 days. The apple's residual is -0.6 days. Did we over or under estimate the shelf-life of the apple? Explain your reasoning.

5.18 Over-under, Part II. Suppose we fit a regression line to predict the number of incidents of skin cancer per 1,000 people from the number of sunny days in a year. For a particular year, we predict the incidence of skin cancer to be 1.5 per 1,000 people, and the residual for this year is 0.5. Did we over or under estimate the incidence of skin cancer? Explain your reasoning.

5.19 Tourism spending. The Association of Turkish Travel Agencies reports the number of foreign tourists visiting Turkey and tourist spending by year.[15] Three plots are provided: scatterplot showing the relationship between these two variables along with the least squares fit, residuals plot, and histogram of residuals.

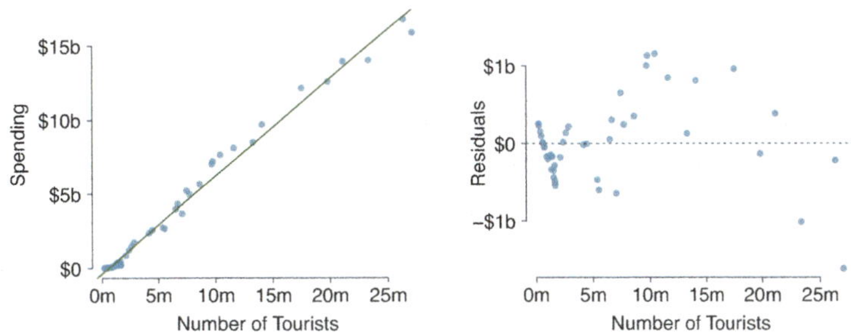

(a) Describe the relationship between number of tourists and spending.

(b) What are the explanatory and response variables?

(c) Why might we want to fit a regression line to these data?

(d) Is a linear model appropriate in this context?

5.20 Nutrition at Starbucks, Part I. The scatterplot below shows the relationship between the number of calories and amount of carbohydrates (in grams) Starbucks food menu items contain.[16] Since Starbucks only lists the number of calories on the display items, we are interested in predicting the amount of carbs a menu item has based on its calorie content.

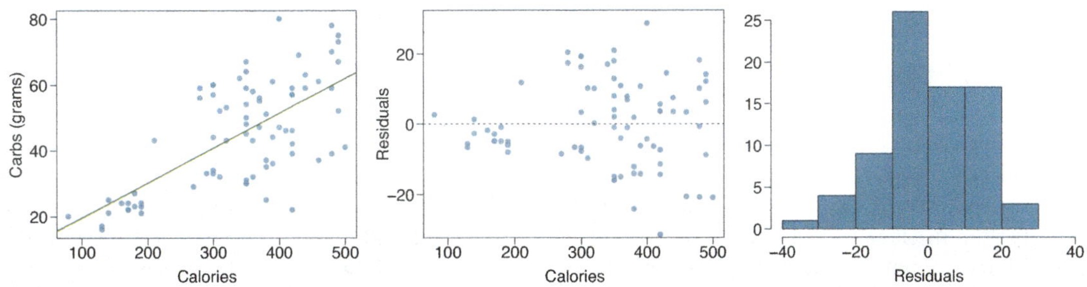

(a) Describe the relationship between number of calories and amount of carbohydrates (in grams) that Starbucks food menu items contain.

(b) In this scenario, what are the explanatory and response variables?

(c) Why might we want to fit a regression line to these data?

(d) Is a linear model appropriate in this context?

[15] Association of Turkish Travel Agencies, Foreign Visitors Figure & Tourist Spendings By Years.

[16] Source: Starbucks.com, collected on March 10, 2011, www.starbucks.com/menu/nutrition.

5.3 Least squares regression

How well can we predict financial aid based on family income for a particular college? How do we measure the fit of a model and compare different models to each other? In this section, we find, interpret, and apply the least-squares regression line and we investigate a new measure that aims to tells us about how well a model "fits" the data.

Learning objectives

1. Calculate the coefficients for the least-squares regression line model using technology or identify them from computer output.
2. Interpret coefficients for the least-squares regression line model.
3. Calculate the coefficient of determination using technology or identify it from from computer output.
4. Interpret the coefficient of determination.

5.3.1 An objective measure for finding the best line

Fitting linear models by eye is open to criticism since it is based on an individual preference. In this section, we use *least squares regression* as a more rigorous approach.

This section considers family income and gift aid data from a random sample of fifty students in the freshman class of Elmhurst College in Illinois. Gift aid is financial aid that does not need to be paid back, as opposed to a loan. A scatterplot of the data is shown in Figure 5.18 along with two linear fits. The lines follow a negative trend in the data; students who have higher family incomes tended to have lower gift aid from the university.

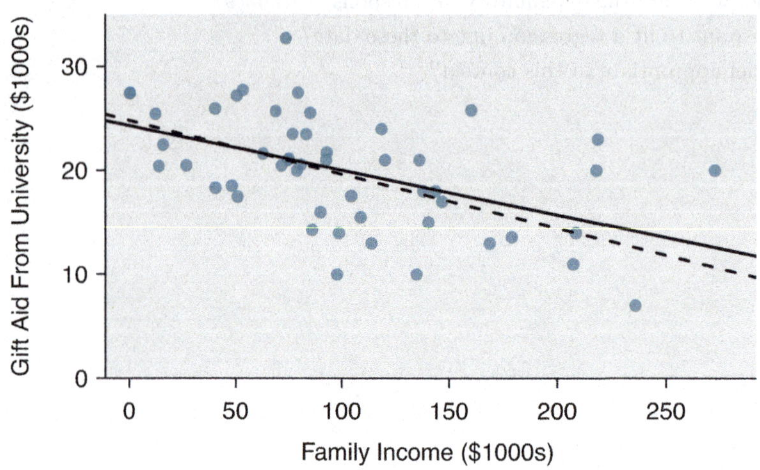

Figure 5.18: Gift aid and family income for a random sample of 50 freshman students from Elmhurst College. Two lines are fit to the data, the solid line being the *least squares line*.

We begin by thinking about what we mean by "best". Mathematically, we want a line that has small residuals. Perhaps our criterion could minimize the sum of the residual magnitudes:

$$|y_1 - \hat{y}_1| + |y_2 - \hat{y}_2| + \cdots + |y_n - \hat{y}_n|$$

We could fit a line using this criteria with a computer program. The resulting dashed line shown in Figure 5.18 demonstrates this fit can be quite reasonable. However, a more common practice is to choose the line that minimizes the sum of the squared residuals:

$$(y_1 - \hat{y}_1)^2 + (y_2 - \hat{y}_2)^2 + \cdots + (y_n - \hat{y}_n)^2$$

The line that minimizes the sum of the squared residuals is represented as the solid line in Figure 5.18. This is commonly called the **least squares line.**

Both lines seem reasonable, so why do data scientists prefer the least squares regression line? One reason is that it is easier to compute by hand and in most statistical software. A more compelling reason is that in many applications, a residual twice as large as another residual is more than twice as bad. For example, being off by 4 is usually more than twice as bad as being off by 2. Squaring the residuals accounts for this discrepancy.

In Figure 5.19, we imagine the squared error about a line as actual squares. The least squares regression line minimizes the sum of the *areas* of these squared errors. In the figure, the sum of the squared error is $4 + 1 + 1 = 6$. There is no other line about which the sum of the squared error will be smaller.

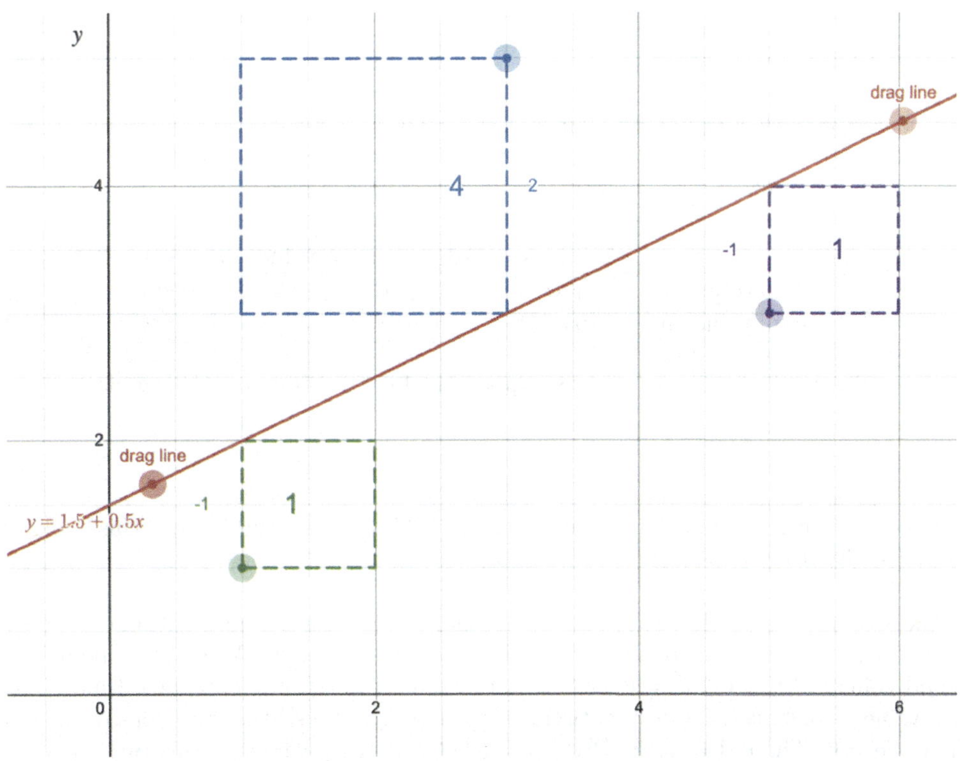

Figure 5.19: A screenshot from the interactive Desmos Activity: Least Squares Demo. Find it at openintro.org/ahss/desmos. The line is the least squares regression line as it makes the sum of squared error (in the y direction) least. Here the smallest possible sum of squared error is 6.

5.3.2 Writing the least squares regression line

For the Elmhurst College data, we could fit a least squares regression line for predicting gift aid based on a student's family income and write the equation as:

$$\widehat{aid} = a + b \times family\_income$$

Here a is the y-intercept of the least squares regression line and b is the slope of the least squares regression line. a and b are both statistics that can be calculated from the data.

We can enter all of the data into a statistical software package and easily find the values of a and b. However, we can also calculate these values by hand, using only the summary statistics.

- The slope of the least squares line is given by

$$b = r\frac{s_y}{s_x}$$

  where r is the correlation coefficient between the variables x and y, and s_x and s_y are the sample standard deviations of x, the explanatory variable, and y, the response variable.

- The point of averages $(\bar{x}, \bar{y})$ is always on the least squares line. Plugging this point in for x and y in the least squares equation and solving for a gives

$$\bar{y} = a + b\bar{x} \qquad\qquad a = \bar{y} - b\bar{x}$$

We mentioned earlier that a computer is usually used to compute the least squares line. A summary table based on computer output is shown in Figure 5.20 for the Elmhurst College data. The first column of numbers provides estimates for a and b, respectively.

	Estimate	Std. Error	t value	Pr(>|t|)
(Intercept)	24.3193	1.2915	18.83	0.0000
family_income	-0.0431	0.0108	-3.98	0.0002

Figure 5.20: Summary of least squares fit for the Elmhurst College data.

EXAMPLE 5.16

Using the second, third, and fourth columns in Figure 5.20 is beyond the scope of this book. However, can you guess what they represent?

Ⓔ Look at the second row, which corresponds to the slope. The first column, Estimate = -0.0431, tells us our best estimate for the slope of the population regression line. We call this point estimate b. The second column, Std. Error = 0.0108, is the standard error of this point estimate. The third column, t value = -3.98, is the T test statistic for the null hypothesis that the slope of the population regression line = 0. The last column, Pr(>|t|) = 0.0002, is the p-value for this two-sided T-test.

EXAMPLE 5.17

Suppose a high school senior is considering Elmhurst College. Can she simply use the linear equation that we have found to calculate her financial aid from the university?

Ⓔ No. Using the equation will provide a prediction or estimate. However, as we see in the scatterplot, there is a lot of variability around the line. While the linear equation is good at capturing the trend in the data, there will be significant error in predicting an individual student's aid. Additionally, the data all come from one freshman class, and the way aid is determined by the university may change from year to year.

5.3.3 Interpreting the coefficients of a regression line

Interpreting the coefficients in a regression model is often one of the most important steps in the analysis.

EXAMPLE 5.18

The slope for the Elmhurst College data for predicting gift aid based on family income was calculated as -0.0431. Intepret this quantity in the context of the problem.

E

You might recall from an algebra course that slope is change in y over change in x. The slope of the regression line tells us about the *average* change in y for each unit change in x. Here, both x and y are in thousands of dollars and the slope is *negative*. So if x is one unit or one thousand dollars higher, the predicted value of y will be 0.0431 thousand dollars *less*. In other words, for each additional thousand dollars of family income, the *predicted* gift aid is 0.0431 thousand, or $43.10 less.

EXAMPLE 5.19

The y-intercept for the Elmhurst College data for predicting gift aid based on family income was calculated as 24.3. Intepret this quantity in the context of the problem.

E

The intercept a describes the predicted value of y when $x = 0$. The predicted gift aid is 24.3 thousand dollars if a student's family has no income. The meaning of the intercept is relevant to this application since the family income for some students at Elmhurst is $0. In other applications, the intercept may have little or no practical value if there are no observations where x is near zero (extrapolation). Here, it would be acceptable to say that the *average* gift aid is 24.3 thousand dollars among students whose family have 0 dollars in income.

INTERPRETING COEFFICIENTS IN A LINEAR MODEL

- The slope, b, describes the *predicted* increase or decreases in the response variable y for a one unit increase of the explanatory variable x.
- The y-intercept, a, describes the *predicted* outcome of y if $x = 0$. The linear model must be valid all the way to $x = 0$ for this to make sense, which in many applications is not the case.

GUIDED PRACTICE 5.20

G

In the previous chapter, we encountered a data set that compared the price of new textbooks for UCLA courses at the UCLA Bookstore and on Amazon. We fit a linear model for predicting price at UCLA Bookstore from price on Amazon and we get:

$$\hat{y} = 1.86 + 1.03x$$

where x is the price on Amazon and y is the price at the UCLA bookstore. Interpret the coefficients in this model and discuss whether the interpretations make sense in this context.[17]

---

[17] The y-intercept is 1.86 and the units of y are in dollars. This tells us that when a textbook costs 0 dollars on Amazon, the *predicted* price of the textbook at the UCLA Bookstore is 1.86 dollars. This does not make sense as Amazon does not sell any $0 textbooks. The slope is 1.03, with units (dollars)/(dollars). For each increase in 1 dollar that a book costs on Amazon, the *predicted* cost at the UCLA Bookstore increases by 1.03 dollars. This interpretation does make sense in this context.

GUIDED PRACTICE 5.21

Can we conclude that if Amazon raises the price of a textbook by 1 dollar, the UCLA Bookstore will raise the price of the textbook by \$1.03?[18]

EXERCISE CAUTION WHEN INTERPRETING COEFFICIENTS OF A LINEAR MODEL

- The slope tells us only the *average* or *predicted* change in y for each unit change in x; it does not tell us how much y might change based on a change in x for any particular *individual.* Moreover, in most cases, the slope cannot be interpreted in a causal way.
- When a value of $x = 0$ doesn't make sense in an application, then the interpretation of the y-intercept won't have any practical meaning. In general, beware of extrapolation.

EXAMPLE 5.22

Use the model $\widehat{aid} = 24.3 - 0.0431 \times family\_income$ to estimate the aid of another freshman student whose family had income of \$1 million.

Recall that the units of family income are in \$1000s, so we want to calculate the aid for *family_income* = 1000:

$$\widehat{aid} = 24.3 - 0.0431 \times family\_income$$
$$\widehat{aid} = 24.3 - 0.431(1000) = -18.8$$

The model predicts this student will have -\$18,800 in aid (!). Elmhurst College cannot (or at least does not) require any students to pay extra on top of tuition to attend.

Using a model to predict y-values for x-values outside the domain of the original data is called **extrapolation**. Generally, a linear model is only an approximation of the real relationship between two variables. If we extrapolate, we are making an unreliable bet that the approximate linear relationship will be valid in places where it has not been analyzed.

[18] No. The slope describes the average or overall trend. This is observational data; a causal conclusion cannot be drawn. Remember, a causal relationship can only be concluded by a well-designed randomized, controlled experiment. Additionally, there may be large variation in the points about the line. The slope does not tell us how much y might change based on a change in x for a *particular* textbook.

5.3.4 Using R^2 to describe the strength of a fit

Consider the scatterplot of gift aid versus family income for the Elmhurst College data, graphed in Figure 5.21. How well does the solid regression line fit the data? And how much better is it than the horizontal dashed line at $\bar{y}$ at fitting the data?

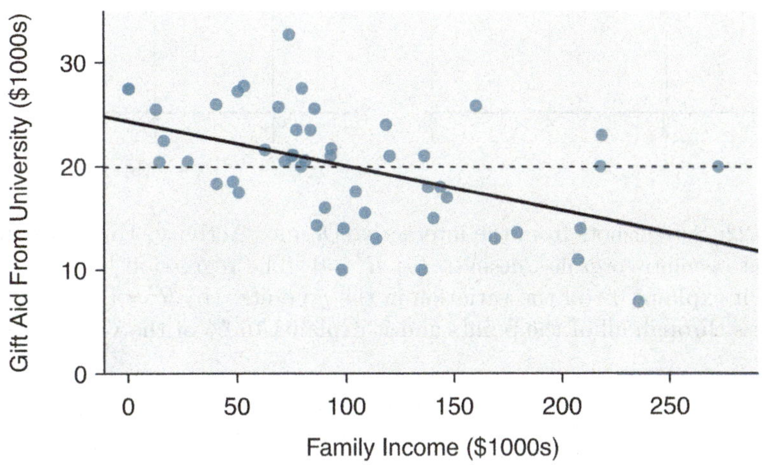

Figure 5.21: Gift aid and family income for a random sample of 50 freshman students from Elmhurst College, shown with the least squares regression line ($\hat{y}$) and the average line ($\bar{y}$).

It is common to explain the fit of a model using R^2 (**R-squared**), the **coefficient of determination**, also called the **explained variance**. In Figure 5.21, the variance of the response variable, aid received, is $s^2_{aid} = 29.8$. However, if we apply our least squares line, then this model reduces our uncertainty in predicting aid using a student's family income. The variability in the residuals describes how much variation remains after using the model: $s^2_{_{RES}} = 22.4$. We could say that the reduction in the variance was:

$$\frac{s^2_{aid} - s^2_{_{RES}}}{s^2_{aid}} = \frac{29.8 - 22.4}{29.8} = \frac{7.5}{29.8} = 0.25$$

If we used the simple standard deviation of the residuals (dividing by $n-1$ instead of $n-2$), this would be exactly R^2. For large data sets, this gives values sufficiently close; however, to get the exact value of R^2 we use a sum of squares method, which is described here in general terms.

Recall that we calculate a residual $(y - \hat{y})$ to see how well a regression line predicts a particular y-value. To measure how well the regression line fits *all* the data, we need to calculate all of the residuals. We saw in the previous section that the sum of squared residuals is a way of measuring overall error about a line. In fact, the sum of squared errors about the regression line is smaller than about any other line. How *much* smaller is this sum than the sum of squared error about the line $\bar{y}$? Note that we if we do not know x, our best prediction of y is simply $\bar{y}$. In a sense, we are asking how much information using x and the regression line provides over ignoring x all together.

Let us take two extreme examples. In Figure 5.22(a), the correlation coefficient is 0 and the regression line is *the same as* the horizontal line at $\bar{y}$; the sum of the squared error about $\hat{y}$ is equal to the sum of the squared error about $\bar{y}$. $R^2 = 0$: the regression line offers no reduction in error; it "explains" 0% of the variation in the y points.

In Figure 5.22(b), the correlation coefficient is -1 and the regression line goes through all of the points. There sum of the squared error about the regression line is 0. $R^2 = 1$: the regression line provides 100% reduction in error and so it "explains" 100% of the variation in the y points.

The exact calculation of R^2 is tedious and is done using technology.[19]

[19]For those interested, the calculations of R^2 in Figure 5.22 can be found as follows. (a) The sum of the squared error about $\hat{y}$ and $\bar{y}$ both equal $(-1)^2 + (2)^2 + (-1)^2 = 6$. $R^2 = \frac{6-6}{6} = 1 - \frac{6}{6} = 0$. (b) $R^2 = \frac{8-0}{8} = 1 - \frac{0}{8} = 1$.

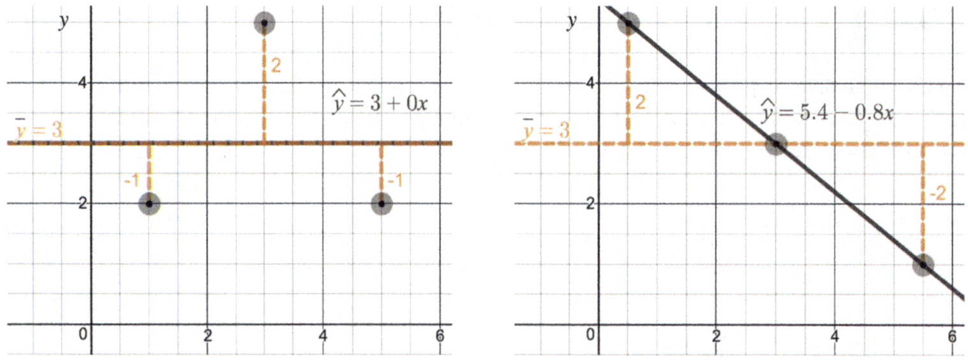

(a) (b)

Figure 5.22: Screenshots from the interactive Desmos Activity: Understanding R^2. Find it at openintro.org/ahss/desmos. (a) $R^2 = 0$. The regression line is equivalent to $\bar{y}$ and it explains 0% of the variation in the y points. (b) $R^2 = 1$. The regression line passes through all of the points and it explains 100% of the variation in the y points.

R^2 IS THE EXPLAINED VARIANCE

R^2 is always between 0 and 1, inclusive. R^2 tells us the proportion of variation in the response variable that is explained by a regression model with one or more explanatory variables. The higher the value of R^2, the better the model "explains" the response variable.

EXAMPLE 5.23

The linear model for predicting gift aid from family income for the Elmhurst College data is given by: $\widehat{aid} = 24.3 - 0.0431 \times family\_income$. $R^2 \approx 0.25$. Interpret this quantity in context.

We can say that about 25% of the variation in gift aid is explained by the linear model with the explanatory variable x = family income.

The value of R^2 is, in fact, equal to r^2, where r is the correlation coefficient. This means that $r = +\sqrt{R^2}$ or $-\sqrt{R^2}$, depending on direction of the association. Use this fact to answer the next two practice problems.

GUIDED PRACTICE 5.24

If a linear model has a very strong negative relationship with a correlation coefficient of -0.97, how much of the variation in the response variable is explained by the linear model?[20]

GUIDED PRACTICE 5.25

If a linear model has an R^2 or explained variance of 0.94, what is the correlation coefficient?[21]

If R^2 is simply the square of the correlation coefficient, why do statisticians prefer using R^2 to assess the fit of a model? The correlation coefficient r only measures *linear* association, so it only has meaning for a linear model. R^2, on the other hand, measures the fit of any model, whether linear, quadratic, exponential, etc. Because of this, R^2 is more useful for measuring and *comparing* the fit of different models.

[20] $R^2 = (-0.97)^2 = 0.94$ or 94%. 94% of the variation in y is explained by the linear model.

[21] We take the square root of R^2 and get 0.97, but we must be careful, because r could be 0.97 *or* -0.97. Without knowing the slope or seeing the scatterplot, we have no way of knowing if r is positive or negative.

5.3.5 Technology: Scatterplots and regression analysis

The data set `loan50`, introduced in Chapter 1, contains information on randomly sampled loans. Download the `loan50` CSV file from openintro.org/data. Open it and perform linear regression analysis for predicting `total_credit_utilized` from `total_credit_limit`.

Desmos:

1. ENTER DATA: Open the CSV file. Copy the two columns that correspond to x and y (the x variable should be on the left). Here we copy the columns `total_credit_limit` and `total_credit_utilized`. In a Desmos cell, paste what you copied. You will see the data presented in a table.

* For small data sets, click + in the upper left, choose `table` and enter x and y values.

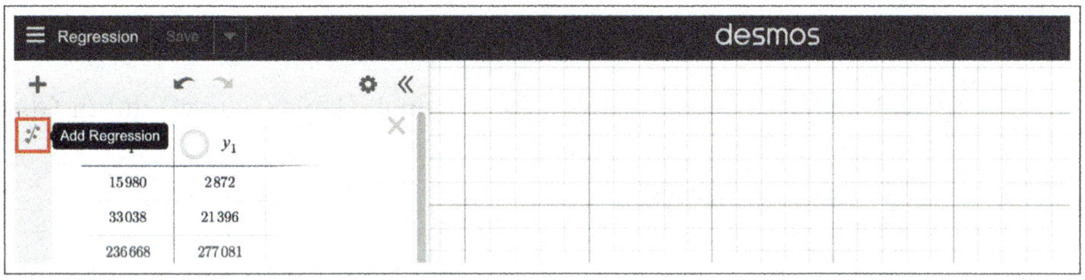

2. SCATTER PLOT and REGRESSION OUTPUT: Click the Add Regression icon shown above. This will return the least squares regression line in the $y = mx + b$ format, R^2, and r as shown below. Click the magnifying glass in the lower left to Zoom Fit the graphing window. Click the wrench icon in the upper right to specify the min, max, and step for the X-Axis and Y-Axis and to add labels to the axes.

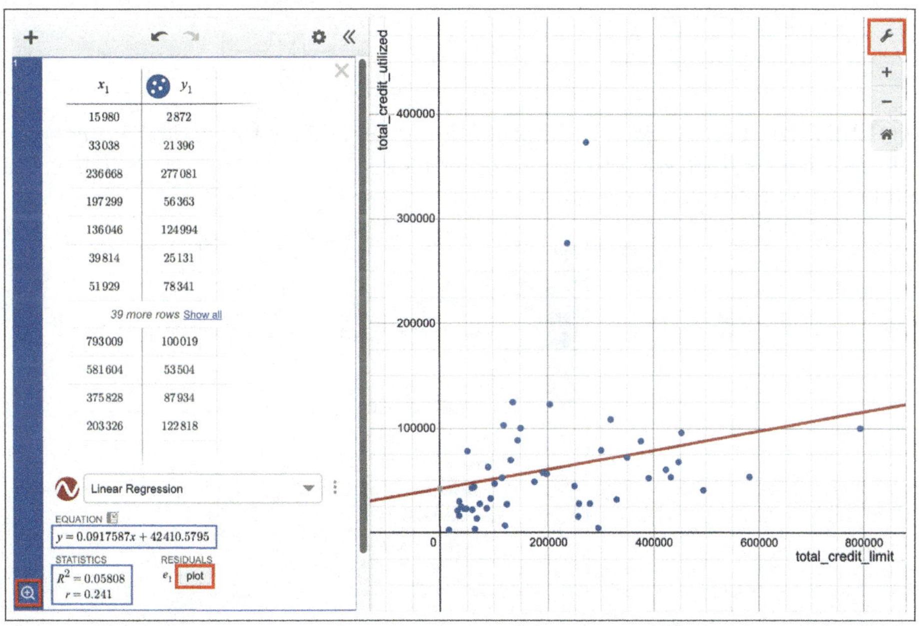

3. RESIDUALS and RESIDUAL PLOT: Click plot below the word RESIDUALS. This will add the residuals to the table and plot them against the x-variable. To make a residual plot, click the circle next to y in the table to deselect it and click the circle next to "Linear Regression" to deselect it. Then click the magnifying glass to recenter. Adjust the Y-axis label to be "residuals".

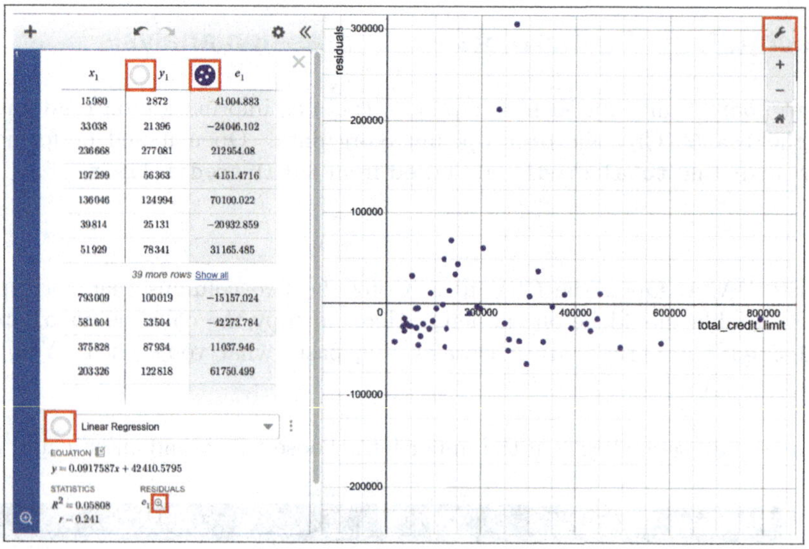

4. SUMMARIZE RESIDUALS: Summarize the residuals, here e_1, as we did in Chapter 1.

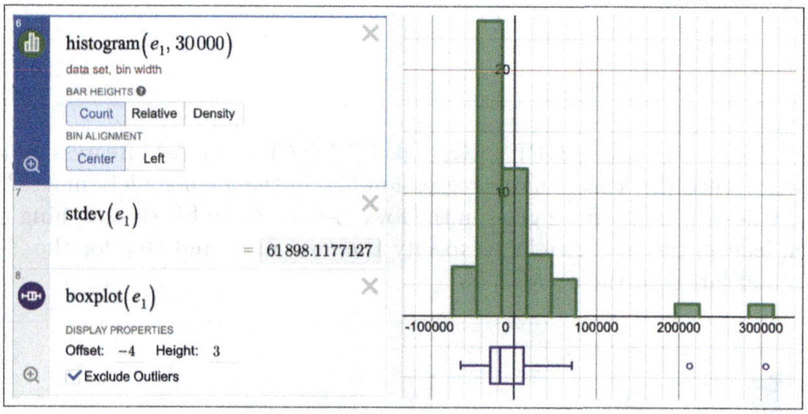

5. NONLINEAR REGRESSION: Make sure the circle next to `Linear Regression` is selected. Click on Linear Regression to choose other models. Note that the residuals will update in the table and on the graph based on the model you select.

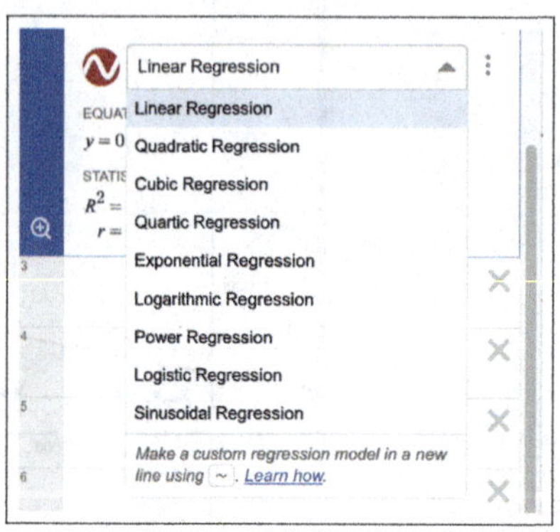

6. MANUALLY CREATE regression model: To see the linear regression model in the form $y = a + bx$, you can type y1 ~ a + bx1. The ~ is in the upper left of keyboard. Adjust the number after y and x to match the subscript on your data table. You can also use this method to manually generate models of your choosing.

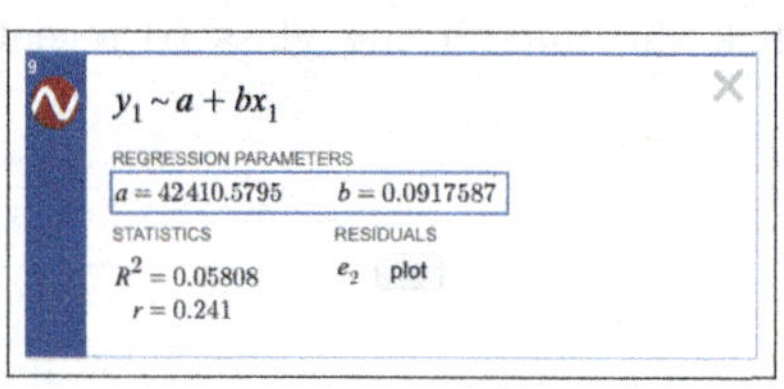

R: Scatterplots and Linear Regression Analysis

First read in the x and y values as described on page 52. For simplicity, we use `scan()`. However, if you have loaded the `openintro` package as described near the bottom of page 53 you can use the `dataset$variable` structure, e.g. `loan50$total_credit_utilized`.

```
> x = scan()
1:  95131
2:  51929
...
50:  390156
51:
Read 50 items
```

```
> y = scan()
1:  32894
2:  78341
...
50:  52534
51:
Read 50 items
```

Use the `lm()` function to build a linear model and `summary()` to summarize it.

```
> model = lm(y ~ x)
> summary(model)
Residuals:
Min      1Q       Median   3Q      Max
-64713  -28247   -16934   10568   305912

Coefficients:
.                Estimate    Std.Error    t value   Pr(>|t|)
(Intercept) 42410.57946   14210.43201    2.984    0.00446 **
.           x     0.09176      0.05333    1.720    0.09179
---
Residual standard error:  62540 on 48 degrees of freedom
Multiple R-squared:  0.05808, Adjusted R-squared:  0.03846
```

Use `plot(x, y)` to make a scatterplot. For fun, type `help(plot)` at the > prompt and investigate the optional `type` argument.

```
> plot(x, y)
> plot(x, y, ylim=c(0,400000), col = "purple", xlab = "Loan Amount ($)",
ylab = "Total Credit Utilized", abline(model, col="red"))
```

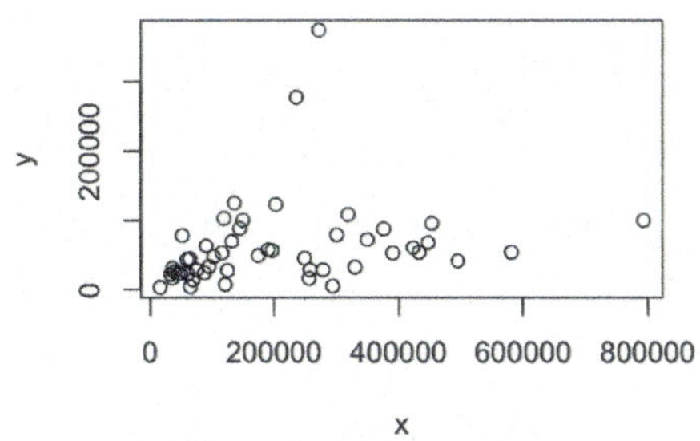

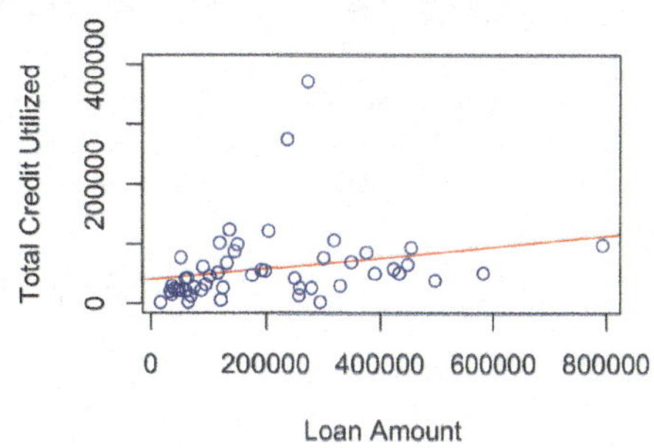

Make a histogram of the residuals and a residual plot, plotting the residuals versus x. Add `xlab = " "` and `ylab = " "` as desired.

```
> hist(model$residuals)
> plot(x, model$residuals, abline(h=0))
```

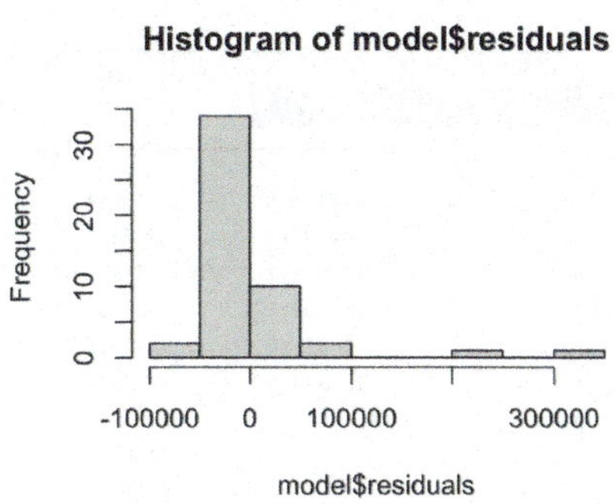

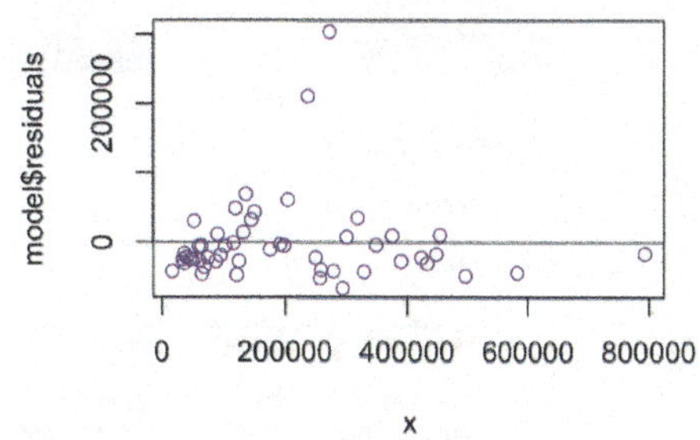

Calculator: Your teacher may give you data files to work with. Alternately, manually enter data into a list as described here. NumWorks calculator instructions are included along with example output. For the TI-83/84 and Casio calculators, general instructions are provided and worked example videos can be accessed via the icon or at **openintro.org/ti** and **openintro.org/casio**.

NUMWORKS: REGRESSION

Use `OK` or `EXE` to make a selection.

1. Click the yellow Home button (above the black Power button) to get to the home screen.
2. Select `Regression` and enter the x and y values.
3. Use the up and right arrow and select `Graph`. This will show you the scatterplot. Use the left and right arrow to toggle among the data points.
4. Press the up arrow and the right arrow to select `Regression`. Here you can choose Linear a+bx format or choose a nonlinear regression. This will draw the regression onto the scatterplot.

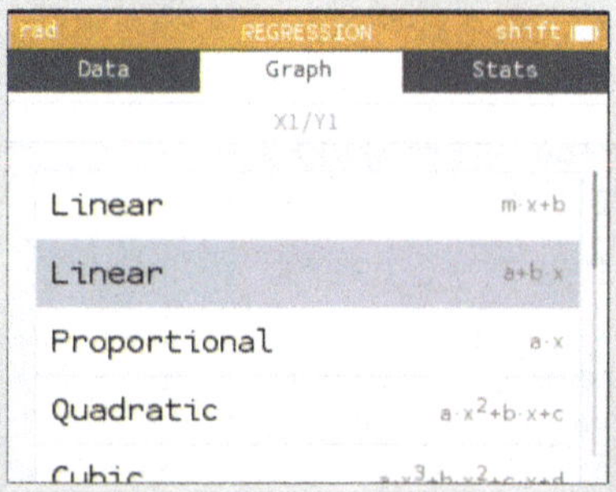

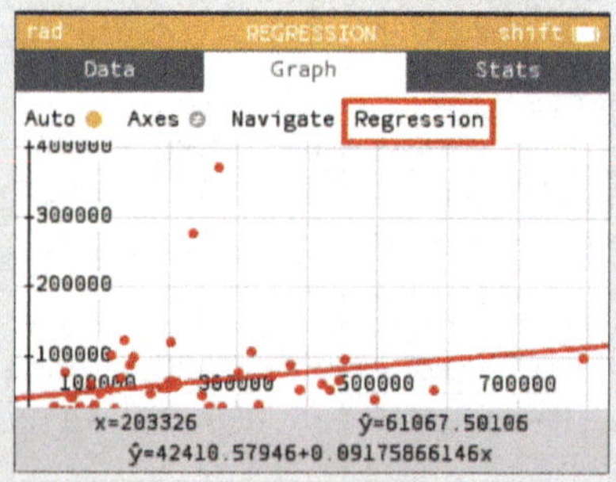

5. Press the up arrow and the right arrow to select `Regression` again. Here you will see the regression output. Press the down arrow to see more options, including `Residual plot`.

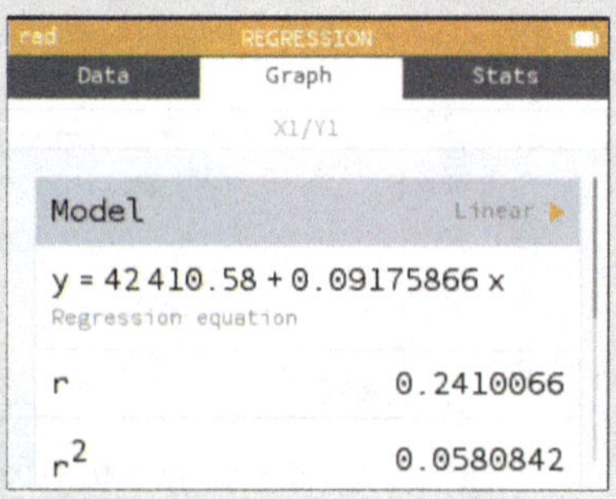

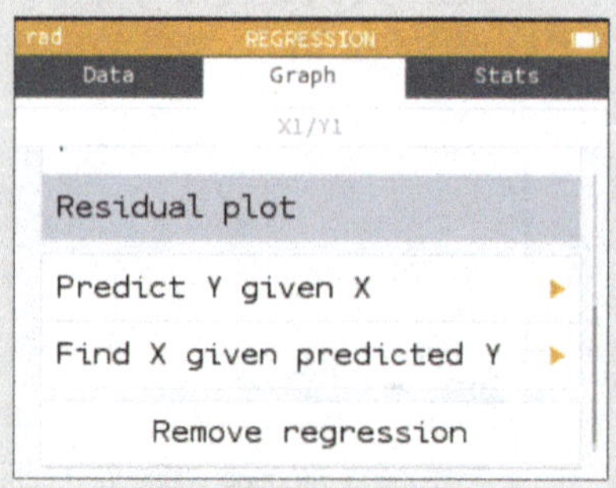

6. On the residual plot, use the left and right arrows to select individual points and see their corresponding residual.
7. Use the arrows to choose `Stats` at the top to see the summary statistics for x and y. Press the right arrow to see the y summary statistics. Press down arrow to see more summary statistics.

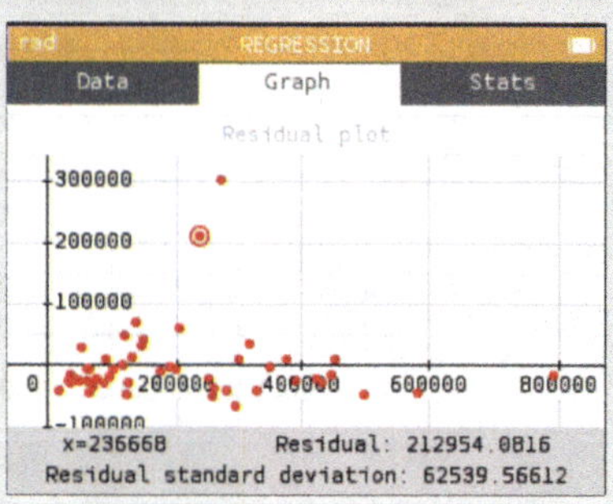

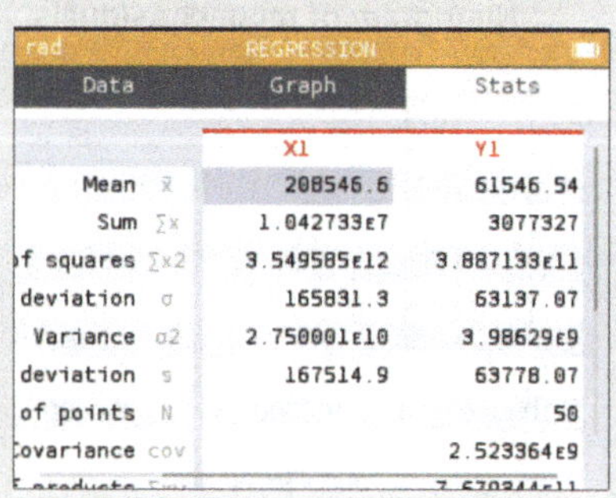

TI-84: FINDING a, b, R^2, AND r FOR A LINEAR MODEL

Use `STAT`, `CALC`, `LinReg(a + bx)`.

1. Choose `STAT`.
2. Right arrow to `CALC`.
3. Down arrow and choose `8:LinReg(a+bx)`.
   - Caution: choosing `4:LinReg(ax+b)` will reverse a and b.
4. Let `Xlist` be `L1` and `Ylist` be `L2` (don't forget to enter the x and y values in L1 and `L2` before doing this calculation).
5. Leave `FreqList` blank.
6. Leave `Store RegEQ` blank.
7. Choose Calculate and hit `ENTER`, which returns:

   `a` a, the y-intercept of the best fit line
   `b` b, the slope of the best fit line
   `r`^{2} R^2, the explained variance
   `r` r, the correlation coefficient

TI-83: Do steps 1-3, then enter the x list and y list separated by a comma, e.g. `LinReg(a+bx) L1, L2`, then hit `ENTER`.

WHAT TO DO IF R^2 AND r DO NOT SHOW UP ON A TI-83/84

If r^2 and r do now show up when doing `STAT`, `CALC`, `LinReg`, the *diagnostics* must be turned on. This only needs to be once and the diagnostics will remain on.

1. Hit `2ND 0` (i.e. `CATALOG`).
2. Scroll down until the arrow points at `DiagnosticOn`.
3. Hit `ENTER` and `ENTER` again. The screen should now say:

```
DiagnosticOn
                    Done
```

WHAT TO DO IF A TI-83/84 RETURNS: ERR: DIM MISMATCH

This error means that the lists, generally L1 and L2, do not have the same length.

1. Choose `1:Quit`.
2. Choose `STAT`, `Edit` and make sure that the lists have the same number of entries.

CASIO FX-9750GII: FINDING a, b, R^2, AND r FOR A LINEAR MODEL

1. Navigate to `STAT` (`MENU` button, then hit the `2` button or select `STAT`).
2. Enter the x and y data into 2 separate lists, e.g. x values in `List 1` and y values in `List 2`. Observation ordering should be the same in the two lists. For example, if $(5,4)$ is the second observation, then the second value in the x list should be 5 and the second value in the y list should be 4.
3. Navigate to `CALC` (`F2`) and then `SET` (`F6`) to set the regression context.
    - To change the `2Var XList`, navigate to it, select `List` (`F1`), and enter the proper list number. Similarly, set `2Var YList` to the proper list.
4. Hit `EXIT`.
5. Select `REG` (`F3`), `X` (`F1`), and `a+bx` (`F2`), which returns:

`a`	a, the y-intercept of the best fit line
`b`	b, the slope of the best fit line
`r`	r, the correlation coefficient
`r`^{2}	R^2, the explained variance
`MSe`	Mean squared error, which you can ignore

If you select `ax+b` (`F1`), the `a` and `b` meanings will be reversed.

5.3.6 Types of outliers in linear regression (special topic)

Outliers in regression are observations that fall far from the "cloud" of points. These points are especially important because they can have a strong influence on the least squares line.

EXAMPLE 5.26

There are six plots shown in Figure 5.23 along with the least squares line and residual plots. For each scatterplot and residual plot pair, identify any obvious outliers and note how they influence the least squares line. Recall that an outlier is any point that doesn't appear to belong with the vast majority of the other points.

E

(1) There is one outlier far from the other points, though it only appears to slightly influence the line.

(2) There is one outlier on the right, though it is quite close to the least squares line, which suggests it wasn't very influential.

(3) There is one point far away from the cloud, and this outlier appears to pull the least squares line up on the right; examine how the line around the primary cloud doesn't appear to fit very well.

(4) There is a primary cloud and then a small secondary cloud of four outliers. The secondary cloud appears to be influencing the line somewhat strongly, making the least squares line fit poorly almost everywhere. There might be an interesting explanation for the dual clouds, which is something that could be investigated.

(5) There is no obvious trend in the main cloud of points and the outlier on the right appears to largely control the slope of the least squares line.

(6) There is one outlier far from the cloud, however, it falls quite close to the least squares line and does not appear to be very influential.

Examine the residual plots in Figure 5.23. You will probably find that there is some trend in the main clouds of (3) and (4). In these cases, the outliers influenced the slope of the least squares lines. In (5), data with no clear trend were assigned a line with a large trend simply due to one outlier (!).

LEVERAGE

Points that fall horizontally away from the center of the cloud tend to pull harder on the line, so we call them points with **high leverage**.

Points that fall horizontally far from the line are points of high leverage; these points can strongly influence the slope of the least squares line. If one of these high leverage points does appear to actually invoke its influence on the slope of the line – as in cases (3), (4), and (5) of Example 5.26 – then we call it an **influential point**. Usually we can say a point is influential if, had we fitted the line without it, the influential point would have been unusually far from the least squares line.

It is tempting to remove outliers. Don't do this without a very good reason. Models that ignore exceptional (and interesting) cases often perform poorly. For instance, if a financial firm ignored the largest market swings – the "outliers" – they would soon go bankrupt by making poorly thought-out investments.

DON'T IGNORE OUTLIERS WHEN FITTING A FINAL MODEL

If there are outliers in the data, they should not be removed or ignored without a good reason. Whatever final model is fit to the data would not be very helpful if it ignores the most exceptional cases.

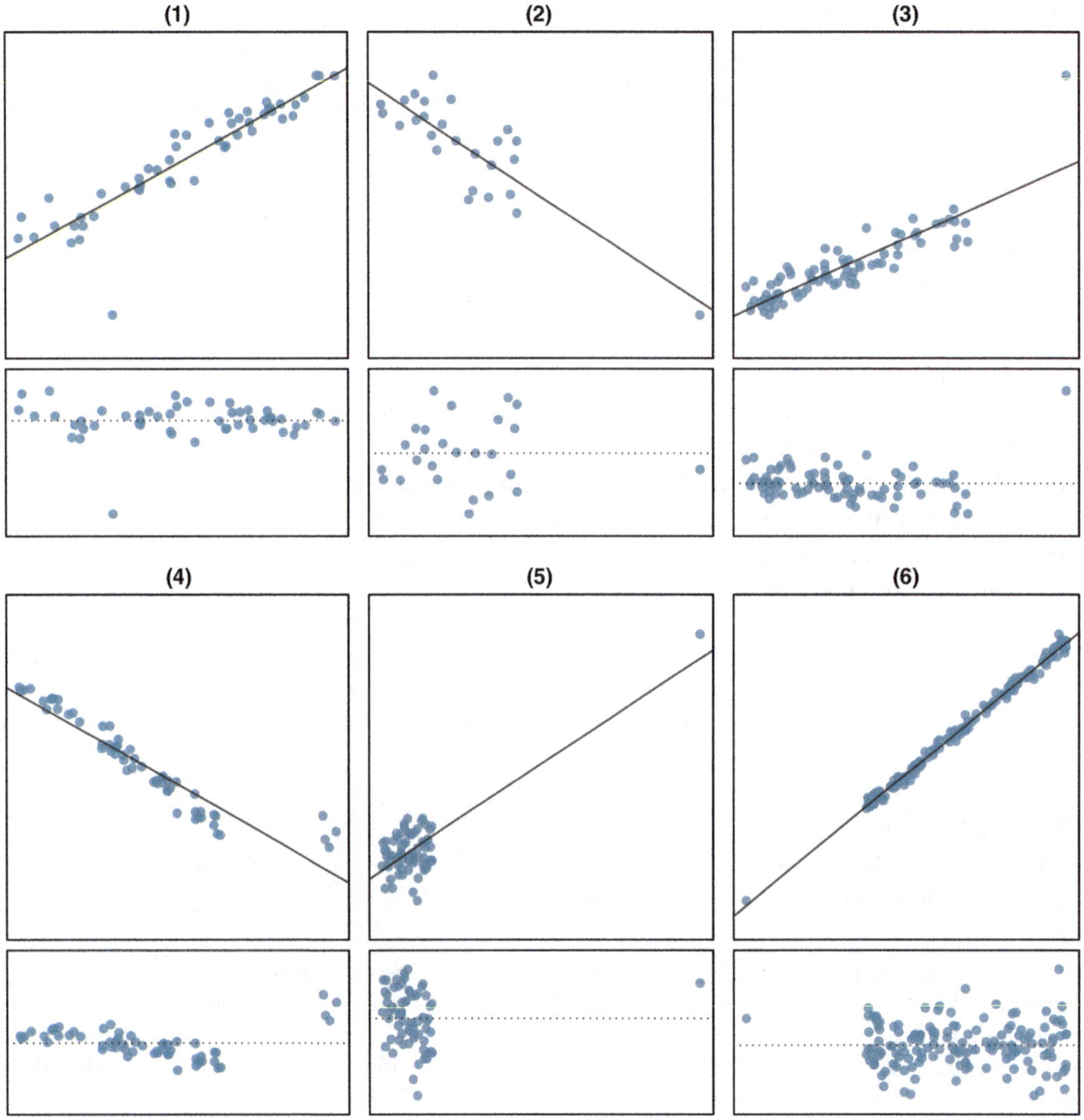

Figure 5.23: Six plots, each with a least squares line and residual plot. All data sets have at least one outlier.

5.3.7 Exploring futher

To explore some regression topics beyond those included in this textbook, visit

openintro.org/ahss/supplements,

where you will find additional content on the following topics:

- Categorical predictors with two levels
- Inference for the slope of a regression line
- Introduction to multiple linear regression
- Introduction to logistic regression

Section summary

- The simple linear regression model is fit to the data by minimizing the sum of the squares of the residuals. Because of this, the resulting equation is often called the **least-squares regression line** (LSRL) and is calculated using technology. This regression line will pass through the point ($\bar{x}$, $\bar{y}$).
- We write the least squares regression line in the form: $\hat{y} = a+bx$. The slope b and the y-intercept a can be calculated using technology or identified from computer output.
- In simple linear regression, the square of the correlation coefficient, R^2, is called the coefficient of determination. R^2 is the proportion of variation in the response variable that is explained by the linear relationship with the explanatory variable.
- R^2 is always between 0 and 1, inclusive, or between 0% and 100%, inclusive. R^2 applies to any type of model, not just a linear model, and can be used to compare the fit among various models. The higher the value of R^2, the better the model "fits" the data.
- The value of R^2 is always positive and cannot tell us the *direction* of the association. If finding r based on R^2, make sure to use either the scatterplot or the slope of the regression line to determine whether $r = +\sqrt{R^2}$ or $-\sqrt{R^2}$.
- The coefficients a and b of the least-squares regression line model (line of best fit) are statistics because they are based on a sample of values.
- The slope b of the least-squares regression line can be interpreted as the *predicted* increase or decrease in the response variable y for a one-unit increase in the explanatory variable x.
- The y-intercept a in the least-squares regression line is the *predicted* value of the response variable y when the explanatory variable x is equal to 0. Sometimes, the y-intercept of the line does not have a reasonable interpretation in context because $x = 0$ might be beyond the interval of x-values used to determine the regression line (extrapolation). At other times, the y-intercept of the line does not have a logical interpretation in context because it might be a negative value for a response variable that has no negative values, such as height.
- When interpreting a, b, and R^2, always interpret them based on the situation, referencing the relevant variables in context.

Exercises

5.21 The Coast Starlight, Part II. Exercise 5.11 introduces data on the Coast Starlight Amtrak train that runs from Seattle to Los Angeles. The mean travel time from one stop to the next on the Coast Starlight is 129 mins, with a standard deviation of 113 minutes. The mean distance traveled from one stop to the next is 108 miles with a standard deviation of 99 miles. The correlation between travel time and distance traveled is 0.636. A least-squares regression line for predicting travel time from distance traveled is fit to the data and has a slope of 0.726 and a y-intercept of 51.

(a) Write the equation of the regression line for predicting travel time and define any variables used.
(b) Interpret the slope and the intercept in this context.
(c) Calculate R^2 of the regression line for predicting travel time from distance traveled for the Coast Starlight, and interpret R^2 in the context of the application.
(d) The distance between Santa Barbara and Los Angeles is 103 miles. Use the model to estimate the time it takes for the Starlight to travel between these two cities.
(e) It actually takes the Coast Starlight about 168 mins to travel from Santa Barbara to Los Angeles. Calculate the residual and explain the meaning of this residual value.
(f) Suppose Amtrak is considering adding a stop to the Coast Starlight 500 miles away from Los Angeles. Would it be appropriate to use this linear model to predict the travel time from Los Angeles to this point?

5.22 Body measurements. Researchers studying anthropometry collected body girth measurements and skeletal diameter measurements, as well as age, weight, height and gender for 507 physically active individuals. The mean shoulder girth is 107.20 cm with a standard deviation of 10.37 cm. The mean height is 171.14 cm with a standard deviation of 9.41 cm. The correlation between height and shoulder girth is 0.67. A least-squares regression line for predicting height from shoulder girth is fit to the data and has a slope of 0.604 and a y-intercept of 106.39.

(a) Write the equation of the regression line for predicting height and define any variables used.
(b) Interpret the slope and the intercept in this context.
(c) Calculate R^2 of the regression line for predicting height from shoulder girth, and interpret it in the context of the application.
(d) A randomly selected student from your class has a shoulder girth of 100 cm. Predict the height of this student using the model.
(e) The student from part (d) is 160 cm tall. Calculate the residual, and explain what this residual means.
(f) A one year old has a shoulder girth of 56 cm. Would it be appropriate to use this linear model to predict the height of this child?

5.23 Murders and poverty, Part I. The following regression output is for predicting annual murders per million from percentage living in poverty in a random sample of 20 metropolitan areas.

	Estimate	Std. Error	t value	Pr(>|t|)
(Intercept)	-29.901	7.789	-3.839	0.001
poverty%	2.559	0.390	6.562	0.000
$s = 5.512$		$R^2 = 70.52\%$		$R^2_{adj} = 68.89\%$

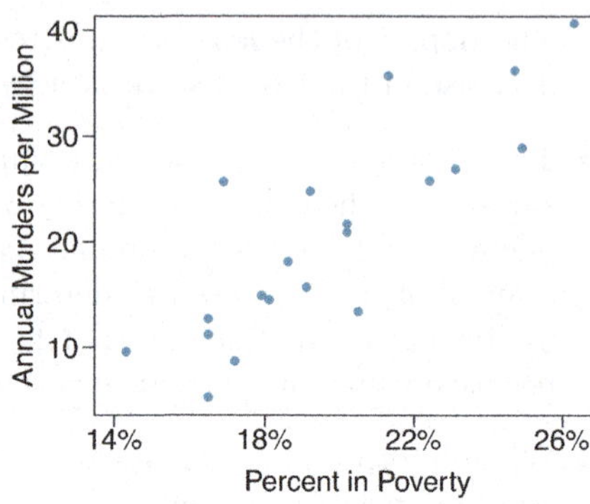

(a) Write out the linear model.
(b) Interpret the intercept.
(c) Interpret the slope.
(d) Interpret R^2.
(e) Calculate the correlation coefficient.

5.24 Cats, Part I. The following regression output is for predicting the heart weight (in g) of cats from their body weight (in kg). The coefficients are estimated using a dataset of 144 domestic cats.

	Estimate	Std. Error	t value	Pr(>|t|)
(Intercept)	-0.357	0.692	-0.515	0.607
body wt	4.034	0.250	16.119	0.000

$s = 1.452$ $R^2 = 64.66\%$ $R^2_{adj} = 64.41\%$

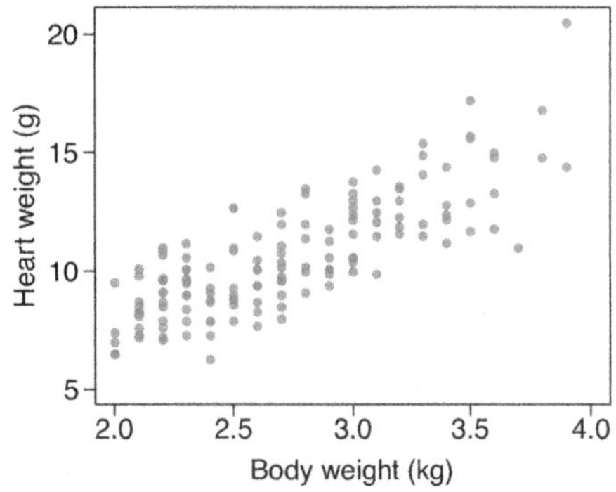

(a) Write out the linear model.

(b) Interpret the intercept.

(c) Interpret the slope.

(d) Interpret R^2.

(e) Calculate the correlation coefficient.

Chapter highlights

This chapter focused on describing the linear association between two numerical variables and fitting a linear model.

- Every analysis should begin with *graphing* the data using a **scatterplot** in order to see the association and any deviations from the trend. A **residual plot** helps us better see patterns in the data.
- When the data show a linear trend, we fit a **least squares regression line** of the form: $\hat{y} = a + bx$, where a is the y-intercept and b is the slope. It is important to be able to interpret a and b in the context of the data.
- A **residual**, $y - \hat{y}$, measures the error for an *individual point*.
- The **correlation coefficient**, r, measures the strength and direction of the linear association between two variables. However, r alone cannot tell us whether data follow a linear trend or whether a linear model is appropriate.
- The **coefficient of determination**, R^2, measures the proportion of variation in the y values explained by a given model. Like r, R^2 alone cannot tell us whether data follow a linear trend or whether a linear model is appropriate.

In this chapter we focused on simple linear models with one explanatory variable. More complex methods of prediction, such as multiple regression (more than one explanatory variable) and non-linear regression can be studied in a future course.

Chapter exercises

5.25 Pet names. The city of Seattle, WA has an open data portal that includes pets registered in the city. For each registered pet, we have information on the pet's name and species. The following visualization plots the proportion of dogs with a given name versus the proportion of cats with the same name. The 20 most common cat and dog names are displayed. The diagonal line on the plot is the $x = y$ line; if a name appeared on this line, the name's popularity would be exactly the same for dogs and cats.

(a) Are these data collected as part of an experiment or an observational study?

(b) What is the most common dog name? What is the most common cat name?

(c) What names are more common for cats than dogs?

(d) Is the relationship between the two variables positive or negative? What does this mean in context of the data?

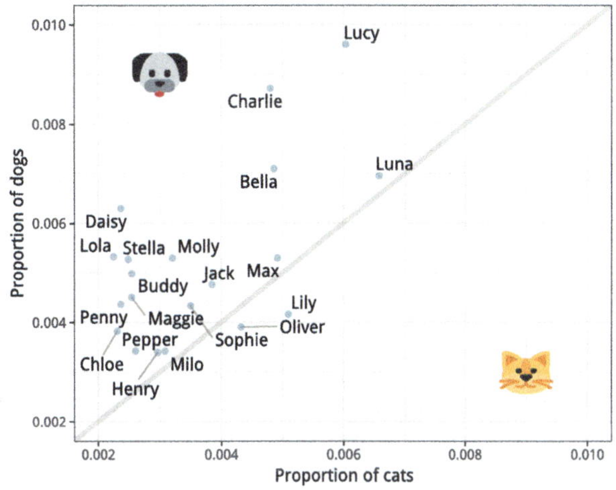

5.26 Trees. The scatterplots below show the relationship between height, diameter, and volume of timber in 31 felled black cherry trees. The diameter of the tree is measured 4.5 feet above the ground.[22]

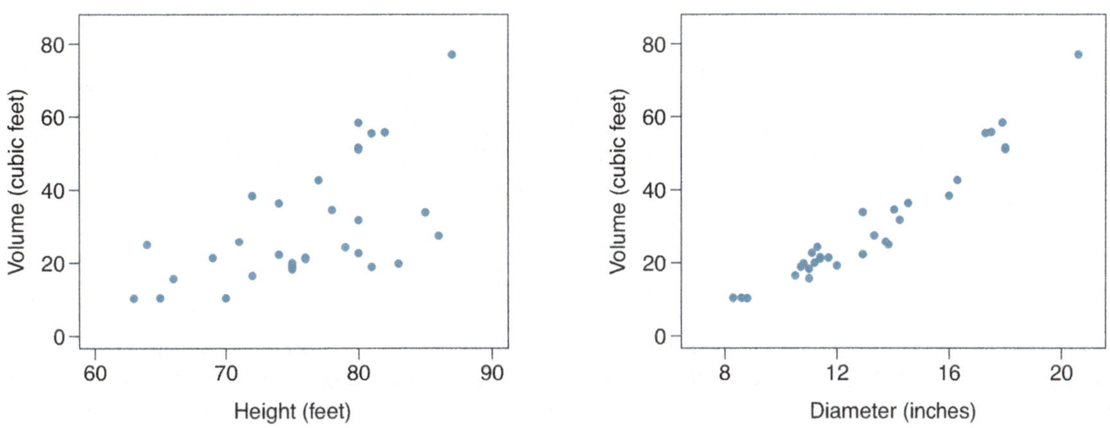

(a) Describe the relationship between volume and height of these trees.

(b) Describe the relationship between volume and diameter of these trees.

(c) Suppose you have height and diameter measurements for another black cherry tree. Which of these variables would be preferable to use to predict the volume of timber in this tree using a simple linear regression model? Explain your reasoning.

5.27 True / False. Determine if the following statements are true or false. If false, explain why.

(a) A correlation coefficient of -0.90 indicates a stronger linear relationship than a correlation of 0.5.

(b) Correlation is a measure of the association between any two variables.

[22]Source: R Dataset, stat.ethz.ch/R-manual/R-patched/library/datasets/html/trees.html.

5.28 Urban homeowners. The scatterplot shows the percent of families who own their home vs. the percent of the population living in urban areas.[23] There are 51 observations, each corresponding to a state and Puerto Rico. A residual plot is also shown.

(a) For these data, $R^2 = 0.28$. What is the correlation? How can you tell if it is positive or negative?

(b) Examine the residual plot. What do you observe? Is a simple least squares fit appropriate for these data?

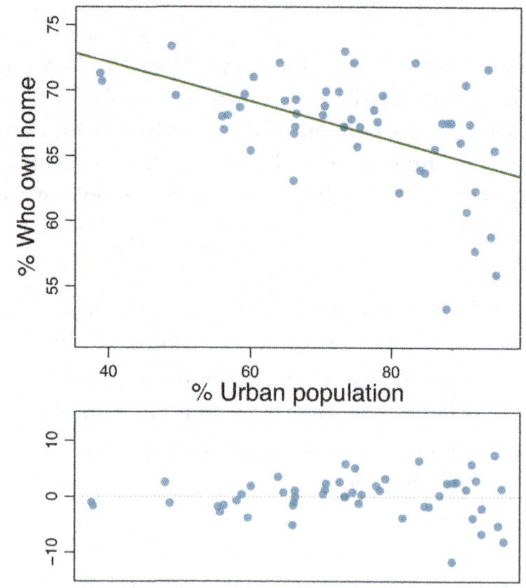

5.29 Nutrition at Starbucks, Part II. Exercise 5.20 introduced a data set on nutrition information on Starbucks food menu items. Based on the scatterplot and the residual plot provided, describe the relationship between the protein content and calories of these menu items, and determine if a simple linear model is appropriate to predict amount of protein from the number of calories.

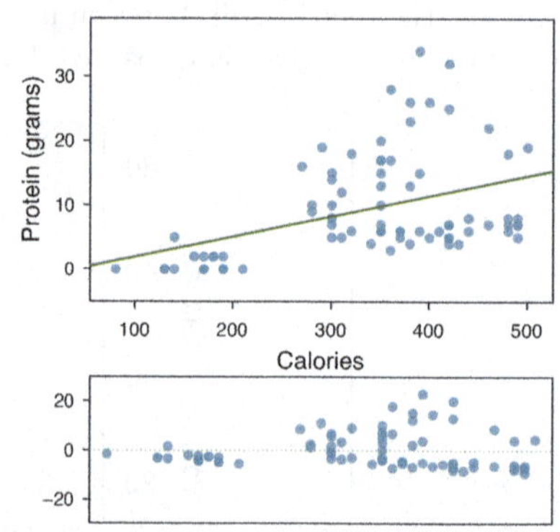

5.30 Helmets and lunches. The scatterplot shows the relationship between socioeconomic status measured as the percentage of children in a neighborhood receiving reduced-fee lunches at school (`lunch`) and the percentage of bike riders in the neighborhood wearing helmets (`helmet`). The regression equation is given by: $\hat{y} = 55.34 - 0.537x$.

(a) If the R^2 for the least-squares regression line for these data is 72%, what is the correlation between `lunch` and `helmet`?

(b) Interpret the intercept of the least-squares regression line in the context of the application.

(c) Interpret the slope of the least-squares regression line in the context of the application.

(d) What would the value of the residual be for a neighborhood where 40% of the children receive reduced-fee lunches and 40% of the bike riders wear helmets? Interpret the meaning of this residual in the context of the application.

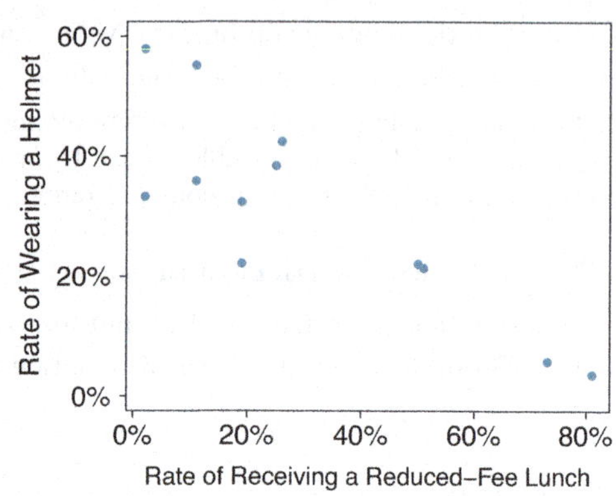

[23] United States Census Bureau, 2010 Census Urban and Rural Classification and Urban Area Criteria and Housing Characteristics: 2010.

5.31 Match the correlation, Part III. Match each correlation to the corresponding scatterplot.

(a) $r = -0.72$
(b) $r = 0.07$
(c) $r = 0.86$
(d) $r = 0.99$

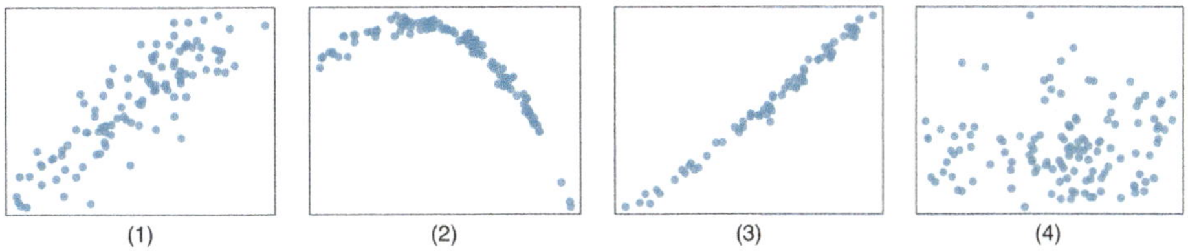

5.32 MCU, predict US theater sales. The Marvel Comic Universe movies were an international movie sensation, containing 23 movies at the time of this writing. Here we consider a model predicting an MCU film's gross theater sales in the US based on the first weekend sales performance in the US. The data are presented below in both a scatterplot and the model in a regression table. Scientific notation is used below, e.g. 42.5e6 corresponds to 42.5×10^6.

	Estimate	Std. Error	t value	Pr(>|t|)
(Intercept)	42.5e6	26.6e6	1.60	0.1251
opening_weekend_us	2.4361	0.1739	14.01	0.0000

(a) Describe the relationship between gross theater sales in the US and first weekend sales in the US.

(b) Write the equation of the regression line. Interpret the slope and intercept in context.

(c) The correlation coefficient for gross sales and first weekend sales is 0.950. Calculate R^2 and interpret it in context.

(d) Suppose we consider a set of all films ever released. Do you think the relationship between opening weekend sales and total sales would have as strong of a relationship as what we see with the MCU films?

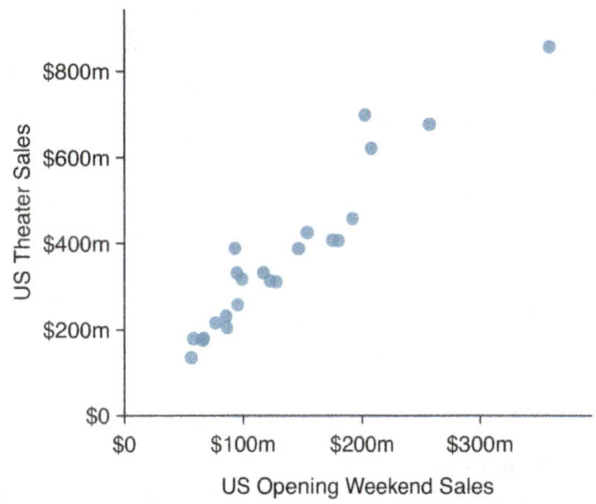

Appendix A

Exercise solutions

1 Exploring one-variable data and collecting data

1.1 (a) "Is there an association between air pollution exposure and preterm births?" (b) 143,196 births in Southern California between 1989 and 1993. (c) Measurements of carbon monoxide, nitrogen dioxide, ozone, and particulate matter less than $10\mu g/m^3$ (PM_{10}) collected at air-quality-monitoring stations as well as length of gestation. Continous numerical variables.

1.3 (a) "Does explicitly telling children not to cheat affect their likelihood to cheat?". (b) 160 children between the ages of 5 and 15. (c) Four variables: (1) age (numerical, continuous), (2) sex (categorical), (3) whether they were an only child or not (categorical), (4) whether they cheated or not (categorical).

1.5 (a) Observation. (b) Variable. (c) Sample statistic (mean). (d) Population parameter (mean).

1.7 (a) $50 \times 3 = 150$. (b) Four continuous numerical variables: sepal length, sepal width, petal length, and petal width. (c) One categorical variable, species, with three levels: *setosa*, *versicolor*, and *virginica*.

1.9 (a) We see the order of the categories and the relative frequencies in the bar plot. (b) There are no features that are apparent in the pie chart but not in the bar plot. (c) We usually prefer to use a bar plot as we can also see the relative frequencies of the categories in this graph.

1.11 (a)

```
0 | 000003333333
0 | 7779
1 | 0011
Legend:  1 | 0 = 10%
```

(b)

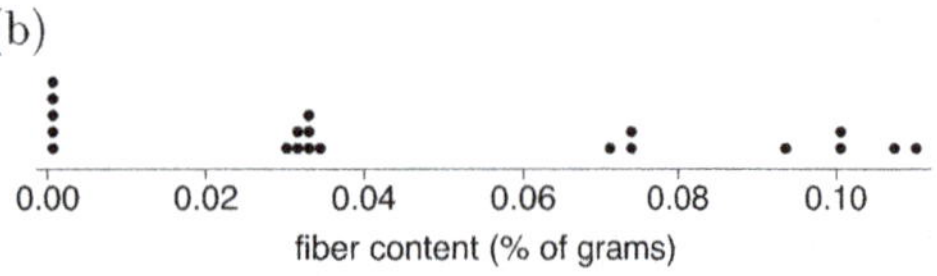

(c)

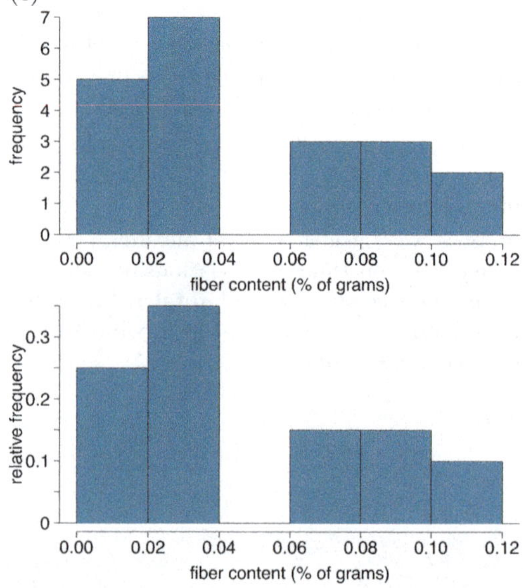

(d) 40% (Note: if using only rel. freq. histogram, you can only get an estimate because 7 is in the middle of the bin. Use the dot plot to get a more accurate answer.)

1.13 Both distributions are right skewed and bimodal with modes at 10 and 20 cigarettes; note that people may be rounding their answers to half a pack or a whole pack. The median of each distribution is between 10 and 15 cigarettes. The middle 50% of the data (the IQR) appears to be spread equally in each group and have a width of about 10 to 15. There are potential outliers above 40 cigarettes per day. It appears that respondents who smoke only a few cigarettes (0 to 5) smoke more on the weekdays than on weekends.

1.15 (a) $\bar{x}_{amtWeekends} = 20$, $\bar{x}_{amtWeekdays} = 16$. (b) $s_{amtWeekends} = 0$, $s_{amtWeekdays} = 4.18$. In this very small sample, higher on weekdays.

1.17 Any 10 employees whose average number of days off is between the minimum and the mean number of days off for the entire workforce at this plant.

1.19 (a) Dist 2 has a higher mean since 20 > 13, and a higher standard deviation since 20 is further from the rest of the data than 13. (b) Dist 1 has a higher mean since −20 > −40, and Dist 2 has a higher standard deviation since -40 is farther away from the rest of the data than -20. (c) Dist 2 has a higher mean since all values in this distribution are higher than those in Dist 1, but both distribution have the same standard deviation since they are equally variable around their respective means. (d) Both distributions have the same mean since they're both centered at 300, but Dist 2 has a higher standard deviation since the observations are farther from the mean than in Dist 1.

1.21 (a) About 30. (b) Since the distribution is right skewed the mean is higher than the median. (c) Q1: between 15 and 20, Q3: between 35 and 40, IQR: about 20. (d) Values that are considered to be unusually low or high lie more than 1.5×IQR away from the quartiles. Upper fence: Q3 + 1.5 × IQR = 37.5 + 1.5 × 20 = 67.5; Lower fence: Q1 - 1.5 × IQR = 17.5 + 1.5 × 20 = −12.5; The lowest AQI recorded is not lower than 5 and the highest AQI recorded is not higher than 65, which are both within the fences. Therefore none of the days in this sample would be considered to have an unusually low or high AQI.

1.23 The histogram shows that the distribution is bimodal, which is not apparent in the box plot. The box plot makes it easy to identify more precise values of observations outside of the whiskers.

1.25 (a) The distribution of number of pets per household is likely right skewed as there is a natural boundary at 0 and only a few people have many pets. Therefore the center would be best described by the median, and variability would be best described by the IQR. (b) The distribution of number of distance to work is likely right skewed as there is a natural boundary at 0 and only a few people live a very long distance from work. Therefore the center would be best described by the median, and variability would be best described by the IQR. (c) The distribution of heights of males is likely symmetric. Therefore the center would be best described by the mean, and variability would be best described by the standard deviation.

1.27 (a) The median is a much better measure of the typical amount earned by these 42 people. The mean is much higher than the income of 40 of the 42 people. This is because the mean is an arithmetic average and gets affected by the two extreme observations. The median does not get effected as much since it is robust to outliers. (b) The IQR is a much better measure of variability in the amounts earned by nearly all of the 42 people. The standard deviation gets affected greatly by the two high salaries, but the IQR is robust to these extreme observations.

1.29 The distribution of commute times is unimodal and approximately symmetric with a mean of about 25 minutes and a standard deviation of about 5 minutes. There does not appear to be any counties with unusually high or low mean travel times.

1.31 (a) Population: all births, sample: 143,196 births between 1989 and 1993 in Southern California. (b) If births in this time span at the geography can be considered to be representative of all births, then the results are generalizable to the population of Southern California. However, since the study is observational the findings cannot be used to establish causal relationships.

1.33 (a) Population: all asthma patients aged 18-69 who rely on medication for asthma treatment. Sample: 600 such patients. (b) If the patients in this sample, who are likely not randomly sampled, can be considered to be representative of all asthma patients aged 18-69 who rely on medication for asthma treatment, then the results are generalizable to the population defined above. Additionally, since the study is experimental, the findings can be used to establish causal relationships.

1.35 (a) Observational. (b) Use stratified sampling to randomly sample a fixed number of students, say 10, from each section for a total sample size of 40 students.

1.37 (a) Simple random sampling is okay. In fact, it's rare for simple random sampling to not be a reasonable sampling method! (b) The student opinions may vary by field of study, so the stratifying by this variable makes sense and would be reasonable. (c) Students of similar ages are probably going to have more similar opinions, and we want clusters to be diverse with respect to the outcome of interest, so this would **not** be a good approach. (Additional thought: the clusters in this case may also have very different numbers of people, which can also create unexpected sample sizes.)

1.39 (a) The cases are 200 randomly sampled men and women. (b) The response variable is attitude towards a fictional microwave oven. (c) The explanatory variable is dispositional attitude. (d) Yes, the cases are sampled randomly. (e) This is an observational study since there is no random assignment to treatments. (f) No, we cannot establish a causal link between the explanatory and response variables since the study is observational. (g) Yes, the results of the study can be generalized to the population at large since the sample is random.

1.41 (a) Simple random sample. Non-response bias, if only those people who have strong opinions about the survey responds his sample may not be representative of the population. (b) Convenience sample. His sample may not be representative of the population since it consists only of his friends. It is also possible that the study will have non-response bias if some choose to not bring back the survey. (c) Convenience sample. This will have a similar issues to handing out surveys to friends. (d) Multi-stage sampling. If the classes are similar to each other with respect to student composition this approach should not introduce bias, other than potential non-response bias.

1.43 (a) Exam performance. (b) Light level: fluorescent overhead lighting, yellow overhead lighting, no overhead lighting (only desk lamps). (c) Sex: man, woman.

1.45 (a) Experiment. (b) Light level (overhead lighting, yellow overhead lighting, no overhead lighting) and noise level (no noise, construction noise, and human chatter noise). (c) Since the researchers want to ensure equal representation of graduate and undergraduate students, program type will be a blocking variable.

1.47 Need randomization and blinding. One possible outline:

(1) Prepare two cups for each participant, one containing regular Coke and the other containing Diet Coke. Make sure the cups are identical and contain equal amounts of soda. Label the cups A (regular) and B (diet). (Be sure to randomize A and B for each trial!)

(2) Give each participant the two cups, one cup at a time, in random order, and ask the participant to record a value that indicates how much she liked the beverage. Be sure that neither the participant nor the person handing out the cups knows the identity of the beverage to make this a double-blind experiment.

(Answers may vary.)

1.49 (a) Decrease: the new score is smaller than the mean of the 24 previous scores. (b) Calculate a weighted mean. Use a weight of 24 for the old mean and 1 for the new mean: $(24 \times 74 + 1 \times 64)/(24 + 1) = 73.6$. (c) The new score is more than 1 standard deviation away from the previous mean, so increase.

1.51 No, we would expect this distribution to be right skewed. There are two reasons for this: (1) there is a natural boundary at 0 (it is not possible to watch less than 0 hours of TV), (2) the standard deviation of the distribution is very large compared to the mean.

1.53 The distribution of ages of best actress winners are right skewed with a median around 30 years. The distribution of ages of best actress winners is also right skewed, though less so, with a median around 40 years. The difference between the peaks of these distributions suggest that best actress winners are typically younger than best actor winners. The ages of best actress winners are more variable than the ages of best actor winners. There are potential outliers on the higher end of both of the distributions.

1.55

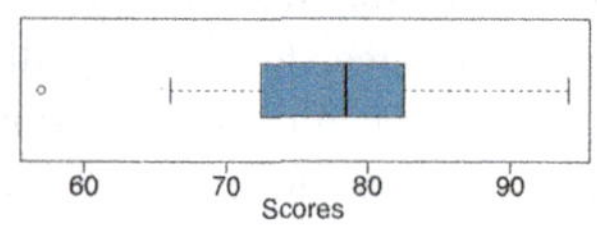

1.57 (a) Experiment. (b) Treatment: 25 grams of chia seeds twice a day, control: placebo. (c) Yes, gender. (d) Yes, single blind since the patients were blinded to the treatment they received. (e) Since this is an experiment, we can make a causal statement. However, since the sample is not random, the causal statement cannot be generalized to the population at large.

1.59 (a) Non-responders may have a different response to this question, e.g. parents who returned the surveys likely don't have difficulty spending time with their children. (b) It is unlikely that the women who were reached at the same address 3 years later are a random sample. These missing responders are probably renters (as opposed to homeowners) which means that they might have a lower socio-economic status than the respondents. (c) There is no control group in this study, this is an observational study, and there may be confounding variables, e.g. these people may go running because they are generally healthier and/or do other exercises.

1.61 (a) Randomized controlled experiment. (b) Explanatory: treatment group (categorical, with 3 levels). Response variable: Psychological well-being. (c) No, because the participants were volunteers. (d) Yes, because it was an experiment. (e) The statement should say "evidence" instead of "proof".

1.63 (a) Categorical, non-ordinal: County, State, Driver's race. Numerical, discrete: No. of stops per year. Numerical, continuous: % searched, % drivers arrested. (b) All categorical, non-ordinal. (c) Response: whether the car was searched or not. Explanatory: race of the driver.

2 Probability, random variables, and probability distributions

2.1 The vertical locations at which the ideological groups break into the Yes, No, and Not Sure categories differ, which indicates that likelihood of supporting the DREAM act varies by political ideology. This suggests that the two variables may be dependent.

2.3 (a) False. Instead of comparing counts, we should compare percentages of people in each group who suffered cardiovascular problems. (b) True. (c) False. Association does not imply causation. We cannot infer a causal relationship based on an observational study. The difference from part (ii) is subtle. (d) True.

2.5 (a) Number of participants in each group. (b) Proportion of survival. (c) The bar chart showing proportions should be displayed as a way to visualize the survival improvement in the treatment versus the control group.

2.7 (a) False. These are independent trials. (b) False. There are red face cards. (c) True. A card cannot be both a face card and an ace.

2.9 (a) 10 tosses. Fewer tosses mean more variability in the sample fraction of heads, meaning there's a better chance of getting at least 60% heads. (b) 100 tosses. More flips means the observed proportion of heads would often be closer to the average, 0.50, and therefore also above 0.40. (c) 100 tosses. With more flips, the observed proportion of heads would often be closer to the average, 0.50. (d) 10 tosses. Fewer flips would increase variability in the fraction of tosses that are heads.

2.11 (a) $0.5^{10} = 0.00098$. (b) $0.5^{10} = 0.00098$. (c) $P(\text{at least one tails}) = 1 - P(\text{no tails}) = 1 - (0.5^{10}) \approx 1 - 0.001 = 0.999$.

2.13 (a) No, but we could if A and B are independent. (b-i) 0.21. (b-ii) 0.79. (b-iii) 0.3. (c) No, because $0.1 \neq 0.21$, where 0.21 was the value computed under independence from part (a). (d) 0.143.

2.15 (a) No, 0.18 of respondents fall into this combination. (b) $0.60 + 0.20 - 0.18 = 0.62$. (c) $0.18/0.20 = 0.9$. (d) $0.11/0.33 \approx 0.33$. (e) No, otherwise the answers to (c) and (d) would be the same. (f) $0.06/0.34 \approx 0.18$.

2.17 (a) No, there are voters who are both independent and swing voters.

(b)

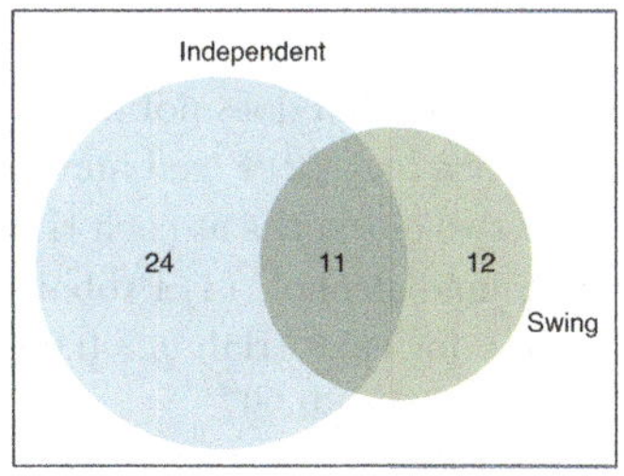

(c) Each Independent voter is either a swing voter or not. Since 35% of voters are Independents and 11% are both Independent and swing voters, the other 24% must not be swing voters. (d) 0.47. (e) 0.53. (f) P(Independent) × P(swing) = $0.35 \times 0.23 = 0.08$, which does not equal P(Independent and swing) = 0.11, so the events are dependent.

2.19 (a) 0.3. (b) 0.3. (c) 0.3. (d) $0.3 \times 0.3 = 0.09$. (e) Yes, the population that is being sampled from is identical in each draw.

2.21 (a) $2/9 \approx 0.22$. (b) $3/9 \approx 0.33$. (c) $\frac{3}{10} \times \frac{2}{9} \approx 0.067$. (d) No, e.g. in this exercise, removing one chip meaningfully changes the probability of what might be drawn next.

2.23 0.0714. Even when a patient tests positive for lupus, there is only a 7.14% chance that he actually has lupus. House may be right.

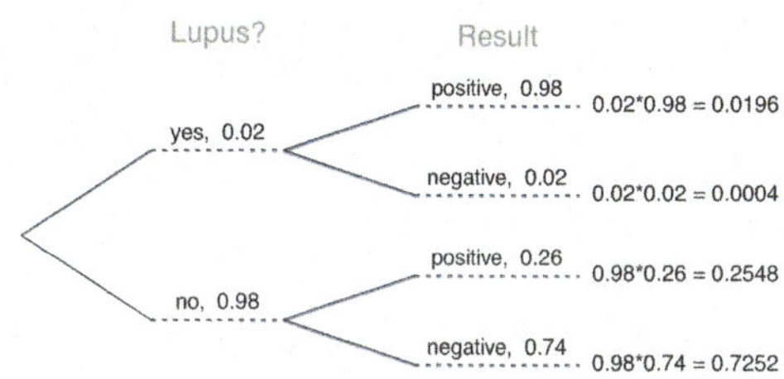

2.25 (a) Mean: $\$3 * 0.5 + \$5 * 0.3 + \$10 * 0.15 + \$25 * 0.05 = \$5.75$. (b) To compute the SD, it is easier to first compute the variance: $(3 - 5.75)^2 * 0.5 + (5 - 5.75)^2 * 0.3 + (10 - 5.75)^2 * 0.15 + (25 - 5.75)^2 * 0.05 = 25.1875$. The SD is then the square root of this value: \$5.02.

2.27 (a) E(X) = 3.59. SD(X) = 9.64. (b) E(X) = -1.41. SD(X) = 9.64. (c) No, the expected net profit is negative, so on average you expect to lose money.

2.29 5% increase in value.

2.31 (a) $\binom{5}{1} = 5$. (b) $\binom{5}{4} = 5$. (c) $\binom{5}{3} = 10$. (d) $\binom{5}{3} + \binom{5}{4} + \binom{5}{5} = 10 + 5 + 1 = 16$.

2.33 (a) Binomial conditions are met: (1) Independent trials: In a random sample, whether or not one 18-20 year old has consumed alcohol does not depend on whether or not another one has. (2) Fixed number of trials: $n = 10$. (3) Only two outcomes at each trial: Consumed or did not consume alcohol. (4) Probability of a success is the same for each trial: $p = 0.697$. (b) 0.203. (c) 0.203. (d) 0.167. (e) 0.997.

2.35 (a) $1 - 0.75^3 = 0.5781$. (b) 0.1406. (c) 0.4219. (d) $1 - 0.25^3 = 0.9844$.

2.37 (a) $\mu = 35$, $\sigma = 3.24$ (b) $Z = \frac{45-35}{3.24} = 3.09$. 45 is more than 3 standard deviations away from the mean, we can assume that it is an unusual observation. Therefore yes, we would be surprised. (c) Using the normal approximation, 0.0010. With 0.5 correction, 0.0017.

2.39 (a) 8.85%. (b) 6.94%. (c) 58.86%. (d) 4.56%.

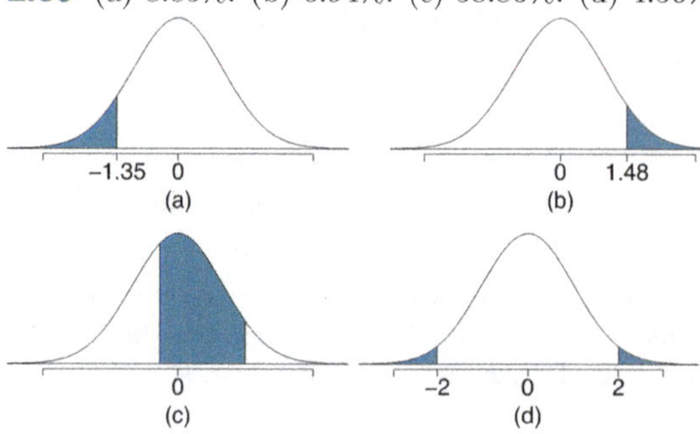

2.41 (a) $Z_{VR} = 1.29$, $Z_{QR} = 0.52$.

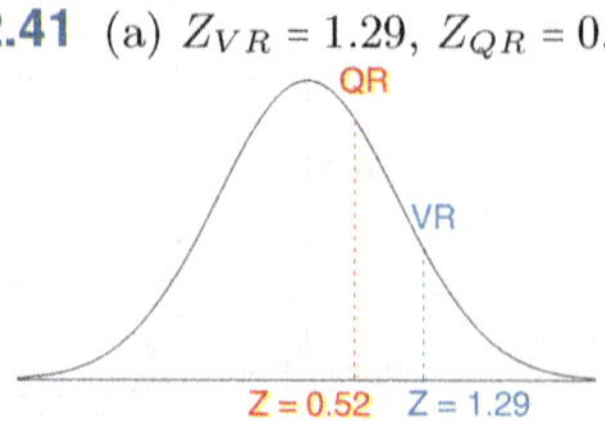

(b) She scored 1.29 standard deviations above the mean on the Verbal Reasoning section and 0.52 standard deviations above the mean on the Quantitative Reasoning section. (c) She did better on the Verbal Reasoning section since her Z-score on that section was higher. (d) $Perc_{VR} = 0.9007 \approx 90\%$, $Perc_{QR} = 0.6990 \approx 70\%$. (e) $100\% - 90\% = 10\%$ did better than her on VR, and $100\% - 70\% = 30\%$ did better than her on QR. (f) We cannot compare the raw scores since they are on different scales. Comparing her percentile scores is more appropriate when comparing her performance to others. (g) Answer to part (b) would not change as Z-scores can be calculated for distributions that are not normal. However, we could not answer parts (c)-(e) since we cannot use the normal probability table to calculate probabilities and percentiles without a normal model.

2.43 (a) $Z = 0.84$, which corresponds to approximately 159 on QR. (b) $Z = -0.52$, which corresponds to approximately 147 on VR.

2.45 (a) $Z = 1.2$, $P(Z > 1.2) = 0.1151$. (b) $Z = -1.28 \rightarrow 70.6°$F or colder.

2.47 $14/20 = 70\%$ are within 1 SD. Within 2 SD: $19/20 = 95\%$. Within 3 SD: $20/20 = 100\%$. They follow this rule closely.

2.49 (a) (i) False. Instead of comparing counts, we should compare percentages of people in each group who suffered cardiovascular problems. (ii) True. (iii) False. Association does not imply causation. We cannot infer a causal relationship based on an observational study. The difference from part (ii) is subtle. (iv) True.
(b) Proportion of all patients who had cardiovascular problems: $\frac{7,979}{227,571} \approx 0.035$
(c) The expected number of heart attacks in the rosiglitazone group, if having cardiovascular problems and treatment were independent, can be calculated as the number of patients in that group multiplied by the overall cardiovascular problem rate in the study: $67,593 * \frac{7,979}{227,571} \approx 2370$.
(d) (i) H_0: The treatment and cardiovascular problems are independent. They have no relationship, and the difference in incidence rates between the rosiglitazone and pioglitazone groups is due to chance. H_A: The treatment and cardiovascular problems are not independent. The difference in the incidence rates between the rosiglitazone and pioglitazone groups is not due to chance and rosiglitazone is associated with an increased risk of serious cardiovascular problems. (ii) A higher number of patients with cardiovascular problems than expected under the assumption of independence would provide support for the alternative hypothesis as this would suggest that rosiglitazone increases the risk of such problems. (iii) In the actual study, we observed 2,593 cardiovascular events in the rosiglitazone group. In the 1,000 simulations under the independence model, we observed somewhat less than 2,593 in every single simulation, which suggests that the actual results did not come from the independence model. That is, the variables do not appear to be independent, and we reject the independence model in favor of the alternative. The study's results provide convincing evidence that rosiglitazone is associated with an increased risk of cardiovascular problems.

2.51 (a) Invalid. Sum is greater than 1. (b) Valid. Probabilities are between 0 and 1, and they sum to 1. In this class, every student gets a C. (c) Invalid. Sum is less than 1. (d) Invalid. There is a negative probability. (e) Valid. Probabilities are between 0 and 1, and they sum to 1. (f) Invalid. There is a negative probability.

2.53 0.8247.

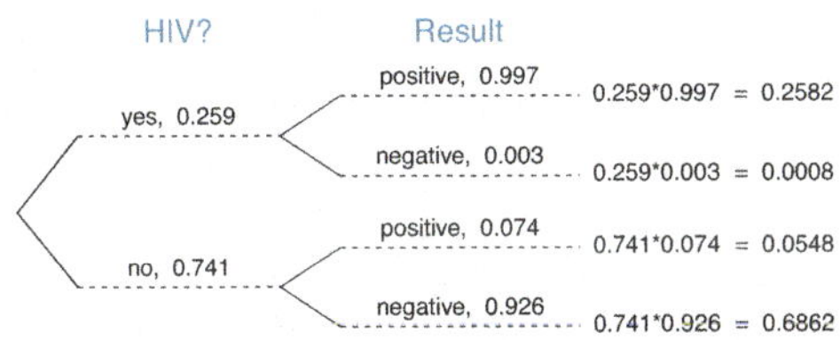

2.55 (a) If the class is not graded on a curve, they are independent. If graded on a curve, then neither independent nor disjoint – unless the instructor will only give one A, which is a situation we will ignore in parts (b) and (c). (b) They are probably not independent: if you study together, your study habits would be related, which suggests your course performances are also related. (c) No. See the answer to part (a) when the course is not graded on a curve. More generally: if two things are unrelated (independent), then one occurring does not preclude the other from occurring.

2.57 (a) $280/792 = 0.354$. (b) $445/792 = 0.562$. (c) $231/792 = 0.292$, (d) $0.354 \times 0.562 = 0.199 \neq 0.292$. The events are not independent, so you cannot just multiply the unconditional probabilities.

2.59 (a) 13. (b) No, these 27 students are not a random sample from the university's student population. For example, it might be argued that the proportion of smokers among students who go to the gym at 9 am on a Saturday morning would be lower than the proportion of smokers in the university as a whole.

2.61 (a) 0.1406. (b) 0.1025. (c) 0.1035.

2.63 (a) $Z = 0.67$. (b) $\mu = \$1650$, $x = \$1800$. (c) $0.67 = \frac{1800-1650}{\sigma} \rightarrow \sigma = \$223.88$.

2.65 $P(^1\text{leggings}, {}^2\text{jeans}, {}^3\text{jeans}) = \frac{5}{24} \times \frac{7}{23} \times \frac{6}{22} = 0.0173$. However, the person with leggings could have come 2nd or 3rd, and these each have this same probability, so $3 \times 0.0173 = 0.0519$.

3 Inference for categorical data: proportions

3.1 (a) Mean. Each student reports a numerical value: a number of hours. (b) Mean. Each student reports a number, which is a percentage, and we can average over these percentages. (c) Proportion. Each student reports Yes or No, so this is a categorical variable and we use a proportion. (d) Mean. Each student reports a number, which is a percentage like in part (b). (e) Proportion. Each student reports whether or not s/he expects to get a job, so this is a categorical variable and we use a proportion.

3.3 (a) The sample is from all computer chips manufactured at the factory during the week of production, so that is the population of interest. We might be tempted to generalize the population to represent all weeks, but we should exercise caution here since the rate of defects may change over time. (b) The proportion of computer chips manufactured at the factory during the week of production that had severe defects. (c) The point estimator for the population proportion is the sample proportion: $\hat{p} = \frac{27}{212} = 0.127$. (d) Answers may vary. An example: it is possible that the rate of chips with severe defects is increasing over time. Therefore, if we sample the first 212 chips during the week of production, we would underestimate the parameter as our estimate for the proportion of computer chips with severe defects manufactured at the factory during the week of production would be lower than the true rate for the week. (e) To get an unbiased estimator of the parameter, we should take a *random* sample from all computer chips manufactured at the factory during the week of production.

3.5 (a) Each observation in each of the distributions represents the sample proportion ($\hat{p}$) from samples of size $n = 20$, $n = 100$, and $n = 500$, respectively. (b) The centers for all three distributions are at 0.95, the true population parameter. When n is small, the distribution is skewed to the left and not smooth. As n increases, the variability of the distribution (standard deviation) decreases, and the shape of the distribution becomes more unimodal and symmetric.

3.7 (a) False. Doesn't satisfy large counts (success-failure) condition. (b) True. The large counts condition is not satisfied. In most samples we would expect $\hat{p}$ to be close to 0.08, the true population proportion. While $\hat{p}$ can be much above 0.08, it is bound below by 0, suggesting it would take on a right skewed shape. Plotting the sampling distribution would confirm this suspicion. (c) False. $\sigma_{\hat{p}} = 0.0243$, and $\hat{p} = 0.12$ is only $\frac{0.12-0.08}{0.0243} = 1.65$ SDs away from the mean, which would not be considered unusual. (d) True. $\hat{p} = 0.12$ is 2.32 standard deviations away from the mean, which is often considered unusual. (e) False. Decreases the SD by a factor of $1/\sqrt{2}$.

3.9 (a) $SD_{\hat{p}} = \sqrt{p(1-p)/n} = 0.0707$. This describes the typical distance that the sample proportion will deviate from the true proportion, $p = 0.5$. (b) $\hat{p}$ approximately follows $N(0.5, 0.0707)$. $Z = (0.55 - 0.50)/0.0707 \approx 0.71$. This corresponds to an upper tail of about 0.2389. That is, $P(\hat{p} > 0.55) \approx 0.24$.

3.11 (a) First we need to check that the necessary conditions are met. There are $200 \times 0.08 = 16$ expected successes and $200 \times (1 - 0.08) = 184$ expected failures, therefore the success-failure condition is met. Then the binomial distribution can be approximated by $N(\mu = 16, \sigma = 3.84)$. $P(X < 12) = P(Z < -1.04) = 0.1492$. (b) Since the success-failure condition is met the sampling distribution of $\hat{p} \sim N(\mu = 0.08, \sigma = 0.0192)$. $P(\hat{p} < 0.06) = P(Z < -1.04) = 0.1492$. (c) As expected, the two answers are the same.

3.13 The sampling distribution is the distribution of sample proportions from samples of the same size randomly sampled from the same population. As the same size increases, the shape of the sampling distribution (when $p = 0.1$) will go from being right-skewed to being more symmetric and resembling the normal distribution. With larger sample sizes, the spread of the sampling distribution gets smaller. Regardless of the sample size, the center of the sampling distribution is equal to the true mean of that population, provided the sampling isn't biased.

3.15 Recall that the general formula is *point estimate* $\pm z^\star \times SE$. First, identify the three different values. The point estimate is 45%, $z^\star = 1.96$ for a 95% confidence level, and $SE = 1.2\%$. Then, plug the values into the formula: $45\% \pm 1.96 \times 1.2\% \rightarrow (42.6\%, 47.4\%)$ We are 95% confident that the proportion of US adults who live with one or more chronic conditions is between 42.6% and 47.4%.

3.17 (a) False. Inference is made on the population parameter, not the point estimate. The point estimate is always in the confidence interval. (b) True. (c) False. The confidence interval is not about a sample mean. (d) False. To be more confident that we capture the parameter, we need a wider interval. Think about needing a bigger net to be more sure of catching a fish in a murky lake. (e) True. Optional explanation: This is true since the normal model was used to model the sample mean. The margin of error is half the width of the interval, and the sample mean is the midpoint of the interval. (f) False. In the calculation of the standard error, we divide the standard deviation by the square root of the sample size. To cut the SE (or margin of error) in half, we would need to sample $2^2 = 4$ times the number of people in the initial sample.

3.19 (a) This claim is reasonable, since the entire interval lies above 50%. (b) The value of 70% lies outside of the interval, so we have convincing evidence that the researcher's conjecture is wrong. (c) A 90% confidence interval will be narrower than a 95% confidence interval. Even without calculating the interval, we can tell that 70% would not fall in the interval, and we would reject the researcher's conjecture based on a 90% confidence level as well.

3.21 (a) True. See the reasoning of 6.1(b). (b) True. We take the square root of the sample size in the SE formula. (c) True. The independence and success-failure conditions are satisfied. (d) True. The independence and success-failure conditions are satisfied.

3.23 (a) False. A confidence interval is constructed to estimate the population proportion, not the sample proportion. (b) True. 95% CI: 82% ± 2%. (c) True. By the definition of the confidence level. (d) True. Quadrupling the sample size decreases the SE and ME by a factor of $1/\sqrt{4}$. (e) True. The 95% CI is entirely above 50%.

3.25 With a random sample, independence is satisfied. The success-failure condition is also satisfied. $ME = z^{\star}\sqrt{\frac{\hat{p}(1-\hat{p})}{n}} = 1.96\sqrt{\frac{0.56\times 0.44}{600}} = 0.0397 \approx 4\%$

3.27 (a) No. The sample only represents students who took the SAT, and this was also an online survey. (b) (0.5289, 0.5711). We are 90% confident that 53% to 57% of high school seniors who took the SAT are fairly certain that they will participate in a study abroad program in college. (c) 90% of such random samples would produce a 90% confidence interval that includes the true proportion. (d) Yes. The interval lies entirely above 50%.

3.29 Because a sample proportion ($\hat{p} = 0.55$) is available, we use this for the sample size calculations. The margin of error for a 90% confidence interval is $1.6449 \times SE = 1.6449 \times \sqrt{\frac{p(1-p)}{n}}$. We want this to be less than 0.01, where we use $\hat{p}$ in place of p:

$$1.6449 \times \sqrt{\frac{0.55(1-0.55)}{n}} \leq 0.01$$

$$1.6449^2 \frac{0.55(1-0.55)}{0.01^2} \leq n$$

From this, we get that n must be at least 6697.

3.31 (a) $H_0 : p = 0.5$ (Neither a majority nor minority of students' grades improved) $H_A : p \neq 0.5$ (Either a majority or a minority of students' grades improved)

(b) $H_0 : \mu = 15$ (The average amount of company time each employee spends not working is 15 minutes for March Madness.) $H_A : \mu \neq 15$ (The average amount of company time each employee spends not working is different than 15 minutes for March Madness.)

3.33 (1) The hypotheses should be about the population proportion (p), not the sample proportion. (2) The null hypothesis should have an equal sign. (3) The alternative hypothesis should have a not-equals sign, and (4) it should reference the null value, $p_0 = 0.6$, not the observed sample proportion. The correct way to set up these hypotheses is: $H_0 : p = 0.6$ and $H_A : p \neq 0.6$.

3.35 (a) Identify: Use a one-sample Z-test for p, where p: proportion of all Independents (US adults) that support National Health Plan; $\alpha = 0.05$.

$$H_0 : p = 0.5 \qquad\qquad H_A : p \neq 0.5$$

Check: We have a sample proportion of $\hat{p} = 0.55$ and a sample size of $n = 617$ independents.
Since this is a random sample and 617 < 10% of all Independents in US, independence is satisfied. The success-failure condition is also satisfied: 617×0.5 and $617 \times (1 - 0.5)$ are both at least 10 (we use the null proportion $p_0 = 0.5$ for this check in a one-proportion hypothesis test). Therefore, we can model $\hat{p}$ using a normal distribution.
Calculate:

$$SE = \sqrt{\frac{0.5(1-0.5)}{617}} = 0.02$$

(We use the null proportion $p_0 = 0.5$ to compute the standard error for a one-proportion hypothesis test.)
Next, we compute the test statistic:

$$Z = \frac{0.55 - 0.5}{0.02} = 2.5$$

This yields a one-tail area of 0.0062, and a p-value of $2 \times 0.0062 = 0.0124$.
Conclude: Because the p-value is smaller than 0.05, we reject the null hypothesis. We have strong evidence that the support is different from 0.5, and since the data provide a point estimate above 0.5, we have strong evidence to support this claim by the TV pundit.

(b) No. Generally we expect a hypothesis test and a confidence interval to align, so we would expect the confidence interval to show a range of plausible values entirely above 0.5. However, if the confidence level is misaligned (e.g. a 99% confidence level and a $\alpha = 0.05$ significance level), then this is no longer generally true.

3.37 (a) H_0: Anti-depressants do not affect the symptoms of Fibromyalgia. H_A: Anti-depressants do affect the symptoms of Fibromyalgia (either helping or harming). (b) Concluding that anti-depressants either help or worsen Fibromyalgia symptoms when they actually do neither. (c) Concluding that anti-depressants do not affect Fibromyalgia symptoms when they actually do.

3.39 (a) Scenario I is higher. Recall that a sample mean based on less data tends to be less accurate and have larger standard errors. (b) Scenario I is higher. The higher the confidence level, the higher the corresponding margin of error.

3.41 (a) $\mu_{\hat{p}_{NE}} = 0.01$. $\sigma_{\hat{p}_{NE}} = 0.0031$. (b) $\mu_{\hat{p}_{NY}} = 0.06$. $\sigma_{\hat{p}_{NY}} = 0.0075$. (c) $\mu_{\hat{p}_{NY}-\hat{p}_{NB}} = 0.06 - 0.01 = 0.05$. $\sigma_{\hat{p}_{NY}-\hat{p}_{NB}} = 0.0081$. (d) We can think of $\hat{p}_{NE}$ and $\hat{p}_{NY}$ as being random variables, and we are considering the standard deviation of the difference of these two random variables, so we square each standard deviation, add them together, and then take the square root of the sum:

$$SD_{\hat{p}_{NY}-\hat{p}_{NE}} = \sqrt{SD^2_{\hat{p}_{NY}} + SD^2_{\hat{p}_{NE}}}$$

3.43 This is not a randomized experiment, and it is unclear whether people would be affected by the behavior of their peers. That is, independence may not hold. Additionally, there are only 5 interventions under the provocative scenario, so the success-failure condition does not hold. Even if we consider a hypothesis test where we pool the proportions, the success-failure condition will not be satisfied. Since one condition is questionable and the other is not satisfied, the difference in sample proportions will not follow a nearly normal distribution.

3.45 (a) False. The entire confidence interval is above 0. (b) True. (c) True. (d) True. (e) False. It is simply the negated and reordered values: (-0.06,-0.02).

3.47 (a) Standard error:

$$SE = \sqrt{\frac{0.79(1-0.79)}{347} + \frac{0.55(1-0.55)}{617}} = 0.03$$

Using $z^\star = 1.96$, we get:

$$0.79 - 0.55 \pm 1.96 \times 0.03 \rightarrow (0.181, 0.299)$$

We are 95% confident that the proportion of Democrats who support the plan is 18.1% to 29.9% higher than the proportion of Independents who support the plan. (b) True.

3.49 Identify: Use a two-sample Z-test for $p_1 - p_2$: rate of sleep deprivation for non-transportation workers – rate of sleep deprivation for truck drivers. Let $\alpha = 0.05$. $H_0: p_C = p_T$. $H_A: p_C \neq p_T$. $\alpha = 0.05$. Check: Independence is satisfied (random samples that are independent), as is the success-failure condition, which we check using the pooled proportion ($\hat{p}_{pool} = 70/495 = 0.141$). Calculate: $Z = -1.65 \rightarrow$ p-value = 0.0989. Conclude: Since the p-value is $> \alpha$, we fail to reject H_0. The data do not provide strong evidence that the rates of sleep deprivation are different for non-transportation workers and truck drivers.

3.51 (a) Identify: Use a two-proportion Z-test for $p_1 - p_2$; p_1 proportion of children that would contract malaria if given malaria vaccine; p_2 proportion of children that would contract malaria if given control vaccine. Let $\alpha = 0.05$. Check: $H_0: p_1 = p_2$ and $H_A: p_1 < p_2$. Use $\hat{p}_c = \frac{89+106}{292+147} = 0.444$ to verify the four success-failure outcomes are at least 10. Also, there were 2 randomly assigned treatments. Calculate: $Z = -8.285 \rightarrow$ p-value $\approx 0 < \alpha$. Reject H_0; Evidence that a lower proportion of children like those in the study would contract malaria if given the malaria vaccine than if given the control vaccine. (b) There is close to a 0% chance that we would get a Z statistic as small as we got (≤ -8.285) if the malaria vaccine and control vaccine were equally effective at reducing malaria rates in children.

3.53 (a) H_0: The distribution of the format of the book used by the students follows the professor's predictions. H_A: The distribution of the format of the book used by the students does not follow the professor's predictions. (b) $E_{hard\ copy} = 126 \times 0.60 = 75.6$. $E_{print} = 126 \times 0.25 = 31.5$. $E_{online} = 126 \times 0.15 = 18.9$. (c) Independence: The sample is not random. However, if the professor has reason to believe that the proportions are stable from one term to the next and students are not affecting each other's study habits, independence could be reasonable. Sample size: All expected counts are at least 5. (d) $\chi^2 = 2.32$, $df = 2$, p-value = 0.313. (e) Since the p-value is large, we fail to reject H_0. The data do not provide strong evidence indicating the professor's predictions were statistically inaccurate.

3.55 (a) False. The chi-square distribution has one parameter called degrees of freedom. (b) True. (c) True. (d) False. As the degrees of freedom increases, the shape of the chi-square distribution becomes more symmetric.

3.57 (a) Two-way table:

Treatment	*Quit* Yes	No	Total
Patch + support group	40	110	150
Only patch	30	120	150
Total	70	230	300

(b-i) $E_{row_1,col_1} = \frac{(row\ 1\ total)\times(col\ 1\ total)}{table\ total} = 35$. This is lower than the observed value.
(b-ii) $E_{row_2,col_2} = \frac{(row\ 2\ total)\times(col\ 2\ total)}{table\ total} = 115$. This is lower than the observed value.

3.59 Identify: Use a chi-square test for independence, with $\alpha = 0.05$. H_0: Opinions regarding offshore drilling for oil and having a college degree are independent. H_A: Opinions regarding offshore drilling for oil and having a college degree are dependent. Check:

$$E_{row\ 1,col\ 1} = 151.5 \qquad E_{row\ 1,col\ 2} = 134.5$$
$$E_{row\ 2,col\ 1} = 162.1 \qquad E_{row\ 2,col\ 2} = 143.9$$
$$E_{row\ 3,col\ 1} = 124.5 \qquad E_{row\ 3,col\ 2} = 110.5$$

Independence: The sample is random, and from less than 10% of the population, so independence between observations is reasonable. Expected counts: All expected counts are at least 5. Calculate: $\chi^2 = 11.47$, $df = 2 \rightarrow$ p-value $= 0.003$. Conclude: Since the p-value $< \alpha$, we reject H_0. There is strong evidence that there is an association between support for offshore drilling and having a college degree.

3.61 No. The samples at the beginning and at the end of the semester are not independent since the survey is conducted on the same students.

3.63 (a) H_0: The age of Los Angeles residents is independent of shipping carrier preference variable. H_A: The age of Los Angeles residents is associated with the shipping carrier preference variable. (b) The conditions are not satisfied since some expected counts are below 5.

3.65 (a) Independence is satisfied (random sample), as is the success-failure condition (40 smokers, 160 non-smokers). The 95% CI: (0.145, 0.255). We are 95% confident that 14.5% to 25.5% of all students at this university smoke. (b) We want $z^\star SE$ to be no larger than 0.02 for a 95% confidence level. We use $z^\star = 1.96$ and plug in the point estimate $\hat{p} = 0.2$ within the SE formula: $1.96\sqrt{0.2(1-0.2)/n} \leq 0.02$. The sample size n should be at least 1,537.

3.67 (a) Proportion of graduates from this university who found a job within one year of graduating. $\hat{p} = 348/400 = 0.87$. (b) This is a random sample, so the observations are independent. Success-failure condition is satisfied: 348 successes, 52 failures, both well above 10. (c) (0.8371, 0.9029). We are 95% confident that approximately 84% to 90% of graduates from this university found a job within one year of completing their undergraduate degree. (d) 95% of such random samples would produce a 95% confidence interval that includes the true proportion of students at this university who found a job within one year of graduating from college. (e) (0.8267, 0.9133). Similar interpretation as before. (f) 99% CI is wider, as we are more confident that the true proportion is within the interval and so need to cover a wider range.

3.69 (a) $H_0 : p = 0.38$. $H_A : p \neq 0.38$. Independence (random sample) and the success-failure condition are satisfied. $Z = -20.5 \rightarrow$ p-value ≈ 0. Since the p-value is very small, we reject H_0. The data provide strong evidence that the proportion of Americans who only use their cell phones to access the internet is different than the Chinese proportion of 38%, and the data indicate that the proportion is lower in the US. (b) If in fact 38% of Americans used their cell phones as a primary access point to the internet, the probability of obtaining a random sample of 2,254 Americans where 17% or less or 59% or more use their only their cell phones to access the internet would be approximately 0. (c) (0.1545, 0.1855). We are 95% confident that approximately 15.5% to 18.6% of all Americans primarily use their cell phones to browse the internet.

3.71 (a) Chi-squared test of independence. (b) H_0: Caffeinated coffee consumption and depression in women are independent. H_A: Caffeinated coffee consumption and clinical are in women dependent. (c) Depression: $2607/50739 = 0.0514$. No depression: $1-0.0514 = 0.9486$ (d) $E = \frac{2607*6617}{50739} = 339.9854 \approx 340$. $\frac{(O-E)^2}{E} = \frac{(373-340)^2}{340} = 3.20$ (e) $df = (R-1)\times(C-1) = 1 \times 4 = 4$, and $p-value < 0.001$. (f) p-value is small and we reject H_0. The data provide convincing evidence to suggest that caffeinated coffee consumption and depression in women are associated. (g) Yes, this is an observational study. Based on this study we can't deduce that drinking more coffee leads to less depression. There may be other factors, lurking variables, that cause decreased depression in women who drink more coffee.

4 Inference for numerical data: means

4.1 (a) The distribution is unimodal and strongly right skewed with a median between 5 and 10 years old. Ages range from 0 to slightly over 50 years old, and the middle 50% of the distribution is roughly between 5 and 15 years old. There are potential outliers on the higher end. (b) When the sample size is small, the sampling distribution is right skewed, just like the population distribution. As the sample size increases, the sampling distribution gets more unimodal, symmetric, and approaches normality. The variability also decreases. This is consistent with the Central Limit Theorem.

4.3 (a) Right skewed. There is a long tail on the higher end of the distribution but a much shorter tail on the lower end. (b) Less than, as the median would be less than the mean in a right skewed distribution. (c) We should not. (d) Even though the population distribution is not normal, the conditions for inference are reasonably satisfied, with the possible exception of skew. If the skew isn't very strong (we should ask to see the data), then we can use the Central Limit Theorem to estimate this probability. For now, we'll assume the skew isn't very strong, though the description suggests it is at least moderate to strong. Use $N(1.3, 0.3/\sqrt{60})$: $Z = 2.58$ $\rightarrow 0.0049$. (e) It would decrease it by a factor of $\sqrt{2}$.

4.5 The centers are the same in each plot, and each data set is from a nearly normal distribution, though the histograms may not look very normal since each represents only 100 data points. The only way to tell which plot corresponds to which scenario is to examine the variability of each distribution. Plot B is the most variable, followed by Plot A, then Plot C. This means Plot B will correspond to the original data, Plot A to the sample means with size 5, and Plot C to the sample means with size 25.

4.7 (a) $Z = -3.33 \rightarrow 0.0004$. (b) The population SD is known and the data are nearly normal, so the sample mean will be nearly normal with distribution $N(\mu, \sigma/\sqrt{n})$, i.e. $N(2.5, 0.0095)$. (c) $Z = -10.54 \rightarrow$ ≈ 0. (d) See below:

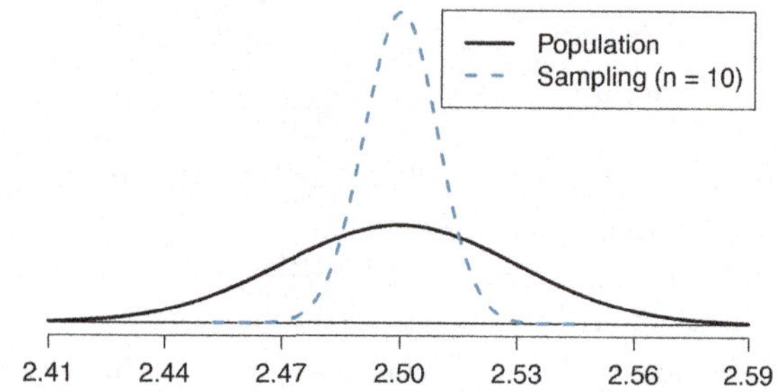

(e) We could not estimate (a) without a nearly normal population distribution. We also could not estimate (c) since the sample size is not sufficient to yield a nearly normal sampling distribution if the population distribution is not nearly normal.

4.9 (a) We cannot use the normal model for this calculation, but we can use the histogram. About 500 songs are shown to be longer than 5 minutes, so the probability is about $500/3000 = 0.167$. (b) Two different answers are reasonable. *Option 1* Since the population distribution is only slightly skewed to the right, even a small sample size will yield a nearly normal sampling distribution. We also know that the songs are sampled randomly and the sample size is less than 10% of the population, so the length of one song in the sample is independent of another. We are looking for the probability that the total length of 15 songs is more than 60 minutes, which means that the average song should last at least $60/15 = 4$ minutes. Using $SD_{\bar{x}} = 1.63/\sqrt{15}$, $Z = 1.31 \rightarrow 0.0951$. *Option 2* Since the population distribution is not normal, a small sample size may not be sufficient to yield a nearly normal sampling distribution. Therefore, we cannot estimate the probability using the tools we have learned so far. (c) We can now be confident that the conditions are satisfied. $Z = 0.92 \rightarrow 0.1788$.

4.11 (a) $SD_{\bar{x}} = \frac{25}{\sqrt{75}} = 2.89$. (b) $Z = 1.73$, which indicates that the two values are not unusually distant from each other when accounting for the uncertainty in John's point estimate.

4.13 (a) $df = 6 - 1 = 5$, $t^\star_5 = 2.02$ (column with two tails of 0.10, row with $df = 5$). (b) $df = 21 - 1 = 20$, $t^\star_{20} = 2.53$ (column with two tails of 0.02, row with $df = 20$). (c) $df = 28$, $t^\star_{28} = 2.05$. (d) $df = 11$, $t^\star_{11} = 3.11$.

4.15 The mean is the midpoint: $\bar{x} = 20$. Identify the margin of error: $ME = 1.015$, then use $t^\star_{35} = 2.03$ and $SE = s/\sqrt{n}$ in the formula for margin of error to identify $s = 3$.

4.17 Independence: it is a random sample, so we can assume that the students in this sample are independent of each other with respect to number of exclusive relationships they have been in. Notice that there are no students who have had no exclusive relationships in the sample, which suggests some student responses are likely missing (perhaps only positive values were reported). The sample size is at least 30. 90% CI: (2.97, 3.43), with df = 202. We are 90% confident that undergraduate students have been in 2.97 to 3.43 exclusive relationships, on average.

4.19 (a) Since it's the same students at the beginning and the end of the semester, there is a pairing between the datasets, for a given student their beginning and end of semester grades are dependent. (b) Since the subjects were sampled randomly, each observation in the men's group does not have a special correspondence with exactly one observation in the other (women's) group. (c) Since it's the same subjects at the beginning and the end of the study, there is a pairing between the datasets, for a subject student their beginning and end of semester artery thickness are dependent. (d) Since it's the same subjects at the beginning and the end of the study, there is a pairing between the datasets, for a subject student their beginning and end of semester weights are dependent.

4.21 False. While it is true that paired analysis requires equal sample sizes, only having the equal sample sizes isn't, on its own, sufficient for doing a paired test. Paired tests require that there be a special correspondence between each pair of observations in the two groups.

4.23 (a) The sample size of 197 is well over 30, so this condition is met. However, we must assume that locations were randomly sampled and that there are more than 1970 total locations from which we could sample. (b) (0.87, 4.93), with df = 196 (c) We are 90% confident that there was an increase of 0.87 to 4.93 in the average number of days that hit 90°F in 2018 relative to 1948 for NOAA stations. (d) Yes, since the interval lies entirely above 0.

4.25 (a) 0.085, do not reject H_0. (b) 0.003, reject H_0. (c) 0.438, do not reject H_0. (d) 0.042, reject H_0.

4.27 First, the hypotheses should be about the population mean (μ), not the sample mean. Second, the null hypothesis should have an equal sign and the alternative hypothesis should be about the null hypothesized value, not the observed sample mean. The correct way to set up these hypotheses is shown below:

$$H_0 : \mu = 10 \ hours$$
$$H_A : \mu \neq 10 \ hours$$

A two-sided test allows us to consider the possibility that the data show us something that we would find surprising.

4.29 (a) H_0: $\mu = 8$ (New Yorkers sleep 8 hrs per night on average.) H_A: $\mu \neq 8$ (New Yorkers sleep less or more than 8 hrs per night on average.) (b) Independence: The sample is random. The min/max suggest there are no concerning outliers. $T = -1.75$. $df = 25 - 1 = 24$. (c) p-value = 0.093. If in fact the true population mean of the amount New Yorkers sleep per night was 8 hours, the probability of getting a random sample of 25 New Yorkers where the average amount of sleep is 7.73 hours per night or less (or 8.27 hours or more) is 0.093. (d) Since p-value > 0.05, do not reject H_0. The data do not provide strong evidence that New Yorkers sleep more or less than 8 hours per night on average. (e) Yes, since we did not rejected H_0.

4.31 (a) Identify: Use a one-sample t-test for μ: true average number of years a child takes piano lessons in this city. Let $\alpha = 0.05$. H_0: $\mu = 5$. H_A: $\mu < 5$. Check: This is a random sample of less than 10% of children in this city, so the observations are independent. To proceed, we assume the distribution of years of piano lessons is approximately normal. Calculate: $SE = 2.2/\sqrt{20} = 0.4919$. The test statistic is $T = (4.6 - 5)/SE = -0.81$. $df = 20 - 1 = 19$. The one-tail p-value is about 0.21. Conclude: p-value $> \alpha = 0.05$, so we do not reject H_0. That is, we do not have sufficiently strong evidence to reject Georgianna's claim.

(b) Identify: Use a one-sample t-interval for μ: true average number of years a child takes piano lessons in this city, with 95% confidence. Check: same as in part (a). Calculate: Using $SE = 0.4919$ and $t^{\star}_{df=19} = 2.093$, the confidence interval is (3.57, 5.63). Conclude: We are 95% confident that the interval (3.57, 5.63) contains the true average number of years a child takes piano lessons in this city. Do not have evidence that average is not 5 because 5 is in the interval.

(c) They agree, since we did not reject the null hypothesis and the null value of 5 was in the t-interval.

4.33 (a) For each observation in one data set, there is exactly one specially corresponding observation in the other data set for the same geographic location. The data are paired. (b) $H_0 : \mu_d = 0$ (There is no difference in average number of days exceeding 90°F in 1948 and 2018 for NOAA stations.) $H_A : \mu_d \neq 0$ (There is a difference.) (c) Locations were randomly sampled, so independence is reasonable. The sample size is at least 30, so we're just looking for particularly extreme outliers: none are present (the observation off left in the histogram would be considered a clear outlier, but not a particularly extreme one). Therefore, the conditions are satisfied. (d) $SE = 17.2/\sqrt{197} = 1.23$. $T = \frac{2.9-0}{1.23} = 2.36$ with degrees of freedom $df = 197 - 1 = 196$. This leads to a one-tail area of 0.0096 and a p-value of about 0.019. (e) Since the p-value is less than 0.05, we reject H_0. The data provide strong evidence that NOAA stations observed more 90°F days in 2018 than in 1948. (f) Type 1 Error, since we may have incorrectly rejected H_0. This error would mean that NOAA stations did not actually observe a decrease, but the sample we took just so happened to make it appear that this was the case. (g) No, since we rejected H_0, which had a null value of 0.

4.35 (a) $\mu_{\bar{x}_1} = 15$, $\sigma_{\bar{x}_1} = 20/\sqrt{50} = 2.8284$. (b) $\mu_{\bar{x}_2} = 20$, $\sigma_{\bar{x}_1} = 10/\sqrt{30} = 1.8257$. (c) $\mu_{\bar{x}_2-\bar{x}_1} = 20 - 15 = 5$, $\sigma_{\bar{x}_2-\bar{x}_1} = \sqrt{\left(20/\sqrt{50}\right)^2 + \left(10/\sqrt{30}\right)^2} = 3.3665$. (d) Think of $\bar{x}_1$ and $\bar{x}_2$ as being random variables, and we are considering the standard deviation of the difference of these two random variables, so we square each standard deviation, add them together, and then take the square root of the sum:

$$SD_{\bar{x}_2-\bar{x}_1} = \sqrt{SD^2_{\bar{x}_2} + SD^2_{\bar{x}_1}}$$

4.37 Paired, data are recorded in the same cities at two different time points. The air quality in a city at one point is not independent of the air quality in the same city at another time point.

4.39 Use a two-sample t-interval for $\mu_1 - \mu_2$: true average weight of chickens that would be fed casein – true average weight of chickens that would be fed soybean. This is an experiment with two randomly assigned treatments/feeds. The treatment group sizes are less than 30, but the distribution of weights for those fed casein and fed soybean are not excessively skewed and have no outliers, so we will assume that the population distributions could be approximately normal. The 95% confidence interval is (28.24, 126.1), with df = 21.64. We are 95% confident that the interval (28.24, 126.1) contains the difference in true average weight of chickens that would be fed casein versus fed soybean (casein – soybean). Because this entire interval is above 0, we have evidence that chickens fed casein would have a higher average weight than chickens fed soybean.

4.41 (a) Chicken fed linseed weighed an average of 218.75 grams while those fed horsebean weighed an average of 160.20 grams. Both distributions are relatively symmetric with no apparent outliers. There is more variability in the weights of chicken fed linseed.
(b) $H_0 : \mu_{ls} = \mu_{hb}$. $H_A : \mu_{ls} \neq \mu_{hb}$.
We leave the conditions to you to consider.
$T = 3.017$, $df = 19.77 \rightarrow$ p-value = 0.007. Since p-value < 0.05, reject H_0. The data provide strong evidence that there is a significant difference between the average weights of chickens that were fed linseed and horsebean.
(c) Type 1 Error, since we rejected H_0.
(d) Yes, since p-value > 0.01, we would not have rejected H_0.

4.43 $H_0: \mu_T = \mu_C$. $H_A: \mu_T \neq \mu_C$. $T = 2.24$, $df = 21$ $\rightarrow$ p-value = 0.036. Since p-value < 0.05, reject H_0. The data provide strong evidence that the average food consumption by the patients in the treatment and control groups are different. Furthermore, the data indicate patients in the distracted eating (treatment) group consume more food than patients in the control group.

4.45 (a) We are building a distribution of sample statistics, in this case the sample mean. Such a distribution is called a sampling distribution for the sample mean. (b) Because we are dealing with the distribution of sample means, we need to check to see if the Central Limit Theorem applies. Our sample size is greater than 30, and we are told that random sampling is employed. With these conditions met, we expect that the distribution of the sample mean will be nearly normal and therefore symmetric. (c) Because we are dealing with a sampling distribution of a sample mean, we measure its variability with the standard deviation of $\bar{x}$. $\sigma_{\bar{x}} = 18.2/\sqrt{45} = 2.713$. (d) The sample means will be more variable with the smaller sample size.

4.47 Paired, data are recorded in the same cities at two different time points. The air quality in a city at one point is not independent of the air quality in the same city at another time point.

4.49 $t^{\star}_{19}$ is 1.73 for a one-tail. We want the lower tail, so set -1.73 equal to the T-score, then solve for $\bar{x}$: 56.91.

4.51 (a) These data are paired. For example, the Friday the 13th in say, September 1991, would probably be more similar to the Friday the 6th in September 1991 than to Friday the 6th in another month or year.
(b) Let $\mu_{diff} = \mu_{sixth} - \mu_{thirteenth}$. $H_0: \mu_{diff} = 0$. $H_A: \mu_{diff} \neq 0$.
(c) Independence: The months selected are not random. However, if we think these dates are roughly equivalent to a simple random sample of all such Friday 6th/13th date pairs, then independence is reasonable. To proceed, we must make this strong assumption, though we should note this assumption in any reported results. Normality: With fewer than 10 observations, we would need to see clear outliers to be concerned. There is a borderline outlier on the right of the histogram of the differences, so we would want to report this in formal analysis results.
(d) $T = 4.93$ for $df = 10 - 1 = 9 \rightarrow$ p-value = 0.001.
(e) Since p-value < 0.05, reject H_0. The data provide strong evidence that the average number of cars at the intersection is higher on Friday the 6th than on Friday the 13th. (We should exercise caution about generalizing the interpetation to all intersections or roads.)
(f) If the average number of cars passing the intersection actually was the same on Friday the 6th and 13th, then the probability that we would observe a test statistic so far from zero is less than 0.01.
(g) We might have made a Type 1 Error, i.e. incorrectly rejected the null hypothesis.

4.53 (a) $H_0: \mu_{diff} = 0$. $H_A: \mu_{diff} \neq 0$. $T = -2.71$. $df = 5$. p-value = 0.042. Since p-value < 0.05, reject H_0. The data provide strong evidence that the average number of traffic accident related emergency room admissions are different between Friday the 6th and Friday the 13th. Furthermore, the data indicate that the direction of that difference is that accidents are lower on Friday the 6th relative to Friday the 13th.
(b) (-6.49, -0.17).
(c) This is an observational study, not an experiment, so we cannot so easily infer a causal intervention implied by this statement. It is true that there is a difference. However, for example, this does not mean that a responsible adult going out on Friday the 13th has a higher chance of harm than on any other night.

5 Regression analysis

5.1 (a) There is a weak and positive relationship between age and income. With so few points it is difficult to tell the form of the relationship (linear or not) however the relationship does look somewhat curved.

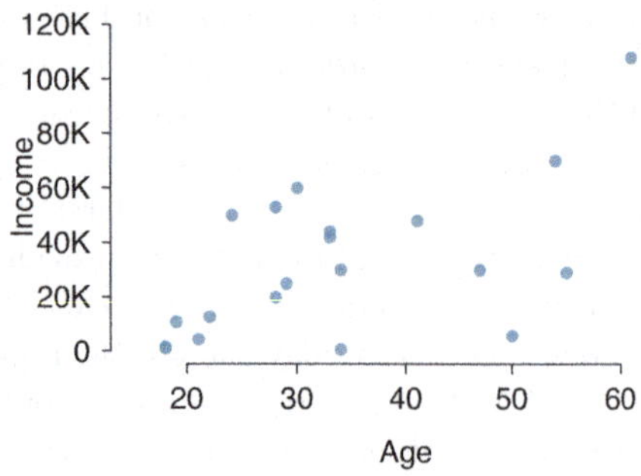

(b)

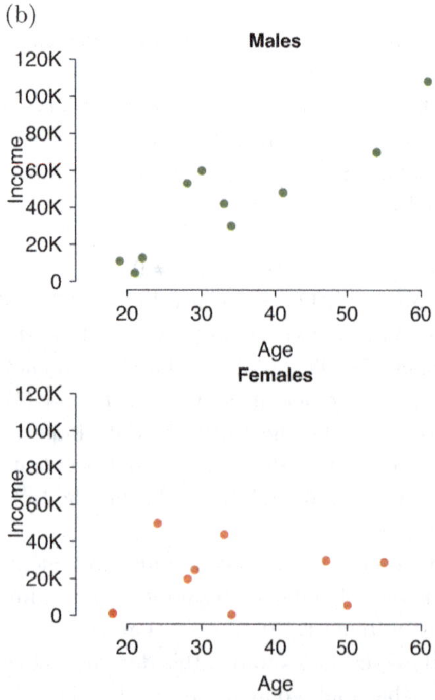

(c) For males as age increases so does income, however this pattern is not apparent for females.

5.3 (a) Positive association: mammals with longer gestation periods tend to live longer as well. (b) Association would still be positive. (c) No, they are not independent. See part (a).

5.5 (a) Exam 2 since there is less of a scatter in the plot of final exam grade versus exam 2. Notice that the relationship between Exam 1 and the Final Exam appears to be slightly nonlinear. (b) Exam 2 and the final are relatively close to each other chronologically, or Exam 2 may be cumulative so has greater similarities in material to the final exam. Answers may vary.

5.7 (a) $r = -0.7 \rightarrow (4)$. (b) $r = 0.45 \rightarrow (3)$. (c) $r = 0.06 \rightarrow (1)$. (d) $r = 0.92 \rightarrow (2)$.

5.9 (a) The relationship is positive, weak, and possibly linear. However, there do appear to be some anomalous observations along the left where several students have the same height that is notably far from the cloud of the other points. Additionally, there are many students who appear not to have driven a car, and they are represented by a set of points along the bottom of the scatterplot. (b) There is no obvious explanation why simply being tall should lead a person to drive faster. However, one confounding factor is gender. Males tend to be taller than females on average, and personal experiences (anecdotal) may suggest they drive faster. If we were to follow-up on this suspicion, we would find that sociological studies confirm this suspicion. (c) Males are taller on average and they drive faster. The gender variable is indeed an important confounding variable.

5.11 (a) There is a somewhat weak, positive, possibly linear relationship between the distance traveled and travel time. There is clustering near the lower left corner that we should take special note of. (b) Changing the units will not change the form, direction or strength of the relationship between the two variables. If longer distances measured in miles are associated with longer travel time measured in minutes, longer distances measured in kilometers will be associated with longer travel time measured in hours. (c) Correlation has no units. (d) Changing units doesn't affect correlation: $r = 0.636$. (e) Correlation will not change.

5.13 (a) The residual plot will show randomly distributed residuals around 0. The variance is also approximately constant. (b) The residuals will show a fan shape, with higher variability for smaller x. There will also be many points on the right above the line. There is trouble with the model being fit here.

5.15 Correlation: no units. Intercept: kg. Slope: kg/cm.

5.17 Over-estimate. Since the residual is calculated as *observed* − *predicted*, a negative residual means that the predicted value is higher than the observed value.

5.19 (a) There is a positive, very strong, linear association between the number of tourists and spending. (b) Explanatory: number of tourists (in thousands). Response: spending (in millions of US dollars). (c) We can predict spending for a given number of tourists using a regression line. This may be useful information for determining how much the country may want to spend in advertising abroad, or to forecast expected revenues from tourism. (d) No, a linear model is not appropriate as seen by the U-shape in the residual plot.

5.21 (a) The regression line can be written as
$\hat{y} = 51 + 0.726 \times x$
where $\hat{y}$ is the predicted travel time and x is the distance traveled. (b) b: For each additional mile in distance, the model predicts an additional 0.726 minutes in travel time. a: When the distance traveled is 0 miles, the travel time is expected to be 51 minutes. It does not make sense to have a travel distance of 0 miles in this context. Here, the y-intercept serves only to adjust the height of the line and is meaningless by itself. (c) $R^2 = 0.636^2 = 0.40$. About 40% of the variation in travel time is explained by the linear model with x=distance traveled. (d) $\widehat{travel\ time} = 51 + 0.726 \times distance = 51 + 0.726 \times 103 \approx 126$ minutes. (Note: we should be cautious in our predictions with this model since we have not yet evaluated whether it is a well-fit model.) (e) $y - \hat{y} = 168 - 126 = 42$ minutes. A positive residual means that the model underestimates the travel time. (f) No, this calculation would require extrapolation.

5.23 (a) $\widehat{murder} = -29.901 + 2.559 \times poverty\%$. (b) Expected murder rate in metropolitan areas with no poverty is -29. 901 per million. This is obviously not a meaningful value, it just serves to adjust the height of the regression line. (c) For each additional percentage increase in poverty, we expect murders per million to be higher on average by 2.559. (d) 70.52% of the variation in murder rates in metropolitan areas is explained by the linear model with x=poverty. (e) $\sqrt{0.7052} = 0.8398$.

5.25 (a) Observational study. (b) Dog: Lucy. Cat: Luna. (c) Oliver and Lily. (d) Positive, as the popularity of a name for dogs increases, so does the popularity of that name for cats.

5.27 (a) True. (b) False, correlation is a measure of the linear association between any two numerical variables.

5.29 There is an upwards trend. However, the variability is higher for higher calorie counts, and it looks like there might be two clusters of observations above and below the line on the right, so we should be cautious about fitting a linear model to these data.

5.31 (a) $r = -0.72 \rightarrow (2)$ (b) $r = 0.07 \rightarrow (4)$ (c) $r = 0.86 \rightarrow (1)$ (d) $r = 0.99 \rightarrow (3)$

Appendix B

Data sets within the text

Each data set within the text is described in this appendix. For those data sets that are in multiple sections in a chapter, only the first section is listed in that chapter. If a data set is not listed here, e.g. Section 2.3.5 lists imagined probabilities for whether a parking garage will fill up and whether there is a sporting event that same evening for an unnamed college, it may not be listed in this data appendix. When a raw data set is available versus just a description, there is a corresponding page for the data set at **openintro.org/data**. That webpage also includes many more data sets than are covered in this textbook, and each data set on the website includes a description, it's source, a detailed overview of each data set's variables, and download options.

Chapter 1: Exploring one-variable data and collecting data

1.1 `loan50`, `loans_full_schema` → This data comes from Lending Club (lendingclub.com), which provides a large set of data on the people who received loans through their platform. The data used in the textbook comes from a sample of the loans made in Q1 (Jan, Feb, March) 2018.

1.4 `loan50` → This data set is described in the data for Section 1.3.

1.4 `possum` → The brushtail possum statistics are based on a sample of possums from Australia and New Guinea. The original source of this data is as follows:
Lindenmayer DB, et al. 1995. *Morphological variation among columns of the mountain brushtail possum, Trichosurus caninus Ogilby (Phalangeridae: Marsupiala).* Australian Journal of Zoology 43: 449-458.

1.6 The study in mind regarding chocolate and heart attack patients:
Janszky et al. 2009. Chocolate consumption and mortality following a first acute myocardial infarction: the Stockholm Heart Epidemiology Program. Journal of Internal Medicine 266:3, p248-257.

1.6 The Nurses' Health Study was mentioned. For more information on this data set, see www.channing.harvard.edu/nhs

1.7 `stent30`, `stent365` → The stent data is split across two data sets, one for the 0-30 day and one for the 0-365 day results.
Chimowitz MI, Lynn MJ, Derdeyn CP, et al. 2011. Stenting versus Aggressive Medical Therapy for Intracranial Arterial Stenosis. New England Journal of Medicine 365:993-1003. www.nejm.org/doi/full/10.1056/NEJMoa1105335.
NY Times article: www.nytimes.com/2011/09/08/health/research/08stent.html.

1.7 The study we had in mind during the introduction of Section 1.7.2 was
Anturane Reinfarction Trial Research Group. 1980. *Sulfinpyrazone in the prevention of sudden death after myocardial infarction.* New England Journal of Medicine 302(5):250-256.

Chapter 2: **Probability, random variables, and probability distributions**

2.2 `email` → These data represent emails sent to David Diez. Each data set includes 21 variables.

2.2 `playing_cards` → A table describing the 52 cards in a standard deck.

2.3 Machine learning on fashion. → This is a simulated data set, not based on any specific machine learning classifier.

2.3 `smallpox` → Fenner F. 1988. Smallpox and Its Eradication (History of International Public Health, No. 6). Geneva: World Health Organization. ISBN 92-4-156110-6.

2.3 `family_college` → A simulated data set based on real population summaries at nces.ed.gov/pubs2001/2001126.pdf.

2.4 `stocks_18` → Monthly returns for Caterpillar, Exxon Mobil Corp, and Google for November 2015 to October 2018.

2.5 Blood type prevalence. → The fraction of people with O+ blood is about 38% according to www.redcrossblood.org/donate-blood/blood-types/o-blood-type.html
We used 35% for simplicity in the examples.

2.6 SAT and ACT score distributions → The SAT score data comes from the 2018 distribution, which is provided at
reports.collegeboard.org/pdf/2018-total-group-sat-suite-assessments-annual-report.pdf
The ACT score data is available at
act.org/content/dam/act/unsecured/documents/cccr2018/P_99_999999_N_S_N00_ACT-GCPR_National.pdf
We also acknowledge that the actual ACT score distribution is *not* nearly normal. However, since the topic is very accessible, we decided to keep the context and examples.

2.7 `malaria` → Lyke et al. 2017. PfSPZ vaccine induces strain-transcending T cells and durable protection against heterologous controlled human malaria infection.
PNAS 114(10):2711-2716. www.pnas.org/content/114/10/2711

Chapter 3: **Inference for categorical data: proportions**

3.2 Blood type prevalence. → This data set is described in the data for Chapter 2.

3.3 `pew_energy_2018` → See the details for this data set above in the Section 3.1 data section.

3.3 `ebola_survey` → In New York City on October 23rd, 2014, a doctor who had recently been treating Ebola patients in Guinea went to the hospital with a slight fever and was subsequently diagnosed with Ebola. Soon thereafter, an NBC 4 New York/The Wall Street Journal/Marist Poll found that 82% of New Yorkers favored a "mandatory 21-day quarantine for anyone who has come in contact with an Ebola patient". This poll included responses of 1,042 New York adults between Oct 26th and 28th, 2014. Poll ID NY141026 on maristpoll.marist.edu.

3.3 Supreme Court → The Gallup organization began measuring the public's view of the Supreme Court's job performance in 2000, and has measured it every year since then with the question: "Do you approve or disapprove of the way the Supreme Court is handling its job?". In 2025, the Gallup poll randomly sampled 1,033 adults in the U.S. and found that 53% of them approved.
news.gallup.com/poll/4732/supreme-court.aspx

3.3 Life on other planets → A February 2018 Marist Poll reported: "Many Americans (68%) think there is intelligent life on other planets". This is up from 52% in 2005. The results were based on a random sample of 1,033 adults in the U.S.
maristpoll.marist.edu/212-are-americans-poised-for-an-alien-invasion

3.4 `transplant` → This is a made up data set about the health outcomes for a hypothetical medical consultant. Note that the data set on the website has 62 patients, not 142 patients, so there will a difference for what is covered in this book vs the data set on the website.

3.4 Alaska residents under 5 years old. → The 2020 statistic comes from the US census: www.census.gov/library/visualizations/interactive/exploring-age-groups-in-the-2020-census.html.

3.4 Toohey poll. → This is a hypothetical scenario not based on a real person or real data.

3.6 `cpr` → Böttiger et al. *Efficacy and safety of thrombolytic therapy after initially unsuccessful cardiopulmonary resuscitation: a prospective clinical trial.* The Lancet, 2001.

3.6 `gear_company` → This is a hypothetical scenario not based on real data.

3.7 `healthcare_law_survey` → Pew research survey on the Affordable Care Act (aka Obamacare) that ran the survey question with two variants.
www.people-press.org/2012/03/26/public-remains-split-on-health-care-bill-opposed-to-mandate/

3.7 `fish_oil_18` → Manson JE, et al. 2018. Marine n-3 Fatty Acids and Prevention of Cardiovascular Disease and Cancer. NEJMoa1811403.

3.8 `jury` → Simulated data set of registered voter proportions and representation on juries from a population.

3.8 M&Ms → Rick Wicklin collected a sample of 712 candies, or about 1.5 pounds, and counted how many there were of each color.
qz.com/918008/the-color-distribution-of-mms-as-determined-by-a-phd-in-statistics

3.9 `gsearch` → Simulated (fake) data set for Google search experiment.

3.9 `ask` → Experiment results from asking about iPods, where the original source is:
Minson JA, Ruedy NE, Schweitzer ME. *There is such a thing as a stupid question: Question disclosure in strategic communication.*
opim.wharton.upenn.edu/DPlab/papers/workingPapers/
Minson_working_Ask%20(the%20Right%20Way)%20and%20You%20Shall%20Receive.pdf

3.9 EVs_region → Pew research results regarding consideration of buying electric vehicle. Percents standardized to add up to 100.
www.pewresearch.org/science/2025/06/05/energy-2025-appendix/
Proportions in Urban/Suburban/Rural taken from:
www.pewresearch.org/social-trends/2018/05/22/demographic-and-economic-trends-in-urban-suburban-and-rural-communities/

3.9 generation_climate_action → A Pew Research poll published in May of 2021 looks at how Americans' attitudes about climate change differ by generation, party and other factors.
www.pewresearch.org/fact-tank/2021/05/26/key-findings-how-americans-attitudes-about-climate-change-differ-by-generation-party-and-other-factors/

Chapter 4: Inference for numerical data: means

4.1 run17, run17samp → These data set represent the full population and a sample of the runners and their run times in the 2017 Cherry Blossom Run in Washington, DC. For more details, see www.cherryblossom.org.

4.1 poker → The full data set includes poker winnings (and losses) for 50 days by a professional poker player, which represents their first 50 days trying to play for a living. Anonymity has been requested by the player.

4.2 Risso's dolphins → Endo T and Haraguchi K. 2009. High mercury levels in hair samples from residents of Taiji, a Japanese whaling town. Marine Pollution Bulletin 60(5):743-747.

Taiji was featured in the movie *The Cove*, and it is a significant source of dolphin and whale meat in Japan. Thousands of dolphins pass through the Taiji area annually, and we will assume these 19 dolphins represent a simple random sample from those dolphins.

4.2 Croaker white fish → www.fda.gov/food/foodborneillnesscontaminants/metals/ucm115644.htm

4.3 run17samp → This data set is described in the data for Chapter 4.

4.2 textbooks, ucla_textbooks_f18 → Data were collected by OpenIntro staff in 2010 and again in 2018. For the 2018 sample, we sampled 201 UCLA courses. Of those, 68 required books could be found on Amazon. The websites where information was retrieved: sa.ucla.edu/ro/public/soc, ucla.verbacompare.com, and amazon.com.

4.2 sat_improve → This is a hypothetical (fake) data set for SAT improvement from an SAT preparation company.

4.5 Jennifer-John study. → Bertrand M, Mullainathan S. 2004. *Science faculty's subtle gender biases favor male students.* PNAS October 9, 2012 109 (41) 16474-16479. www.pnas.org/content/109/41/16474

4.5 resume → Study for racial bias in hiring, where the study's data is available in the resume data set. This data set is explored in great detail in the logistic regression section of the OpenIntro Statistics textbook (free PDF). The original source for this data is:
Bertrand M, Mullainathan S. 2004. *Are Emily and Greg More Employable than Lakisha and Jamal? A Field Experiment on Labor Market Discrimination.* The American Economic Review 94:4 (991-1013). www.nber.org/papers/w9873

4.5 Exams variants. → This is a simulated (fake) data set for exam performance of students for two different exam variations.

4.6 ncbirths → A random sample of 1000 NC births. A sample of that random sample was used for the example in the section.

4.6 stem_cells → Menard C, et al. 2005. Transplantation of cardiac-committed mouse embryonic stem cells to infarcted sheep myocardium: a preclinical study. The Lancet: 366:9490, p1005-1012. thelancet.com/journals/lancet/article/PIIS0140-6736(05)67380-1/fulltext

Chapter 5: **Regression Analysis**

5.2 `county_2023`, `county_complete` → The county_2023 data comes from: American Community Survey 2023 5-year estimates (census.gov/programs-surveys/acs/data.html) and was collected using the "tidycensus" package. The county_complete data set includes additional variables and data from multiple years. This data comes from several government sources, including: USDA (ers.usda.gov), Bureau of Labor Statistics (bls.gov/lau),
and SAIPE (census.gov/did/www/saipe).

5.2 `simulated_scatter` → Fake data used for the first three plots. The perfect linear plot uses group 4 data, where `group` variable in the data set (Figure 5.10). The group of 3 imperfect linear plots use groups 1-3 (Figure 5.11). The sinusoidal curve uses group 5 data (Figure 5.12). The group of 3 scatterplots with residual plots use groups 6-8 (Figure 5.17). The correlation plots uses groups 9-19 data (Figures 5.7 and 5.8).

5.2 `possum` → This data is described in the data for Chapter 1.

5.3 `elmhurst` → These data were sampled from a table of data for all freshman from the 2011 class at Elmhurst College that accompanied an article titled *What Students Really Pay to Go to College* published online by *The Chronicle of Higher Education*: chronicle.com/article/What-Students-Really-Pay-to-Go/131435.

5.3 `loan50` → This data set is described in the data for Chapter 1.

5.3 `simulated_scatter` → The plots for types of outliers uses groups 24-29 (Figure 5.23).

Appendix C

Distribution tables

C.1 Random Number Table

Row	Column 1-5	6-10	11-15	16-20	21-25	26-30	31-35	36-40
1	44394	76100	85973	26853	07080	91603	00476	19681
2	61578	75037	54792	74216	31952	31235	31258	57886
3	18529	73285	95291	49606	67174	95905	33679	75811
4	81238	18321	71085	08284	39318	31434	26173	07440
5	11173	58878	25516	15058	48639	52723	95864	89673
6	96737	95194	14419	22202	92867	73525	94382	29927
7	63514	55066	65162	96016	91723	21160	24285	33264
8	35087	57036	10001	39424	50536	77380	45042	48180
9	00148	73933	49369	32403	53850	16291	93619	27557
10	28999	76232	32637	95697	63679	54506	11299	94294
11	37911	50834	10927	74075	26558	42311	36483	71820
12	33624	82379	03625	58336	27390	00586	06344	89625
13	93282	63059	10830	89432	26917	31555	51793	18718
14	57429	71933	80329	56521	97594	92651	14819	86546
15	65029	24328	06826	61448	54760	09351	73930	99564
16	14779	23173	97183	59835	69580	94653	55095	80666
17	52072	12187	35360	82925	44923	44532	18251	96991
18	76282	91849	17138	59554	35476	67007	02484	10122
19	46561	33015	04577	02178	32915	35912	48974	92985
20	70623	36097	48780	06921	60683	22461	36175	61281
21	03605	08541	17546	85790	48413	69382	89785	80206
22	46147	07603	92057	87609	52670	96255	96660	83167
23	09547	77804	95099	22158	53279	23161	72675	92804
24	12899	05005	86667	72331	09114	28187	97404	26750
25	21223	38353	56970	48965	58371	02697	61417	54746
26	35770	35697	32281	53514	10854	16778	56447	46965
27	04243	65817	81819	64381	83509	44316	56316	47742
28	56989	05587	79995	36598	02316	81627	50104	47720
29	53233	48698	59304	63566	25352	03322	29938	82306
30	20232	30909	77126	50041	96500	24033	77422	20150

C.2 Standard Normal Probability Table

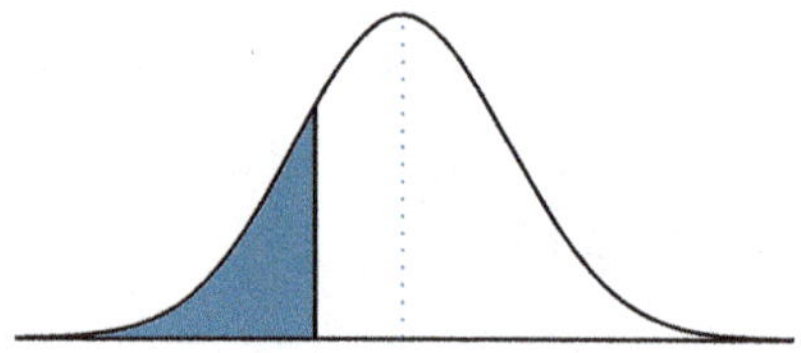

Negative Z

Table entry for Z is the probability lying below Z.

Second decimal place of Z										
0.09	0.08	0.07	0.06	0.05	0.04	0.03	0.02	0.01	0.00	Z
0.0002	0.0003	0.0003	0.0003	0.0003	0.0003	0.0003	0.0003	0.0003	0.0003	−3.4
0.0003	0.0004	0.0004	0.0004	0.0004	0.0004	0.0004	0.0005	0.0005	0.0005	−3.3
0.0005	0.0005	0.0005	0.0006	0.0006	0.0006	0.0006	0.0006	0.0007	0.0007	−3.2
0.0007	0.0007	0.0008	0.0008	0.0008	0.0008	0.0009	0.0009	0.0009	0.0010	−3.1
0.0010	0.0010	0.0011	0.0011	0.0011	0.0012	0.0012	0.0013	0.0013	0.0013	−3.0
0.0014	0.0014	0.0015	0.0015	0.0016	0.0016	0.0017	0.0018	0.0018	0.0019	−2.9
0.0019	0.0020	0.0021	0.0021	0.0022	0.0023	0.0023	0.0024	0.0025	0.0026	−2.8
0.0026	0.0027	0.0028	0.0029	0.0030	0.0031	0.0032	0.0033	0.0034	0.0035	−2.7
0.0036	0.0037	0.0038	0.0039	0.0040	0.0041	0.0043	0.0044	0.0045	0.0047	−2.6
0.0048	0.0049	0.0051	0.0052	0.0054	0.0055	0.0057	0.0059	0.0060	0.0062	−2.5
0.0064	0.0066	0.0068	0.0069	0.0071	0.0073	0.0075	0.0078	0.0080	0.0082	−2.4
0.0084	0.0087	0.0089	0.0091	0.0094	0.0096	0.0099	0.0102	0.0104	0.0107	−2.3
0.0110	0.0113	0.0116	0.0119	0.0122	0.0125	0.0129	0.0132	0.0136	0.0139	−2.2
0.0143	0.0146	0.0150	0.0154	0.0158	0.0162	0.0166	0.0170	0.0174	0.0179	−2.1
0.0183	0.0188	0.0192	0.0197	0.0202	0.0207	0.0212	0.0217	0.0222	0.0228	−2.0
0.0233	0.0239	0.0244	0.0250	0.0256	0.0262	0.0268	0.0274	0.0281	0.0287	−1.9
0.0294	0.0301	0.0307	0.0314	0.0322	0.0329	0.0336	0.0344	0.0351	0.0359	−1.8
0.0367	0.0375	0.0384	0.0392	0.0401	0.0409	0.0418	0.0427	0.0436	0.0446	−1.7
0.0455	0.0465	0.0475	0.0485	0.0495	0.0505	0.0516	0.0526	0.0537	0.0548	−1.6
0.0559	0.0571	0.0582	0.0594	0.0606	0.0618	0.0630	0.0643	0.0655	0.0668	−1.5
0.0681	0.0694	0.0708	0.0721	0.0735	0.0749	0.0764	0.0778	0.0793	0.0808	−1.4
0.0823	0.0838	0.0853	0.0869	0.0885	0.0901	0.0918	0.0934	0.0951	0.0968	−1.3
0.0985	0.1003	0.1020	0.1038	0.1056	0.1075	0.1093	0.1112	0.1131	0.1151	−1.2
0.1170	0.1190	0.1210	0.1230	0.1251	0.1271	0.1292	0.1314	0.1335	0.1357	−1.1
0.1379	0.1401	0.1423	0.1446	0.1469	0.1492	0.1515	0.1539	0.1562	0.1587	−1.0
0.1611	0.1635	0.1660	0.1685	0.1711	0.1736	0.1762	0.1788	0.1814	0.1841	−0.9
0.1867	0.1894	0.1922	0.1949	0.1977	0.2005	0.2033	0.2061	0.2090	0.2119	−0.8
0.2148	0.2177	0.2206	0.2236	0.2266	0.2296	0.2327	0.2358	0.2389	0.2420	−0.7
0.2451	0.2483	0.2514	0.2546	0.2578	0.2611	0.2643	0.2676	0.2709	0.2743	−0.6
0.2776	0.2810	0.2843	0.2877	0.2912	0.2946	0.2981	0.3015	0.3050	0.3085	−0.5
0.3121	0.3156	0.3192	0.3228	0.3264	0.3300	0.3336	0.3372	0.3409	0.3446	−0.4
0.3483	0.3520	0.3557	0.3594	0.3632	0.3669	0.3707	0.3745	0.3783	0.3821	−0.3
0.3859	0.3897	0.3936	0.3974	0.4013	0.4052	0.4090	0.4129	0.4168	0.4207	−0.2
0.4247	0.4286	0.4325	0.4364	0.4404	0.4443	0.4483	0.4522	0.4562	0.4602	−0.1
0.4641	0.4681	0.4721	0.4761	0.4801	0.4840	0.4880	0.4920	0.4960	0.5000	−0.0

*For $Z \leq -3.50$, the probability is less than or equal to 0.0002.

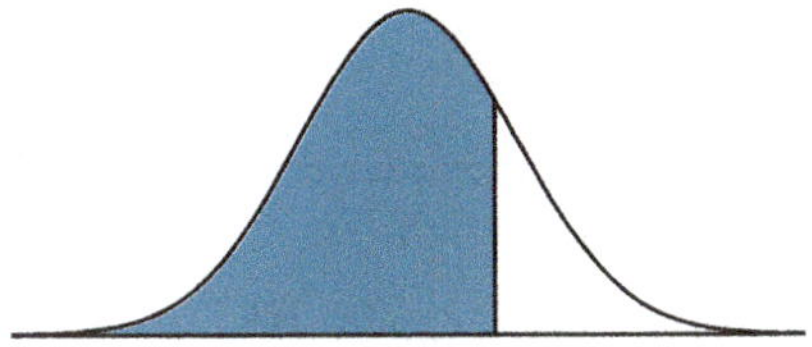

Positive Z

Table entry for Z is the probability lying below Z.

Z	0.00	0.01	0.02	0.03	0.04	0.05	0.06	0.07	0.08	0.09
	Second decimal place of Z									
0.0	0.5000	0.5040	0.5080	0.5120	0.5160	0.5199	0.5239	0.5279	0.5319	0.5359
0.1	0.5398	0.5438	0.5478	0.5517	0.5557	0.5596	0.5636	0.5675	0.5714	0.5753
0.2	0.5793	0.5832	0.5871	0.5910	0.5948	0.5987	0.6026	0.6064	0.6103	0.6141
0.3	0.6179	0.6217	0.6255	0.6293	0.6331	0.6368	0.6406	0.6443	0.6480	0.6517
0.4	0.6554	0.6591	0.6628	0.6664	0.6700	0.6736	0.6772	0.6808	0.6844	0.6879
0.5	0.6915	0.6950	0.6985	0.7019	0.7054	0.7088	0.7123	0.7157	0.7190	0.7224
0.6	0.7257	0.7291	0.7324	0.7357	0.7389	0.7422	0.7454	0.7486	0.7517	0.7549
0.7	0.7580	0.7611	0.7642	0.7673	0.7704	0.7734	0.7764	0.7794	0.7823	0.7852
0.8	0.7881	0.7910	0.7939	0.7967	0.7995	0.8023	0.8051	0.8078	0.8106	0.8133
0.9	0.8159	0.8186	0.8212	0.8238	0.8264	0.8289	0.8315	0.8340	0.8365	0.8389
1.0	0.8413	0.8438	0.8461	0.8485	0.8508	0.8531	0.8554	0.8577	0.8599	0.8621
1.1	0.8643	0.8665	0.8686	0.8708	0.8729	0.8749	0.8770	0.8790	0.8810	0.8830
1.2	0.8849	0.8869	0.8888	0.8907	0.8925	0.8944	0.8962	0.8980	0.8997	0.9015
1.3	0.9032	0.9049	0.9066	0.9082	0.9099	0.9115	0.9131	0.9147	0.9162	0.9177
1.4	0.9192	0.9207	0.9222	0.9236	0.9251	0.9265	0.9279	0.9292	0.9306	0.9319
1.5	0.9332	0.9345	0.9357	0.9370	0.9382	0.9394	0.9406	0.9418	0.9429	0.9441
1.6	0.9452	0.9463	0.9474	0.9484	0.9495	0.9505	0.9515	0.9525	0.9535	0.9545
1.7	0.9554	0.9564	0.9573	0.9582	0.9591	0.9599	0.9608	0.9616	0.9625	0.9633
1.8	0.9641	0.9649	0.9656	0.9664	0.9671	0.9678	0.9686	0.9693	0.9699	0.9706
1.9	0.9713	0.9719	0.9726	0.9732	0.9738	0.9744	0.9750	0.9756	0.9761	0.9767
2.0	0.9772	0.9778	0.9783	0.9788	0.9793	0.9798	0.9803	0.9808	0.9812	0.9817
2.1	0.9821	0.9826	0.9830	0.9834	0.9838	0.9842	0.9846	0.9850	0.9854	0.9857
2.2	0.9861	0.9864	0.9868	0.9871	0.9875	0.9878	0.9881	0.9884	0.9887	0.9890
2.3	0.9893	0.9896	0.9898	0.9901	0.9904	0.9906	0.9909	0.9911	0.9913	0.9916
2.4	0.9918	0.9920	0.9922	0.9925	0.9927	0.9929	0.9931	0.9932	0.9934	0.9936
2.5	0.9938	0.9940	0.9941	0.9943	0.9945	0.9946	0.9948	0.9949	0.9951	0.9952
2.6	0.9953	0.9955	0.9956	0.9957	0.9959	0.9960	0.9961	0.9962	0.9963	0.9964
2.7	0.9965	0.9966	0.9967	0.9968	0.9969	0.9970	0.9971	0.9972	0.9973	0.9974
2.8	0.9974	0.9975	0.9976	0.9977	0.9977	0.9978	0.9979	0.9979	0.9980	0.9981
2.9	0.9981	0.9982	0.9982	0.9983	0.9984	0.9984	0.9985	0.9985	0.9986	0.9986
3.0	0.9987	0.9987	0.9987	0.9988	0.9988	0.9989	0.9989	0.9989	0.9990	0.9990
3.1	0.9990	0.9991	0.9991	0.9991	0.9992	0.9992	0.9992	0.9992	0.9993	0.9993
3.2	0.9993	0.9993	0.9994	0.9994	0.9994	0.9994	0.9994	0.9995	0.9995	0.9995
3.3	0.9995	0.9995	0.9995	0.9996	0.9996	0.9996	0.9996	0.9996	0.9996	0.9997
3.4	0.9997	0.9997	0.9997	0.9997	0.9997	0.9997	0.9997	0.9997	0.9997	0.9998

*For $Z \geq 3.50$, the probability is greater than or equal to 0.9998.

C.3 *t*-Distribution Critical Values Table

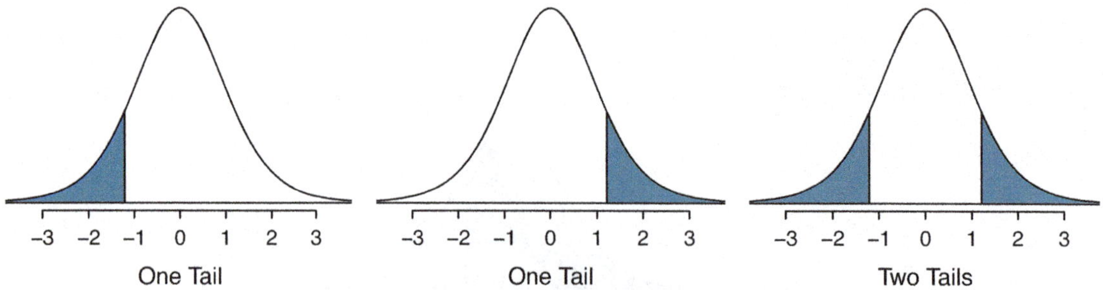

	one tail	0.100	0.050	0.025	0.010	0.005
df	1	3.08	6.31	12.71	31.82	63.66
	2	1.89	2.92	4.30	6.96	9.92
	3	1.64	2.35	3.18	4.54	5.84
	4	1.53	2.13	2.78	3.75	4.60
	5	1.48	2.02	2.57	3.36	4.03
	6	1.44	1.94	2.45	3.14	3.71
	7	1.41	1.89	2.36	3.00	3.50
	8	1.40	1.86	2.31	2.90	3.36
	9	1.38	1.83	2.26	2.82	3.25
	10	1.37	1.81	2.23	2.76	3.17
	11	1.36	1.80	2.20	2.72	3.11
	12	1.36	1.78	2.18	2.68	3.05
	13	1.35	1.77	2.16	2.65	3.01
	14	1.35	1.76	2.14	2.62	2.98
	15	1.34	1.75	2.13	2.60	2.95
	16	1.34	1.75	2.12	2.58	2.92
	17	1.33	1.74	2.11	2.57	2.90
	18	1.33	1.73	2.10	2.55	2.88
	19	1.33	1.73	2.09	2.54	2.86
	20	1.33	1.72	2.09	2.53	2.85
	21	1.32	1.72	2.08	2.52	2.83
	22	1.32	1.72	2.07	2.51	2.82
	23	1.32	1.71	2.07	2.50	2.81
	24	1.32	1.71	2.06	2.49	2.80
	25	1.32	1.71	2.06	2.49	2.79
	26	1.31	1.71	2.06	2.48	2.78
	27	1.31	1.70	2.05	2.47	2.77
	28	1.31	1.70	2.05	2.47	2.76
	29	1.31	1.70	2.05	2.46	2.76
	30	1.31	1.70	2.04	2.46	2.75
Confidence level C		80%	90%	95%	98%	99%

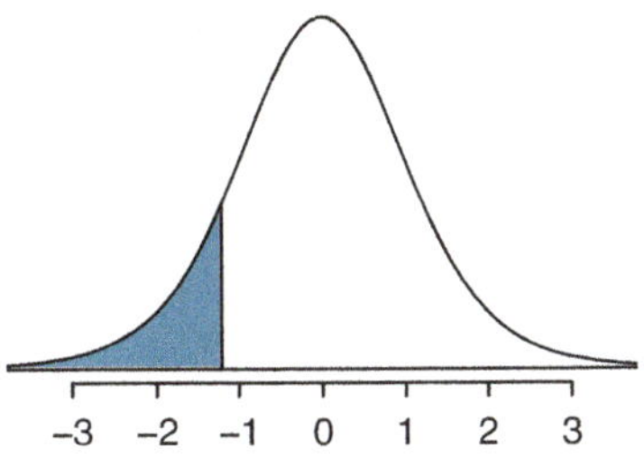

One Tail

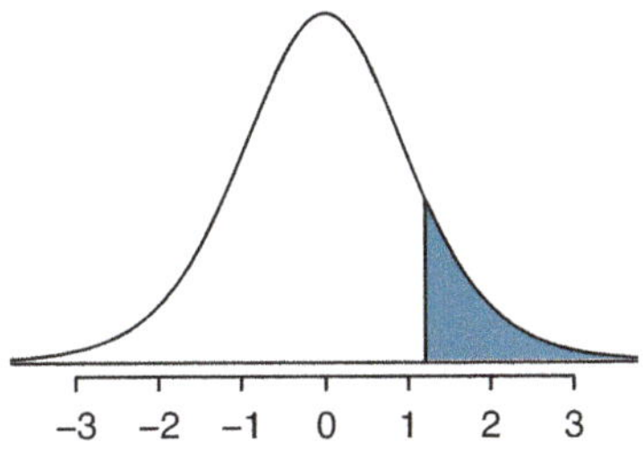

One Tail

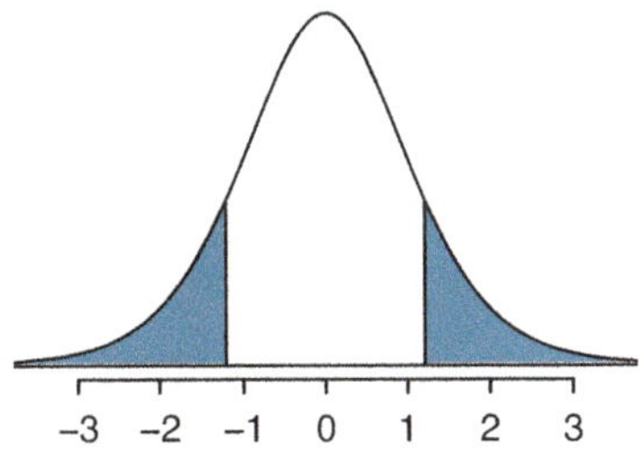

Two Tails

	one tail	0.100	0.050	0.025	0.010	0.005
df	31	1.31	1.70	2.04	2.45	2.74
	32	1.31	1.69	2.04	2.45	2.74
	33	1.31	1.69	2.03	2.44	2.73
	34	1.31	1.69	2.03	2.44	2.73
	35	1.31	1.69	2.03	2.44	2.72
	36	1.31	1.69	2.03	2.43	2.72
	37	1.30	1.69	2.03	2.43	2.72
	38	1.30	1.69	2.02	2.43	2.71
	39	1.30	1.68	2.02	2.43	2.71
	40	1.30	1.68	2.02	2.42	2.70
	41	1.30	1.68	2.02	2.42	2.70
	42	1.30	1.68	2.02	2.42	2.70
	43	1.30	1.68	2.02	2.42	2.70
	44	1.30	1.68	2.02	2.41	2.69
	45	1.30	1.68	2.01	2.41	2.69
	46	1.30	1.68	2.01	2.41	2.69
	47	1.30	1.68	2.01	2.41	2.68
	48	1.30	1.68	2.01	2.41	2.68
	49	1.30	1.68	2.01	2.40	2.68
	50	1.30	1.68	2.01	2.40	2.68
	60	1.30	1.67	2.00	2.39	2.66
	70	1.29	1.67	1.99	2.38	2.65
	80	1.29	1.66	1.99	2.37	2.64
	90	1.29	1.66	1.99	2.37	2.63
	100	1.29	1.66	1.98	2.36	2.63
	150	1.29	1.66	1.98	2.35	2.61
	200	1.29	1.65	1.97	2.35	2.60
	300	1.28	1.65	1.97	2.34	2.59
	400	1.28	1.65	1.97	2.34	2.59
	500	1.28	1.65	1.96	2.33	2.59
	∞	1.28	1.65	1.96	2.33	2.58
Confidence level C		80%	90%	95%	98%	99%

C.4 Chi-Square Critical Values Table

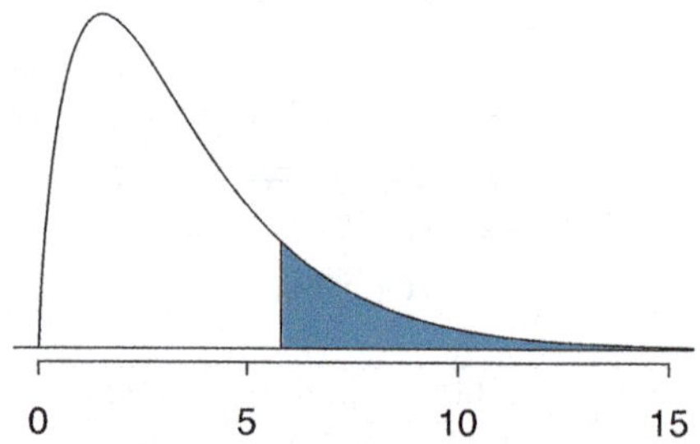

Figure C.1: Areas in the chi-square table always refer to the right tail.

Upper tail	0.3	0.2	0.1	0.05	0.02	0.01	0.005	0.001
df 1	1.07	1.64	2.71	3.84	5.41	6.63	7.88	10.83
2	2.41	3.22	4.61	5.99	7.82	9.21	10.60	13.82
3	3.66	4.64	6.25	7.81	9.84	11.34	12.84	16.27
4	4.88	5.99	7.78	9.49	11.67	13.28	14.86	18.47
5	6.06	7.29	9.24	11.07	13.39	15.09	16.75	20.52
6	7.23	8.56	10.64	12.59	15.03	16.81	18.55	22.46
7	8.38	9.80	12.02	14.07	16.62	18.48	20.28	24.32
8	9.52	11.03	13.36	15.51	18.17	20.09	21.95	26.12
9	10.66	12.24	14.68	16.92	19.68	21.67	23.59	27.88
10	11.78	13.44	15.99	18.31	21.16	23.21	25.19	29.59
11	12.90	14.63	17.28	19.68	22.62	24.72	26.76	31.26
12	14.01	15.81	18.55	21.03	24.05	26.22	28.30	32.91
13	15.12	16.98	19.81	22.36	25.47	27.69	29.82	34.53
14	16.22	18.15	21.06	23.68	26.87	29.14	31.32	36.12
15	17.32	19.31	22.31	25.00	28.26	30.58	32.80	37.70
16	18.42	20.47	23.54	26.30	29.63	32.00	34.27	39.25
17	19.51	21.61	24.77	27.59	31.00	33.41	35.72	40.79
18	20.60	22.76	25.99	28.87	32.35	34.81	37.16	42.31
19	21.69	23.90	27.20	30.14	33.69	36.19	38.58	43.82
20	22.77	25.04	28.41	31.41	35.02	37.57	40.00	45.31
25	28.17	30.68	34.38	37.65	41.57	44.31	46.93	52.62
30	33.53	36.25	40.26	43.77	47.96	50.89	53.67	59.70
40	44.16	47.27	51.81	55.76	60.44	63.69	66.77	73.40
50	54.72	58.16	63.17	67.50	72.61	76.15	79.49	86.66

Index

Appendix D

Technology reference, Formulas, and Inference guide

D.1 Technology reference

Instructions for Desmos, R, and the NumWorks, TI-83/84, and Casio fx-9750GII calculators are included for the following topics.

D.2 Inference Guide

INFERENCE GUIDE

CONFIDENCE INTERVALS

Use **confidence intervals** to **estimate** a parameter with a particular **confidence level, C.**

IDENTIFY: Interval name, parameter, and C level

CHECK: Check that conditions for the procedure are met.

CALCULATE:
CI: point estimate $\pm$ critical value $\times$ *SE* of estimate

df = (if applicable)
(_____ , _____)

CONCLUDE:
We are C% confident that the interval (_ , _) contains the true [parameter]. (Put the parameter in *context.*)

We have evidence that [...], because [...]. OR
We do not have evidence that [...], because [...].

HYPOTHESIS TESTS

Use **hypothesis tests** to **test** H_0 versus H_A at a particular **significance level, α.**

IDENTIFY: Test name, parameter, hypotheses, and α

CHECK: Check that conditions for the procedure are met.

CALCULATE:

$$\textbf{standardized test statistic} = \frac{\textbf{point estimate} - \textbf{null value}}{\textbf{\textit{SE} of estimate}}$$

df = (if applicable)
p-value =

CONCLUDE:
p-value $\leq$ α, so we reject H_0.
We have evidence that $[H_A]$. (Put H_A in *context.*)
OR
p-value > α, so we do NOT reject H_0.
We do NOT have evidence that $[H_A]$. (Put H_A in *context.*)

When the parameter is: a single proportion p

IDENTIFY: **1-Sample Z-Interval** to estimate p, or
1-Sample Z-Test to test $H_0: p = p_0$

CHECK:
- Data come from a random sample or process
- If sampling without replacement, n $\leq$ 10% of N
- For CI: $n\hat{p} \geq 10$ and $n(1-\hat{p}) \geq 10$
  For Test: $np_0 \geq 10$ and $n(1-p_0) \geq 10$

CALCULATE:
point estimate: sample proportion $\hat{p}$

***SE* of estimate:** for CI: use $\sqrt{\frac{\hat{p}(1-\hat{p})}{n}}$; for Test: use $\sqrt{\frac{p_0(1-p_0)}{n}}$

When the parameter is: a single mean μ

IDENTIFY: **1-Sample T-Interval** to estimate μ, or
1-Sample T-Test to test $H_0: \mu = \mu_0$

CHECK:
- Data come from a random sample or process
- If sampling without replacement, n $\leq$ 10% of N
- $n \geq 30$ OR population distribution is nearly normal OR sample data is free from strong skewness and outliers

CALCULATE:
point estimate: sample mean $\bar{x}$

***SE* of estimate**: $\frac{s}{\sqrt{n}}$

$df = n - 1$

When the parameter is: a difference of proportions $p_1 - p_2$

IDENTIFY: **2-Sample Z-Interval** to estimate $p_1 - p_2$, or
2-Sample Z-Test to test $H_0: p_1 = p_2$.

CHECK:
- Data come from 2 independent random samples or 2 randomly assigned treatments.
- If sampling w/o repl., $n_1 \leq$ 10% of N_1 and $n_2 \leq$ 10% of N_2
- For CI: $n_1\hat{p}_1 \geq 10$, $n_1(1-\hat{p}_1) \geq 10$,
  $n_2\hat{p}_2 \geq 10$, $n_2(1-\hat{p}_2) \geq 10$
  For Test: use $\hat{p}_c$, the pooled proportion, in place of $\hat{p}_1$ and $\hat{p}_2$ above

CALCULATE:
point estimate: difference of sample proportions $\hat{p}_1 - \hat{p}_2$
***SE* of estimate:**

CI: use $\sqrt{\frac{\hat{p}_1(1-\hat{p}_1)}{n_1} + \frac{\hat{p}_2(1-\hat{p}_2)}{n_2}}$; Test: use $\sqrt{\hat{p}_c(1-\hat{p}_c)\left(\frac{1}{n_1} + \frac{1}{n_2}\right)}$

When the parameter is: a difference of means $\mu_1 - \mu_2$

IDENTIFY: **2-Sample T-Interval** to estimate $\mu_1 - \mu_2$, or
2-Sample T-Test to test $H_0: \mu_1 = \mu_2$.

CHECK:
- Data come from 2 independent random samples or 2 randomly assigned treatments.
- If sampling w/o repl., $n_1 \leq$ 10% of N_1 and $n_2 \leq$ 10% of N_2
- $n_1 \geq 30$ and $n_2 \geq 30$ OR both population distributions nearly normal OR both sample data sets are free from strong skewness and outliers

CALCULATE:
point estimate: difference of sample means $\bar{x}_1 - \bar{x}_2$

***SE* of estimate**: $\sqrt{\frac{s_1^2}{n_1} + \frac{s_2^2}{n_2}}$

df: use technology

The χ2 Hypothesis Tests for categorical variables: **chi-square statistic** $= \sum \frac{(\text{observed} - \text{expected})^2}{\text{expected}}$

*When comparing the distribution of **one categorical variable to a fixed/specified population distribution**

IDENTIFY: **χ2 Goodness of Fit Test**

CHECK:
- Data come from a random sample or process.
- If sampling without replacement, n $\leq$ 10% of N
- All expected counts ≥ 5. (To calculate expected counts for each category, multiply the sample size by the expected proportion under H_0.)

CALCULATE:
$\chi^2 =$
df = # of categories – 1

When comparing the distribution of **a categorical variable across 2 or more populations/treatments**

IDENTIFY: **χ2 Test for Homogeneity**

CHECK:
- Data come from 2 or more independent random samples or 2 or more randomly assigned treatments.
- If sampling without replacement, each n should $\leq$ 10% of each corresponding N
- All expected counts ≥ 5. (Calculate expected counts and verify this to be true.)

CALCULATE:
$\chi^2 =$
df = (# of rows – 1)(# of cols – 1)

When looking for **association or dependence between two categorical variables**

IDENTIFY: **χ2 Test for Independence**

CHECK:
- Data come from a random sample or process.
- If sampling without replacement, n $\leq$ 10% of N
- All expected counts ≥ 5. (Calculate expected counts and verify this to be true.)

CALCULATE:
$\chi^2 =$
df = (# of rows – 1)(# of cols – 1)

*Not tested on the AP® Exam.

D.3 Formulas

Descriptive Statistics

$$\bar{x} = \frac{1}{n}\sum x_i = \frac{\sum x_i}{n}$$

$$s_x = \sqrt{\frac{1}{n-1}\sum (x_i - \bar{x})^2}$$

$$r = \frac{1}{n-1}\sum \left(\frac{x_i - \bar{x}}{s_x}\right)\left(\frac{y_i - \bar{y}}{s_y}\right)$$

$$\hat{y} = a + bx$$

Probability

$$P(A \cup B) = P(A) + P(B) - P(A \cap B)$$

$$P(A \mid B) = \frac{P(A \cap B)}{P(B)}$$

$$\mu_X = E(X) = \sum x_i \cdot P(x_i)$$

$$\sigma_X = \sqrt{\sum (x_i - \mu_x)^2 \cdot P(x_i)}$$

$$\mu_{\bar{x}} = \mu \qquad \sigma_{\bar{x}} = \frac{\sigma}{\sqrt{n}}$$

If X has a binomial distribution with parameters n and p, then:

$$P(X = x) = \binom{n}{x} p^x (1-p)^{n-x}$$

$$\mu_X = np \qquad \sigma_X = \sqrt{np(1-p)}$$

$$\mu_{\hat{p}} = p \qquad \sigma_{\hat{p}} = \sqrt{\frac{p(1-p)}{n}}$$

Inferential Statistics

standardized test statistic: $\frac{\text{point estimate} - \text{null value}}{SE \text{ of estimate}}$

confidence interval: point estimate $\pm$ critical value $\times$ SE of estimate

	parameter	point estimate	SE of estimate	
single proportion	p	$\hat{p}$	$\sqrt{\frac{\hat{p}(1-\hat{p})}{n}}$	when H_0: $p = p_0$, use $\sqrt{\frac{p_0(1-p_0)}{n}}$
diff. of proportions	$p_1 - p_2$	$\hat{p}_1 - \hat{p}_2$	$\sqrt{\frac{\hat{p}_1(1-\hat{p}_1)}{n_1} + \frac{\hat{p}_2(1-\hat{p}_2)}{n_2}}$	when H_0: $p_1 = p_2$, use $\sqrt{\hat{p}_c(1-\hat{p}_c)\left(\frac{1}{n_1}+\frac{1}{n_2}\right)}$
single mean	μ	$\bar{x}$	$\frac{s}{\sqrt{n}}$	
mean of differences	μ_d	$\bar{x}_d$	$\frac{s_d}{\sqrt{n_d}}$	
difference of means	$\mu_1 - \mu_2$	$\bar{x}_1 - \bar{x}_2$	$\sqrt{\frac{s_1^2}{n_1} + \frac{s_2^2}{n_2}}$	

Chi-square test statistic $= \sum \frac{(\text{observed}-\text{expected})^2}{\text{expected}}$

Made in the USA
Monee, IL
11 September 2026

62816371R10262